Understanding Thermodynamics

Understanding Thermodynamics

Edited by **Lucy Flynn**

New Jersey

Published by Clanrye International,
55 Van Reypen Street,
Jersey City, NJ 07306, USA
www.clanryeinternational.com

Understanding Thermodynamics
Edited by Lucy Flynn

International Standard Book Number: 978-1-63240-507-4 (Hardback)

Printed in the United States of America.

Contents

Preface

The term 'thermodynamics' usually describes a macroscopic description of bodies and its related processes. Thermodynamics is a branch of physical chemistry, which plays a great role in modern science. It essentially deals with heat and temperature, and how these two parameters are related to energy and work. If we turn the pages of history all the world over, the origins of thermodynamics can be traced to the early steam engines. Scientists attempted for an increased efficiency and power output in steam engines which led to work in this field. Since its invention, great theories have been derived by experts of thermodynamics. Prominent scientists who have made contributions in the development of thermodynamics include Maxwell, Boltzman, Bernoulli, Josiah Willard Gibbs, Sadi Carnot and Clapeyron Claussius amongst numerous others.

In all studies related to thermodynamics, there are three kinds of physical entities namely; states of a system, thermodynamic processes of a system, and thermodynamic operations. Also, thermodynamic equilibrium happens to be one of the most crucial concepts in this field. One of the essential features of thermodynamics is the fact that it incorporates the concept of heat or thermal energy as an important component in the energy systems. This is also a feature in which thermodynamics differs from the other branches of science. Today, Thermodynamics has expanded to studies related to energy transfers in chemical processes.

I would like to express my sincere appreciation to the authors of the chapters in this book, for their excellent contributions and efforts involved in the publication process. Not only do I appreciate their participation, but also their adherence as a group to the time parameters set for this publication. This book has been profusely illustrated to make the concepts clear to the readers. I do believe that the contents in this book will be helpful to many researchers in this field around the world.

Editor

Conductometric Studies of Thermodynamics of Complexation of Co^{2+}, Ni^{2+}, Cu^{2+}, and Zn^{2+} Cations with Aza-18-crown-6 in Binary Acetonitrile-Methanol Mixtures

Mehdi Taghdiri, Mahmood Payehghadr, Reza Behjatmanesh-Ardakani, and Homa Gha'ari

Department of Chemistry, Payame Noor University, P.O. Box 19395-3697, Tehran, Iran

Correspondence should be addressed to Mehdi Taghdiri, mehditaghdiri@yahoo.com

Academic Editor: Perla B. Balbuena

The complexation reactions between aza-18-crown-6 (A18C6) and Co^{2+}, Ni^{2+}, Cu^{2+}, and Zn^{2+} ions were studied conductometrically in different acetonitrile-methanol mixtures at various temperatures. The formation constants of the resulting 1 : 1 complexes were calculated from the computer fitting of the molar conductance-mole ratio data at different temperatures. Selectivity of A18C6 for Co^{2+}, Ni^{2+}, Cu^{2+}, and Zn^{2+} cations is sensitive to the solvent composition. At 20°C and in acetonitrile solvent, the stability of the resulting complexes varied in the order $Zn^{2+} > Cu^{2+} > Co^{2+} \sim Ni^{2+}$ but the order was reversed by adding 20% methanol. The enthalpy and entropy changes of the complexation reactions were evaluated from the temperature dependence of formation constants. It was found that the stability of the resulting complexes decreased with increasing methanol in the solvent mixture. The $T\Delta S^\circ$ versus ΔH° plot of thermodynamic data obtained shows a fairly good linear correlation indicating the existence of enthalpy-entropy compensation in the complexation reactions. In addition, binding energies of Ni^{2+}, Cu^{2+}, and Zn^{2+} complexes with A18C6 were calculated at B3LYP/6-31G level of theory.

1. Introduction

Macrocyclic polyethers (crown ethers), first prepared in the 1960s [1], constitute an important class of host molecules that have found broad application to studies of molecular recognition and inclusion phenomena [2]. They are macrocycles capable of ion encapsulation due to their cage like structures. The metal ion is held in the crown ether cavity by electrostatic attraction between the charged cation and dipoles created by the nonbonding electrons of donor atoms [3, 4]. The selectivity and stability of crown ethers are also influenced by their structural flexibility, the number and type of donor atoms on the cavity of the crown, and the solvation energy of the metal ion. Thus, compounds of this type have been used extensively for selective complexation and transport of cations, anions, and neutral molecules [5]. Macrocyclic crown compounds have gained a great deal of attention due to their wide applications in chemistry [6] such as microanalysis, sensing and separation of metal ions [7], and extraction of biogenic amines [8].

Aza-crown ethers have especially been focused on as useful ligands because of their versatility and applicability [8–11]. It was of interest to us to study the interaction of the various cations with aza-crown ethers. Since the nature of these compounds, the size of cavity, and solvent may strongly influence the stoichiometry and complexation of transition metal complexes in solution [6], the complexation reactions between aza-18-crown-6 (A18C6) (Figure 1) with Co^{2+}, Ni^{2+}, Cu^{2+}, and Zn^{2+} ions were studied by conductometric in acetonitrile (AN) and different acetonitrile-methanol (MeOH) mixtures at various temperatures. The formation constants of the resulting 1 : 1 complexes were calculated from the computer fitting of the molar conductance-mole ratio data at different temperatures. Little work has been reported on the study of the stability of transition and heavy metal ion complexes with A18C6, mainly in binary

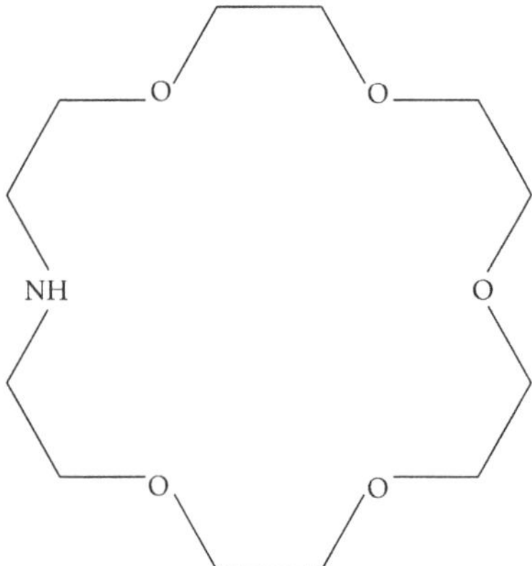

Figure 1: Structure of aza-18-crown-6.

solvent mixtures [12–14]. Thus, studies of complexation reactions as a function of the solvent composition in binary solvent mixtures might be used to investigate the factors contributing to the changes in complexation stability and selectivity of the resulting complexes [11–13]. It should be noted that AN and MeOH have different Gutmann donor numbers, (DN) (i.e., for AN, DN = 14.1, and for MeOH, DN = 19.0) [15]. The Gutmann donor numbers are a measure of the electron-donating properties of a solvent. The donor number is defined as the negative enthalpy value for the 1 : 1 adduct formation between a given electron-pair donor solvent and the standard Lewis acid $SbCl_5$, in dilute solution in the non-coordinating solvent 1,2-dichloroethane, for which a DN of zero is assigned. DN reflects the ability of the solvent to solvate cations and other Lewis acids. DN values range from zero, for solvents like hexane or tetrachloromethane, to 61.0 for triethylamine. In general, it is observed that the smaller the values of DN, the more stable the cation-crown ether complex [6].

2. Experimental

Reagent-grade nitrate salts of cobalt, nickel, copper, and zinc, dried acetonitrile ($H_2O < 0.005\%$) and methanol (all from Merck) were of the highest purity available and used as received. A18C6 was purchased from Fluka. All the AN-MeOH mixtures used were prepared by weight.

Conductance measurements were carried out with a Metrohm 712 conductivity meter. A dip-type conductivity cell made of platinum black was used. The cell constant at the different temperatures used was determined by conductivity measurements of a 0.010 M solution of analytical-grade KCl (Merck) in triply distilled deionized water. The specific conductance of this solution at various temperatures has been reported in the literature [16]. In all measurements, the cell was thermostated at the desired temperature ±0.1°C using a Haake D1 thermostat-circulator water bath.

In a typical experiment, 10 mL of the desired metal ion (5.0×10^{-5} M) was placed in the titration cell, thermostated to the desired temperature and the conductance of solution was measured. Then, a known amount of a concentrated A18C6 solution (5.0×10^{-3} M) was added in a stepwise manner using a calibrated micropipette. The conductance of the solution was measured after each addition. The ligand solution was continually added until the desired ligand to cation mole ratio was achieved. The dilution of the salt solution during the titration is negligible because the total volume of added concentrated A18C6 solution is 300 μL. Therefore, the conductance changes during the titration due to the dilution are very negligible and are neglected.

NMR spectra in deuterated acetonitrile (CD_3CN) were recorded on a Bruker Avance 300 spectrometer, and all chemical shifts are reported in δ units downfield from Me_4Si. In a typical measurement, 0.5 mL of ligand solution (0.02 M) in CD_3CN was placed in the 5 mm BBO NMR tube and the spectrum was recorded. Then a known amount of a concentrated solution of zinc nitrate (0.2 M) in CD_3CN was added in a stepwise manner using a 10-μL Hamilton syringe and the spectrum of the solution was recorded after each addition. The zinc ion solution was continually added until the desired cation to ligand molar ratio was achieved.

The formation constants, K_f, and the limiting molar conductances, Λ_o, of the resulting 1 : 1 complexes between A18C6 and the cations used, in different AN-MeOH mixtures and at various temperatures, were calculated by fitting the observed molar conductance, Λ_{obs}, at varying $[A18C6]/[M^{2+}]$ mole ratios to a previously derived equation [6, 17, 18] which expresses the Λ_{obs} as a function of the free and complexed metal ions. A nonlinear least squares curve fitting using Microsoft Excel Solver (version 11.0) was applied for the evaluation of formation constant and limiting molar conductance of the resulting 1 : 1 complexes. The large formation constant values ($\log K_{ML} > 6$) can be determined by this method.

Chi-square statistic was used to evaluate the fitness of equation to the experimental data. The Chi-square test statistic is the sum of the squares of the differences between the experimental data and data obtained by calculating from equation. The equivalent mathematical statement is

$$\chi^2 = \sum_{i=1}^{N} \frac{\left(\Lambda_{\text{exp.}} - \Lambda_{\text{calc.}}\right)^2}{\Lambda_{\text{calc.}}}, \quad (1)$$

where $\Lambda_{\text{exp.}}$ and $\Lambda_{\text{calc.}}$ are the molar conductance obtained by experimental data and calculating from the equation, respectively. If calculated data are similar to the experimental data, χ^2 will be a small number; if they are different, χ^2 will be a large number.

3. Results and Discussion

The conductometric method has been extensively used for obtaining the formation constants of complexes of crown ethers with metal cations [6]. In order to evaluate the influence of adding A18C6 on the molar conductance of Co^{2+}, Ni^{2+}, Cu^{2+}, and Zn^{2+} ions in different AN-MeOH mixtures, the molar conductance at a constant salt concentration (5.0×10^{-5} M) was monitored while increasing the crown ether concentration at various temperatures. Some of the resulting molar conductances versus A18C6/cation mole ratio plots are shown in Figures 2 and 3.

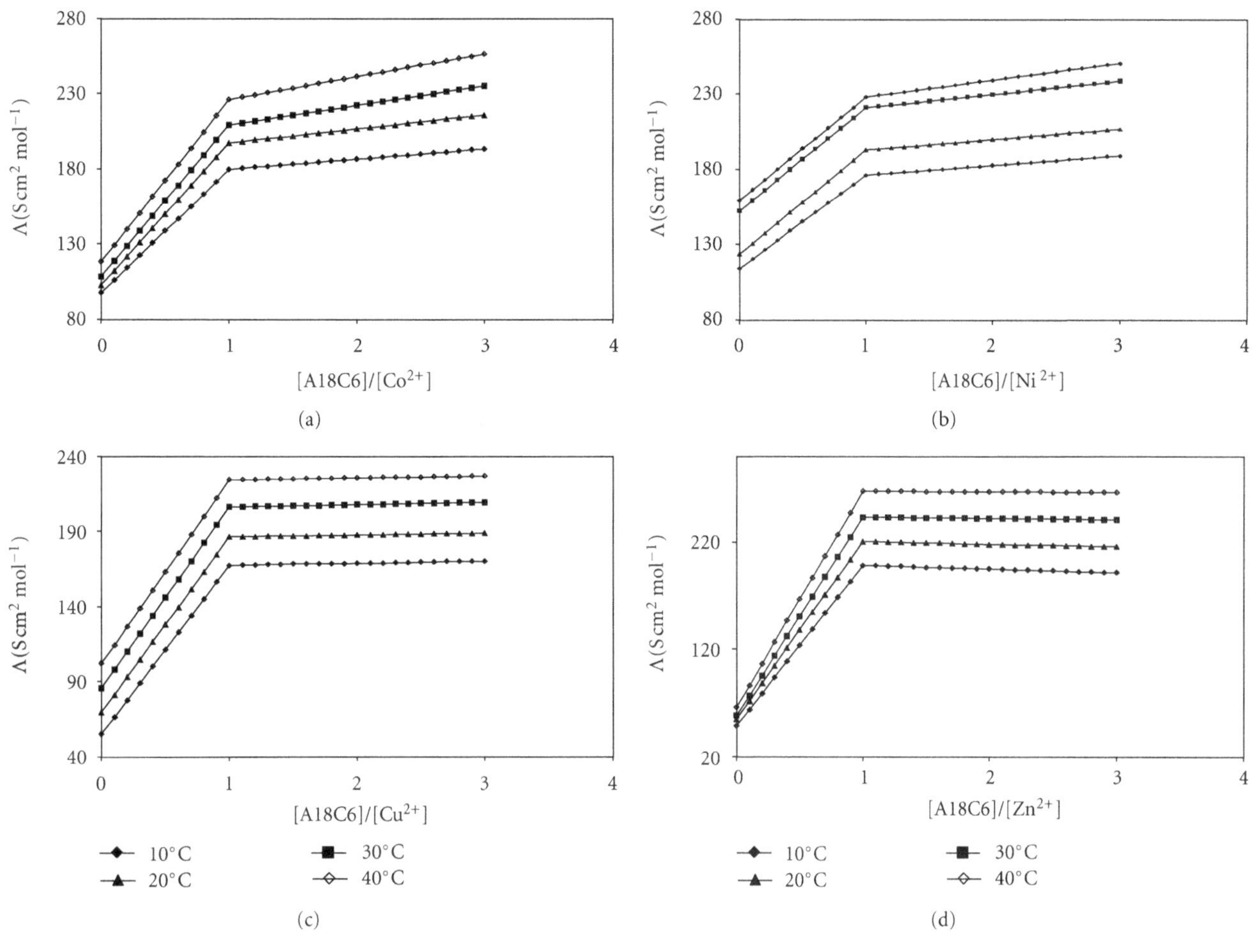

Figure 2: Molar conductance (S cm^2 mol^{-1}) versus [A18C6]/[M^{2+}] mole ratio plots in 100% AN at various temperatures. The M^{2+} cations are (a) Co^{2+}, (b) Ni^{2+}, (c) Cu^{2+} and (d) Zn^{2+}.

As it is seen, in all cases, there is a gradual increase in the molar conductance with an increase in the crown ether concentration. Figure 2 shows that, in the case of all M^{2+}-A18C6 system in AN solvent, the addition of A18C6 to the M^{2+} ion solution causes a continuous increase in the molar conductance, which begins to level off at a mole ratio greater than one. Such a conductance behavior is indicative of the formation of fairly stable 1 : 1 complexes in AN.

The formation constants of all A18C6-M^{2+} complexes in different solvent mixtures at various temperatures, obtained by computer fitting of the molar conductance-mole ratio data and Chi-square values (χ^2), are listed in Table 1. The computer fits of the molar ratio data are shown as solid lines. The small numbers of Chi-square indicate that our assumption of 1 : 1 stoichiometry seems reasonable in the light of the fair agreement between the observed and calculated molar conductances.

In order to have a better understanding of the thermodynamics of complexation reactions of Co^{2+}, Ni^{2+}, Cu^{2+}, and Zn^{2+} ions with A18C6, it is useful to investigate the enthalpic and entropic contributions to these reactions. The ΔH° and ΔS° of the complexation reactions in different AN-MeOH mixtures were evaluated from the temperature dependence of the formation constants by applying a linear least-squares analysis according to the van't Hoff equation. The van't Hoff plots of log K_f versus $1/T$ are shown in Figure 4. The enthalpies and entropies of complexation were determined in the usual manner from the slopes and intercepts of the plots, respectively and the results are listed in Table 2.

In the case of complexation of macrocyclic ligands, there are many factors which can make significant contributions to the stability of their metal ion complexes: the cation size, the ionic solvation of the charged species involved, conformations of the free and complexed crown ethers, the electronic structure of metal ion and the binding strengths of solvent-ion and crown-ion. Moreover, the changes of the enthalpy and entropy influence the stability constant. Many factors contribute to changes in enthalpy and entropy of complexation reactions. It seems that solvent properties such as donor number, relative permittivity (dielectric constant) and their dipole moments influence upon the enthalpy and entropy.

The data given in Table 1 clearly illustrate the fundamental role of the solvent properties in the M^{2+}-A18C6 complexation reactions studied. The stability of the resulting complexes decreases strongly with increasing weight percent

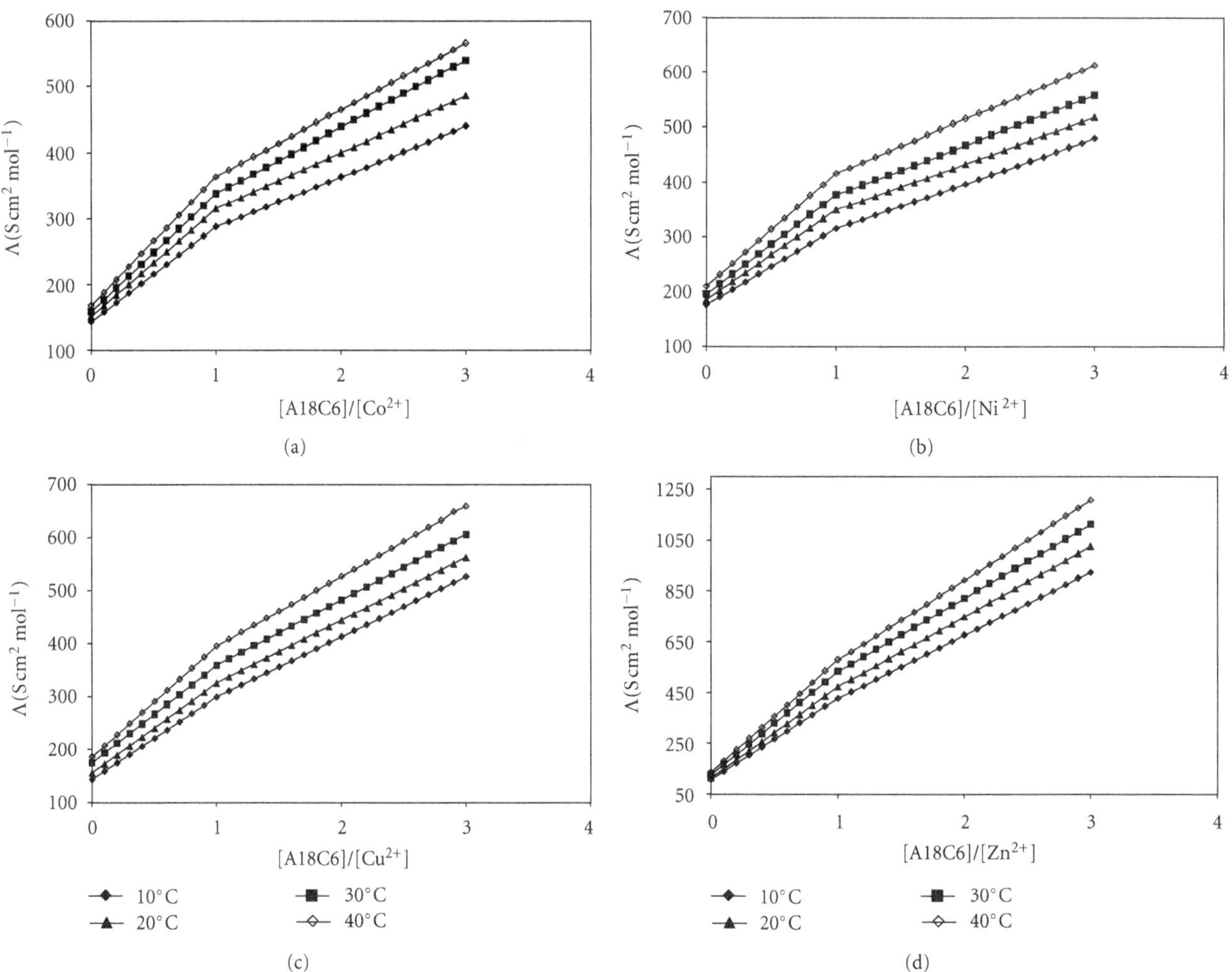

FIGURE 3: Molar conductance (S cm^2 mol^{-1}) versus [A18C6]/[M^{2+}] mole ratio plots in AN-MeOH (80 : 20) mixture at various temperatures. The M^{2+} cations are (a) Co^{2+}, (b) Ni^{2+}, (c) Cu^{2+}, and (d) Zn^{2+}.

TABLE 1: Formation constants for different M^{2+}-A18C6 complexes in various AN-MeOH mixtures.

Cation[a]	wt% AN	Log K_f							
		10°C	χ^2	20°C	χ^2	30°C	χ^2	40°C	χ^2
Co^{2+} (1.50)	100	5.66	0.76	5.50	1.11	5.27	1.48	5.19	1.69
	80	3.87	1.60	3.89	1.95	3.79	1.87	3.90	2.41
	60	2.82	0.10	2.90	0.15	2.98	0.25	2.91	1.35
	40	2.83	0.11	2.87	0.14	2.83	0.15	2.86	0.25
Ni^{2+} (1.38)	100	5.47	0.54	5.49	0.60	5.28	0.63	5.05	0.72
	80	3.73	1.09	3.91	1.81	3.95	2.14	4.00	2.72
	60	2.90	0.12	2.94	0.18	3.06	0.69	3.14	0.44
	40	2.96	0.17	2.99	0.21	3.01	0.23	3.06	0.30
Cu^{2+} (1.54)	100	7.44	0.09	7.77	0.07	7.48	0.09	7.63	0.07
	80	3.39	0.71	3.47	0.94	3.53	1.16	3.63	1.73
	60	3.11	0.31	3.14	0.37	3.17	0.45	3.20	0.77
	40	3.08	0.30	3.06	0.29	3.03	0.27	3.02	0.28
Zn^{2+} (1.48)	100	15.27	0.64	14.60	0.29	14.33	0.07	14.07	0.01
	80	3.25	1.43	3.31	1.72	3.44	2.99	3.46	3.17
	60	2.96	0.75	3.04	0.79	3.13	1.31	3.18	1.66
	40	1.44	1.82	1.93	0.008	2.44	0.12	2.82	0.75

[a]The values in parenthesis are the ionic sizes in Å [19].

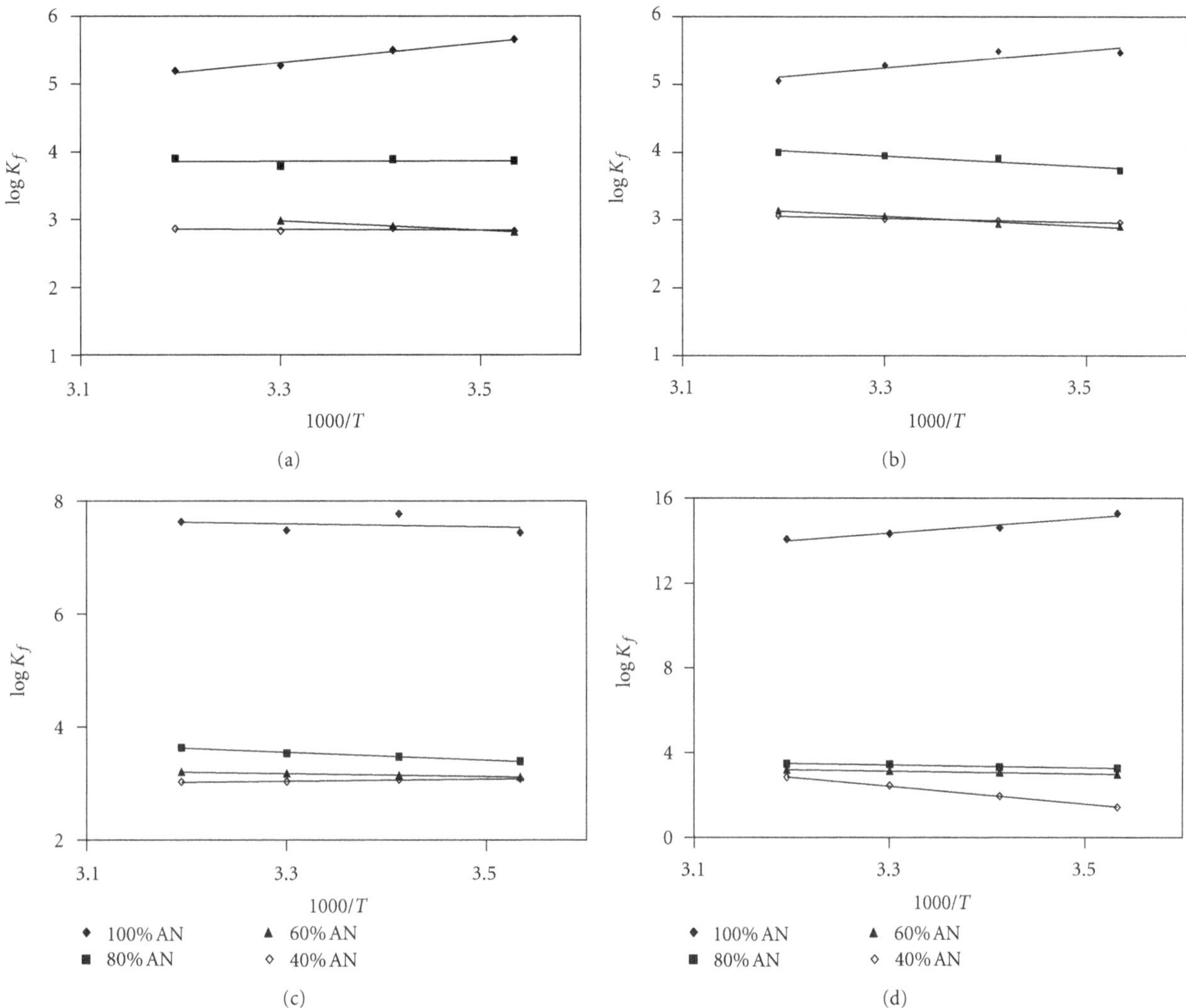

FIGURE 4: van't Hoff plots for the 1 : 1 complexation of Co^{2+} (a), Ni^{2+} (b), Cu^{2+} (c), and Zn^{2+} (d) with A18C6 in different solvent mixtures.

of MeOH in the solvent mixture (Figure 5). It is well known that the solvating ability of the solvent, as expressed by the Gutmann donor number [15], plays an important role in different complexation reactions [4, 16, 18, 20, 21]. There is actually an inverse relationship between the stabilities of the complexes and the solvating abilities of the solvents. Methanol has a higher donicity (DN = 19.0) than acetonitrile (DN = 14.1) and, therefore, shows more competition with the crown ether for mentioned ions; thus, it is not unexpected to observe that addition of methanol to acetonitrile will decrease the stability of the complexes. In addition, selectivity for certain cations over others may be altered according to the nature of solvent. The order of stability of complexes in pure AN at 20°C is $Zn^{2+} > Cu^{2+} > Co^{2+} \sim Ni^{2+}$. It is interesting to note that the order of the stability of the complexes formed between A18C6 and these metal cations in AN-MeOH (wt% MeOH = 20) binary mixtures is reversed. This reversal of stabilities indicates the selectivity of macrocyclic ligands is sensitive to the solvent composition and may change in certain composition of the mixed solvent systems. A reversal in stabilities has been observed for M^{n+}-A18C6 (M^{n+} = Ag^{+}, Hg^{2+} and Pb^{2+}) in DMSO-H_2O binary mixture [13]. The selectivity of complexation not only depends on the ratio of the cation diameter and the diameter of the crown ether cavity, but also on the solvent composition. The change of solvent composition influences upon the changes in complexation enthalpies and entropies and hence the selectivity of the crown ethers towards cations. It seems that solvent properties such as donor number, dielectric constant and the isosolvation point (the point at which both solvents participate equally in the inner salvation shell of the cation [22]) contribute to the changes of enthalpy and entropy as a function of solvent composition.

Table 2 shows that, as expected, for M^{2+}-A18C6 systems studied, the thermodynamic data vary significantly with the solvent properties. However, the observed increase (or decrease, depending on the nature of the metal ion) in ΔH° value upon addition of MeOH to the solvent mixture will be compensated by an increase (or decrease) in the corresponding ΔS° value. The existence of such a compensating effect (Figure 6) between ΔH° and ΔS° values, which has been frequently reported for a variety of metal-ligand systems

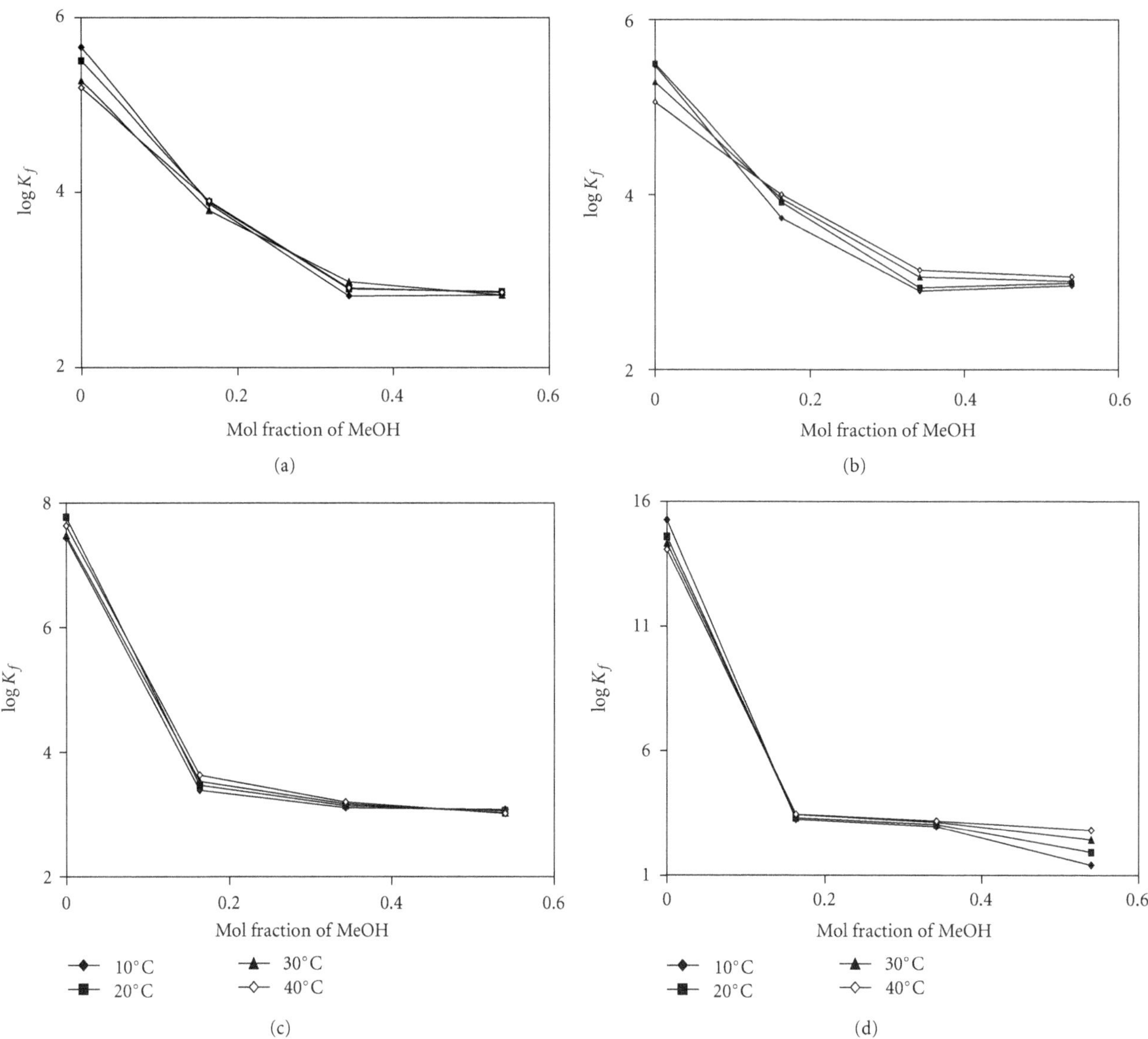

FIGURE 5: Variation of the stability constants of various A18C6-M^{2+} systems with the composition of the AN-MeOH binary mixture at different temperatures. The M^{2+} cations are (a) Co^{2+}, (b) Ni^{2+}, (c) Cu^{2+}, and (d) Zn^{2+}.

[23–25], would cause the overall change in the ΔG° value of the complex to be smaller than might be expected from the change in either ΔH° or ΔS° independently. There is an entropy-driven stabilization for Co^{2+}, Ni^{2+}, Cu^{2+}, and Zn^{2+} complexation in AN-MeOH mixtures. These cations have small ionic sizes (Table 1) and are strongly solvated by MeOH. Therefore, complete or partial desolvation of cations and ligand are important steps of complexation process and seems to be mainly responsible for increasing of entropy.

The stability constant between Zn^{2+} and A18C6 is much greater in AN than other studied cations. It also shows a slight decrease in the molar conductance after the mole ratio of one (Figure 2(d)). Therefore, the ^{1}H NMR spectra of a 0.02 M solution of A18C6 in deuterated AN in the presence of increasing concentration of zinc nitrate were recorded (Figure 7(a)). The resulting chemical shift versus metal ion to ligand mole ratio (shown in Figure 7(b)) revealed a distinct inflection point at a mole ratio of 1 : 2, indicating the formation of a ML_2 complex in solution. A sandwich type complex has been proposed for biphenyl containing A18C6 with Zn^{2+} from ^{1}HNMR, UV visible and fluorescence observations in AN [9]. The ML_2 complexes have been also proposed for 18C6-Cu^{2+} in methanol [26], A18C6-Pb^{2+} in H_2O, A18C6-Ag^+ in DMSO-H_2O [13], and immobilized DB18C6-Zn^{2+} [27].

An attempt was made to obtain more information from the quantum chemical calculations about the structures of both free and complex forms of A18C6. To do this, the initial structures of compounds were built with HyperChem 7.0 program [28]. All calculations were done with Gaussian 2003 suit of programs [29] on an Intel SMP computer with 8 processors and 8 GB of RAM. B3LYP/6-31G level of theory was used for all optimization and frequency calculations. All computations were done in the gas phase. Full geometry

TABLE 2: Thermodynamic parameters for different M^{2+}-A18C6 complexes in various AN-MeOH mixtures.

Cation	wt% AN	ΔH° (kJ mol^{-1})	ΔS° (J mol^{-1} K^{-1})	ΔG° (kJ mol^{-1})
Co^{2+}	100	−27.9 ± 3.3	9.8 ± 11.1	−30.8 ± 4.7
	80[a]	1.6 ± 0.6	79.8 ± 2.0	−21.8 ± 0.8
	60[b]	13.1 ± 0.3	100.4 ± 1.1	−16.3 ± 0.5
	40[c]	1.6 ± 0.8	59.6 ± 2.7	−15.9 ± 1.1
Ni^{2+}	100	−24.6 ± 7.5	19.2 ± 25.3	−30.2 ± 10.7
	80	14.6 ± 3.7	123.5 ± 12.4	−21.6 ± 5.2
	60	14.2 ± 2.0	105.3 ± 6.7	−16.7 ± 2.8
	40	5.4 ± 0.8	75.7 ± 2.5	−16.8 ± 1.1
Cu^{2+}	100	—[d]	—[d]	—[d]
	80	13.2 ± 1.0	111.5 ± 3.2	−19.5 ± 1.4
	60	5.1 ± 0.1	77.5 ± 0.3	−17.6 ± 0.1
	40	−3.6 ± 0.4	46.3 ± 1.5	−17.2 ± 0.6
Zn^{2+}	100[e]	−46.5 ± 0.5	120.8 ± 1.6	−81.9 ± 0.7
	80	12.9 ± 2.4	107.9 ± 8.1	−18.7 ± 3.4
	60	12.7 ± 0.9	101.7 ± 3.1	−17.1 ± 1.3
	40	80.5 ± 3.3	311.7 ± 11.0	−10.8 ± 4.6

[a] Without considering Log K_f at 30°C.
[b] Without considering Log K_f at 40°C.
[c] Without considering Log K_f at 20°C.
[d] High uncertainty.
[e] Without considering Log K_f at 10°C.

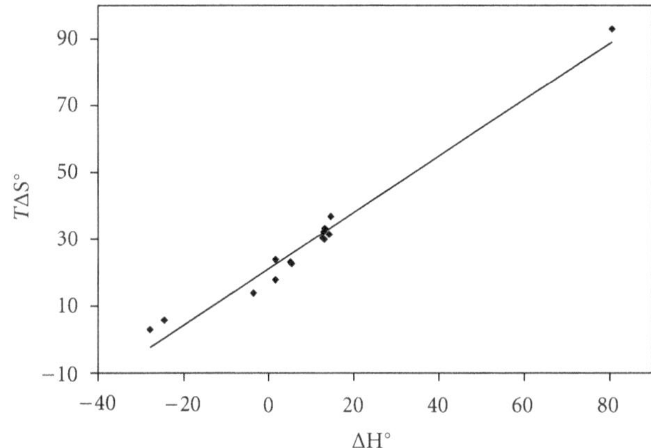

FIGURE 6: Plot of TΔS° (kJ mol^{-1}) versus ΔH° (kJ mol^{-1}) for 1 : 1 complexation of Co^{2+}, Ni^{2+}, Cu^{2+}, and Zn^{2+} ions with A18C6 in different AN-MeOH binary mixtures ($R = 0.987$).

TABLE 3: Calculated electronic energies of different species and binding energies of complexes formed by interaction between A18C6 and Ni^{2+}, Cu^{2+}, and Zn^{2+}. All calculations have been done with B3LYP/6-31G.

Compounds	Electronic energy (a.u.)	Binding energy (kcal/mol)[a]
A18C6	−903.07162832	
Ni^{2+}	−1507.14512292	
Cu^{2+}	−1639.396656	
Zn^{2+}	−1778.31513190	
A18C6-Ni^{2+}	−2410.87730681	−414.50
A18C6-Cu^{2+}	−2543.10355707	−398.64
$(A18C6)_2$-Zn^{2+}	−3585.13086395	−422.56

[a] Binding energy (ΔE) is defined as ΔE = E(complex)−E(Ligand)−E(Ion). 1 a.u. is equal to 627.5095 kcal/mol.

optimization was used with no symmetry or any constraints about bond lengths, bond angles, or bond torsions. Figure 8 shows the optimized geometries of both free and metal ion complexes of A18C6. Frequency analyses were done to test whether the optimized structures are stationary points or saddle points. In all outputs, number of imaginary frequency was equal to zero. This indicates that all structures are true optimized forms. Table 3 shows binding energies for A18C6-cation complexation for three metal cations of Ni^{2+}, Cu^{2+}, and Zn^{2+}. Despite many attempts, we could not obtain binding energies for A18C6-Co^{2+}. All binding energies are less than zero. It shows that all cations tend to form complex with A18C6 in the gas phase. It should be noted that these cations have relatively close ionic radii. The ML_2 structure for A18C6-Zn^{2+} (Figure 8(d)) indicates the strong interaction of Zn^{2+} with nitrogen atoms. The ^{1}HNMR study (Figure 7(a)) confirms this strong interaction. The ^{1}HNMR signal at δ = 2.17 ppm corresponding to N–H, disappears upon addition of Zn^{2+} to A18C6.

4. Conclusions

From the conductometric results obtained on the thermodynamics of complexation of A18C6 with some transition metal ions in different AN-MeOH binary mixtures and

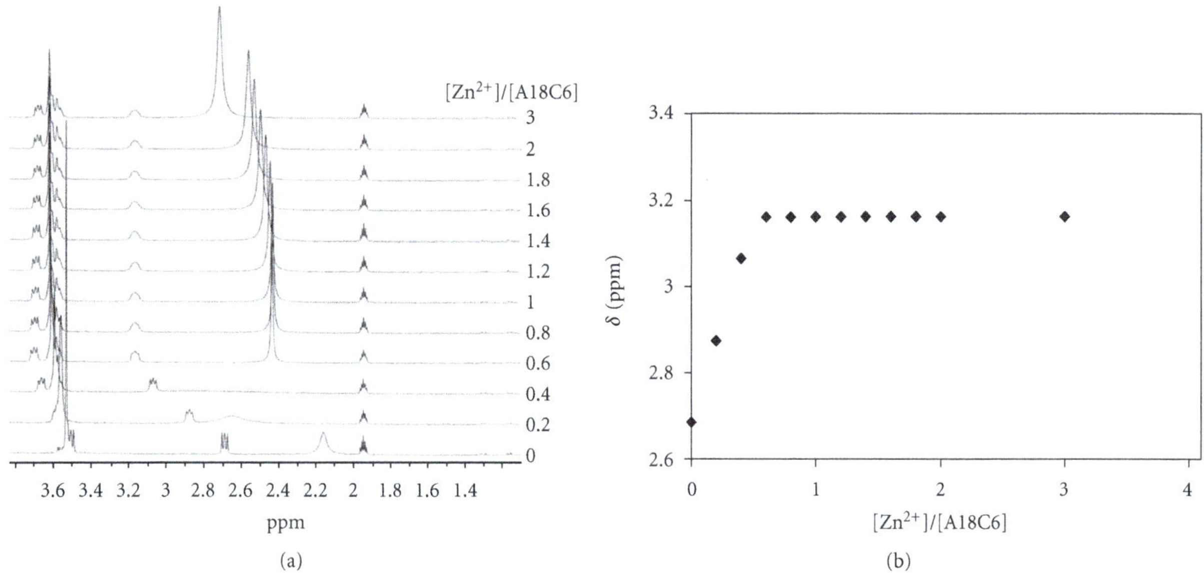

FIGURE 7: (a) The ^{1}HNMR spectra of A18C6 ligand in the presence of increasing concentration of Zn^{2+} ions in deuterated AN and (b) chemical shift versus $[Zn^{2+}]/[A18C6]$ plot.

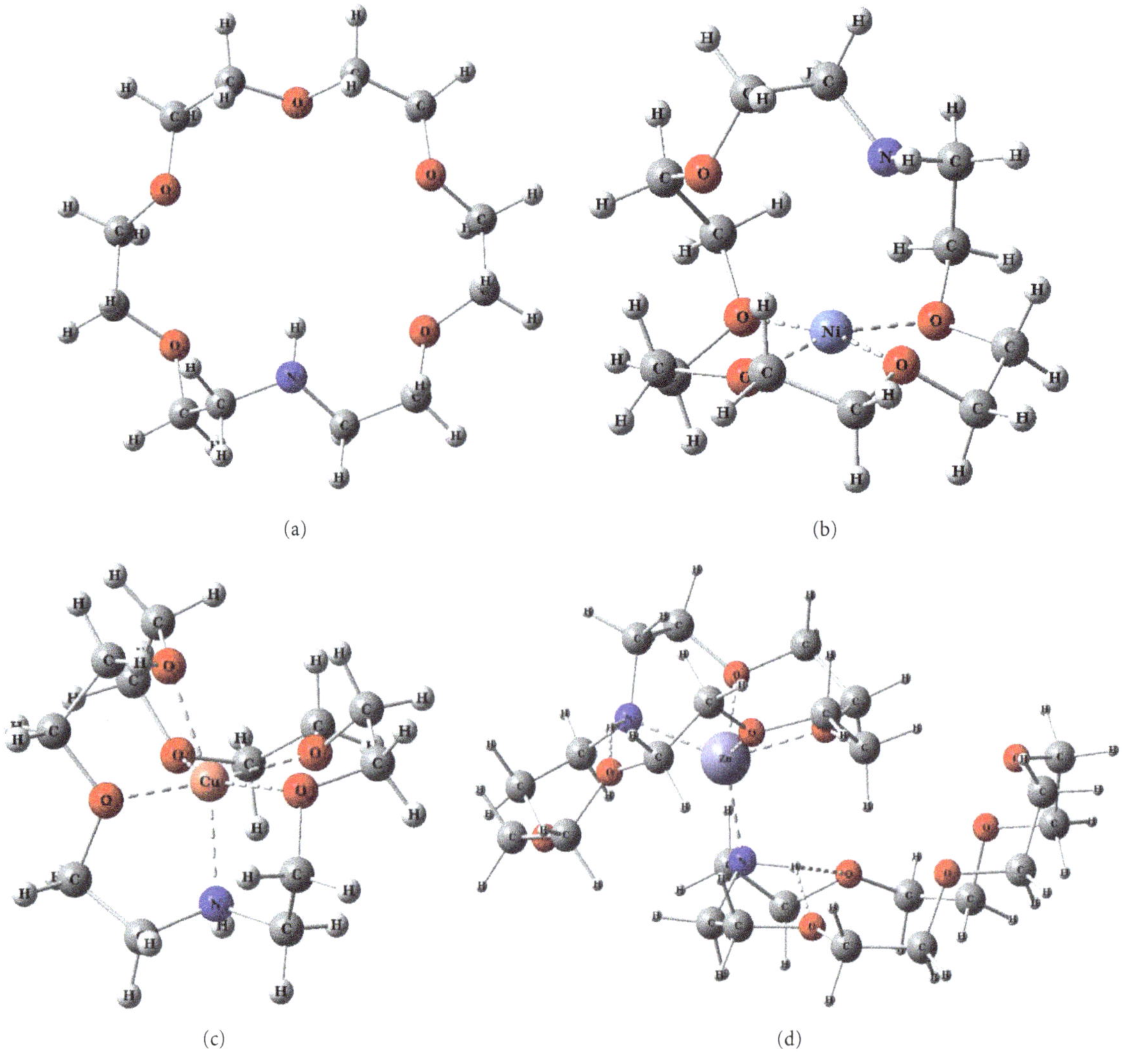

(a) (b)

(c) (d)

FIGURE 8: The optimized geometries of free A18C6 (a) and its complexes with Ni^{2+} (b), Cu^{2+} (c), and Zn^{2+}(d).

quantum chemical calculations, the following can be concluded.

(1) The solvent illustrate the fundamental role in the M^{2+}-A18C6 complexation reactions. In the case of all metal ions, the stability of the resulting complexes with A18C6 decreases and the order of stability vary with increasing MeOH in the solvent mixture.

(2) Although the enthalpy and entropy changes are strongly solvent dependent, the observed increase (or decrease, depending on the nature of the metal ion) in ΔH° value upon addition of MeOH to AN will be compensated by an increase (or decrease) in the corresponding ΔS° value.

(3) The quantum chemical calculations confirm the formation of stable A18C6-M^{2+} (M^{2+} = Ni^{2+} and Cu^{2+}) and $(A18C6)_2$- Zn^{2+} complexes.

References

[1] C. J. Pedersen, "Cyclic polyethers and their complexes with metal salts," *Journal of the American Chemical Society*, vol. 89, no. 26, pp. 7017–7036, 1967.

[2] E. Blasius and K. Janzen, "Host guest complex chemistry," in *Macrocycles: Synthesis, Structures, Applications*, F. Vögtle and E. Weber, Eds., pp. 189–216, Springer, Berlin, Germany, 1985.

[3] J. L. Atwood and J. M. Lehn, "Comprehensive supramolecular chemistry," in *Molecular Recognition: Receptors For Cationic Guests*, vol. 1, Pergamon, 1996.

[4] R. M. Izatt, K. Pawlak, J. S. Bradshaw, and R. L. Bruening, "Thermodynamic and kinetic data for macrocycle interaction with cations, anions, and neutral molecules," *Chemical Reviews*, vol. 95, no. 7, pp. 2529–2586, 1995.

[5] J. L. Atwood, *Inclusion Phenomena and Molecular Recognition*, Plenum, 1990.

[6] F. A. Christy and P. S. Shrivastav, "Conductometric studies on cation-crown ether complexes: a review," *Critical Reviews in Analytical Chemistry*, vol. 41, no. 3, pp. 236–269, 2011.

[7] M. C. Aragoni, M. Arca, F. Demartin et al., "Fluorometric chemosensors. Interaction of toxic heavy metal ions PbII, CdII, and HgII with novel mixed-donor phenanthroline-containing macrocycles: spectrofluorometric, conductometric, and crystallographic studies," *Inorganic Chemistry*, vol. 41, no. 25, pp. 6623–6632, 2002.

[8] M. Saaid, B. Saad, I. A. Rahman, A. S. M. Ali, and M. I. Saleh, "Extraction of biogenic amines using sorbent materials containing immobilized crown ethers," *Talanta*, vol. 80, no. 3, pp. 1183–1190, 2010.

[9] A. M. Costero, S. Gil, J. Sanchis, S. Peransí, V. Sanz, and J. A. Gareth Williams, "Conformationally regulated fluorescent sensors. Study of the selectivity in Zn^{2+} versus Cd^{2+} sensing," *Tetrahedron*, vol. 60, no. 30, pp. 6327–6334, 2004.

[10] A. M. Costero, J. Sanchis, S. Peransi, S. Gil, V. Sanz, and A. Domenech, "Bis(crown ethers) derived from biphenyl: extraction and electrochemical properties," *Tetrahedron*, vol. 60, no. 21, pp. 4683–4691, 2004.

[11] G. Dubois, R. Tripier, S. Brandès, F. Denat, and R. Guilard, "Cyclam complexes containing silica gels for dioxygen adsorption," *Journal of Materials Chemistry*, vol. 12, no. 8, pp. 2255–2261, 2002.

[12] J. Bradshaw, S. Nielsen, J. Lamb, J. Christensen, and D. Sen, "Thermodynamic and kinetic data for cation-macrocycle interaction," *Chemical Reviews*, vol. 85, pp. 271–339, 1985.

[13] G. H. Rounaghi, A. Soleamani, and K. R. Sanavi, "Conductance studies on complex formation between aza-18-crown-6 with Ag^+, Hg^{2+} and Pb^{2+} cations in DMSO-H_2O binary solutions," *Journal of Inclusion Phenomena and Macrocyclic Chemistry*, vol. 58, no. 1-2, pp. 43–48, 2007.

[14] M. Shamspur and H. R. Pouretedal, "Conductance study of the thermodynamics of some transition and heavy metal complexes with Aza-18-crown-6 in dimethylformamide solution," *Journal of Chemical Society of Pakistan*, vol. 21, no. 1, pp. 14–20, 1999.

[15] V. Gutmann, *The Donor-Acceptor Approach to Molecular Interactions*, Plenum Press, New York, NY, USA, 1978.

[16] Y. C. Wu and W. F. Koch, "Absolute determination of electrolytic conductivity for primary standard KCl solutions from 0 to 50°C," *Journal of Solution Chemistry*, vol. 20, no. 4, pp. 391–401, 1991.

[17] S. Katsuta, Y. Ito, and Y. Takeda, "Stabilities in nitromethane of alkali metal ion complexes with dibenzo-18-crown-6 and dibenzo-24-crown-8 and their transfer from nitromethane to other polar solvents," *Inorganica Chimica Acta*, vol. 357, no. 2, pp. 541–547, 2004.

[18] D. P. Zollinger, E. Bulten, A. Christenhusz, M. Bos, and W. E. van der Linden, "Computerized conductometric determination of stability constants of complexes of crown ethers with alkali metal salts and with neutral molecules in polar solvents," *Analytica Chimica Acta*, vol. 198, pp. 207–222, 1987.

[19] R. Shannon, "Revised effective ionic radii and systematic studies of interatomic distances in halides and chalcogenides," *Acta Crystallographica Section A*, vol. 32, pp. 751–767, 1976.

[20] M. A. Bush and M. R. Truter, "Crystal structures of complexes between alkali-metal salts and cyclic polyethers. Part IV. The crystal structures of dibenzo-30-crown-10 (2,3:17,18-dibenzo-1,4,7,10,13,16,19,22,25,28-decaoxacyclotriaconta-2,17-diene) and of its complex with potassium iodide," *Journal of the Chemical Society, Perkin Transactions 2*, no. 3, pp. 345–350, 1972.

[21] R. M. Izatt, K. Pawlak, J. S. Bradshaw, and R. L. Bruening, "Thermodynamic and kinetic data for macrocycle interaction with cations and anions," *Chemical Reviews*, vol. 91, no. 8, pp. 1721–2085, 1991.

[22] M. K. Rofouei, M. Taghdiri, M. Shamsipur, and K. Alizadeh, "^{133}Cs NMR study of Cs^+ ion complexes with dibenzo-24-crown-8, dicyclohexano-24-crown-8 and dibenzo-30-crown-10 in binary acetonitrile-nitromethane mixtures," *Journal of Solution Chemistry*, vol. 39, no. 9, pp. 1350–1359, 2010.

[23] E. Grunwald and C. Steel, "Solvent reorganization and thermodynamic enthalpy-entropy compensation," *Journal of the American Chemical Society*, vol. 117, no. 21, pp. 5687–5692, 1995.

[24] Y. Inoue, Y. Liu, and T. Hakushi, *Thermodynamics of Cation-Macrocycle Complexation: Enthalpy-Entropy Compensation*, Dekker, New York, NY, USA, 1990.

[25] G. Khayatian and F. S. Karoonian, "Conductance and thermodynamic study of the complexation of ammonium ion with different crown ethers in binary nonaqueous solvents," *Journal of the Chinese Chemical Society*, vol. 55, no. 2, pp. 377–384, 2008.

[26] L. Chen, M. Bos, P. D. J. Grootenhuis et al., "Stability constants for some divalent metal ion/crown ether complexes in methanol determined by polarography and conductometry," *Analytica Chimica Acta*, vol. 201, pp. 117–125, 1987.

[27] B. Gao, S. Wang, and Z. Zhang, "Study on complexation adsorption behavior of dibenzo-18-crown-6 immobilized on CPVA microspheres for metal ions," *Journal of Inclusion Phenomena and Macrocyclic Chemistry*, vol. 68, no. 3-4, pp. 475–483, 2010.

[28] R. Hyperchem, 7.0, Hypercube Inc., Gainesville, Fla, USA, 2002.

[29] M. Frisch, G. Trucks, H. Schlegel et al., "Program Gaussian 03," Revision b 3, 2003.

2

Thermodynamic Equilibrium Analysis of Methanol Conversion to Hydrocarbons Using Cantera Methodology

Duminda A. Gunawardena and Sandun D. Fernando

Biological and Agricultural Engineering Department, Texas A&M University, College Station, TX 77843, USA

Correspondence should be addressed to Sandun D. Fernando, sfernando@tamu.edu

Academic Editor: Krzysztof J. Ptasinski

Reactions associated with removal of oxygen from oxygenates (deoxygenation) are an important aspect of hydrocarbon fuels production process from biorenewable substrates. Here we report the equilibrium composition of methanol-to-hydrocarbon system by minimizing the total Gibbs energy of the system using Cantera methodology. The system was treated as a mixture of 14 components which had CH_3OH, C_6H_6, C_7H_8, C_8H_{10} (ethyl benzene), C_8H_{10} (xylenes), C_2H_4, C_2H_6, C_3H_6, CH_4, H_2O, C, CO_2, CO, H_2. The carbon in the equilibrium mixture was used as a measure of coke formation which causes deactivation of catalysts that are used in aromatization reaction(s). Equilibrium compositions of each species were analyzed for temperatures ranging from 300 to 1380 K and pressure at 0–15 atm gauge. It was observed that when the temperature increases the mole fractions of benzene, toluene, ethylbenzene, and xylene pass through a maximum around 1020 K. At 300 K the most abundant species in the system were CH_4, CO_2, and H_2O with mole fractions 50%, 16.67%, and 33.33%, respectively. Similarly at high temperature (1380 K), the most abundant species in the system were H_2 and CO with mole fractions 64.5% and 32.6% respectively. The pressure in the system shows a significant impact on the composition of species.

1. Introduction

Methanol is the simplest alcohol which has a tremendous importance as an industrial feedstock [1, 2]. As a fuel, methanol does not have high enough specific heat value to compete with gasoline and therefore its not attractive as a substitute but as a motor fuel additive it is said to be improving the fuel quality. The prospect of methanol being used as raw material for fuel processing actually started with the accidental discovery by Chang and Silvestry in the early 70s [3]. With the use of newly discovered ZSM-5 it was found that methanol can be transformed to gasoline grade products. Methanol conversion process in the industry has branched into two paths, namely, methanol to olefins (MTO) and methanol to gasoline (MTG). Even though MTG got the global attention as an alternative route to produce fuel, it was unable to make the process economically viable [4, 5]. To make the process economical the process parameters has to be optimized. Catalyst upgrading to make deoxygenation reaction more selective toward gasoline products such as benzene, toluene, ethylbenzene, and xylene (BTEX) is one such approach [6, 7]. Another approach is to alter the reaction conditions such as temperature, pressure, and residence time to augment the desired product spectrum [6, 8]. For this purpose, understanding the energetics of the MTG reaction pathway by thermodynamic analysis is also an important step.

The reaction pathway of MTG process is not yet completely resolved. However, from the available information in the literature, it is clear that it involves a series of reactions [9]. How the first C–C bond formation occurs is still under debate [10–12]. The widely accepted model so far is based on the hydrocarbon pool method where it is described as a catalytic scaffold with organic molecules adsorbed on to the zeolite structure [13, 14].

As given in Figure 1, MTG process takes place in a series of steps where formation of dimethyl ether is said to be the first step. Olefinic products produced at the secondary stage is significant for the MTO process while at tertiary stage is gasoline grade products are obtained. The products of MTG

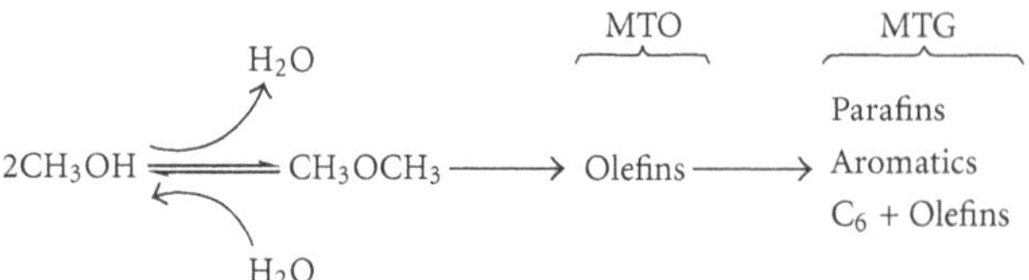

FIGURE 1: Reaction pathway of MTO and MTG processes as proposed in the literature [15].

process consists of a mixture of paraffins, aromatics, and olefins, and for the thermodynamic system, are treated as equilibrium products in this analysis.

2. Thermodynamic Equilibrium Analysis

If the stoichiometric equation of methanol conversion to aromatic hydrocarbons is known, the thermodynamic equilibrium analysis can easily be performed using the equilibrium constant. However, when the reaction stoichiometry is not known, the nonstoichiometric method should be used. Equilibrium composition for nonstichiometric systems can be determined by different methods such as energy minimization technique, kinetic/dynamic model, or by neural network technique. Since our system involves temperature and pressure as variables, it is much convenient to select Gibbs free energy minimization technique. In this study, a code written in Python (an open source programming language) to run the Cantera software library was used to analyze the equilibrium composition of the model mixture. Cantera is an object oriented software tool developed by a team from California Institute of Technology for solving chemical kinetics, thermodynamics, and transport processes [16]. Equilibrium composition was determined by Gibbs energy minimization by the Villars-Cruise-Smith algorithm [17]. Thermodynamic data for the species were calculated using the nine-coefficient NASA polynomial [18]. All the coefficients for the polynomial were obtained from Burcats online database [19].

3. Results and Discussion

The outcome of the optimization routine is the mole fraction of each of the fourteen compounds in the mixture at each pressure and temperature. Since it is a gas phase system, the partial pressure of each component is proportional to mole fraction and results are analyzed in those terms. As expected, the system with the fourteen components reaches equilibrium at different temperatures and pressures.

The key input to the model was the standard formation enthalpies of different species considered. Table 1 gives the formation enthalpies as obtained from [20].

TABLE 1: Standard formation enthalpies of the species considered in the system.

Compound name	ΔH_f^o (kJ/kmol)
CH_3OH	−205
C_6H_6	82.8
C_7H_8	50.1
C_8H_{10} (ethyl benzene)	49.0
C_8H_{10} (xylenes)	17.9
C_2H_4	52.47
C_2H_6	−83.8
C_3H_6	20.41
CH_4	−74.87
H_2O	−241.83
C	0.0
CO_2	−393.52
CO	−110.53
H_2	0.0

3.1. Methanol Conversion to Aromatics. According to the thermodynamic analysis, equilibrium aromatics yield at different temperatures is extremely low. Figure 2(a) depicts that the BTEX mole fraction trend in the equilibrium mixture passes through a maximum around 1100 K. The mole fraction of BTEX clearly increases as pressure increases. The reaction model considered in this study resembles a thermochemical conversion of methanol at different pressures. It is interesting to note that all the aromatics considered in the model initiated formation around 850 K and peaked approximately around 1110 K. The formation started to disappear beyond 1300 K.

This analysis suggests that to maximize gasoline fraction aromatic products formation, the reactions should be carried out at a narrow temperature regime. This temperature window broadens as pressure increases but is quite narrow at lower pressures. It appears that the best reaction temperature for gasoline range aromatics formation is around 800–850°C.

In order to better approximate experimental BTX yields the model was forced to reach equilibrium with eleven hydrocarbon products such as CH_3OH, C_6H_6, C_7H_8, C_8H_{10} (ethyl benzene), C_8H_{10} (xylenes), C_2H_4, C_2H_6, C_3H_6, CH_4, H_2O, H_2 neglecting thermodynamically more stable species such as CO_2, CO, and C. This forced equilibrium conditions resulted in changing the aromatic mole fraction in the system significantly as shown in Figure 2(b). According to this analysis, the highest total aromatic mole percentage of 9.41% was observed at 0 psi and 1380 K. Thermodynamics favor benzene formation in both methods in the aromatic fraction and the highest yield is reported at 0 psi and 1380 K.

3.2. Methanol Conversion to Paraffins. Methane, ethane, and propane are the alkanes considered in this model. It is quite

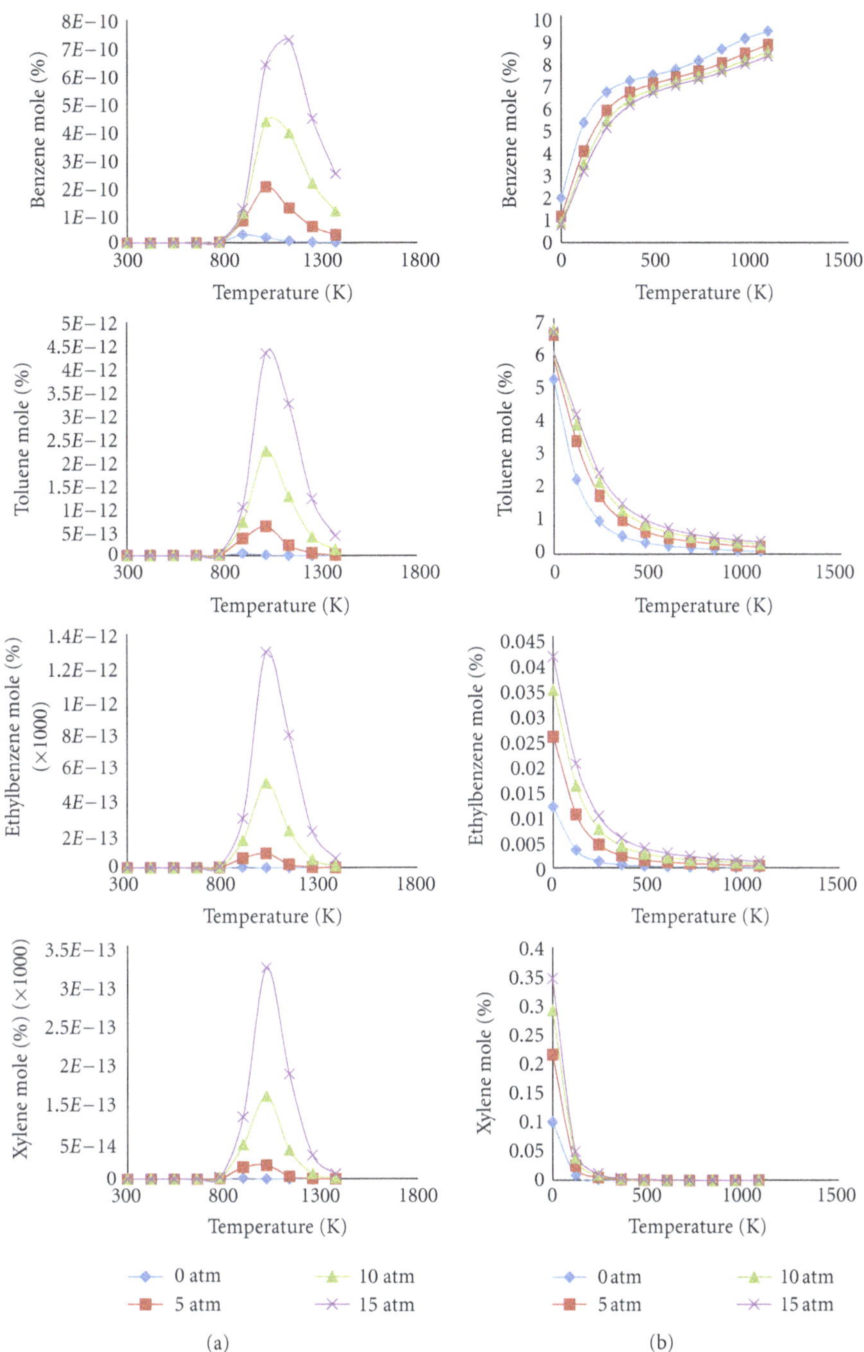

Figure 2: (a) Equilibrium composition of gasoline-range aromatic hydrocarbons in the fourteen-component model. (b) The equilibrium composition of aromatics after the system was forced to equilibrate with eleven species removing CO_2, CO, and C.

interesting that methane is the most abundant alkane in the equilibrium mixture at 300 K under both models that were tested.

Molar faction of ethane is relatively low, but, shows an interesting behavior with respect to temperature and pressure (Figure 3). Increasing pressure increases the mole fraction of ethane and with increasing temperature, the trends is to pass through a maximum around 900 K–1120 K. According to the unrestricted model, propane mole percentage in the equilibrium mixture is very low—including the forced equilibrium model. The analysis indicates that methane is the only alkane, that is, thermodynamically favorable.

3.3. Methanol Conversion to Olefins. In this thermodynamic model, we have considered ethylene which has been reported to be present when methanol is catalytically processed [6]. It is widely believed that ethylene is one of the preliminary

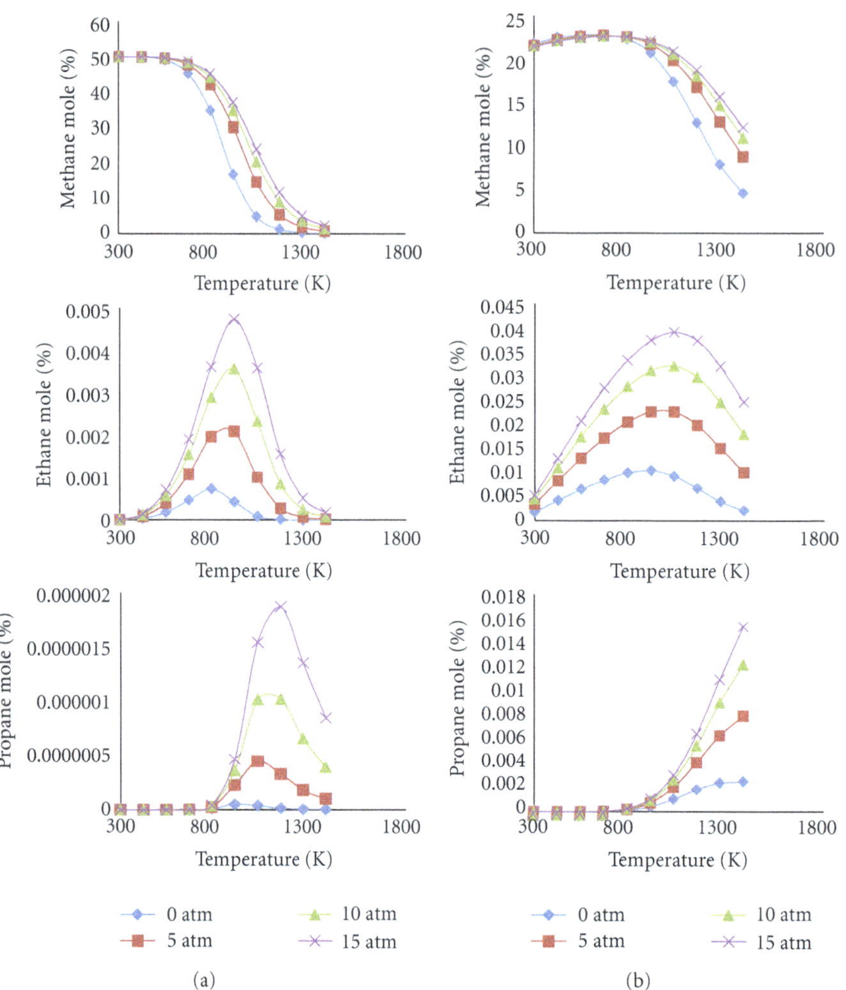

FIGURE 3: Equilibrium mole fractions of paraffins (a) with thermodynamically stable species considered in the model, (b) when CO_2, CO, and C dropped from the model.

products when methanol is converted in the presence of catalyst such as ZSM-5. However, according to Figure 4(a), it is clear that in the equilibrium mixture, the mole percentage of ethylene is quite low but under the forced equilibrium ethylene shows a significant presence than ethane and propane. It is clear that under both models, the increase of temperature clearly favored formation of ethylene.

3.4. Formation of Solid Carbon and Syngas. Carbon that is produced in the reaction is commonly attributed as a cause for catalyst deactivation. It can be seen from Figure 5(a) that carbon formation varies significantly as the temperature increases but the equilibrium mole percents are extremely low. Accordingly, the effect of pressure on carbon formation is not significant. It should be noted that in practice, the coke formation can happen not only as a result of direct carbon deposition but also by deposition of larger molecular weight carbonaceous compounds on active sites of the catalyst.

Under unrestricted equilibrium conditions, two reaction schemes can be proposed for the system. At low temperatures (300–500 K) the most abundant species in the system are methane, water, and carbon dioxide. All the other species in the model can be neglected since these are present in extremely low concentrations at these temperatures. This reaction is depicted in stoichiometric form in (1). The heat of reaction which amounts to −281.79 kJ/mol indicates that it is highly exothermic and the products methane, water, and carbon dioxide are the most stable at low temperatures. Similarly, at higher temperatures the most abundant and the most stable species are carbon monoxide and hydrogen. Hydrogen is present in the highest concentrations at temperatures > 1000 K in the mixture which amounts to 66.1% in the unrestricted equilibrium model. The products (formed at high temperature environment) can be represented in a stoichiometric form as shown in (2):

$$4CH_3OH \longrightarrow 3CH_4 + 2H_2O + CO_2, \quad \Delta H_f = -281.79\,\text{kJ/mol}, \tag{1}$$

$$CH_3OH \longrightarrow CO + 2H_2, \quad \Delta H_f = 94.47\,\text{kJ/mol}. \tag{2}$$

The system with fourteen species attains thermodynamic equilibrium at different temperatures and pressures. At

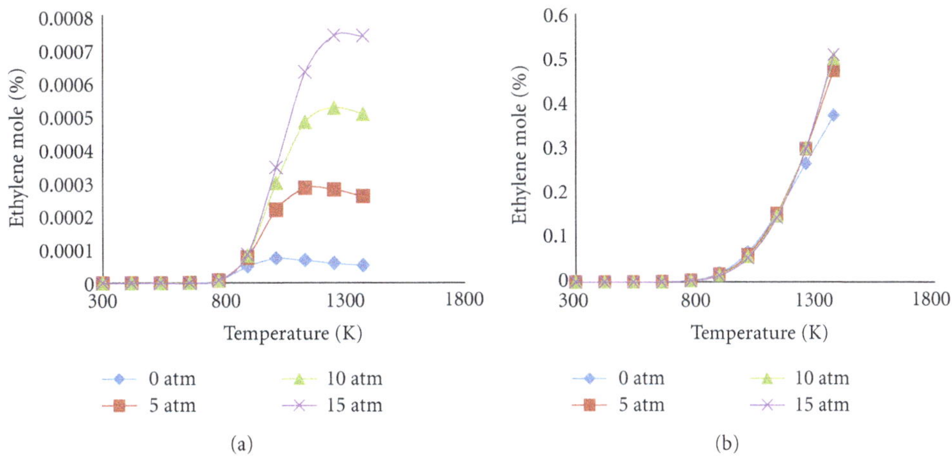

FIGURE 4: Variation of carbon mole fraction with respect to temperature at different pressures.

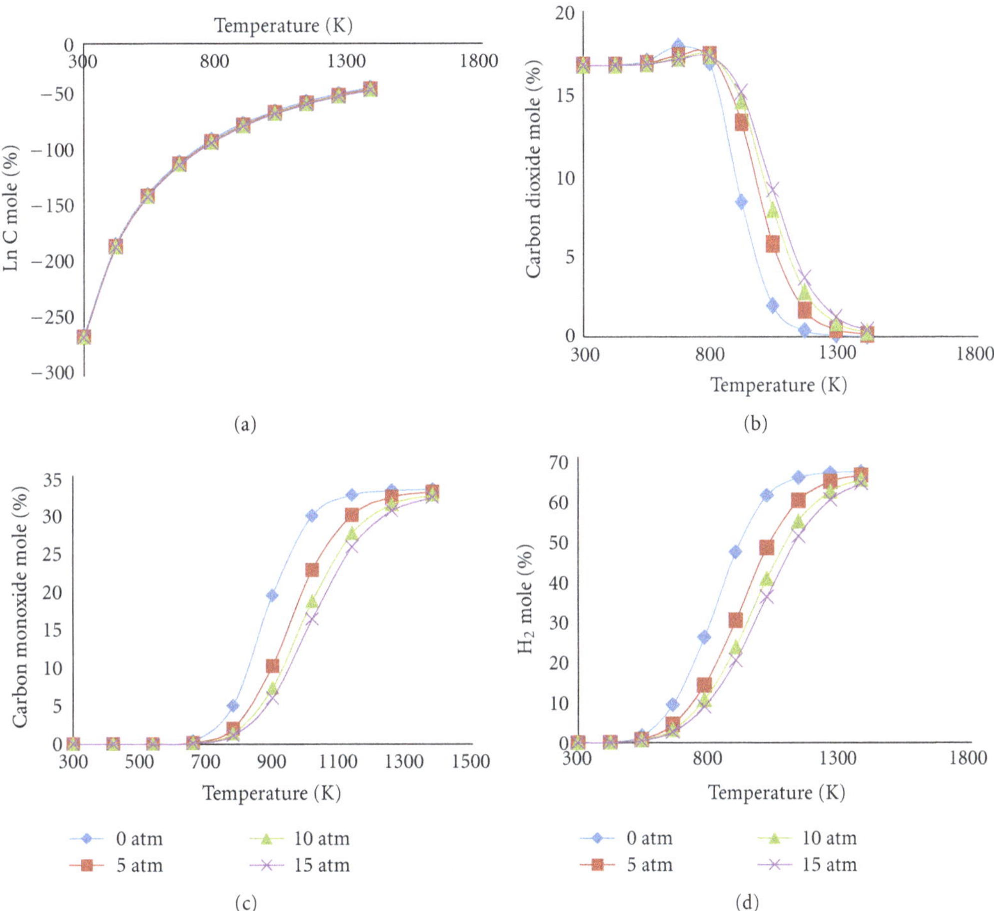

FIGURE 5: Variation of (a) carbon, (b) carbon dioxide, (c) carbon monoxide, (d) hydrogen, mole fraction with respect to temperature at different pressures.

equilibrium, the overall ΔG of the mixture is a measure of how stable the reaction is at respective reaction conditions. According to Figure 6, the Gibbs free energy of the system stays in negative region implying that the reaction is spontaneous. As the temperature increases the negative value of ΔG increases and reaches a minimum around 780 K.

For this system under equilibrium, an equilibrium constant can be calculated using (3). According to Figure 7, the system has a large equilibrium constant. A reaction with

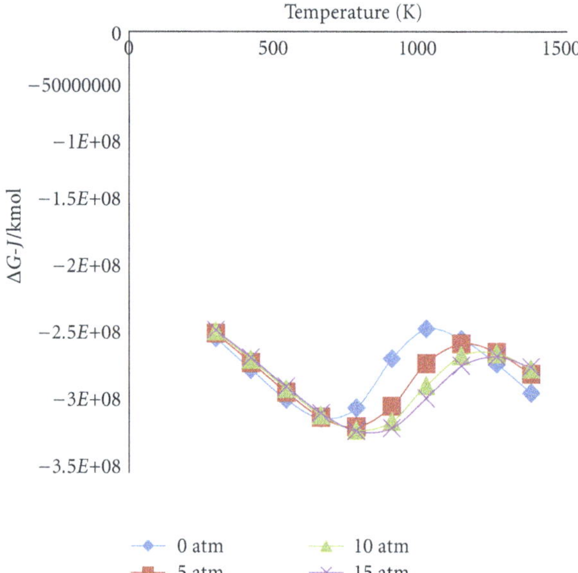

FIGURE 6: Variation of the total Gibbs free energy of the reaction (ΔG^o_{rx}) with respect to temperature at different pressures.

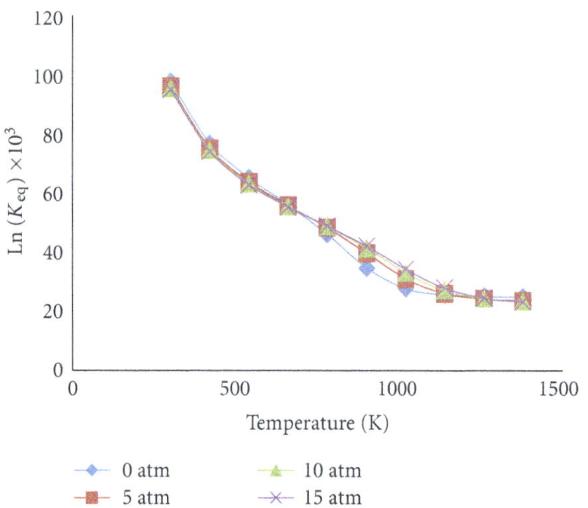

FIGURE 7: Variation of the equilibrium constant of the system with respect to temperature at different pressures.

a large equilibrium constant can take a long time to reach the equilibrium depending on the kinetics of the reaction. As the temperature increases the equilibrium constant reduces exponentially:

$$K_a = \exp\left(\frac{-\Delta G}{RT}\right). \quad (3)$$

The overall analysis suggests that when this mixture is left to equilibrate at the temperature and pressure conditions specified, the tendency of the system is to favor formation of low molecular weight compounds (such as CH_3, CO, CO_2, H_2, and H_2O). Although in a liquid fuel production point of view, we anticipate obtaining high yields of hydrocarbons, this analysis suggests that the equilibrium mole percent of BETX can reach up to 9.41%. This observation could be reinforced by the fact that ΔH_f for BTEX are all positive, the system tends to favor formation of more stable compounds with more negative ΔH_f values [21]. Accordingly, it is clear that at higher temperatures, in the absence of a catalyst, the tendency of methanol is to produce a Syngas containing primarily CO and H_2. However, literature suggests that the dynamics of the reaction changes in the presence of a catalyst such as ZSM-5. It has been shown that in the presence of a catalyst under analogous conditions, significant amounts of BTEX (with ethylene and xylenes having the highest selectivities) are produced [6]. The inference of this analysis is that, in the presence of a catalyst, simultaneous formation of all considered molecules does not occur from methanol; rather, independent reactions that comprise of only a few selected intermediate products may occur. However, with sufficient time, these energetically favorable intermediates will break down to more stable low molecular weight compounds.

4. Conclusions

Two models were tested to understand the behavior of aromatics, alkanes, and alkenes production when methanol is thermochemically deoxygenated. Under the forced equilibrium conditions it shows that the aromatic mole percentage is 9.41%. At low temperature the most abundant stable species were methane, carbon dioxide, and water, while hydrogen and carbon monoxide were the most dominant stable species at higher temperatures. Methane was the most abundant species at low temperature with a mole fraction of 50%. At high temperature the most abundant species was hydrogen with mole fraction 66.1%. The system under equilibrium produces negligible amounts of elemental carbon. The negative values of the free energy (ΔG) of the system indicate that the reaction is spontaneous for the entire range of temperatures pressures that were analyzed. Further, the analysis indicates that increasing temperature would increase the negative value of free energy making the system stable. The overall equilibrium constant of the system drops exponentially with increasing temperature.

Acknowledgment

This paper is based upon work supported by the National Science Foundation under Grant no. CBET 0965772.

References

[1] X. Yin, D. Y. C. Leung, J. Chang, J. Wang, Y. Fu, and C. Wu, "Characteristics of the synthesis of methanol using biomass-derived syngas," *Energy & Fuels*, vol. 19, no. 1, pp. 305–310, 2005.

[2] P. Reubroycharoen, T. Yamagami, T. Vitidsant, Y. Yoneyama, M. Ito, and N. Tsubaki, "Continuous low-temperature methanol synthesis from syngas using alcohol promoters," *Energy & Fuels*, vol. 17, no. 4, pp. 817–821, 2003.

[3] C. D. Chang and A. J. Silvestri, "The conversion of methanol and other O-compounds to hydrocarbons over zeolite catalysts," *Journal of Catalysis*, vol. 47, no. 2, pp. 249–259, 1977.
[4] M. W. Anderson and J. Klinowski, "Solid state NMR studies of the shape-selective catalytic conversion of Methanol into gasoline on Zeolite ZSM-5," *Journal of the American Chemical Society*, vol. 112, no. 1, pp. 10–16, 1990.
[5] B. C. Gates, *Catalytic Chemistry*, The Wiley Series in Chemical Engineering, Wiley, 1991.
[6] R. Barthos, T. Bánsági, T. Süli Zakar, and F. Solymosi, "Aromatization of methanol and methylation of benzene over Mo_2C/ZSM-5 catalyst," *Journal of Catalysis*, vol. 247, no. 2, pp. 368–378, 2007.
[7] D. Freeman, R. P. K. Wells, and G. J. Hutchings, "Conversion of methanol to hydrocarbons over Ga_2O_3/H-ZSM-5 and Ga_2O_3/WO_3 catalysts," *Journal of Catalysis*, vol. 205, no. 2, pp. 358–365, 2002.
[8] Ø. Mikkelsen and S. Kolboe, "The conversion of methanol to hydrocarbons over zeolite H-beta," *Microporous and Mesoporous Materials*, vol. 29, no. 1-2, pp. 173–184, 1999.
[9] D. M. McCann, D. Lesthaeghe, P. W. Kletnieks et al., "A complete catalytic cycle for supramolecular methanol-to-olefins conversion by linking theory with experiment," *Angewandte Chemie*, vol. 47, no. 28, pp. 5179–5182, 2008.
[10] S. R. Blaszkowski and R. A. Van Santen, "Theoretical study of C-C bond formation in the methanol-to- gasoline process," *Journal of the American Chemical Society*, vol. 119, no. 21, pp. 5020–5027, 1997.
[11] J. E. Jackson and F. M. Bertsch, "Conversion of methanol to gasoline: a new mechanism for formation of the first carbon-carbon bond," *Journal of the American Chemical Society*, vol. 112, no. 25, pp. 9085–9092, 1990.
[12] S. M. Campbell, X. Z. Jiang, and R. F. Howe, "Methanol to hydrocarbons: spectroscopic studies and the significance of extra-framework aluminium," *Microporous and Mesoporous Materials*, vol. 29, no. 1-2, pp. 91–108, 1999.
[13] U. Olsbye, M. Bjørgen, S. Svelle, K. P. Lillerud, and S. Kolboe, "Mechanistic insight into the methanol-to-hydrocarbons reaction," *Catalysis Today*, vol. 106, no. 1–4, pp. 108–111, 2005.
[14] I. M. Dahl and S. Kolboe, "On the reaction mechanism for propene formation in the MTO reaction over SAPO-34," *Catalysis Letters*, vol. 20, no. 3-4, pp. 329–336, 1993.
[15] D. A. Gunawardena and S. D. Fernando, "Deoxygenation of methanol over ZSM-5 in a high pressure catalytic pyroprobe," *Chemical Engineering and Technology*, vol. 34, no. 2, pp. 173–178, 2011.
[16] D. Goodwin, "Cantera: an object-oriented software toolkit for chemical kinetics, thermodynamics, and transport processes," http://code.google.com/p/cantera/.
[17] M. Baratieri, P. Baggio, L. Fiori, and M. Grigiante, "Biomass as an energy source: thermodynamic constraints on the performance of the conversion process," *Bioresource Technology*, vol. 99, no. 15, pp. 7063–7073, 2008.
[18] A. Burcat and B. Ruscic, "Third millenium ideal gas and condensed phase thermochemical database for combustion with update from active thermochemical tables," Tech. Rep., Argonne National Laboratory, 2005.
[19] A. Burcat and B. Ruscic, New NASA Thermodynamic Polynomials Database With Active Thermochemical Tables updates, ANL 05/20 TAE 960, 2010, ftp://ftp.technion.ac.il/pub/supported/aetdd/ thermodynamics.
[20] A. F. Kazakov, NIST/TRC Web Thermo Tables (WTT)—Professional Edition, Thermodynamics Research Center, 2010.
[21] S.I. Sandler, *Chemical, Biochemical, and Engineering Thermodynamics*, John Wiley & Sons, 2005.

3

Maximum Power Point Characteristics of Generalized Heat Engines with Finite Time and Finite Heat Capacities

Abhishek Khanna[1,2] and Ramandeep S. Johal[1]

[1] *Department of Physical Sciences, Indian Institute of Science Education and Research Mohali, Sector 81, Manauli, Mohali, Punjab 140306, India*
[2] *Elite Course Theoretical and Mathematical Physics, Ludwig Maximillian University, D-80333 Munich, Germany*

Correspondence should be addressed to Ramandeep S. Johal, rsjohal@iisermohali.ac.in

Academic Editor: M. A. Rosen

We revisit the problem of optimal power extraction in four-step cycles (two adiabatic and two heat-transfer branches) when the finite-rate heat transfer obeys a linear law and the heat reservoirs have finite heat capacities. The heat-transfer branch follows a polytropic process in which the heat capacity of the working fluid stays constant. For the case of ideal gas as working fluid and a given switching time, it is shown that maximum work is obtained at Curzon-Ahlborn efficiency. Our expressions clearly show the dependence on the relative magnitudes of heat capacities of the fluid and the reservoirs. Many previous formulae, including infinite reservoirs, infinite-time cycles, and Carnot-like and non-Carnot-like cycles, are recovered as special cases of our model.

1. Introduction

Curzon-Ahlborn efficiency, $\eta_{CA} = 1 - \sqrt{T_2/T_1}$, where T_1 and T_2 are the reservoir temperatures [1], is regarded as a landmark result of finite-time thermodynamics. It models the effect of irreversibilities due to finite rate of heat transfer on the performance of heat engines. Such models are termed as endoreversible. The flow of heat between the working fluid and the reservoirs is assumed to be Newtonian. All the components maintain an internal equilibrium and losses due to friction, and heat leak are assumed to be negligible. Many authors [2–10] extended and clarified the scope of this model. In fact, various other models of heat engines also yield maximum power output at efficiencies very close to CA efficiency [11–15]. This universal-like behaviour of efficiency has been recently analysed using Bayesian probabilities in the case of quantum models of engines [16].

In this paper, we revisit the problem of optimal performance with "linear" irreversibilities of finite time and finite heat reservoirs in classical models of engines. This question was addressed in [7] using a Lagrangian formalism, for a one-component working fluid without assuming an equation of state. However, the generic problem with finite reservoirs could not be solved in a closed form. Then Gordon [8] used an ancillary device of intermediate reservoirs to arrive at a closed form solution. In particular, it was shown [8] that for Carnot-like engines, finiteness of reservoirs has no effect on the efficiency at maximum power, and it is still at Curzon-Ahlborn efficiency. The case of other non-Carnot heat engines (such as Otto cycle, Joule-Brayton cycle, and Atkinson cycle) was discussed by using an infinite chain of Carnot cycles.

The present analysis of the problem is based on the following features: (a) the working fluid is a classical ideal gas; (b) the total cycle time as well as switching time is given. We consider a generic four-step cycle with two adiabatic and two heat-transfer branches, which follow a polytropic process with a constant heat capacity $C > 0$. The process is described by the relation $TV^x = k$. Here, T is the temperature of the working fluid, V is the volume of the working fluid, and k is an arbitrary constant. Consequently, many popular heat engines like Carnot cycle ($x = 0$), Otto cycle ($x \to \infty$), and Joule-Brayton ($x = -1$) can be incorporated in this study.

Our key results may be listed as follows. For ideal gas as the working fluid and the generic four-step cycle (including

Carnot and non-Carnot heat cycles), the power of the engine is maximum at CA efficiency, even for finite reservoirs. Many special cases like infinite time and infinite reservoirs [11], finite time and infinite reservoirs with isothermal branches [1] or polytropic branches can be described. We note that closed form expressions for work can be derived without invoking a device like an intermediate reservoir [8]. Further, the heat capacities of the working fluid and the (finite) reservoirs enter explicitly in the expressions, and their relative sizes appear as the respective ratios of heat capacities. Also, the explicit expressions of work for these special cases clearly show that the irreversibilities of finite time and/or finite reservoirs reduce the maximum amount of work extracted as compared with infinite time and infinite reservoirs cases.

2. The Model

2.1. Temperature Profiles. First of all, we calculate how the temperatures of the two bodies change, when kept in contact with one another for some time. The heat transfer between the bodies is Newtonian, and during heat transfer, their heat capacities remain constant.

Let the two bodies denoted as A and B have heat capacities, C_A and C_B, respectively. The bodies are kept in thermal contact with one another from time, $t = 0$ to $t = t_1$. The temperatures of the two bodies at any time t are $T_A(t)$ and $T_B(t)$, respectively, and the coefficient of heat conductivity is K.

At time t, the heat flow between bodies A and B should be equal to heat gained (lost) by one or heat lost (gained) by the other. Therefore,

$$K[T_A(t) - T_B(t)]dt = -C_A dT_A(t) = C_B dT_B(t). \quad (1)$$

Solving the above equations, temperature profiles of the two bodies at time t are calculated as

$$T_A(t) = T_A(0) + \frac{\gamma_1}{\gamma}(1 - e^{-\gamma t}), \quad (2)$$

$$T_B(t) = T_B(0) + \frac{\gamma_2}{\gamma}(1 - e^{-\gamma t}), \quad (3)$$

where

$$\gamma = K\left(\frac{1}{C_A} + \frac{1}{C_B}\right), \quad (4)$$

$$\gamma_1 = \frac{-K}{C_A}(T_A(0) - T_B(0)), \quad (5)$$

$$\gamma_2 = \frac{K}{C_B}(T_A(0) - T_B(0)). \quad (6)$$

2.2. The Cycle. We have two finite heat reservoirs at initial temperatures $T_1(0)$ and $T_2(0)$ and heat capacities are C_1 and C_2, respectively. A working fluid is chosen to extract work from the reservoirs. The time period for one working cycle is t_2. One working cycle consists of the following steps.

(i) Working fluid is brought in contact with the hot reservoir from time $t = 0$ to t_1. The fluid expands along a path with heat capacity C. Consequently, temperature of the hot reservoir reduces from $T_1(0)$ to $T_1(t_1)$, and that of the working fluid rises from $T(0)$ to $T^-(t_1)$.

(ii) The fluid is allowed to expand adiabatically. This step is assumed to be instantaneous and takes negligible time. As a result, the temperature of the working fluids jumps from $T^-(t_1)$ to $T^+(t_1)$.

(iii) The fluid is brought in contact with the cold reservoir from time t_1 to t_2. The fluid contracts along a path with heat capacity C. Consequently, temperature of the cold reservoir rises from $T_2(t_1)$ to $T_2(t_2)$ and that of the working fluid reduces from $T^+(t_1)$ to $T^-(t_2)$.

(iv) Working fluid is allowed to contract adiabatically. This step is also assumed to take negligible time. As a result, the temperature of the working fluids jumps from $T^-(t_2)$ to $T(0)$, completing the cycle.

Using (2)–(6), the temperature profiles of the reservoirs and the working fluid in different steps can be described as follows:

(i) $0 < t < t_1$:

$$T_1(t) = T_1(0) + \frac{\lambda_1}{\lambda}\left(1 - e^{-\lambda t}\right),$$
$$T(t) = T(0) + \frac{\lambda_2}{\lambda}\left(1 - e^{-\lambda t}\right), \quad (7)$$

where

$$\lambda = K_1\left(\frac{1}{C_1} + \frac{1}{C}\right),$$
$$\lambda_1 = \frac{-K_1}{C_1}[T_1(0) - T(0)], \quad (8)$$
$$\lambda_2 = \frac{K_1}{C}[T_1(0) - T(0)].$$

(ii) $t = t_1$:

$$\lim_{t \to t_1^-} T(t) = T(0) + \frac{\lambda_2}{\lambda}\left(1 - e^{-\lambda t}\right) = T^-(t_1), \quad (9)$$

$$\lim_{t \to t_1^+} T(t) = T^+(t_1), \quad T^+(t_1) < T^-(t_1). \quad (10)$$

(iii) $t_1 < t < t_2$:

$$T(t) = T^+(t_1) + \frac{\beta_1}{\beta}\left(1 - e^{-\beta(t-t_1)}\right),$$
$$T_2(t) = T_2(t_1) + \frac{\beta_2}{\beta}\left(1 - e^{-\beta(t-t_1)}\right). \quad (11)$$

Since $T_2(t_1) = T_2(0)$, this implies

$$T_2(t) = T_2(0) + \frac{\beta_1}{\beta}\left(1 - e^{-\beta(t-t_1)}\right), \quad (12)$$

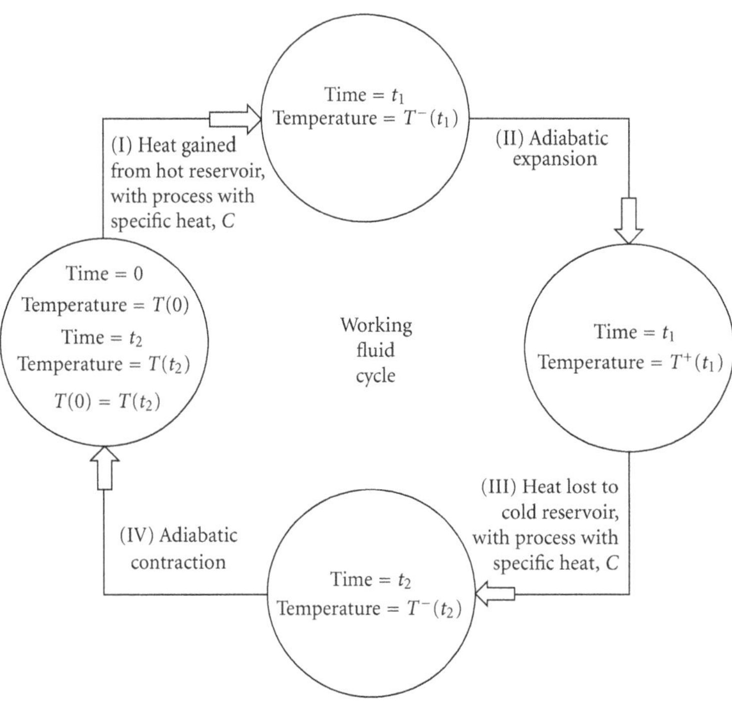

Figure 1: Heat cycle of the working fluid.

where

$$\beta = K_2\left(\frac{1}{C_2} + \frac{1}{C}\right),$$

$$\beta_1 = \frac{-K_2}{C}(T^+(t_1) - T_2(t_1)), \tag{13}$$

$$\beta_2 = \frac{K_2}{C_2}(T^+(t_1) - T_2(t_1)).$$

(iv) $t = t_2$:

$$\lim_{t\to t_2^-} T(t) = T^+(t_1) + \frac{\beta_1}{\beta}\left(1 - e^{-\beta(t-t_1)}\right) = T^-(t_2), \tag{14}$$

$$\lim_{t\to t_2^+} T(t) = T(0). \tag{15}$$

The complete cycle is depicted in Figure 1.

We will now use the above temperatures to evaluate the work performed by the engine. Due to the cyclic process, the first law of thermodynamics implies that the net heat exchanged by the working fluid during the cycle equals the work performed. Thus

$$W = \int_{T(0)}^{T^-(t_1)} C dT + \int_{T^+(t_1)}^{T^-(t_2)} C dT. \tag{16}$$

Since specific heat is not a function of temperature,

$$W = C[T^-(t_1) - T(0)] + C[T^-(t_2) - T^+(t_1)]. \tag{17}$$

Using (9), (10), and (14), we obtain

$$W = C\frac{\lambda_2}{\lambda}\left(1 - e^{\lambda t_1}\right) + C\frac{\beta_1}{\beta}\left(1 - e^{-\beta(t_2-t_1)}\right). \tag{18}$$

Substituting values for λ, λ_2, β and β_1 from (8) and (13), we have

$$W = \frac{CC_1}{C+C_1}[T_1(0) - T(0)]\left(1 - e^{-K_1((1/C_1)+(1/C))t_1}\right) - \frac{CC_2}{C+C_2}[T^+(t_1) - T_2(0)]\left(1 - e^{-K_2((1/C_2)+(1/C))(t_2-t_1)}\right). \tag{19}$$

Heat absorbed from the hot reservoir (Q_1) is given by

$$Q_1 = \int_{T(0)}^{T^-(t_1)} C dT = \frac{CC_1}{C+C_1}[T_1(0) - T(0)] \times \left(1 - e^{-K_1((1/C_1)+(1/C))t_1}\right). \tag{20}$$

There is no change in the entropy of the working fluid after one cycle. Since the change in the entropy of the working fluid occurs only in the nonadiabatic branches ((i) and (iii)),

$$\oint dS = C\int_{T(0)}^{T^-(t_1)} \frac{dT}{T} + C\int_{T^+(t_1)}^{T^-(t_2)} \frac{dT}{T} = 0, \tag{21}$$

which yields

$$T^+(t_1) = \frac{T^-(t_1)T^-(t_2)}{T(0)}. \tag{22}$$

Here, it should be noticed that no heat leakage or entropy production is considered in this model.

3. Case I: Finite Time Studies with Infinite Reservoirs

In this section, we will first optimize the work per cycle over the initial temperature $T(0)$ of the working fluid and then with respect to the switching time t_1.

Under the infinite reservoirs condition ($C_1 \to \infty, C_2 \to \infty$), the temperatures of hot and the cold reservoirs remain fixed at values $T_1(0)$ and $T_2(0)$, respectively. Using (9), (14), and (22) and substituting $\lambda = K_1/C$ and $\beta = K_2/C$, we can calculate temperature of working fluid after the adiabatic step as

$$\begin{aligned}T^+(t_1) = &\left[T_2(0)\left(1 - e^{-(K_2/C)(t_2-t_1)}\right)\right.\\&\left.\times\left(T_1(0) + T(0)e^{-(K_1/C)t_1} - T_1(0)e^{-(K_1/C)t_1}\right)\right]\\&\times\left[T(0) - T_1(0)e^{-(K_2/C)(t_2-t_1)}\right.\\&\left.+ (T_1(0) - T(0))e^{-(K_1/C)t_1-(K_2/C)(t_2-t_1)}\right]^{-1}.\end{aligned} \quad (23)$$

Substituting the above expression for $T^+(t_1)$ in (19), the work obtained is given by

$$\begin{aligned}W = &C(T_1(0) - T(0))\left(1 - e^{-(K_1/C)t_1}\right)\\&\times\left[1 - \frac{T_2(0)\left(1 - e^{-(K_2/C)(t_2-t_1)}\right)}{\alpha}\right],\end{aligned} \quad (24)$$

where for convenience, we have defined

$$\begin{aligned}\alpha = &T(0) - T_1(0)e^{-(K_2/C)(t_2-t_1)}\\&+ (T_1(0) - T(0))e^{-(K_1/C)t_1-(K_2/C)(t_2-t_1)}.\end{aligned} \quad (25)$$

On using (20) with the infinite reservoir condition, and (24), the efficiency of the heat engine ($\eta = W/Q_1$) is given by

$$\eta = 1 - \frac{T_2(0)\left(1 - e^{-(K_2/C)(t_2-t_1)}\right)}{\alpha}. \quad (26)$$

Now we will optimize the work per cycle with respect to initial temperature of the working medium, for a fixed switching time, and later, optimise with respect to the switching time also

$$\left(\frac{\partial W}{\partial T(0)}\right)_{t_1} = 0. \quad (27)$$

Solving the above equation, we observe that at maximum work,

$$\alpha = \left(1 - e^{-(K_2/C)(t_2-t_1)}\right)\sqrt{T_1(0)T_2(0)}. \quad (28)$$

$$\begin{aligned}T(0) = &\left[\sqrt{T_1(0)T_2(0)}\left(1 - e^{-(K_2/C)(t_2-t_1)}\right)\right.\\&\left.+ T_1(0)e^{-(K_2/C)(t_2-t_1)}\left(1 - e^{-(K_1/C)t_1}\right)\right]\\&\times\left[1 - e^{-(K_1/C)t_1-(K_2/C)(t_2-t_1)}\right]^{-1}.\end{aligned} \quad (29)$$

Substituting the optimum values of α and $T(0)$ in (24), we obtain

$$\begin{aligned}W_{\text{opt}} = &C\left(\sqrt{T_1(0)} - \sqrt{T_2(0)}\right)^2\\&\times\frac{\left(1 - e^{-(K_1/C)t_1}\right)\left(1 - e^{-(K_2/C)(t_2-t_1)}\right)}{1 - e^{-(K_1/C)t_1-(K_2/C)(t_2-t_1)}}.\end{aligned} \quad (30)$$

Using (26) and (28), the efficiency at optimum work is found to be

$$\eta = 1 - \sqrt{\frac{T_2(0)}{T_1(0)}} \equiv \eta_{CA}. \quad (31)$$

Note that the efficiency is independent of the switching time t_1.

We can further maximize the work with respect to the switching time:

$$\frac{\partial W_{\text{opt}}}{\partial t_1} = 0. \quad (32)$$

Solving the above equation, we get

$$\begin{aligned}K_1 e^{-(K_1/C)t_1}\left(1 - e^{-(K_2/C)(t_2-t_1)}\right)^2 = &K_2 e^{-(K_2/C)(t_2-t_1)}\\&\times\left(1 - e^{-(K_1/C)t_1}\right)^2.\end{aligned} \quad (33)$$

Though we have not obtained an explicit formula for t_1, (33) can be solved numerically to get the optimum t_1.

3.1. The Isothermal Limit: $C \to \infty$. When we take the nonadiabatic paths to be isothermal, that is, $C \to \infty$, we obtain the following results for the maximum work:

$$T(0) = \kappa\sqrt{T_1(0)}, \quad (34)$$

$$T^+(t_1) = \kappa\sqrt{T_2(0)}, \quad (35)$$

$$\kappa = \frac{\sqrt{K_1 T_1(0)} + \sqrt{K_2 T_2(0)}}{\sqrt{K_1} + \sqrt{K_2}}. \quad (36)$$

The temperature of the working fluid is equal to that given by (34) from time, $t = 0$ to $t = t_1$ and to that given by (35) from time $t = t_1$ to $t = t_2$. Further, with the isothermal condition ($C \to \infty$), we can simplify (33) to calculate the optimum switching time explicitly as

$$t_1 = \frac{t_2}{1 + \sqrt{K_1/K_2}}. \quad (37)$$

Using this switching time and the isothermal condition, the maximum work that can be extracted from the heat engine is calculated to be

$$W_{\max} = K_1 K_2 t_2\left[\frac{\sqrt{T_1(0)} - \sqrt{T_2(0)}}{\sqrt{K_1} + \sqrt{K_2}}\right]^2. \quad (38)$$

The above results are exactly those obtained for an endoreversible cycle [1]. We observe that in the isothermal limit, our heat engine becomes identical to the Curzon-Ahlborn cycle.

4. Case II: Finite Time Studies with Finite Reservoirs

In this section, we carry out the optimisation of work under the constraints of both finite reservoirs and finite cycle time. Solving similarly as in Section 3, the work becomes optimal when the initial temperature of the fluid is given by

$$T(0) = \left[\sqrt{T_2(0)T_1(0)}(C + C_1)\left(1 - e^{-\beta(t_2-t_1)}\right) + C_1 T_1(0)e^{-\beta(t_2-t_1)}\left(1 - e^{-\lambda t_1}\right)\right] \times\left[C\left(1 - e^{-\lambda t_1}\right) + C_1\left(1 - e^{-\lambda t_1-\beta(t_2-t_1)}\right)\right]^{-1}. \quad (39)$$

The optimal work is explicitly given as

$$W_{\mathrm{opt}} = \left[CC_1C_2\left(\sqrt{T_1(0)} - \sqrt{T_2(0)}\right)^2 \times\left(1 - e^{-\lambda t_1}\right)\left(1 - e^{-\beta(t_2-t_1)}\right)\right] \times\left[CC_1\left(1 - e^{-\lambda t_1}\right) + CC_2\left(1 - e^{-\beta(t_2-t_1)}\right) + C_1C_2\left(1 - e^{-\lambda t_1-\beta(t_2-t_1)}\right)\right]^{-1}. \quad (40)$$

The impact of finiteness of time can be seen on the work obtained, in Figure 2. It is shown that when time is large enough, the maximum work obtained is approximately equal to that in infinite time case (43). The efficiency at optimal work is again found to be CA value. To the best of our knowledge, only for Carnot-like cycles with isothermal branches ($C \to \infty$) [8], it has been shown that finiteness of the reservoirs does not affect the efficiency at maximum power. Here, we have seen that the generic heat cycle with two adiabatic and two polytropic branches obtains maximum power at the CA efficiency. This efficiency is still unaffected by the switching time, even with the finite reservoirs.

We may further maximize the work with respect to the switching time:

$$\frac{\partial W_{\mathrm{opt}}}{\partial t_1} = 0. \quad (41)$$

The optimal switching time is the solution to the following equation:

$$\lambda e^{-\lambda t_1}\left(1 - e^{-\beta(t_2-t_1)}\right)^2 = \beta e^{-\beta(t_2-t_1)}\left(1 - e^{-\lambda t_1}\right)^2, \quad (42)$$

where λ and β are given by (8) and (13). The impact of finiteness of reservoirs on the optimal switching time can be seen in Figure 3. The optimal switching time with infinite reservoirs (but finite C for working fluid) for the parameters in Figure 3 is approximately 0.792, whereas the corresponding Curzon-Ahlborn expression, (37) (infinite reservoirs and infinite C working fluid), will yield about 0.828. In short, from numerical results we observe that optimal switching time varies significantly from the above limiting values if the reservoirs and working fluid are finite.

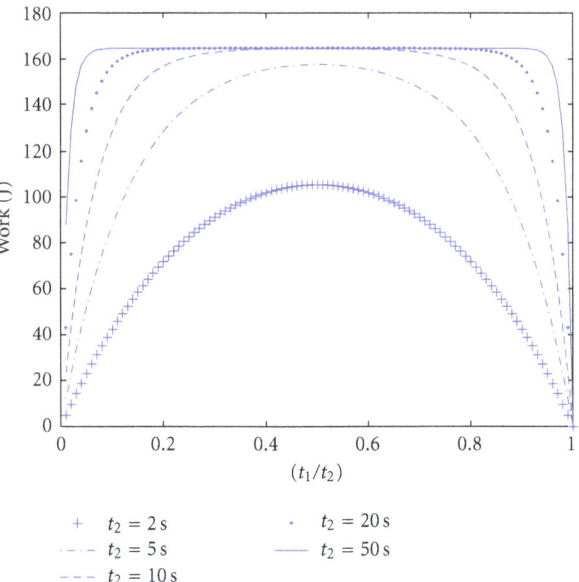

FIGURE 2: Optimal work as function of t_1/t_2. Here, $C = 2$ J/K; $T_1(0) = 700$ K; $T_2(0) = 300$ K; $K_1 = 3$ J/K sec; $K_2 = 3$ J/K sec; $C_1 = 300$ J/K; $C_2 = 500$ J/K.

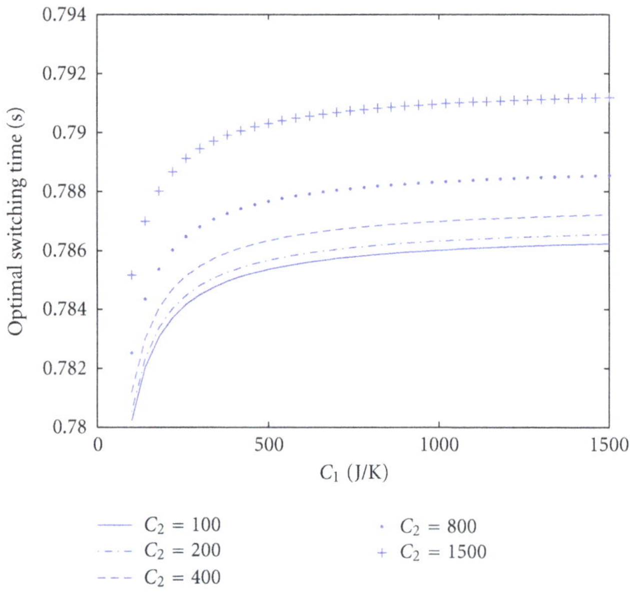

FIGURE 3: Optimal switching time versus the heat capacity of the hot reservoir. Here, $C = 2$ J/K; $T_1(0) = 700$ K; $T_2(0) = 300$ K; $K_1 = 6$ J/K sec; $K_2 = 3$ J/K sec; $t_2 = 2$ sec.

4.1. Infinite Time Limit. Further, if in (33), we keep the constraints of finite reservoirs, but let the nonadiabatic branches proceed very slowly ($t_1 \to \infty$ and $(t_2 - t_1) \to \infty$), we get the following expression for the maximum possible work that can be extracted from the heat engine:

$$\widehat{W}_{\mathrm{opt}} = \frac{C\left(\sqrt{T_1(0)} - \sqrt{T_2(0)}\right)^2}{(1 + (C/C_1) + (C/C_2))}. \quad (43)$$

The factor $1/(1+(C/C_1)+(C/C_2))$ accounts for finiteness of the reservoirs but with an infinite cycle time. In the limit of both infinite reservoirs as well as infinitely slow cycles, the above expression for the maximum work goes back to the one derived by Leff [11] (particularly Otto cycle and Joule-Brayton cycle),

$$\widehat{W}_{\text{opt}} = C\left(\sqrt{T_1(0)} - \sqrt{T_2(0)}\right)^2, \tag{44}$$

where C is the heat capacity of the working fluid.

5. Summary

In this paper, the maximum power point characteristics of a generalized four-step heat engine are studied. The heat cycle has two adiabatic steps and in the remaining two steps the working fluid follows a polytropic process with a constant heat capacity. It is observed that some common heat cycles such as Otto cycle, Joule-Brayton cycle, and Carnot cycle can be incorporated as special cases of this model. Curzon and Ahlborn in their seminal paper derived the expression for efficiency at maximum power for a Carnot-like engine with infinite reservoirs. Later, Gordon proved that even with finite reservoirs, Carnot-like engines show this efficiency at maximum power. Here we have analysed this result for the generalised heat cycle and shown that CA efficiency is obtained by optimizing the initial temperature of the working fluid and is independent of the switching time. However, we can further maximize this work over the switching time. Finally, the effect of finiteness of cycle time alone, the finiteness of the reservoirs alone, and the finiteness of both the cycle time and the reservoirs together on the maximum work per cycle can be respectively attributed to the following factors:

$$\frac{\left(1-e^{-(K_1/C)t_1}\right)\left(1-e^{-(K_2/C)(t_2-t_1)}\right)}{1-e^{-(K_1/C)t_1-(K_2/C)(t_2-t_1)}}, \tag{45}$$

$$\frac{1}{(1+(C/C_1)+(C/C_2))}, \tag{46}$$

$$\frac{C_1C_2\left(1-e^{-\lambda t_1}\right)\left(1-e^{-\beta(t_2-t_1)}\right)}{CC_1(1-e^{-\lambda t_1})+CC_2(1-e^{-\beta(t_2-t_1)})+C_1C_2(1-e^{-\lambda t_1-\beta(t_2-t_1)})}. \tag{47}$$

It can be easily shown (see the appendix) that all these factors are less than unity, and thus, the irreversibilities due to the finiteness of time and/or the reservoirs actually reduce the maximum work that can be performed by these model engines, over the infinite-time and infinite-reservoir models.

Appendix

A. Time- and Reservoir-Irreversibility Factors Are Less Than Unity

A.1. Finite Cycle-Time Irreversibility. The factor in (45) is attributed to the finite cycle-time irreversibility. Let $e^{-(K_1/C)t_1} = a$ and $e^{-(K_2/C)(t_2-t_1)} = b$. Since K_1, K_2, C, t_1, $(t_2 - t_1)$ are all greater than zero, we have $0 < a < 1$ and $0 < b < 1$. It follows that $ab < b$ and $ab < a$. Adding the last two inequalities, we get $2ab < a + b$, which may be rewritten as $(1 - a - b + ab) < 1 - ab$. So finally, we get

$$\frac{(1-a)(1-b)}{1-ab} < 1. \tag{A.1}$$

Thus we see that the finite cycle-time irreversibility factor is less than one.

A.2. Finite-Reservoir Irreversibility. Since C, C_1, C_2 are all greater than zero, we have

$$\frac{1}{(1+(C/C_1)+(C/C_2))} < 1. \tag{A.2}$$

A.3. Finite-Time and Finite-Reservoir Irreversibility. Using similar procedure as to time irreversibility, we can show that

$$\frac{C_1C_2\left(1-e^{-\lambda t_1}\right)\left(1-e^{-\beta(t_2-t_1)}\right)}{C_1C_2(1-e^{-\lambda t_1-\beta(t_2-t_1)})} < 1. \tag{A.3}$$

Since, C_1, C_2, C, λ, β, t_1, $(t_2 - t_1) > 0$, therefore, $(1 - e^{-\lambda t_1}) > 0$ and $(1 - e^{-\beta(t_2-t_1)}) > 0$. So, adding $(CC_1(1 - e^{-\lambda t_1}) + CC_2(1 - e^{-\beta(t_2-t_1)}))$ to the denominator of left hand expression in (A.3) will reduce the expression further and retain the inequality. Thus we conclude

$$\frac{C_1C_2\left(1-e^{-\lambda t_1}\right)\left(1-e^{-\beta(t_2-t_1)}\right)}{CC_1(1-e^{-\lambda t_1})+CC_2(1-e^{-\beta(t_2-t_1)})+C_1C_2(1-e^{-\lambda t_1-\beta(t_2-t_1)})} < 1. \tag{A.4}$$

Acknowledgments

This work was initiated during the summer program at IISER Mohali. A. Khanna expresses his gratitude towards IISER Mohali, for hospitality and financial support. R. S. Johal acknowledges financial support from the Department of Science and Technology, India, under the Research Project no. SR/S2/CMP-0047/2010(G).

References

[1] F. L. Curzon and B. Ahlborn, "Efficiency of a Carnot engine at maximum power output," *American Journal of Physics*, vol. 43, no. 1, p. 22, 1975.

[2] D. Gutkowicz-Krusin, I. Procaccia, and J. Ross, "On the efficiency of rate processes. Power and efficiency of heat engines," *The Journal of Chemical Physics*, vol. 69, no. 9, pp. 3898–3906, 1978.

[3] M. H. Rubin, "Optimal configuration of a class of irreversible heat engines. I," *Physical Review A*, vol. 19, no. 3, pp. 1272–1276, 1979.

[4] M. H. Rubin, "Optimal configuration of a class of irreversible heat engines. II," *Physical Review A*, vol. 19, no. 3, pp. 1277–1289, 1979.

[5] M. J. Ondrechen, B. Andressen, M. Mozurkewich, and R. S. Berry, "Maximum work from a finite reservoir by sequential Carnot cycles," *American Journal of Physics*, vol. 49, p. 681, 1981.

[6] P. Salamon, Y. B. Band, and O. Kafri, "Maximum power from a cycling working fluid," *Journal of Applied Physics*, vol. 53, no. 1, pp. 197–202, 1982.

[7] M. J. Ondrechen, M. H. Rubin, and Y. B. Band, "The generalized Carnot cycle: a working fluid operating in finite time between finite heat sources and sinks," *The Journal of Chemical Physics*, vol. 78, no. 7, pp. 4721–4727, 1983.

[8] J. M. Gordon, "Maximum power point characterstics of heat engines as a general thermodynamic problem," *American Journal of Physics*, vol. 57, p. 1136, 1989.

[9] L. Chen, F. Sun, and C. Wu, "Effect of heat transfer law on the performance of a generalized irreversible Carnot engine," *Journal of Physics D*, vol. 32, no. 2, pp. 99–105, 1999.

[10] L. Chen, S. Zhou, F. Sun, and C. Wu, "Optimal configuration and performance of heat engines with heat leak and finite heat capacity," *Open Systems and Information Dynamics*, vol. 9, no. 1, pp. 85–96, 2002.

[11] H. S. Leff, "Thermal Efficiency at maximum work output: new results for old heat engines," *American Journal of Physics*, vol. 55, p. 602, 1987.

[12] C. Van den Broeck, "Thermodynamic efficiency at maximum power," *Physical Review Letters*, vol. 95, Article ID 190602, 2005.

[13] Z. C. Tu, "Efficiency at maximum power of Feynman's ratchet as a heat engine," *Journal of Physics A*, vol. 41, Article ID 312003, 2008.

[14] M. Esposito, K. Lindenberg, and C. Van den Broeck, "Thermoelectric efficiency at maximum power in a quantum dot," *Physical Review Letters*, vol. 102, Article ID 130602, 2009.

[15] Y. Zhou and D. Segal, "Minimal model of a heat engine: information theory approach," *Physical Review E*, vol. 82, Article ID 011120, 2010.

[16] R. S. Johal, "Universal Efficiency at optimal work with Bayesian statistics," *Physical Review E*, vol. 82, Article ID 061113, 2010.

4

Thermodynamic and Acoustic Study on Molecular Interactions in Certain Binary Liquid Systems Involving Ethyl Benzoate

B. Nagarjun,[1] A. V. Sarma,[2] G. V. Rama Rao,[3] and C. Rambabu[4]

[1] *Department of Physics, G.V.P. College of Engineering (A), Visakhapatnam 530048, Andhra Pradesh, India*
[2] *Department of Physics, Andhra University, Visakhapatnam 530003, Andhra Pradesh, India*
[3] *Department of Physics, DAR College, Nuzvid 521201,Andhra Pradesh, India*
[4] *Department of Chemistry, Acharya Nagarjuna University, Guntur 522510, Andhra Pradesh, India*

Correspondence should be addressed to B. Nagarjun; au.nagarjun@gmail.com

Academic Editor: Felix Sharipov

Speeds of sound and density for binary mixtures of ethyl benzoate (EB) with N,N-dimethylformamide (NNDMF), N,N-dimethyl acetamide (NNDMAc), and N,N-dimethylaniline (NNDMA) were measured as a function of mole fraction at temperatures 303.15, 308.15 K, 313.15 K, and 318.15 K and atmospheric pressure. From the experimental data, adiabatic compressibility (β_{ad}), intermolecular free length (L_f), and molar volume (V) have been computed. The excess values of the above parameters were also evaluated and discussed in light of molecular interactions. Deviation in adiabatic compressibilities and excess intermolecular free length (L_f^E) are found to be negative over the molefraction of ethyl benzoate indicating the presence of strong interactions between the molecules. The negative excess molar volume V^E values are attributed to strong dipole-dipole interactions between unlike molecules in the mixtures. The binary data of $\Delta\beta_{\text{ad}}$, V^E, and L_f^E were correlated as a function of molefraction by using the Redlich-Kister equation.

1. Introduction

The thermophysical study of esters is of increasing interest due to their wide range in flavouring, perfumery, artificial essences, and cosmetics. Esters are also important solvents in pharmaceutical, paint, and plastic industries [1]. Recently, interest in the study of liquid mixtures containing esters as one of the components [2–5] has increased. These studies are of great significance because one can get information regarding structural changes that occur in pure ester because of mixing.

Among the different types of esters, ethyl benzoate (aromatic ester) is a polar solvent (μ = 1.8 D) and is not a strongly associated liquid. On the other hand, NNDMF, to some extent, is associated by means of dipole-dipole interactions. Significant structural effects are absent due to the lack of hydrogen bonds. Therefore, it acts as an aprotic protophilic solvent [6]. NNDMAc is a dipolar aprotic solvent and is moderately structured [7]. The N,N-dialkylamides have no significant intermolecular hydrogen bonding capability but are highly polar with high percentage of ionic character, making oxygen of C=O group strongly negative [8]. NNDMA is a unassociated liquid [9]. The interactions of ethyl benzoate with N,N-dialkylamides and N,N-dimethylaniline are important to enable analysis of their dissimilar geometric structures on mixture properties.

Rao [10] calculated the excess isentropic compressibilities for the binary mixtures of N,N-dimethylformamide and N,N-dimethylacetamide and various normal and branched esters. The negative excess values are attributed to the presence of n-n interactions between unlike molecules. Aminabhavi et al. [11] studied the physicochemical behavior of binary mixtures of ethyl benzoate with diethylene glycol and dimethyl ether. Rathnam et al. [12] reported the densities, viscosities, and refractive indices of ethyl benzoate with o-xylene, m-xylene, p-xylene, and ethyl benzene. Lien et al. [13] have measured the excess molar enthalpies of the binary mixture of ethyl benzoate with 1-Octanol at 298.15 K. Joshi et al. [14] measured

Table 1: Comparison of experimental velocities and densities of pure liquids with the literature values at 303.15 K.

Component	U (m/s)		$\rho \times 10^3$ (kg·m^{-3})	
	Expt	Lit	Expt	Lit
Ethyl benzoate	1347.00	1346.50[a]	1.0456	1.0455[a]
N,N dimethyl formamide	1438.53	1438.32[b]	0.9392	0.9393[b]
N,N dimethyl acetamide	1438.00	1435.55[b]	0.9319	0.9317[b]
N,N dimethyl aniline	1470.50	1470.00[c]	0.9482	0.94815[d]

[a]Data taken from [18]. [b]Data taken from [19]. [c]Data taken from [20]. [d]Data taken from [21].

the densities and viscosities of the binary mixture of bromoform with ethyl benzoate.

We report herein the experimental values of densities and ultrasonic velocities of mixtures of ethyl benzoate with NNDMF, NNDMAc, and NNDMA at temperatures 303.15, 308.15, 313.15, and 318.15 K covering the whole miscibility range to enable analysis of the effect of their dissimilar geometric structures on the mixture properties. The nonrectilinear behavior of ultrasonic velocity, compressibility, and other thermodynamic parameters of liquid mixtures reveal the strength of interactions. To get additional information about the nature and strength of molecular interactions, other parameters such as free length, adiabatic compressibility, and their excess parameters have been calculated in the liquid mixtures. These results were correlated with the Redlich-Kister polynomial equation [15] to derive the coefficients and standard deviation.

2. Materials and Experimental Details

Ethyl benzoate (SD Fine Chemicals, purity > 99%) was used without any further treatment. NNDMF, NNDMAc, and NNDMA were purchased from SD Fine or Merck and were purified by the recommended methods [16, 17]. Further, the purities were ascertained from their ultrasonic speed and density values at 303.15 K which agreed with the literature values (Table 1).

Mixtures were prepared by mixing appropriate volumes of liquids in airtight bottles and weighed in a single-pan Mettler balance to an accuracy of ±0.001 mg. Preferential evaporation losses of solvents from mixtures were kept to a minimum as evidenced by repeated measurement of the physical properties over an interval of 2-3 days, during which no changes in physical properties were observed. The possible error in mole fractions is estimated to be around ±0.0001.

The densities of liquids and their mixtures were measured by a 25 mL specific gravity bottle, calibrated with redistilled water. The average uncertainty in measurement in the measured density is ±0.05 kgm^{-3}. With the fluctuation of ±0.05 K, temperature was controlled by a water thermostat. Ultrasonic velocity was measured using the ultrasonic interferometer (Model M-83) provided by Mittal Enterprises, New Delhi. The values agree closely with the values given in the literature.

The experimental values of ultrasonic speeds (u), densities (ρ), adiabatic compressibilities (β_{ad}), molar volumes (V), and free length (L_f) of pure ethyl benzoate, NNDMF, NNDMAc, NNDMA, and those of their binary mixtures over the entire composition range and at 303.15 K, expressed by mole fraction x_1 of ethyl benzoate, are listed in Table 2.

3. Results and Discussion

The experimental density (ρ) values of binary mixtures were used to calculate the excess molar volumes as

$$V^E = x_1 M_1 \left(\frac{1}{\rho} - \frac{1}{\rho_1}\right) + x_2 M_2 \left(\frac{1}{\rho} - \frac{1}{\rho_2}\right), \quad (1)$$

where M is the molar mass; subscripts 1 and 2 stand for pure components, ethyl benzoate and NNDMF, NNDMAc, and NNDMA, respectively. The uncertainty in V^E is estimated to be within $\pm 2 \times 10^{-6}\ \text{m}^3 \cdot \text{mol}^{-1}$.

Assuming that ultrasonic absorption is negligible, adiabatic compressibility (β_{ad}) can be obtained from the density and velocity of ultrasonic sound (U) using the relation

$$\beta_{ad} = \frac{1}{\rho U^2}. \quad (2)$$

Deviation in adiabatic compressibility was calculated using the relation

$$\Delta\beta = \beta_{ad} - (\beta_1 x_1 + \beta_2 x_2), \quad (3)$$

where β_{ad} is the compressibility of the mixture.

Intermolecular free length (L_f) has been calculated using the relation

$$L_f = K(\beta_{ad})^{1/2}, \quad (4)$$

where K is Jacobson's temperature-dependant constant and is equal to $(93.875 + 0.375T) \cdot 10^{-8}$.

The excess intermolecular free length (L_f^E) at a given mole fraction is the difference between mean free length and the sum of the fractional contributions of the two liquids given by

$$L_f^E = L_f - (L_{f1} x_1 + L_{f2} x_2), \quad (5)$$

where L_{f1} and L_{f2} are the individual intermolecular free length values of pure liquids in the binary mixtures.

The experimental values of ultrasonic velocity (U) and values of density (ρ), at the four temperatures, namely, T = 303.15 K, 308.15 K, 313.15 K, and 318.15 K, along with the derived values of adiabatic compressibility (β_{ad}), intermolecular free length (L_f), and molar volume (V) and their excess parameters are given in Tables 2, 3, and 4.

The excess properties were fitted to a Redlich, Kister-type [14] polynomial equation

$$Y^E = x_1 x_2 \sum_{i=0}^{j} A_i (1 - 2x_1)^i. \quad (6)$$

The optimum number of coefficients, j, was ascertained from an examination of the variation of the standard deviation σ. The values of coefficients, A_i, were evaluated by using the

TABLE 2: Experimental ultrasonic velocities, u, densities, ρ, and related thermodynamic parameters for ethyl benzoate + NNDMF system.

x_1	U ms^{-1}	$\rho \times 10^3$ kgm^{-3}	$\beta_{ad} \times 10^{-12}$ m^{2}N^{-1}	$V \times 10^{-6}$ m^{3}mol^{-1}	L_f Å	$\Delta\beta_{ad} \times 10^{-12}$ m^{2}N^{-1}	$V^E \times 10^{-6}$ m^{3}mol^{-1}	L_f^E Å
				T = 303.15 K				
0.0000	1438.53	0.9392	51.4522	77.8216	0.4497	0.0000	0.0000	0.0000
0.0568	1433.50	0.9529	51.0688	81.2961	0.4481	−0.4549	−0.2620	−0.0019
0.1193	1427.94	0.9663	50.7512	85.1516	0.4467	−0.8511	−0.5200	−0.0037
0.1885	1421.80	0.9793	50.5133	89.4680	0.4456	−1.1761	−0.7540	−0.0051
0.2654	1414.94	0.9922	50.3414	94.2800	0.4449	−1.4448	−1.0030	−0.0063
0.3514	1407.25	1.0045	50.2715	99.7325	0.4446	−1.6230	−1.2130	−0.0071
0.4484	1398.50	1.0158	50.3334	105.9725	0.4448	−1.6831	−1.3510	−0.0073
0.5584	1388.25	1.0249	50.6279	113.3098	0.4461	−1.5271	−1.2520	−0.0066
0.6843	1376.50	1.0327	51.1051	121.8471	0.4482	−1.2084	−1.0000	−0.0052
0.8298	1362.89	1.0395	51.7198	131.8472	0.4512	−0.7048	−0.5770	−0.0030
1.0000	1347.00	1.0456	52.7108	143.6209	0.4552	0.0000	0.0000	0.0000
				T = 308.15 K				
0.0000	1420.69	0.9345	53.0178	78.2129	0.4598	0.0000	0.0000	0.0000
0.0568	1416.24	0.9498	52.4916	81.5612	0.4575	−0.5600	−0.3950	−0.0024
0.1193	1411.27	0.9639	52.0884	85.3661	0.4558	−1.0003	−0.7110	−0.0043
0.1885	1405.75	0.9773	51.7820	89.6557	0.4544	−1.3478	−0.9800	−0.0059
0.2654	1399.58	0.9900	51.5655	94.4878	0.4535	−1.6100	−1.2180	−0.0070
0.3514	1392.69	1.0023	51.4383	99.9475	0.4529	−1.7883	−1.4310	−0.0078
0.4484	1384.81	1.0134	51.4578	106.2300	0.4530	−1.8265	−1.5380	−0.0080
0.5584	1375.55	1.0226	51.6827	113.5634	0.4540	−1.6669	−1.4560	−0.0073
0.6843	1364.87	1.0304	52.0990	122.1266	0.4558	−1.3254	−1.1930	−0.0058
0.8298	1352.50	1.0368	52.7261	132.1880	0.4585	−0.7848	−0.7260	−0.0034
1.0000	1338.00	1.0419	53.6120	144.1309	0.4624	0.0000	0.0000	0.0000
				T = 313.15 K				
0.0000	1398.00	0.9294	55.0533	78.6421	0.4719	0.0000	0.0000	0.0000
0.0568	1394.18	0.9463	54.3682	81.8657	0.4690	−0.6765	−0.5300	−0.0029
0.1193	1389.85	0.9609	53.8719	85.6300	0.4668	−1.1633	−0.8980	−0.0050
0.1885	1385.04	0.9742	53.5109	89.9392	0.4652	−1.5137	−1.1600	−0.0065
0.2654	1379.64	0.9870	53.2289	94.7763	0.4640	−1.7840	−1.4070	−0.0077
0.3514	1373.59	0.9990	53.0536	100.2767	0.4632	−1.9462	−1.5950	−0.0084
0.4484	1366.70	1.0101	52.9994	106.5688	0.4630	−1.9856	−1.7100	−0.0086
0.5584	1358.48	1.0191	53.1703	113.9502	0.4638	−1.7980	−1.6000	−0.0078
0.6843	1349.00	1.0267	53.5206	122.5584	0.4653	−1.4285	−1.3150	−0.0061
0.8298	1337.99	1.0331	54.0707	132.6663	0.4677	−0.8562	−0.8280	−0.0037
1.0000	1325.00	1.0375	54.9010	144.7422	0.4712	0.0000	0.0000	0.0000
				T = 318.15 K				
0.0000	1380.74	0.9299	56.4079	78.5998	0.4810	0.0000	0.0000	0.0000
0.0568	1377.12	0.9487	55.5820	81.6576	0.4775	−0.8276	−0.7380	−0.0035
0.1193	1373.00	0.9638	55.0381	85.3744	0.4752	−1.3733	−1.2000	−0.0059
0.1885	1368.39	0.9770	54.6608	89.6770	0.4735	−1.7527	−1.5200	−0.0075
0.2654	1363.20	0.9888	54.4219	94.6046	0.4725	−1.9938	−1.7337	−0.0085
0.3514	1357.36	0.9998	54.2883	100.2006	0.4719	−2.1300	−1.8900	−0.0091
0.4484	1350.64	1.0098	54.2828	106.5998	0.4719	−2.1383	−1.9700	−0.0092
0.5584	1342.70	1.0184	54.4667	114.0329	0.4727	−1.9577	−1.8900	−0.0084
0.6843	1333.49	1.0256	54.8316	122.6897	0.4743	−1.5965	−1.6500	−0.0069
0.8298	1322.72	1.0310	55.4405	132.9388	0.4769	−0.9919	−1.1300	−0.0043
1.0000	1310.00	1.0325	56.4374	145.4431	0.4812	0.0000	0.0000	0.0000

TABLE 3: Experimental ultrasonic velocities, u, densities, ρ, and related thermodynamic parameters for ethyl benzoate + NNDMAc.

x_1	U ms^{-1}	$\rho \times 10^3$ kgm^{-3}	$\beta_{ad} \times 10^{-12}$ m^{2}N^{-1}	$V \times 10^{-6}$ m^{3}mol^{-1}	L_f Å	$\Delta\beta_{ad} \times 10^{-12}$ m^{2}N^{-1}	$V^E \times 10^{-6}$ m^{3}mol^{-1}	L_f^E Å
				T = 303.15 K				
0.0000	1438.00	0.9319	51.8935	93.4864	0.4517	0.0000	0.0000	0.0000
0.0674	1432.95	0.9480	51.3740	96.3878	0.4494	−0.5747	−0.4800	−0.0025
0.1400	1427.32	0.9635	50.9473	99.5830	0.4475	−1.0606	−0.9200	−0.0047
0.2181	1421.01	0.9782	50.6271	103.1217	0.4461	−1.4447	−1.3000	−0.0063
0.3026	1414.06	0.9925	50.3887	107.0031	0.4451	−1.7521	−1.6553	−0.0077
0.3943	1406.27	1.0058	50.2747	111.3334	0.4446	−1.9410	−1.9200	−0.0085
0.4940	1397.57	1.0175	50.3173	116.2342	0.4448	−1.9800	−2.0200	−0.0087
0.6030	1387.19	1.0274	50.5796	121.7970	0.4459	−1.8068	−1.9200	−0.0079
0.7225	1375.40	1.0358	51.0348	128.0889	0.4479	−1.4492	−1.6200	−0.0063
0.8542	1362.17	1.0419	51.7277	135.3109	0.4510	−0.8639	−1.0000	−0.0037
1.0000	1347.00	1.0456	52.7108	143.6209	0.4552	0.0000	0.0000	0.0000
				T = 308.15 K				
0.0000	1425.00	0.9282	53.0553	93.8591	0.4600	0.0000	0.0000	0.0000
0.0674	1414.43	0.9533	52.4328	96.6298	0.4573	−0.6600	−0.6200	−0.0029
0.1400	1407.10	0.9757	51.7632	99.7749	0.4543	−1.3700	−1.1200	−0.0060
0.2181	1400.06	0.9955	51.2467	103.3044	0.4521	−1.9300	−1.5200	−0.0085
0.3026	1392.49	1.0127	50.9238	107.2326	0.4506	−2.3000	−1.8400	−0.0101
0.39 43	1385.16	1.0275	50.7248	111.5802	0.4498	−2.5500	−2.1000	−0.0112
0.4940	1378.23	1.0390	50.6703	116.4747	0.4495	−2.6600	−2.2200	−0.0117
0.6030	1370.45	1.0463	50.8910	122.0725	0.4505	−2.5000	−2.1000	−0.0109
0.7225	1362.13	1.0499	51.3375	128.3908	0.4525	−2.1200	−1.7900	−0.0093
0.8542	1351.69	1.0489	52.1808	135.6309	0.4562	−1.3500	−1.1700	−0.0059
1.0000	1338.00	1.0419	53.6120	144.1309	0.4624	0.0000	0.0000	0.0000
				T = 313.15 K				
0.0000	1405.50	0.9238	54.7974	94.3061	0.4708	0.0000	0.0000	0.0000
0.0674	1403.10	0.9419	53.9280	97.0079	0.4671	−0.8764	−0.7000	−0.0038
0.1400	1400.00	0.9589	53.2064	100.0549	0.4639	−1.6055	−1.3100	−0.0069
0.2181	1396.00	0.9743	52.6688	103.5372	0.4616	−2.1512	−1.7700	−0.0093
0.3026	1390.60	0.9883	52.3255	107.4594	0.4601	−2.5033	−2.1100	−0.0109
0.3943	1384.40	1.0015	52.0988	111.8120	0.4591	−2.7395	−2.3800	−0.0119
0.4940	1377.50	1.0138	51.9853	116.6629	0.4586	−2.8632	−2.5600	−0.0124
0.6030	1368.80	1.0237	52.1359	122.2385	0.4592	−2.7239	−2.4800	−0.0118
0.7225	1358.10	1.0314	52.5672	128.6366	0.4611	−2.3051	−2.1100	−0.0100
0.8542	1345.00	1.0365	53.3302	136.0082	0.4645	−1.5557	−1.3800	−0.0067
1.0000	1325.00	1.0375	54.9010	144.7422	0.4712	0.0000	0.0000	0.0000
				T = 318.15 K				
0.0000	1385.00	0.9194	56.7017	94.7575	0.4823	0.0000	0.0000	0.0000
0.0674	1384.98	0.9396	55.4839	97.2461	0.4771	−1.2000	−0.9300	−0.0051
0.1400	1381.08	0.9573	54.7647	100.2213	0.4740	−1.9000	−1.6300	−0.0082
0.2181	1376.75	0.9726	54.2441	103.7130	0.4717	−2.4000	−2.1000	−0.0103
0.3026	1371.70	0.9871	53.8417	107.5884	0.4700	−2.7800	−2.5078	−0.0120
0.3943	1366.33	1.0002	53.5575	111.9618	0.4687	−3.0400	−2.7800	−0.0131
0.4940	1359.66	1.0121	53.4475	116.8575	0.4683	−3.1236	−2.9400	−0.0135
0.6030	1352.00	1.0220	53.5323	122.4504	0.4686	−3.0100	−2.8700	−0.0130
0.7225	1342.03	1.0295	53.9307	128.8682	0.4704	−2.5800	−2.5100	−0.0111
0.8542	1330.16	1.0320	54.7659	136.6042	0.4740	−1.7100	−1.4485	−0.0074
1.0000	1310.00	1.0325	56.4374	145.4431	0.4812	0.0000	0.0000	0.0000

TABLE 4: Experimental ultrasonic velocities, u, densities, ρ, and related thermodynamic parameters for ethyl benzoate + NNDMA system.

x_1	U ms^{-1}	$\rho \times 10^3$ kgm^{-3}	$\beta_{ad} \times 10^{-12}$ m^{2}N^{-1}	$V \times 10^{-6}$ m^{3}mol^{-1}	L_f Å	$\Delta\beta_{ad} \times 10^{-12}$ m^{2}N^{-1}	$V^E \times 10^{-6}$ m^{3}mol^{-1}	L_f^E Å
				T = 303.15 K				
0.0000	1470.50	0.9482	48.7719	127.8106	0.4379	0.0000	0.0000	0.0000
0.0900	1463.99	0.9605	48.5763	128.8888	0.4370	−0.5500	−0.3445	−0.0025
0.1820	1455.87	0.9724	48.5187	130.0539	0.4367	−0.9700	−0.6340	−0.0043
0.2761	1446.91	0.9841	48.5394	131.2837	0.4368	−1.3200	−0.8920	−0.0058
0.3724	1436.85	0.9957	48.6486	132.5568	0.4373	−1.5900	−1.1410	−0.0070
0.4709	1425.42	1.0070	48.8766	133.9043	0.4383	−1.7500	−1.3510	−0.0077
0.5717	1413.22	1.0176	49.2038	135.3745	0.4398	−1.8200	−1.4750	−0.0080
0.6750	1398.88	1.0268	49.7705	137.0818	0.4423	−1.6600	−1.4000	−0.0072
0.7807	1383.14	1.0347	50.5169	138.9865	0.4456	−1.3300	−1.1670	−0.0058
0.8890	1366.09	1.0412	51.4636	141.1360	0.4498	−0.8100	−0.7300	−0.0035
1.0000	1347.00	1.0456	52.7108	143.6209	0.4552	0.0000	0.0000	0.0000
				T = 308.15 K				
0.0000	1453.40	0.9439	50.1538	128.3928	0.4472	0.0000	0.0000	0.0000
0.0900	1448.57	0.9571	49.7950	129.3529	0.4456	−0.6700	−0.4560	−0.0029
0.1820	1441.85	0.9691	49.6332	130.4900	0.4449	−1.1500	−0.7670	−0.0051
0.2761	1433.31	0.9810	49.6186	131.6920	0.4448	−1.4900	−1.0460	−0.0066
0.3724	1423.92	0.9925	49.6915	132.9731	0.4452	−1.7500	−1.2800	−0.0077
0.4709	1413.61	1.0038	49.8522	134.3235	0.4459	−1.9300	−1.4800	−0.0085
0.5717	1402.59	1.0144	50.1109	135.8035	0.4470	−2.0200	−1.5870	−0.0089
0.6750	1389.28	1.0236	50.6179	137.5103	0.4493	−1.8700	−1.5050	−0.0082
0.7807	1374.46	1.0316	51.3136	139.4123	0.4524	−1.5400	−1.2670	−0.0067
0.8890	1358.23	1.0383	52.2081	141.5350	0.4563	−1.0200	−0.8490	−0.0044
1.0000	1338.00	1.0419	53.6120	144.1309	0.4624	0.0000	0.0000	0.0000
				T = 313.15 K				
0.0000	1435.20	0.9398	51.6583	128.9530	0.4571	0.0000	0.0000	0.0000
0.0900	1431.88	0.9534	51.1601	129.8537	0.4549	−0.7900	−0.5200	−0.0035
0.1820	1425.32	0.9658	50.9684	130.9465	0.4541	−1.2800	−0.8800	−0.0056
0.2761	1417.25	0.9777	50.9236	132.1433	0.4539	−1.6300	−1.1690	−0.0071
0.3724	1408.35	0.9892	50.9658	133.4173	0.4540	−1.9000	−1.4150	−0.0083
0.4709	1398.80	1.0004	51.0852	134.7768	0.4546	−2.1000	−1.6110	−0.0092
0.5717	1388.45	1.0109	51.3122	136.2689	0.4556	−2.2000	−1.7110	−0.0096
0.6750	1375.74	1.0199	51.8070	138.0100	0.4578	−2.0400	−1.6000	−0.0088
0.7807	1361.81	1.0279	52.4598	139.9134	0.4606	−1.7300	−1.3660	−0.0075
0.8890	1345.54	1.0345	53.3911	142.0506	0.4647	−1.1500	−0.9390	−0.0049
1.0000	1325.00	1.0375	54.9010	144.7422	0.4712	0.0000	0.0000	0.0000
				T = 318.15 K				
0.0000	1412.00	0.9356	53.6093	129.5319	0.4690	0.0000	0.0000	0.0000
0.0900	1409.96	0.9497	52.9638	130.3486	0.4661	−0.9000	−0.6150	−0.0040
0.1820	1404.50	0.9624	52.6740	131.4026	0.4649	−1.4500	−1.0250	−0.0064
0.2761	1397.05	0.9743	52.5901	132.6049	0.4645	−1.8000	−1.3200	−0.0079
0.3724	1389.02	0.9857	52.5824	133.8966	0.4645	−2.0800	−1.5600	−0.0091
0.4709	1380.54	0.9967	52.6410	135.2781	0.4647	−2.3000	−1.7460	−0.0100
0.5717	1371.38	1.0071	52.7962	136.7845	0.4654	−2.4300	−1.8440	−0.0106
0.6750	1359.31	1.0160	53.2681	138.5332	0.4675	−2.2500	−1.7380	−0.0098
0.7807	1345.99	1.0239	53.9072	140.4535	0.4703	−1.9100	−1.5000	−0.0083
0.8890	1330.79	1.0303	54.8035	142.6280	0.4742	−1.3200	−1.0490	−0.0057
1.0000	1310.00	1.0325	56.4374	145.4431	0.4812	0.0000	0.0000	0.0000

TABLE 5: Coefficients, A_i, of (6) and standard deviations, σ, for binary systems at different temperatures.

	T/K	A_0	A_1	A_2	σ
Ethyl benzoate + NNDMF					
$\Delta\beta_{ad} \times 10^{-12}/m^2N^{-1}$	303.15	−6.4467	2.2499	−0.0638	0.0171
	308.15	−6.8775	2.8689	−1.3139	0.0391
	313.15	−7.2181	3.6083	−2.8140	0.0767
	318.15	−7.6186	4.2860	−5.1293	0.1145
$V^E \times 10^{-6}/m^3mol^{-1}$	303.15	−5.1947	0.8178	1.3098	0.0259
	308.15	−5.8404	1.3461	−0.4321	0.0456
	313.15	−6.1729	2.0602	−2.4103	0.0899
	318.15	−7.0606	−2.4146	−5.8222	0.1234
L_f^E/Å	303.15	−0.0285	0.0096	0.0020	0.0001
	308.15	−0.0301	0.0123	−0.0047	0.0002
	313.15	−0.0312	0.0154	−0.0118	0.0003
	318.15	−0.0328	0.0177	−0.0215	0.0005
Ethyl benzoate + NNDMAc					
$\Delta\beta_{ad} \times 10^{-12}/m^2N^{-1}$	303.15	−7.8289	1.3767	−0.1647	0.0155
	308.15	−10.7235	−0.2065	0.0599	0.0681
	313.15	−11.2398	0.4350	−3.1080	0.0260
	318.15	−11.9293	2.3285	−6.8757	0.1496
$V^E \times 10^{-6}/m^3mol^{-1}$	303.15	−8.0967	−0.1873	0.3985	0.0196
	308.15	−8.7131	0.0699	−1.4562	0.0315
	313.15	−9.9358	−0.2638	−1.9087	0.0390
	318.15	−11.1717	1.5314	−3.0602	0.1497
L_f^E/Å	303.15	−0.0342	0.0064	0.0000	0.0001
	308.15	−0.0471	−0.0007	0.0004	0.0003
	313.15	−0.0488	0.0022	−0.0131	0.0001
	318.15	−0.0516	0.0093	−0.0287	0.0007
Ethyl benzoate + NNDMA					
$\Delta\beta_{ad} \times 10^{-12}/m^2N^{-1}$	303.15	−7.0776	−0.9608	−0.6333	0.0256
	308.15	−7.8226	−1.4455	−2.2954	0.0322
	313.15	−8.4985	−1.4000	−3.4138	0.0474
	318.15	−9.3139	−1.6851	−4.5462	0.0618
$V^E \times 10^{-6}/m^3mol^{-1}$	303.15	−5.5381	−2.0172	−0.4794	0.0310
	308.15	−6.0491	−1.9707	−1.6874	0.0364
	313.15	−6.5792	−2.0734	−2.1865	0.0427
	318.15	−7.1185	−2.0844	−3.1230	0.0440
L_f^E/Å	303.15	−0.0311	−0.0032	−0.0031	0.0001
	308.15	−0.0344	−0.0061	−0.0089	0.0001
	313.15	−0.0372	−0.0049	−0.0143	0.0003
	318.15	−0.0404	−0.0064	−0.0203	0.0003

method of least squares, with all points weighted equally. The coefficients A_0, A_1, and A_2 along with standard deviations, σ of fit for all the mixtures are listed in Table 5.

The $\Delta\beta_{ad}$, V^E, and L_f^E values plotted against mole fraction of ethyl benzoate (x_1) are shown in Figures 2, 3, and 4.

In this investigation, the values of ultrasonic velocity decrease with increase in the concentration of ethyl benzoate and decrease with increase in temperature at any particular concentration for all the three systems (Figure 1). The ultrasonic velocity values decrease with increase of temperature due to the breaking of hetero-and homomolecular clusters at high temperatures [22]. Lagemann and Duban [23] were the first to point out the ultrasonic velocity approach for qualitative estimation of the interaction in liquids.

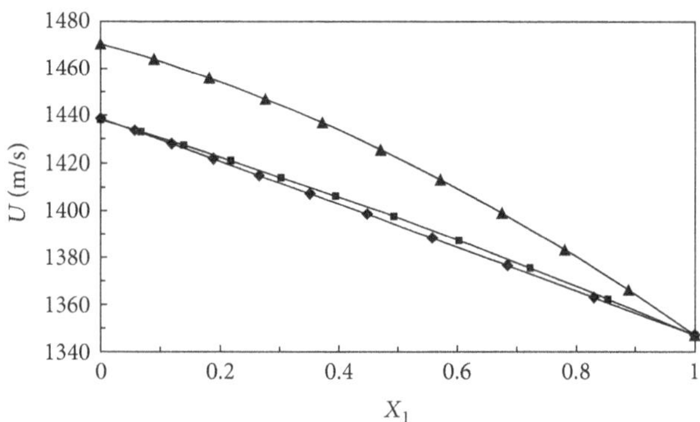

FIGURE 1: Ultrasonic velocities (U) plotted against the mole fractions of ethyl benzoate (X) in binary mixtures of ethyl benzoate + NNDMF (♦), ethyl benzoate + NNDMAc (■), and ethyl benzoate + NNDMA (▲).

The intermolecular free length is the distance between the surfaces of the neighboring molecules. The variation of ultrasonic velocity in a solution depends upon the increase or decrease of intermolecular free length after mixing the components. The interdependence of intermolecular free length and ultrasonic velocity was evolved from a model for sound propagation proposed by Kincaid and Eyring [24]. The ultrasonic velocity should decrease if the intermolecular free length increases as a result of mixing of components. This fact is observed in the present investigation for ethyl benzoate + NNDMF, ethyl benzoate + NNDMAc, and ethyl benzoate + NNDMA systems. Figure 2 represents the variation of excess intermolecular free length with mole fraction of ethyl benzoate. The excess intermolecular free length values are negative, and the curves appear to reach negative peak value at about 0.45 mole fraction of ethyl benzoate. According to Ramamurthy and Sastry [25], the negative L_f^E values indicate that sound wave has to travel a longer distance. This may be attributed to the dominant nature of interactions between unlike molecules.

The compressibility behavior of solutes, which is the second derivative of the Gibbs energy, is a very sensitive indicator of molecular interactions and can provide useful information about this phenomenon [26–28]. The structural change of molecules takes place due to the existence of electrostatic field between interacting molecules. The change in adiabatic compressibility value in liquids and liquid mixtures may be ascribed to the strength of intermolecular attraction. The effect of depolymerization increases the compressibility of the system [29]. Electrostatic attraction and association decrease the compressibility. The compressibility of the mixtures is the result of these two effects depending upon the predominance [30]. According to Jacobson [31, 32],

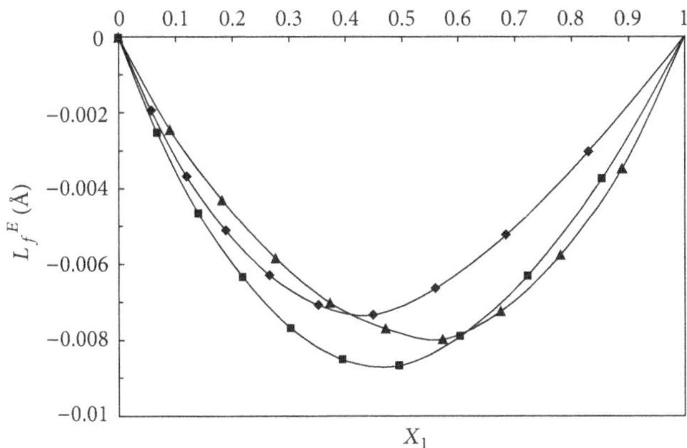

Figure 2: Excess intermolecular free lengths (L_f^E) plotted against the mole fractions of ethyl benzoate (X) in binary mixtures of ethyl benzoate + NNDMF (♦), ethyl benzoate + NNDMAc (■), and ethyl benzoate + NNDMA (▲).

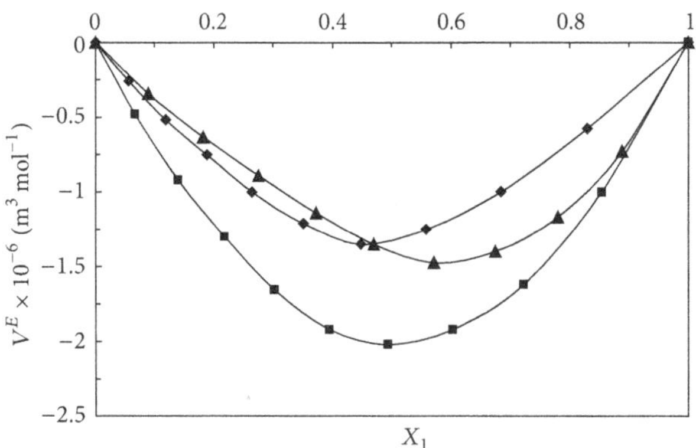

Figure 4: Excess molar volumes (V^E) plotted against the mole fractions of ethyl benzoate (X) in binary mixtures of ethyl benzoate + NNDMF (♦), ethyl benzoate + NNDMAc (■), and ethyl benzoate + NNDMA (▲).

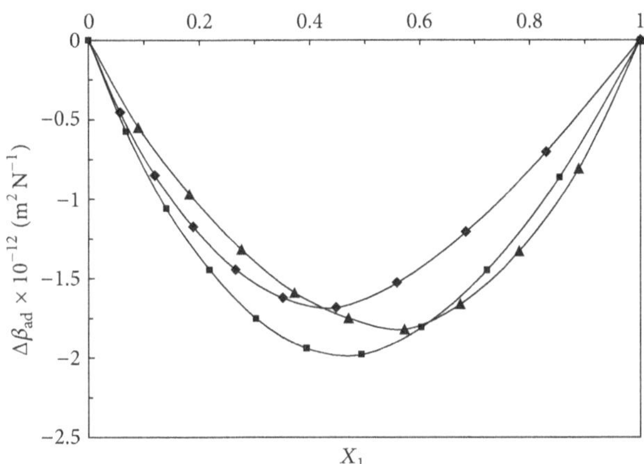

Figure 3: Deviation in adiabatic compressibilities ($\Delta\beta_{ad}$) plotted against the mole fractions of ethyl benzoate (X) in binary mixtures of ethyl benzoate + NNDMF (♦), ethyl benzoate + NNDMAc (■), and ethyl benzoate + NNDMA (▲).

the adiabatic compressibility can be studied more through intermolecular free length.

Figure 3 represents the deviation in adiabatic compressibility with the mole fraction of ethyl benzoate. In the present investigation, deviation in adiabatic compressibilities is found to be negative over the mole fraction of ethyl benzoate indicating the presence of strong interactions between the molecules. The negative deviation in adiabatic compressibility reaches a peak at about 0.45 mole fraction of ethyl benzoate in all the three systems chosen.

The deviation in adiabatic compressibility can be explained by taking into consideration the following factors.

(a) Loss of dipolar association and difference in size and shape of component molecules which lead to decrease in velocity and increase in compressibility.

(b) Dipole-dipole interaction or hydrogen-bonded complex formation between unlike molecules which lead to increase in sound velocity and decrease of compressibility.

The actual deviation depends on the resultant effect. The strength of the interaction between the components increases when excess values tend to become increasingly negative. This may be qualitatively interpreted in terms of closer approach of unlike molecules leading to reductions in compressibility and volume [33, 34]. This type of interactions for the binary mixtures has been already reported previously [35]. The deviations in adiabatic compressibilities are found to increase with increasing temperature which is in agreement with the previously reported results [36]. The $\Delta\beta_{ad}$ and L_f^E minima occur at the same concentrations further strengthen the occurrence of molecular associations [37].

The values of V^E calculated by using (1) are included in Tables 2–4.

The sign of V^E of a system depends upon the relative magnitude of expansion and contraction of the two liquids due to mixing [38].

The negative V^E arises due to dominance of the following factors.

(a) Chemical interaction between constituent molecules such as heteromolecular association through the formation of H-bond, often termed as strong specific interaction.

(b) Association through weaker physical forces such as dipolar force or any other forces of this kind.

(c) Accommodation of molecules of one component into the interstitial positions of the structural network of molecules of the other component.

(d) Geometry of the molecular structure that favors fitting of the component molecules with each other.

The V^E values are negative over the entire mole fraction range and at all temperatures investigated for all binary systems under study (Figure 4). The observed negative values of V^E for the three mixtures indicate the presence of specific interactions between ethyl benzoate and amide molecules. The negative V^E values are attributed to strong dipole-dipole interactions between unlike molecules in the mixtures. The

V^E values are more negative for ethyl benzoate + DMAc than those for other two systems.

The electron density at oxygen atom of the carbonyl atom of DMAc is greater than that of DMF due to the presence of methyl group at carbon atom of carbonyl group in DMF resulting in stronger interaction in thesystem [39, 40]. The V^E values of ethyl benzoate + DMA are in between these two. The strength of interactions between ethyl benzoate + DMA is stronger than ethyl benzoate + DMF due to the fact that negative charge on nitrogen in DMA is more than that of nitrogen in DMF due to conjugation with C=O group (this is also evident from the fact that the DMA is more basic than DMF).

Rao and Reddy [41] reported increased negative values of excess molar volumes with the increase in carbon chain length of the ester, in the binary mixtures of DMF, and aliphatic esters. The observed higher negative values of V^E ($-1.35\,\text{cm}^3\,\text{mol}^{-1}$) for aromatic ester with DMF in the present investigation, when compared to that of ethyl acetate and DMF ($V^E = -0.375$) at the same temperature (303.15 K), indicate much stronger interactions between the unlike molecules of components in this binary mixture due to the formation of not only dipole-dipole interactions but also of induced polar interaction like dipole-induced dipole and dipole interactions between aromatic ester and DMF. Similar interactions are reported in the mixtures of DMF and polycyclic aromatic hydrocarbons by Nikam and Kharat [42, 43] and by Ramadevi and Prabhakara Rao [44]. Also, formation of induced polar interactions like dipole-induced dipole interactions between polycyclic aromatic hydrocarbons and NMP (acyclic amide) is reported by Sugiura and Ogawa [45]. This further supports that amides react with aromatic compounds more strongly than the corresponding aliphatic counter parts.

The V^E values decrease (become more negative) with increase in temperature for all the 3 systems. This is attributed due to the dissociation of self-associated amide molecules resulting in the more favorable fitting of the smaller molecules into large voids of bigger ethyl benzoate molecules leading to contraction in volume and hence resulting in more negative V^E values with rise in temperature.

4. Conclusions

The adiabatic compressibility (β_{ad}) and intermolecular free length (L_f) both have an inverse relationship with ultrasonic velocity (U). Occurrence of U maxima, β_{ad}, and L_f minima at the same concentrations indicates the strong interaction through dipole-dipole interactions between the components. The negative V^E values are attributed to strong dipole-dipole interactions between unlike molecules in the mixtures.

Acknowledgment

The authors sincerely thank the University Grants Commission, India, for funding the current research work under UGC Scholarship Assistance Program (SAP) in the Department of Physics, Andhra University.

References

[1] Y.-W. Sheu and C. H. Tu, "Densities and viscosities of binary mixtures of ethyl acetoacetate, ethyl isovalerate, methyl benzoate, benzyl acetate, ethyl salicylate, and benzyl propionate with ethanol at T = (288.15, 298.15, 308.15, and 318.15) K," *Journal of Chemical and Engineering Data*, vol. 51, no. 2, pp. 545–553, 2006.

[2] J. N. Nayak, M. I. Aralaguppi, and T. M. Aminabhavi, "Density, viscosity, refractive index, and speed of sound in the binary mixtures of 1,4-dioxane + ethyl acetoacetate, + diethyl oxalate, + diethyl phthalate, or + dioctyl phthalate at 298.15, 303.15, and 308.15 K," *Journal of Chemical and Engineering Data*, vol. 48, no. 6, pp. 1489–1494, 2003.

[3] W. V. Steele, R. D. Chirico, A. B. Cowell, S. E. Knipmeyer, and A. Nguyen, "Thermodynamic properties and ideal-gas enthalpies of formation for methyl benzoate, ethyl benzoate, (R)-(+)-limonene, tert-amyl methyl ether, trans-crotonaldehyde, and diethylene glycol," *Journal of Chemical and Engineering Data*, vol. 47, no. 4, pp. 667–688, 2002.

[4] N. Indrswati, Mudjiati, F. Wicaksana, H. Hindarso, and S. IsmadjiS, "Measurements of density and viscosity of binary mixtures of several flavor compounds with 1-butanol and 1-pentanol at 293.15 K, 303.15 K, 313.15 K, and 323.15 K," *Journal of Chemical and Engineering Data*, vol. 46, no. 3, pp. 696–702, 2001.

[5] M. Hasan, A. P. Hiray, U. B. Kadam, D. F. Shirude, K. J. Kurhe, and A. B. Sawant, "Densities, sound speed, and IR studies of (methanol + 1-acetoxybutane) and (methanol + 1,1-dimethylethyl ester) at (298.15, 303.15, 308.15, and 313.15) K," *Journal of Chemical and Engineering Data*, vol. 55, no. 1, pp. 535–538, 2010.

[6] M. M. H. Bhuiyan and M. H. Uddin, "Excess molar volumes and excess viscosities for mixtures of N,N-dimethylformamide with methanol, ethanol and 2-propanol at different temperatures," *Journal of Molecular Liquids*, vol. 138, no. 1–3, pp. 139–146, 2008.

[7] P. J. Victor, D. Das, and D. K. Hazra, "Excess molar volumes, viscosity deviations and isentropic compressibility changes in binary mixtures of N,N-dimethylacetamide with 2-methoxyethanol and water in the temperature range 298.15 to 318.15 K," *Journal of the Indian Chemical Society*, vol. 81, no. 12, pp. 1045–1050, 2004.

[8] R. K. Bhardwaj and A. Pal, "Spectroscopic studies of binary liquid mixtures of alkoxyethanols with substituted and cyclic amides at 298.15 K," *Journal of Molecular Liquids*, vol. 118, no. 1–3, pp. 37–39, 2005.

[9] R. Palepu and J. H. MacNeil, "Viscosities and densities of binary-mixtures of 2,2,2-trichloroethanol with substituted anilines," *Australian Journal of Chemistry*, vol. 41, no. 5, pp. 791–797, 1988.

[10] K. P. C. Rao, "Excess isentropic compressibilities of N,N-dimethylformamide and N,N-dimethylacetamide with aliphatic esters at 303.15 K," *Ultrasonics*, vol. 28, no. 2, pp. 120–124, 1990.

[11] T. M. Aminabhavi, H. T. S. Phayde, R. S. Khinnavar, B. Gopalakrishna, and K. C. Hansen, "Densities, refractive indices, speeds of sound, and shear viscosities of diethylene glycol dimethyl ether with ethyl acetate, methyl benzoate, ethyl benzoate, and diethyl succinate in the temperature range from 298.15 to 318.15 K," *Journal of Chemical and Engineering Data*, vol. 39, no. 2, pp. 251–260, 1994.

[12] M. V. Rathnam, S. Mohite, and M. S. S. Kumar, "Viscosity, density, and refractive index of some (ester + hydrocarbon)

binary mixtures at 303.15 K and 313.15 K," *Journal of Chemical and Engineering Data*, vol. 50, no. 2, pp. 325–329, 2005.

[13] P.-J. Lien, S. T. Wu, M. J. Lee, and H. M. Lin, "Excess molar enthalpies for dimethyl carbonate with o-xylene, m-xylene, p-xylene, ethylbenzene, or ethyl benzoate at 298.15 K and 10.2 MPa," *Journal of Chemical and Engineering Data*, vol. 48, no. 3, pp. 632–636, 2003.

[14] S. S. Joshi, T. M. Aminabhavi, and S. S. Shukla, "Densities and viscosities of binary mixtures of bromoform with anisole, acetophenone, ethyl benzoate, 1,2- dichloroethane and 1,1,2,2-tetrachloroethane from 298.15 to 313.15 K," *Indian Journal of Technology*, vol. 29, pp. 319–326, 1991.

[15] O. Redlich and A. T. Kister, "Algebraic representation of thermodynamic properties and the classification of solutions," *Industrial and Engineering Chemistry*, vol. 40, no. 2, pp. 345–348, 1948.

[16] A. I. Vogel, *A Text Book of Practical Organic Chemistry*, John Willey, New York, NY, USA, 5th edition, 1989.

[17] J. A. Riddick, W. B. Bunger, and T. K. Sokano, *Techniques in Chemistry, Vol 2, Organic Solvents*, John Willey, NewYork, NY, USA, 4th edition, 1986.

[18] G. V. Rama Rao, A. V. Sarma, and C. Rambabu, "Evaluation of excess thermodynamic properties in some binary mixtures of o-chlorophenol," *Indian Journal of Chemistry A*, vol. 43, no. 12, pp. 2518–2528, 2004.

[19] D. Papamatthaiakis, F. Aroni, and V. Havredaki, "Isentropic compressibilities of (amide + water) mixtures: a comparative study," *Journal of Chemical Th rmodynamics*, vol. 40, no. 1, pp. 107–118, 2008.

[20] R. Kumar, *Ultrasonic and spectroscopic investigations of molecular interactions in liquid mixtures [Ph.D. thesis]*, University of Madras, Chennai, India, 2009.

[21] M. Katz, P. W. Lobo, A. S. Minano, and H. Solimo, "Viscosities, densities, and refractive indices of binary liquid mixtures," *Canadian Journal of Chemistry*, vol. 49, no. 15, pp. 2605–2609, 1971.

[22] P. S. Nikam, B. S. Jagdale, A. B. Sawant, and M. Hasan, *Journal of Pure and Applied Ultrasonics*, vol. 22, p. 115, 2000.

[23] R. T. Lagemann and W. S. Duban, *Journal of Chemical Physics*, vol. 49, p. 423, 1945.

[24] J. F. Kincaid and H. Eyring, "Free volumes and free angle ratios of molecules in liquids," *The Journal of Chemical Physics*, vol. 6, no. 10, pp. 620–629, 1938.

[25] M. Ramamurthy and O. S. Sastry, *Indian Journal of Pure and Applied Physics*, vol. 21, p. 579, 1983.

[26] I. Mozo, J. A. Gonzalez, I. G. de la Fuente, J. C. Cobos, N. Riesco, and N. Riesco, "Thermodynamics of mixtures containing alkoxyethanols. Part XXV. Densities, excess molar volumes and speeds of sound at 293.15, 298.15 and 303.15 K, and isothermal compressibilities at 298.15 K for 2-alkoxyethanol + 1-butanol systems," *Journal of Molecular Liquids*, vol. 140, no. 1–3, pp. 87–100, 2008.

[27] V. Ulagendran, R. Kumar, S. Jayakumar, and V. Kannappan, "Ultrasonic and spectroscopic investigations of charge-transfer complexes in ternary liquid mixtures," *Journal of Molecular Liquids*, vol. 148, no. 2-3, pp. 67–72, 2009.

[28] P. J. Victor, P. K. Muhuri, B. Das, and D. K. Hazra, "Thermodynamics of ion association and solvation in 2-methoxyethanol: behavior of tetraphenylarsonium, picrate, and tetraphenylborate ions from conductivity and ultrasonic data," *Journal of Physical Chemistry B*, vol. 103, no. 50, pp. 11227–11232, 1999.

[29] M. V. Kaulgud and K. J. Patil, *Acustica*, vol. 28, p. 130, 1973.

[30] K. Vijayalakshmi, V. Lalitha, and R. Gandhimathi, *Journal of Pure and Applied Ultrasonics*, vol. 28, pp. 154–156, 2006.

[31] B. Jacobson, "Intermolecular free lengths in liquids in relation to compressibility, surface tension and viscosity," *Acta Chemica Scandanavia*, vol. 5, pp. 1214–1216, 1951.

[32] B. Jacobson, *Acta Chemica Scandanavia*, vol. 34, p. 121, 1975.

[33] S. S. Yadava and A. Yadav, "Ultrasonic study on binary liquid mixtures between some bromoalkanes and hydrocarbons," *Ultrasonics*, vol. 43, no. 9, pp. 732–735, 2005.

[34] R. J. Fort and W. R. Moore, "Adiabatic compressibilities of binary liquid mixtures," *Transactions of the Faraday Society*, vol. 61, pp. 2102–2111, 1965.

[35] M. Yasmin, K. P. Singh, S. Parveen, M. Gupta, and J. P. Shukla, "Thermoacoustical excess properties of binary liquid mixtures-a comparative experimental and theoretical study," *Acta Physica Polonica A*, vol. 115, no. 5, pp. 890–900, 2009.

[36] K. Subbarangaiah, N. M. Murthy, and S. V. Subrahmanyam, *Acustica*, vol. 55, p. 105, 1985.

[37] A. M. E. Raj, L. B. Resmi, V. B. Jothy, M. Jayachandran, and C. Sanjeeviraja, "Ultrasonic study on binary mixture containing dimethylformamide and methanol over the entire miscibility range ($0 < x < 1$) at temperatures 303–323 K," *Fluid Phase Equilibria*, vol. 281, no. 1, pp. 78–86, 2009.

[38] M. M. H. Bhuiyan and M. H. Uddin, "Excess molar volumes and excess viscosities for mixtures of N,N-dimethylformamide with methanol, ethanol and 2-propanol at different temperatures," *Journal of Molecular Liquids*, vol. 138, no. 1–3, pp. 139–146, 2008.

[39] A. Ali and A. K. Nain, "Ultrasonic and volumetric study of binary mixtures of benzyl alcohol with amides," *Bulletin of the Chemical Society of Japan*, vol. 75, no. 4, pp. 681–687, 2002.

[40] N. K. Kim, H. J. Lee, K. H. Choi et al., "Substituent effect of N,N-dialkylamides on the intermolecular hydrogen bonding with thioacetamide," *Journal of Physical Chemistry A*, vol. 104, no. 23, pp. 5572–5578, 2000.

[41] K. P. C. Rao and K. S. Reddy, "Excess volumes of N,N-dimethyl formamide and N,N-dimethyl acetamide with aliphatic esters at room temperature," *Physics and Chemistry of Liquids*, vol. 18, no. 1, pp. 75–79, 1988.

[42] P. S. Nikam and S. J. Kharat, "Densities, viscosities, and thermodynamic properties of (N,N-dimethylformamide + benzene + chlorobenzene) ternary mixtures at (298.15, 303.15, 308.15, and 313.15) K," *Journal of Chemical and Engineering Data*, vol. 48, no. 5, pp. 1202–1207, 2003.

[43] P. S. Nikam and S. J. Kharat, "Density and viscosity studies of binary mixtures of N,N-dimethylformamide with toluene and methyl benzoate at (298.15, 303.15, 308.15, and 313.15) K," *Journal of Chemical and Engineering Data*, vol. 50, no. 2, pp. 455–459, 2005.

[44] R. S. Ramadevi and M. V. Prabhakara Rao, "Excess volumes of substituted benzenes with N,N-dimethylformamide," *Journal of Chemical and Engineering Data*, vol. 40, no. 1, pp. 65–67, 1995.

[45] T. Sugiura and H. Ogawa, "Enthalpies of solution, partial molar volumes and isentropic compressibilities of polycyclic aromatic hydrocarbons in 1-methyl-2-pyrrolidone at 298.15 K," *Fluid Phase Equilibria*, vol. 277, no. 1, pp. 29–34, 2009.

5

Nonequilibrium Thermodynamics and Distributions Time to Achieve a Given Level of a Stochastic Process for Energy of System

V. V. Ryazanov

Institute for Nuclear Research, Prospect Nauki 47, Kiev 252028, Ukraine

Correspondence should be addressed to V. V. Ryazanov, vryazan@kinr.kiev.ua

Academic Editor: L. De Goey

In a previous paper (Ryazanov (2011)) with the joint statistical distribution for the energy and lifetime (time to achieve a given level of a stochastic process for energy of system) to derive thermodynamic relationships, clarifying similar expressions of extended irreversible thermodynamics we used an exponential distribution of lifetime. In this paper, we explore a more realistic expression for the distribution of time to achieve a given level of a stochastic process for energy of system (or relaxation times or lifetimes), and we analyse how such distribution affects the corresponding expressions of nonequilibrium entropy, temperature, and entropy production.

1. Introduction

In [1] expressions for a nonequilibrium entropy S, entropy production σ, nonequilibrium temperature, and generalized transport equations for thermodynamic fluxes were proposed, taking into account a given distribution of internal relaxation times (or lifetimes) of the variables of the system. In the theory of random processes such values are called time to achieve a given level of the variables of the system (or the first-passage-time problems or escape time). In [1] for this quantity we used the term lifetime. These expressions generalize similar expressions of the extended irreversible thermodynamics (EIT) [2], a theory which takes into account the nonvanishing character of the relaxation time of the heat flux and other thermodynamic fluxes. In [1] we proposed to take into consideration not only a single or a few relaxation time, but a whole distribution of relaxation times, generalizing in this way the previous proposals of EIT.

There are many motivations for taking a full distribution of relaxation times; for instance, in heat transfer, the heat flux is the sum of contributions of molecules moving at different speeds, or of phonons having different frequencies, and the relaxation times of the mentioned contributions usually depend on the speed or the frequency, thus yielding a relaxation time distribution. Glassy materials have also a wide distribution of relaxation times. Collisions of heavy nuclei yield different products having widely different decay times. Thus, the consideration of relaxation time distributions seems natural in the analysis of nonequilibrium systems.

Approach [1] was based on the distribution containing lifetime as a thermodynamic parameter of the form

$$\rho(z; E = v, \Gamma = y) = \exp\frac{\{-\beta v - \gamma y\}}{Z(\beta, \gamma)}, \tag{1}$$

where E is energy of the system, z are dynamical variables, coordinates and momenta of the system particles, Γ is random variables of lifetime, a first-passage time till the random process $v(t) = E(t)$ reaches its zero value, β and γ are Lagrange multipliers conjugated to thermodynamic parameters E and Γ, and $Z(\beta, \gamma)$ is the partition function. Reference [1] uses an exponential distribution for the lifetime

$$p_\Gamma(y) = \Gamma_0^{-1} \exp\left\{-\frac{y}{\Gamma_0}\right\}, \tag{2}$$

where Γ_0 is average lifetime. The expression (2) can be obtained from the algorithm of phase coarsening of the complex systems (Appendix 2 [1]). The distribution in the form (2) is valid for the existence of the weak ergodicity in a system. Mixing the system states at long times will lead to the distribution (2). For systems in the vicinity of phase transitions or in the chaotic regime, distribution (2) is no longer valid. The aim of this paper is to generalize the formalism proposed in [1] to this more general situation.

2. Generalized Lifetime Thermodynamics with Lifetime Distribution in (3)

Setting the form of the function $p_\Gamma(y)$ reflects not only the internal properties of a system, but also the influence of the environment on an open system and the particular character of its interaction with the environment. The following physical interpretation of the exponential distribution for the function $p_\Gamma(y)$ is given: a system evolves freely like an isolated system governed by the Liouville operator. Besides that the system undergoes random transitions and the phase point representing the system switches from one trajectory to another one with an exponential probability under the influence of the "thermostat." The exponential distribution describes completely random systems. The influence of the environment on a system can have organized character as well; for example, this is the case of systems in a nonequilibrium state with input and output nonstationary fluxes. The character of the interaction with the environment can also vary; therefore different forms of the function $p_\Gamma(y)$ can be used.

In [3] the maximum entropy principle for Liouville equations with source for the determination of the function $p_\Gamma(y)$ was applied. The result obtained in [3] can be written as

$$p_\Gamma(y,t) = \frac{p_\Gamma(0)e^{-C_i y/F_i}}{1+(p_\Gamma(0)/F_i)e^{-C_i y/F_i}(R(t)-R(t_0))}, \quad y = t - t_0, \tag{3}$$

where t_0 is some initial moment of birth of the system, t is the current time,

$$R(t) = \sum_j\sum_m F_m(t_0)F_j(t_0) \times \sum_k \frac{\langle P_k P_j P_m\rangle - \langle P_j P_m\rangle\langle P_k\rangle}{\langle P_i P_k\rangle - \langle P_i\rangle\langle P_k\rangle} + F_i \ln Z_\beta(t_0) - \sum_m\sum_j F_j(t_0)\frac{\langle P_j P_m\rangle - \langle P_j\rangle\langle P_m\rangle}{\langle P_i P_m\rangle - \langle P_i\rangle\langle P_m\rangle}. \tag{4}$$

In (4) it is assumed that, along with the energy $E = P_0$ system is described and by M other physical quantities; for example, $P_1 = p$ is density of momentum, $P_{i+1} = n_i$ are the number of particles ith component, and so forth and conjugate thermodynamic parameters F_j. If we restrict ourselves, $P_0 = E$, $F_j = \beta$, $R(t)$ becomes

$$R(t) = \beta^2(t_0)\frac{\langle E^3\rangle - \langle E^2\rangle\langle E\rangle}{\langle E^2\rangle - \langle E\rangle^2} + \beta(t)\ln Z_\beta(t_0) - \beta(t_0), \quad Z_\beta = \int \omega(E=v)e^{-\beta v}dv. \tag{5}$$

In deriving (3) we used the Zubarev-Peletminsky rule [4, 5]

$$\frac{d\vec{z}}{dt} = \vec{w}(\vec{z}),$$
$$\vec{w}\vec{\nabla}P_i = \sum_{j=1}^{M} C_{ij}P_j; \quad i = 1,\ldots,M, \tag{6}$$
$$C_i = \sum_j C_{ji}F_j(t_0),$$

where C_{ij} are c-numbers. When the local density of dynamic variables is considered, the value of P_m may depend on the spatial variables. Since then the value C_{ij} may also depend on the spatial variables or may be differential operators. The equation for the specific energy u has the form $\rho du/dt + \vec{\nabla}\vec{J}q = 0$; $E = \int_V \rho u dV$; $\vec{J}_q = \vec{J}_u = \vec{q}$ is the heat flux. In this paper we consider what changes in the thermodynamic behavior of the system results with replacement of the form (2), used in [1], the function of the form (3).

From the normalization of the distribution (3) it is seen that

$$p_\Gamma(0) = \frac{\beta\left(1 - e^{-rC_i/\beta^2}\right)}{r}, \quad r = R(t_0) - R(t). \tag{7}$$

The value of $(C_i/\beta)^{-1}$ is close to the average lifetime Γ_0 of the relation (2). It may be shown that in the linear approximation to r

$$p_\Gamma(0,t_0) = a = \frac{C_i}{F_i},$$
$$p_\Gamma(y,t) = ae^{-ay}\left(1 + \frac{are^{-ay}}{\beta}\right). \tag{8}$$

In the expression for the partition function

$$Z(\beta,\gamma) = \int \exp\{-\beta v - \gamma y\}dz = \iint dv dy \omega(v,y)\exp\{-\beta v - \gamma y\} \tag{9}$$

is the structure factor $\omega(E,\Gamma)$ which has a meaning of the joint probability density of values E, Γ as it was shown in [1]. It may be shown that for the distribution (3) the function $\omega(E,\Gamma)$ takes on the form:

$$\omega(E,\Gamma = y) = \omega(E)\frac{p_\Gamma(0)e^{-(C_i y/F_i)}}{1+(p_\Gamma(0)/F_i)e^{-(C_i y/F_i)}(R(t)-R(t_0))}. \tag{10}$$

Substituting (10) into the partition function (9) yields

$$Z(\beta,\gamma) = Z(\beta)Z(\gamma),$$

$$\begin{aligned} Z(\gamma) &= \int_0^\infty e^{-\gamma y} p_\Gamma(y,t)\,dy \\ &= \frac{p_\Gamma(0)}{a(1+(\gamma/a))}\, {}_2F_1\left(1, 1+\frac{\gamma}{a}, 2+\frac{\gamma}{a}, w\right), \\ w &= \frac{p_\Gamma(0) r}{\beta}, \end{aligned} \tag{11}$$

where $Z(\beta) = \int \omega(E = v)\exp\{-\beta v\}dv$ is the Gibbs partition function, ${}_2F_1(.,.,.,.)$ is ordinary hypergeometric function [6].

We have from (11) when $\Gamma\gamma = -\partial \ln Z(\beta,\gamma)/\partial\gamma$; $\Gamma_0(V) = \Gamma_\gamma(V)/\gamma = 0$;

$$\begin{aligned} \gamma\Gamma\gamma &= \frac{x}{1+x} - x\frac{w\left(1/(2+x)^2\right) + \cdots + w^n\left(n/(x+n+1)^2\right) + \cdots}{1 + w((x+1)/(x+2)) + \cdots + w^n((x+1)/(x+n+1)) + \cdots}, \quad x = \frac{\gamma}{a} \approx \gamma\Gamma_0, \\ {}_2F_1(1, x+1, x+2, w) &= 1 + w\frac{x+1}{x+2} + \cdots + w^n\frac{x+1}{x+n+1} + \cdots. \end{aligned} \tag{12}$$

where

The expression for $x = \gamma/a$ is close to that considered in [1], the expression $x = \gamma\Gamma_0$, but not identical to it. At $Z(\beta,\gamma) = Z(\beta)Z(\gamma)$ the entropy is equal to

$$\frac{S}{k_B} = -\langle \ln \rho(z; E, \Gamma)\rangle = \frac{S_\beta}{k_B} + \frac{S_\gamma}{k_B}, \tag{13}$$

$$\begin{aligned} S_\beta &= \beta\langle E\rangle + \ln Z(\beta), \qquad S_\gamma = \gamma\langle\Gamma\rangle + \ln Z(\gamma), \\ S_\gamma &= \frac{x}{1+x} - \ln(1+x) + \ln\frac{p_\Gamma(0)}{a} - x\frac{w\left(1/(2+x)^2\right) + \cdots + w^n\left(n/(x+n+1)^2\right) + \cdots}{1 + w((1+x)/(2+x)) + \cdots + w^n((1+x)/(1+n+x)) + \cdots} + \\ &\quad + \ln\left[1 + w\frac{(1+x)}{(2+x)} + \cdots + w^n\frac{(1+x)}{(1+n+x)} + \cdots\right]. \end{aligned} \tag{14}$$

The first two terms on the right side of (14) coincide with the expression for S_γ of [1]. The remaining terms are a supplement to this expression. They depend on w and tend to 0 if a value w (or a value r in (11), (3), (7)) tends to 0.

From (14) treating E and Γ as variables,

$$dS = k_B\beta dE + k_B\gamma d\Gamma. \tag{15}$$

From (14)

$$\begin{aligned} \frac{\partial S_\gamma}{\partial x} &= -\frac{D}{x}, \\ D &= x\left\{\left(\frac{x}{(1+x)^2}\right) - \frac{x}{1 + w((1+x)/(2+x)) + \cdots + w^n((1+x)/(1+n+x)) + \cdots}\right. \\ &\quad \times\left[2\left(\frac{w}{(2+x)^3} + \cdots + \frac{w^n n}{(1+n+x)^3} + \cdots\right) + \right. \\ &\quad \left.\left. + \frac{\left(\left(w/(2+x)^2\right) + \cdots + \left(w^n n/(1+n+x)^2\right) + \cdots\right)^2}{1 + w((1+x)/(2+x)) + \cdots + w^n((1+x)/(1+n+x)) + \cdots}\right]\right\}. \end{aligned} \tag{16}$$

In [1] for the case of thermal conductivity obtained the expression

$$x_q = y_q \Gamma_0 \approx \frac{t_{0q} S_a q}{E} = \frac{t_{0q} q}{\rho u L} = t_{0q} y_q, \qquad y_q = \frac{q}{\rho u L}, \; E \approx \rho u V, \tag{17}$$

where $q = \pm|\vec{q}| = (\vec{q}\vec{q})^{1/2}$, $u = E/\rho V$ is energy density in a system, L is its linear size, V is its volume $\sim L^3$, and the surface $S_a \sim L^2$; $q_+ = (\vec{q}_+ \vec{s}) = q = (\vec{q}\vec{q})^{1/2}$, where q_- and q_+ are projections of the outcoming and incoming heat flux density vectors on the surface normal vector $\vec{s}$ (ordered chronologically [1]); the signs of q_- and q_+ depend on whether we are heating or cooling the system. Then $q_+ = -q_- - R\rho\partial u/\partial t$; in the stationary state $q_+ = -q_-$.

Thermodynamic parameters that determine the value of x will be the energy E, the heat flux q, and the size of L. From (17) we obtain

$$dx = x\left(-\frac{dE}{E} + \frac{dq}{q} + \frac{2dL}{L}\right). \tag{18}$$

Expression (15) can be rewritten as

$$dS = k_B \beta dE + \frac{\partial S_y}{\partial x} dx. \tag{19}$$

The value r in (11) is small, although it is possible to specify a physical situation where the value of r is not small. In the linear approximation to r (and w) the relationship (16) takes the form

$$\frac{\partial S_y}{\partial x} \approx -\frac{x}{(1+x)^2} + w\frac{2x}{(2+x)^3}. \tag{20}$$

In [1] value of x was determined by comparing the obtained expression for the nonequilibrium entropy to the same expression of the EIT [2]. It is possible to determine the value of x in the other ratios, for example, by comparing the expression for the inverse nonequilibrium temperature with corresponding expression of EIT in the form

$$\theta^{-1} = \frac{1}{T} + q^2 \frac{\alpha_q}{u}, \quad \alpha_q = \frac{\tau_q}{\rho \lambda T^2}, \tag{21}$$

where τ_q is the time of the flux correlation, T is equilibrium temperature, and λ is the heat conductivity coefficient.

The expression of $\theta^{-1} = (\partial S/\partial E_{|q,R})$, obtained from (14), (16), (18) in the linear approximation to w is given by

$$\theta^{-1} = \frac{\partial S}{\partial E}_{|q,R} = k_B\left[\beta + \frac{x}{E}\left(\frac{x}{(1+x)^2} - w\frac{2x}{(2+x)^3}\right)\right]. \tag{22}$$

Equating the expressions (21) and (22), we obtain the equation for x. Neglecting the powers of x higher than the second, we find the solution of the resulting quadratic equation

$$x = \frac{-28c \pm (16c(30c+4-w))^{1/2}}{2(38c-8-2w)}, \quad c = \frac{q^2 \alpha_q E}{u k_B}. \tag{23}$$

The exact expression of θ^{-1} is more cumbersome ((20) than linear approximation of (16))

$$\theta^{-1} = k_B\left(\beta + \frac{D}{E}\right), \tag{24}$$

where D is from (16). Here you can substitute the value of x from (17). From (16)–(19) we obtain

$$dS = k_B\left(\beta + \frac{D}{E}\right)dE - k_B\frac{D}{q}dq - k_B\frac{2D}{L}dL. \tag{25}$$

Determining in (25) the value of dS/dt, we find the entropy balance equation $(1/V)(dS/dt) = -\vec{\nabla}\vec{j}_S + \sigma_S$, $S = \int_V \rho s d\vec{r}$, where for $\rho = \text{const}$, $\vec{j}_S = \theta^{-1}\vec{q}$,

$$\sigma_S = \vec{q}\vec{\nabla}\theta^{-1} - \frac{k_B D(dq/dt)}{qV} + \frac{k_B(dL/dt)[3u\beta\rho + V^{-1}D]}{L}. \tag{26}$$

Comparing entropy production σ_S (26) and $\sigma_S = \vec{q}\vec{q}/\lambda T^2$ of EIT [2], we shall find

$$\lambda T^2 k_B\left(\frac{d\vec{q}}{dt}\right)q^{-2}V^{-1}D + \vec{q} = \lambda T^2 \nabla\theta^{-1} + \left(\frac{dL}{dt}\right)\lambda T^2 k_B \frac{[3u\beta\rho + DV^{-1}]}{\vec{q}L}. \tag{27}$$

The first two terms of the expressions (24)–(27) coincide with the expressions of [1]. The remaining terms are a supplement to these expressions. They depend on w and tends to 0 if a value w tends to 0.

Just as in [1], we can write the conditions of thermodynamic stability and also consider not only by heat transfer, but by the internal friction of the system which is represented by the dissipative part of the stress tensor, the processes with variable mass, and so forth.

3. Conclusion

In this paper instead of the exponential distribution for the lifetime of [1] we examined the lifetime distribution of the form (3). The expressions derived here generalize expressions for the nonequilibrium entropy, entropy production, inverse nonequilibrium temperature, and Maxwell-Cattaneo equation derived in [1]. Expressions have the following structure: the terms of [1] and supplements, depending on the value of w (11).

There are other possibilities to generalize the results of [1]. Thus, in [7] the linearised dual-phase-lag model of heat transfer is based on the equation

$$\tau \frac{\partial q}{\partial t} + q = -\lambda \nabla T - \varepsilon \frac{\partial \nabla T}{\partial t}, \quad (28)$$

where T is the equilibrium temperature, λ the heat conductivity, q the heat flux, τ and ε are relaxation times. In [7] it was supposed that the entropy density s depends not only on the internal energy density u but also on $\dot{u}$ and $\ddot{u} = (\partial^2 u)/(\partial t^2)$, $s = s(u, \dot{u}, \ddot{u})$. The use of these variables could be extended to higher-order time derivatives; such formalism is compatible with our proposal of taking a distribution of relaxation times, because it allows for rather general dynamical decays, much more complicated than an exponential decay with a single relaxation time.

It is possible to suggest some actual physical systems to which these ideas may be applicable. Those are systems which take into account the nonvanishing character of the relaxation time of the heat flux and other thermodynamic fluxes. Examples include solids with phonons having different frequencies and the relaxation times of the mentioned contributions usually depend on the speed or the frequency. A more detailed study of glass materials with a wide range of relaxation times, different times of the decay in collisions of heavy nuclei—those and other examples—can serve as objects of description of the proposed theory. The proposed description also applies to the processes of deformation of a continuous medium, to the chemical reactions, and so forth. The proposed description characterizes open systems describing the interaction with the environment.

References

[1] V. V. Ryazanov, "Nonequilibrium thermodynamics based on the distributions containing lifetime as thermodynamic parameter," *Journal of Thermodynamics*, vol. 2011, Article ID 203203, 10 pages, 2011.

[2] D. Jou, J. Casas-Vazquez, and G. Lebon, *Extended Irreversible Thermodynamics*, Springer, Berlin, Germany, 1993.

[3] V. V. Ryazanov, "Maximum entropy principle and the form of source in non-equilibrium statistical operator method," http://arxiv.org/abs/0910.4490v1.

[4] D. N. Zubarev, *Nonequilibrium Statistical Thermodynamics*, Plenum-Consultants Bureau, New York, NY, USA, 1974.

[5] A. I. Akhiezer and S. V. Peletminskii, *Methods of Statistical Physics*, Pergamon Press, Oxford, UK, 1981.

[6] M. Abramowitz and I. A. Stegun, Eds., *Handbook of Mathematical Functions with Formulas, Graphs, and Mathematical Tables*, Dover, New York, NY, USA, 1972.

[7] S. I. Serdyukov, "A new version of extended irreversible thermodynamics and dual-phase-lag model in heat transfer," *Physics Letters A*, vol. 281, no. 1, pp. 16–20, 2001.

6

Water Desorption Process in Room Temperature Ionic Liquid-H_2O Mixtures: *N, N*-diethyl-*N*-methyl-*N*-(2-methoxyethyl) Ammonium Tetrafluoroborate

Hiroshi Abe,[1] Tomohiro Mori,[1] Yusuke Imai,[1] and Yukihiro Yoshimura[2]

[1] *Department of Materials Science and Engineering, National Defense Academy, Yokosuka 239-8686, Japan*
[2] *Department of Applied Chemistry, National Defense Academy, Yokosuka 239-8686, Japan*

Correspondence should be addressed to Hiroshi Abe, ab@nda.ac.jp

Academic Editor: Ramesh Gardas

A water desorption process of a mixture of room temperature ionic liquid (*N*, *N*-diethyl-*N*-methyl-*N*-(2-methoxyethyl) ammonium tetrafluoroborate) and water was investigated via simultaneous X-ray diffraction and differential scanning calorimetry (DSC) measurements, in which relative humidity was controlled by a water vapor generator. In these measurements, H_2O concentration was estimated by the peak position of the *principal* peak in X-ray diffraction patterns, and the thermal property associated with a mixing state was detected by a DSC thermograph. In addition, the density of the mixture was measured as a macroscopic property. *In situ* observations revealed that the thermally unstable mixing state in the water-rich region has an important correlation with density and thermal and structural properties.

1. Introduction

Numerous investigations of room temperature ionic liquids (RTILs) have been performed to produce green solvents [1]. These RTILs are organic salts, which consist only of a cation and an anion. Nonmeasurable vapor pressure is considered as an outstanding feature of RTILs.

To study physicochemical properties, chemical potential and enthalpy are investigated using RTILs and water mixtures [2]. RTILs include 1-butyl-3-methylimidazolium tetrafluoroborate, $[C_4mim][BF_4]$ and 1-butyl-3-methylimidazolium iodide, $[C_4mim]I$. Here a series of cations such as 1-alkyl-3-methyl-imidazolium is expressed as $[C_nmim]$ using the alkyl chain length, *n*. Anomalies in the water-rich region were observed in excess partial molar enthalpies. Moreover, the excess molar enthalpy was examined systematically in various aqueous solutions [3]. For example, in $[C_4mim][BF_4]$, the maximum value of the excess isobaric molar heat capacity appeared at around 80 mol% H_2O. Considering hydrophilicity and hydrophobicity, which are determined by the chemical structure of the anion, the hydrogen bonding of water was discussed. Solubilities of various types of RTILs and additives have been recently summarized [4]. The liquid-liquid equilibrium (LLE) and the solid-liquid equilibrium (SLE) in a binary liquid system revealed the mixing states as a result of molecular interactions including a hydrogen-bond donor and acceptor. From the perspective thermodynamics, activity and enthalpy were estimated by solute heat capacity and melting temperature. Moreover, thermomorphic phase separation in RTIL-organic liquid systems was investigated by electrical conductivity, FT-Raman, and NMR measurements [5], and the LLE relative to the phase separation in binary liquids was determined. Proton dynamics in the vicinity of the phase separation were clearly distinguished in NMR spectra. Macroscopic solubilities and conductivity were correlated with molecular interactions described by FT-Raman and NMR spectroscopy. Moreover, LLE and SLE in quaternary phosphonium RTIL-based mixtures were determined [6]. RTIL was regarded as

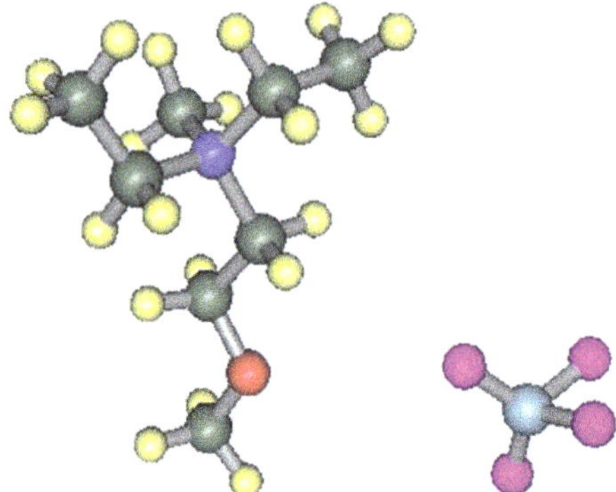

Figure 1: A [DEME] cation and a $[BF_4]$ anion.

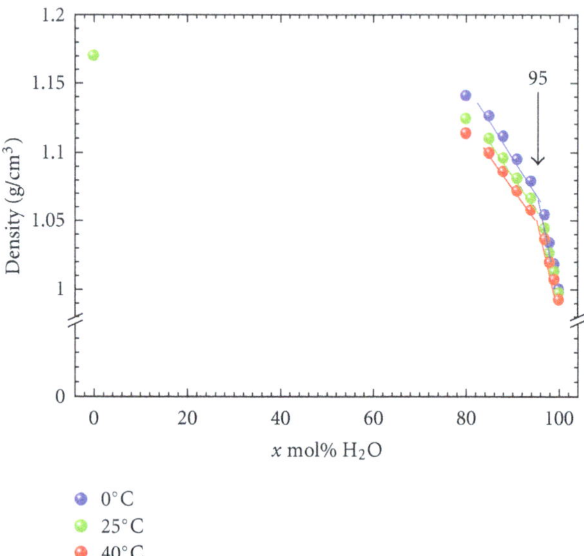

Figure 2: Temperature and concentration dependences of density in $[DEME][BF_4]$-x mol% H_2O mixtures.

a nonimidazolium system, and the solubility was interpreted by the polar anion-solvent interaction.

Recently, a different type of RTIL *N*, *N*-diethyl-*N*-methyl-*N*-(2-methoxyethyl) ammonium tetrafluoroborate, $[DEME][BF_4]$, was synthesized having a high potential for electrochemical capacitor applications [7]. The complicated SLE in the $[DEME][BF_4]$-H_2O mixture, caused by various types of hydrogen bonding, was determined by simultaneous X-ray diffraction and differential scanning calorimetry (DSC) measurements [8, 9]. The bonding nature strongly depends on the local environment, that is, coordination number, molecular orientational order, and molecular packing efficiency. We recently determined that an outstanding hierarchy structure of $[DEME][BF_4]$-H_2O mixtures is formed even in the liquid state [10]. On each scale, various types of anomalies appeared at respective water concentrations that included anomalous optical absorption in the UV-vis region, density fluctuation by small angle X-ray scattering, a network-forming property by the *prepeak* in the X-ray diffraction patterns, and the liquid structure by X-ray diffraction. We recently summarized a series of our researches on $[DEME][BF_4]$-H_2O mixtures from the perspective of hydrogen bonding and discussed water-mediated glassy states and double glass transition [11].

In the present study, we investigate the desorption process of the $[DEME][BF_4]$-H_2O mixture. *In situ* observations reveal that the mixing state in the water-rich region became thermally unstable, and the mixing state has an important correlation with density and structural anomalies.

2. Experiments

We selected $[DEME][BF_4]$ (Kanto Chemical Co.) as an RTIL (Figure 1). Because this RTIL is hydrophilic, 126 ppm H_2O was included in the as-received sample by the Karl-Fischer titration method (870 KT Titrino plus, Metrohm AG). For H_2O mixtures, we used distilled water (Wako Pure Chemical Co.). The mixtures were prepared by dissolving RTILs in a dry box under a flow of helium gas to exclude atmospheric H_2O. The sample solutions were simply prepared by dissolving H_2O in $[DEME][BF_4]$.

In situ observations were carried out using simultaneous X-ray diffraction and DSC measurements. The DSC was attached to a vertical goniometer with 2 kW X-ray generator (RINT-Ultima III, Rigaku Co.). For *in situ* observations of a liquid state, a sample stage was fixed horizontally with a sealed X-ray tube and a scintillation counter moving simultaneously. A parallel beam was obtained by a parabolic multilayer mirror. A long Soller slit was placed in front of the scintillation counter. Cu Kα radiation (λ = 0.1542 nm) was selected for the simultaneous measurements. Here the scattering vector, Q, is defined as $4\pi(\sin\theta)/\lambda$ (nm^{-1}). Relative humidity (RH) was controlled by a water vapor generator (HUM-1, Rigaku Co.), which was directly connected to the DSC. Moisture inside the DSC cell was stable within ±2%RH throughout the entire desorption process.

Liquid density of the mixtures was measured with a density/specific gravity meter (DA-645, Kyoto Electronics Manufacturing Co.). The temperature range was 0–90°C. Accuracies of density and temperature were estimated to be $\pm 5 \times 10^{-5}$ g/cm^3 and ±0.03°C, respectively.

3. Results

The liquid density of $[DEME][BF_4]$-H_2O mixtures was measured from 40 to 0°C in the water-rich region (Figure 2). At a fixed temperature, density as a function of water concentration (x mol% H_2O), $\rho(x)$, exhibited a sharp bend at 95 mol%. The two different gradients $(\partial\rho/\partial x)_T$ between $\rho(x < 95\,\text{mol\%})$ and $\rho(x > 95\,\text{mol\%})$ suggest that molecular packing is not described by a simple superposition of $[DEME][BF_4]$ and H_2O molecules. On the other hand, at fixed concentrations, a substantially different temperature dependence of density, $(\partial\rho/\partial T)_x$, was observed below and above 95 mol%. For example, $\rho(T)$ below 95 mol% depended remarkably on temperature, while the temperature dependence above 95 mol% was small. The volumetric thermal expansion coefficient, β, can enhance these density changes.

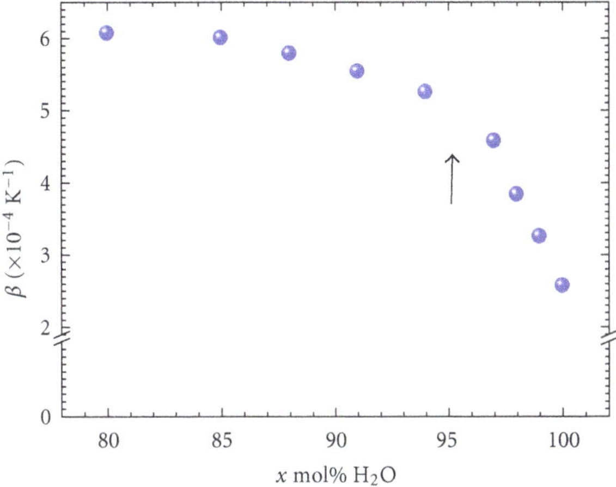

FIGURE 3: Concentration dependence of volumetric thermal expansion coefficient, β, at 25°C.

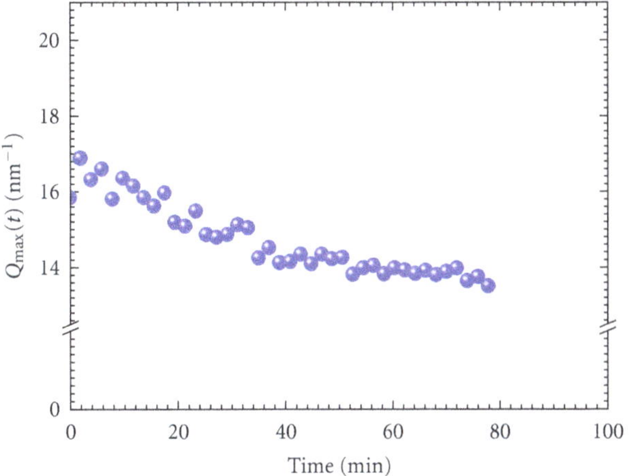

FIGURE 4: Time dependence of Q_{max} position.

In general, β is provided by $-(1/\rho)(\partial\rho/\partial T)_x$. Concentration dependence of β at 25°C is shown in Figure 3. Note that the distinct change at 95 mol% is revealed more clearly.

Simultaneous X-ray diffraction and DSC measurements were performed at the fixed temperature (33°C) and humidity (30%RH). The initial and final water concentrations were determined to be 93.0 and 18.1 mol%, respectively, by an electric balance. In the same manner as that detailed in a previous study [12], Q positions at the maximum in X-ray diffraction patterns, Q_{max}, were calculated by the least square fitting method using the pseudo-Voigt function. Time dependence of Q_{max} at the fixed temperature is shown in Figure 4. Apart from $Q_{max}(t)$ in the present study, the Q_{max}-x mol% plot values in [DEME][BF_4]-H_2O mixtures were previously obtained (Figure 5) [12]. For further analysis, we represented the Q_{max}-x relationship by a simple equation:

$$x = [1 - \exp\{-1.7(Q_{max} - Q_0)\}] \times 100. \quad (1)$$

Q_0 is the constant value determined to be 13.5 nm^{-1} by the nonlinear fitting method. The solid curve in the

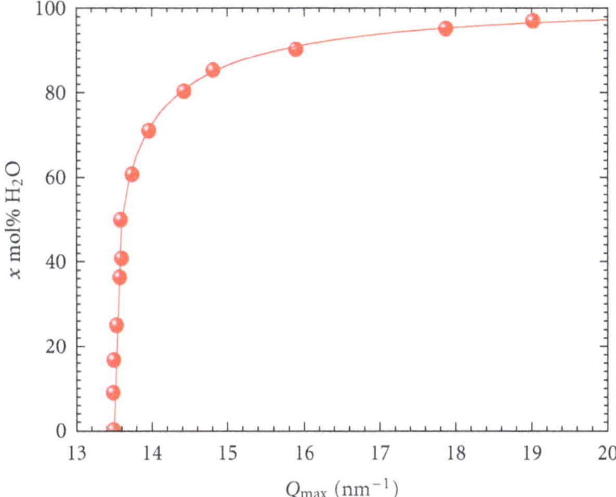

FIGURE 5: x-Q_{max} relationship. Solid curve is obtained by the least square fitting method.

figure indicates the fitted values obtained using (1). To obtain time-dependent concentrations, the experimentally obtained $Q_{max}(t)$ values were transformed into $x(t)$ using (1) (Figure 6(a)). Here the solid curve in Figure 6(a) was obtained by the least square fitting method of a polynomial. For the quantitative analysis of the DSC thermograph, time dependence of mass during the desorption process is required. Therefore, we assumed that the mass of RTIL, m_{IL}, in the mixture could not decrease during the desorption process. Thus, the mass of water as a function of time, $m_w(t)$, was evaluated easily from $x(t)$ at fixed m_{IL}. In fact, even under vacuum (400 Pa), the mass of [DEME][BF_4] including various types of additives did not decrease for 48 hours [13]. Therefore, the total mass curve, $m(t)$, is given by

$$m(t) = m_{IL} + m_w(t). \quad (2)$$

Figure 6(b) shows the total mass curve, $m(t)$, which was obtained by satisfying (2). The final water concentration (18.1 mol%), which was measured by an electric balance after the desorption process, remarkably coincided with the calculated $m(t)$ within the experimental error.

In addition to the X-ray diffraction measurement for the estimation of water concentration, the DSC thermogram with time evolution was measured via *in situ* observations. As mentioned previously, the total mass decreased with increasing desorption time. Thus, we corrected the thermogram trace by $m(t)$ in Figure 6(b). As a result, we obtained a normalized heat flow as shown in Figure 7. Here x_c (85 mol% H_2O) was the crossover concentration, which was determined by the hierarchy structure in [DEME][BF_4]-H_2O mixtures [10]. At around x_c, the heat flow apparently exhibited the minimum value on the thermogram. To check an experimental error in the heat flow, the DSC trace having a little mass change, Δm, is plotted in Figure 8, where $\Delta m/m$ during desorption process was estimated to be less than 10%. Data as background were collected at 30°C and 85%RH. Both intensity and Q position of the principal peak were almost constant during the measurements. Compared with the background, it is clear that the heat flow on desorption

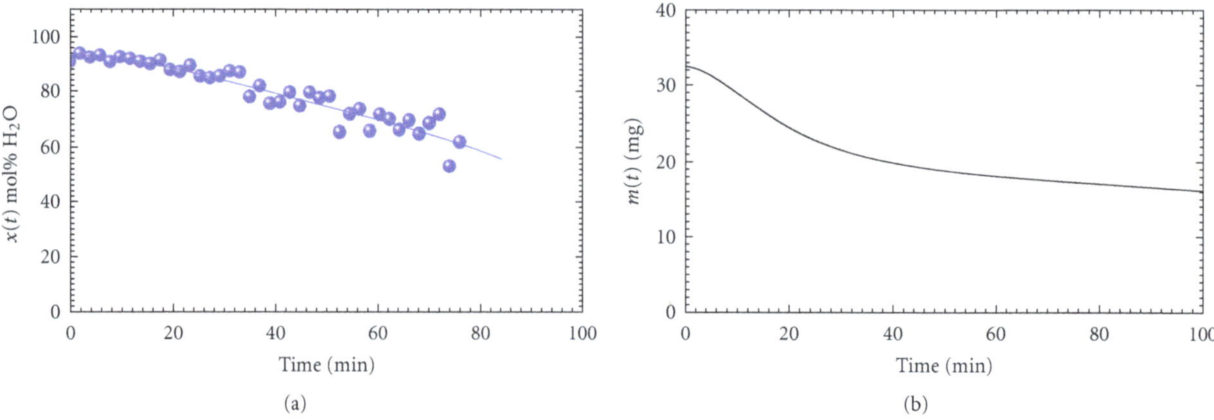

FIGURE 6: Time dependences of (a) water concentrations and (b) mass.

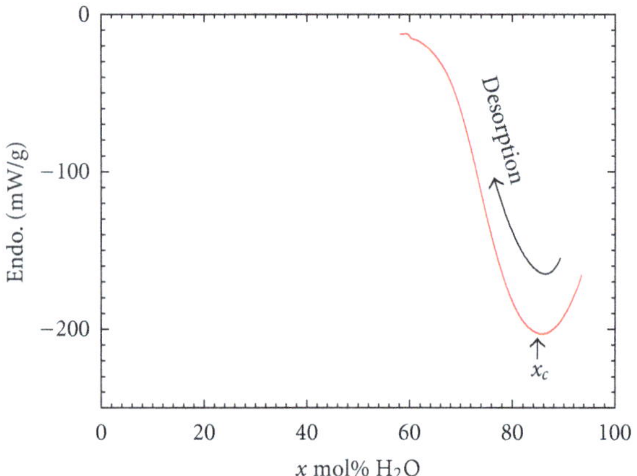

FIGURE 7: Water concentration dependence of normalized heat flow.

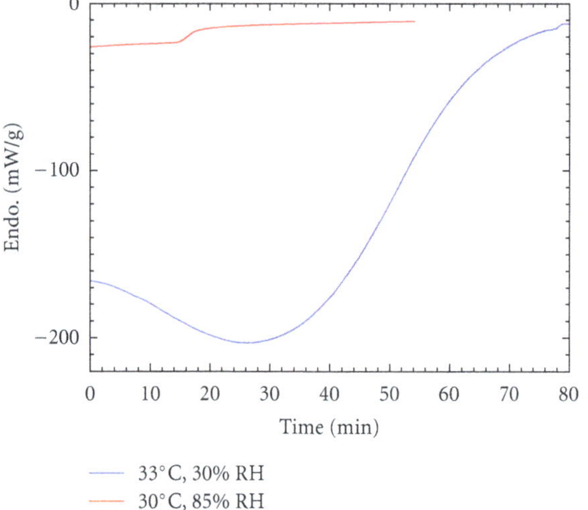

FIGURE 8: Time dependence of the normalized heat flow. For a comparison, the normalized heat flow at 30°C and 85%RH is plotted as background, where a little mass change was measured after the DSC scan.

process (33°C and 30%RH) has the distinct endothermic peak.

4. Discussion

We have shown the anomalous behavior of [DEME][BF_4]-H_2O mixtures with respect to density and the DSC trace in Figures 2 and 7, respectively. Including the structural property obtained in a previous study [10, 12], these anomalies are summarized as follows: (i) medium-range order (MRO) develops at 85 mol% $< x <$ 95 mol% as a structural property; (ii) a crossover point from RTIL to water is observed in 85 mol% (x_c) by X-ray diffraction patterns; (iii) density as a function of concentration is decomposed into two regions such as $\rho(x < 95\,\text{mol\%})$ and $\rho(x > 95\,\text{mol\%})$; (iv) by fixing x below 95 mol%, density, $\rho(T)$, highly depends on temperature; (v) at the fixed temperature, the DSC trace exhibits an endothermal peak at around x_c.

To explain these anomalies, we introduced the concept of specific aggregation, which consists of an inner core and an outer shell [14]. If aggregation exists at 85 mol% $< x <$ 95 mol%, structural anomalies (i) and (ii) can be well recognized. Aggregation can occur when a cation is isolated completely from an anion above 85 mol% and the MRO is developed by aggregation as an intermediate state. At the same time, anomaly (iii) supports the theory that $\rho(x < 95\,\text{mol\%})$ is influenced by the existence of aggregation. The formation of an aggregation-mediated network suppresses the density changes as a function of water concentration. On the other hand, $\rho(x > 95\,\text{mol\%})$ without aggregation tends to converge to the water density with increasing x. For example, the liquid structure in $ZnCl_2$-KCl mixtures is characterized by heterogeneous density fluctuations [15]. In molecular dynamics of several alkali halides, the low-density part is described by a void that exhibits MRO [16]. Next, we focused on $\rho(T)$ below 95 mol% (anomaly (iv)). In general, heterogeneous density fluctuations, which are composed of dense aggregations and voids, cause volume expansion. In the case of $\rho(T)$ below 95 mol%, appearance and disappearance of heterogeneity provide the highly changing density according to temperature. At high temperatures, a pore part easily appears by thermal expansion. To compensate for the drastic change in density, the thermally activated aggregation (a dense part) could increase with increasing temperature. In contrast, the aggregation-assisted heterogeneity could disappear at low temperatures. In accordance with the population of the aggregation, we confirm that $\rho(T)$ below 95 mol% shows a high temperature dependence.

As a result, under isothermal holding, the DSC trace in the desorption process can also support the aforementioned theory. Anomaly (v) indicates that the endothermal peak at the fixed temperature is derived from the unstable mixing state. From 93 to 85 mol%, aggregation occurs as an intermediate state between water and RTIL. We determine that the anomalous thermal property of [DEME][BF_4]-H_2O mixtures agrees with the density and structural anomalies such as the hierarchy structure.

This new methodology during the desorption process using simultaneous X-ray and DSC measurements was introduced and applied for the first time. Time dependence of concentration, $x(t)$, in the mixture is not generally evaluated by conventional DSC measurements because mass transfer, $m(t)$, should be determined during the desorption process for the quantitative thermal analysis. We have successfully estimated concentration in real time from X-ray diffraction patterns in the simultaneous measurements. From the perspective of thermodynamics, thermal property as a response to concentration is important for obtaining information of the stable/unstable mixing states at fixed temperatures.

5. Summary

[DEME][BF_4]-H_2O mixtures have an anomalous mixing state in the water-rich region. Combined with the structural anomalies, we determined that specific aggregation could occur at 85 mol% $< x <$ 95 mol%. This water concentration is regarded as an intermediate state between water and RTIL. The thermally activated aggregation can explain the time and concentration dependences of density and the endothermic peak at around x_c. We conclude that energetically unstable aggregation contributes to the hierarchy structure in the [DEME][BF_4]-H_2O system at around x_c.

Acknowledgments

The authors appreciate Ms. M. Yasaka and Mr. A. Kishi of Rigaku Co. for experimental support and helpful discussions. Also, they thank Dr. T. Takekiyo, Professor H. Matsumoto, and Professor T. Arai of National Defense Academy for helpful discussions.

References

[1] M. J. Earle and K. R. Seddon, "Ionic liquids. Green solvents for the future," *Pure and Applied Chemistry*, vol. 72, no. 7, pp. 1391–1398, 2000.

[2] H. Katayanagi, K. Nishikawa, H. Shimozaki, K. Miki, P. Westh, and Y. Koga, "Mixing schemes in ionic liquid—H_2O systems: a thermodynamic study," *Journal of Physical Chemistry B*, vol. 108, no. 50, pp. 19451–19457, 2004.

[3] G. García-Miaja, J. Troncoso, and L. Romaní, "Excess enthalpy, density, and heat capacity for binary systems of alkylimidazolium-based ionic liquids + water," *Journal of Chemical Thermodynamics*, vol. 41, no. 2, pp. 161–166, 2009.

[4] U. Domańska, "Solubilities and thermophysical properties of ionic liquids," *Pure and Applied Chemistry*, vol. 77, no. 3, pp. 543–557, 2005.

[5] A. Riisager, R. Fehrmann, R. W. Berg, R. Van Hal, and P. Wasserscheid, "Thermomorphic phase separation in ionic liquid-organic liquid systems—conductivity and spectroscopic characterization," *Physical Chemistry Chemical Physics*, vol. 7, no. 16, pp. 3052–3058, 2005.

[6] U. Domańska and L. M. Casás, "Solubility of phosphonium ionic liquid in alcohols, benzene, and alkylbenzenes," *Journal of Physical Chemistry B*, vol. 111, no. 16, pp. 4109–4115, 2007.

[7] T. Sato, G. Masuda, and K. Takagi, "Electrochemical properties of novel ionic liquids for electric double layer capacitor applications," *Electrochimica Acta*, vol. 49, no. 21, pp. 3603–3611, 2004.

[8] Y. Imai, H. Abe, T. Goto, Y. Yoshimura, Y. Michishita, and H. Matsumoto, "Structure and thermal property of *N*, *N*-diethyl-*N*-methyl-*N*-2-methoxyethyl ammonium tetrafluoroborate-H_2O mixtures," *Chemical Physics*, vol. 352, no. 1–3, pp. 224–230, 2008.

[9] H. Abe, Y. Yoshimura, Y. Imai, T. Goto, and H. Matsumoto, "Phase behavior of room temperature ionic liquid—H_2O mixtures: *N*, *N*-diethyl-*N*-methyl-*N*-2-methoxyethyl ammonium tetrafluoroborate," *Journal of Molecular Liquids*, vol. 150, no. 1–3, pp. 16–21, 2009.

[10] M. Aono, Y. Imai, Y. Ogata et al., "Anomalous mixing state in room-temperature ionic liquid-water mixtures: *N*, *N*-diethyl-*N*-methyl-*N*-(2-methoxyethyl) ammonium tetrafluoroborate," *Metallurgical and Materials Transactions A*, vol. 42, no. 1, pp. 37–40, 2011.

[11] A. Kokorin, Ed., *Ionic Liquids: Theory, Properties, New Approaches*, InTech, 2011.

[12] H. Abe, Y. Imai, T. Takekiyo, and Y. Yoshimura, "Deuterated water effect in a room temperature ionic liquid: *N*, *N*-diethyl-*N*-methyl-*N*-2-methoxyethyl ammonium tetrafluoroborate," *Journal of Physical Chemistry B*, vol. 114, no. 8, pp. 2834–2839, 2010.

[13] H. Abe, T. Mori, R. Abematsu et al., submitted to *Journal of Molecular Liquids*. In press.

[14] M. Aono, Y. Imai, H. Abe, H. Matsumoto, and Y. Yoshimura, *Thermochimica Acta*. In press.

[15] D. A. Allen, R. A. Howe, N. D. Wood, and W. S. Howells, "The structure of molten zinc chloride and potassium chloride mixtures," *Journal of Physics Condensed Matter*, vol. 4, no. 6, article 005, pp. 1407–1418, 1992.

[16] M. Salanne, C. Simon, P. Turq, and P. A. Madden, "Intermediate range chemical ordering of cations in simple molten alkali halides," *Journal of Physics Condensed Matter*, vol. 20, no. 33, Article ID 332101, 2008.

Thermo Physical Properties for Binary Mixture of Dimethylsulfoxide and Isopropylbenzene at Various Temperatures

Maninder Kumar and V. K. Rattan

University Institute of Chemical Engineering and Technology, Panjab University, Chandigarh 160014, India

Correspondence should be addressed to Maninder Kumar; maninderbhatoy@gmail.com

Academic Editor: K. A. Antonopoulos

Density, refractive index, speed of sound, and viscosity have been measured of binary mixture dimethylsulfoxide (DMSO) + isopropylbenzene (CUMENE) over the whole composition range at 298.15, 303.15, 308.15, and 313.15 K and atmospheric pressure. From these experimental measurements the excess molar volume, deviations in viscosity, molar refractivity, speed of sound, and isentropic compressibility have been calculated. These deviations have been correlated by a polynomial Redlich-Kister equation to derive the coefficients and standard error. The viscosities have furthermore been correlated with two or three parameter models, that is, herric correlation and McAllister model, respectively.

1. Introduction

This paper contributes in part to our ongoing research on the solution properties. In the present study, data on density, viscosity, refractive index and speed of sound of binary mixture dimethylsulfoxide (DMSO) + isopropylbenzene at 298.15, 303.15, 308.15, and 313.15 K have been measured experimentally. From these results the excess molar volumes, viscosity deviations, and deviations in molar refraction and isentropic compressibility have been derived. Dimethylsulfoxide is a versatile nonaqueous dipolar aprotic solvent having wide range of applications. It is used as a solvent in many nucleophilic substitutions reactions. It has the ability to pass through membranes, an ability that has been verified by numerous subsequent researchers. It can penetrate through living tissues without damaging them. Therefore local anesthetic or penicillin can be carried through the skin without using a needle which makes it a paramount in medicinal field.

Isopropylbenzene is a naturally occurring substance present in coal tar and petroleum, insoluble in water, but is soluble in many organic solvents. It is used as a feedback for the production of Phenol and its coproduct acetone. It is also used as a solvent for fats and raisins.

The study of the thermodynamic properties of DMSO + isopropylbenzene mixtures is of interest in industrial fields where solvent mixtures could be used as selective solvents for numerous reactions.

2. Experimental Section

2.1. Materials. The chemicals used are of AR grade, dimethylsulfoxide (DMSO) and isopropylbenzene (CUMENE) are from Riedel, Germany. The chemicals are purified using standard procedure [1] and are stored over molecular sieves. The purity of the chemicals was verified by comparing viscosity, density, and refractive index with the known values reported in the literature as shown in Table 1. All the compositions are prepared by using SARTOIS balance. The possible error in the mole fraction is estimated to be less than $\pm 1 \times 10^{-4}$.

2.2. Density and Speed of Sound. Density and Speed of sound were measured by ANTON PAAR densimeter (DSA 5000) to

TABLE 1: Physical properties of components at 298.15 K.

Component	T/K	ρ/g·cm^{-3}		η/mPa·s		n_D	
		exptl	lit	exptl	lit	exptl	lit
DMSO	298.15	1.0940	1.09537[9]	1.9834	1.9910[9]	1.4798	1.4775[9]
CUMENE	298.15	0.8581	0.85743[9]	0.7388	0.7390[9]	1.4928	1.4889[9]

an accuracy of ±0.000005 g/cm^{-3} and ±0.5 m/s, respectively. The densimeter was calibrated with bi-distilled degassed water.

2.3. Viscosity. Viscosities were measured by using calibrated modified Ubbelohde viscometer [2] as described earlier. The calibration of viscometer was done at each temperature in order to determine constants A and B of equation

$$\nu = \frac{\eta}{\rho} = At + \frac{B}{t}. \quad (1)$$

Flow time was measured with an electronic stop watch with precision of ±0.01 s. For each measurement viscometer was kept vertically in water bath for half an hour at constant temperature in order to attain thermodynamic equilibrium. The temperature of the bath was maintained constant with the help of circulating type cryostat where the temperature is controlled to ±0.02 K. The efflux time was repeated at least three times for each composition. The uncertainty in the values is within ±0.003 mPa·s.

2.4. Refractive Index. Refractive indices were measured for sodium D-line by ABBE-3L refractometer having Bausch and Lomb lenses. The temperature was maintained constant with the water bath as described for the viscosity measurement. A minimum of three independent readings were taken for each composition, and the average value was considered in all the calculations. Refractive index data are accurate to ±0.0001 units.

3. Experimental Results and Correlations

At least three independent readings of all the physical property measurements on ρ, η, n_D, and u were taken for each composition and the averages of these experimental values are presented in Table 2. The experimentally determined values are used for the deviation calculations.

3.1. Excess Molar Volume. Density is used to evaluate excess molar volume calculated by the equation

$$V^E = \frac{x_1M_1 + x_2M_2}{\rho} - \frac{x_1M_1}{\rho_1} - \frac{x_2M_2}{\rho_2}, \quad (2)$$

where ρ_1, ρ_2 are the densities of pure components and ρ is the density of mixture. M_1, M_2 are the molecular weight of the two components. x_1, x_2 are the mole fraction of DMSO.

Excess gibb's free energy of activation has been also calculated using the viscosity and density of the mixture by the equation

$$\Delta G^E = RT\left[\ln(\eta V) - \sum_{i=1}^{2} x_i \ln(\eta_i V_i)\right] \quad (3)$$

R is a universal gas constant; T is the temperature of the mixture. η, η_i are the viscosity of the mixture and pure compound, respectively. V, V_i refers to the molar volume of the mixture and pure components, respectively.

3.2. Viscosity Calculations. The deviation in viscosity is obtained by equation

$$\Delta\eta = \eta - x_1\eta_1 - x_2\eta_2. \quad (4)$$

η_1, η_2 refers to the viscosity of pure components and η is the viscosity of mixture. McAllister [3] model

$$\begin{aligned}\ln\nu &= x_1^3\ln\nu_1 + x_2^3\ln\nu_2 + 3x_1^2x_2\ln\eta_{12} + 3x_1x_2^2\ln\eta_{21} \\ &\quad - \ln\left[x_1 + x_2\frac{M_2}{M_1}\right] + 3x_1^2x_2\ln\left[\frac{2+(M_2/M_1)}{3}\right] \\ &\quad + 3x_1x_2^2\ln\left[\frac{1+(2M_2/M_1)}{3}\right] + x_2^3\ln\left[\frac{M_2}{M_1}\right]\end{aligned} \quad (5)$$

and herric correlation

$$\begin{aligned}\ln\nu &= x_1\ln\nu_1 + x_2\ln\nu_2 + x_1x_2\left[\alpha_{12} + \alpha'_{12}(x_1 - x_2)\right] \\ &\quad - \ln M_{\text{mix}} + x_1\ln M_1 + x_2\ln M_2\end{aligned} \quad (6)$$

have been fitted to viscosity data and it was found that both have the same standard errors at each temperature.

3.3. Isentropic Compressibility. The experimental results for the speed of sound of binary mixture are listed in Table 2. The isentropic compressibility was evaluated by $K_S = u^{-2}\rho^{-1}$ and the deviation in isentropic compressibility is calculated using the below equation:

$$K_S^E = K_S - K_S^{id} \quad (7)$$

and deviation in speed of sound by

$$\Delta u = u - x_1u_1 - x_2u_2, \quad (8)$$

TABLE 2: Density, ρ, viscosity, η, speed of Sound, u, and refractive indices, n_D, for DMSO(1) + CUMENE(2) at different temperatures.

x_1	ρ g·cm^{-3}	η mPa·s	u m·s^{-1}	n_D
		Dimethylsulfoxide(1) + isopropylbenzene(2)		
		298.15 K		
0.0000	0.8581	0.7388	1325	1.4928
0.1047	0.8721	0.8025	1336	1.4918
0.1866	0.8842	0.8603	1348	1.4909
0.3013	0.9031	0.9610	1365	1.4897
0.3835	0.9180	1.0397	1378	1.4887
0.5032	0.9421	1.1661	1400	1.4871
0.5858	0.9610	1.2692	1417	1.4859
0.6972	0.9895	1.4337	1442	1.4841
0.7926	1.0178	1.5896	1467	1.4829
0.8929	1.0519	1.7648	1493	1.4811
1.0000	1.0940	1.9834	1523	1.4798
		303.15 K		
0.0000	0.8538	0.6806	1313	1.4915
0.1047	0.8676	0.7519	1327	1.4905
0.1866	0.8797	0.8143	1339	1.4898
0.3013	0.8984	0.9064	1356	1.4885
0.3835	0.9132	0.9814	1369	1.4875
0.5032	0.9372	1.1086	1391	1.4859
0.5858	0.9559	1.2075	1408	1.4848
0.6972	0.9843	1.3691	1434	1.4829
0.7926	1.0122	1.5230	1459	1.4815
0.8929	1.0459	1.6943	1485	1.4796
1.0000	1.0890	1.8949	1512	1.4781
		308.15 K		
0.0000	0.8495	0.6167	1301	1.4896
0.1047	0.8632	0.6962	1317	1.4888
0.1866	0.8751	0.7616	1329	1.4881
0.3013	0.8936	0.8565	1346	1.4869
0.3835	0.9083	0.9294	1359	1.4859
0.5032	0.9322	1.0493	1381	1.4844
0.5858	0.9508	1.1459	1398	1.4833
0.6972	0.9788	1.3059	1425	1.4814
0.7926	1.0064	1.4565	1449	1.4799
0.8929	1.0404	1.6131	1473	1.4784
1.0000	1.0840	1.7979	1498	1.4765
		313.15 K		
0.0000	0.8452	0.5588	1289	1.4871
0.1047	0.8587	0.6434	1306	1.4864
0.1866	0.8705	0.7103	1319	1.4857
0.3013	0.8887	0.8063	1337	1.4846
0.3835	0.9033	0.8756	1349	1.4836
0.5032	0.9270	0.9947	1372	1.4821
0.5858	0.9454	1.0872	1389	1.4809
0.6972	0.9732	1.2459	1416	1.4792
0.7926	1.0008	1.3888	1439	1.4778
0.8929	1.0348	1.5402	1462	1.4762
1.0000	1.0786	1.7054	1486	1.4742

TABLE 3: Derived parameters of Redlich-Kister equation (12) and standard deviation (13) for various functions of the binary mixtures at different temperatures.

T/K	A_0	A_1	A_2	A_3	σ
			V^E/cm^3·mol^{-1}		
298.15	−1.6771	−0.2071	−0.0446	−0.6836	0.00561
303.15	−1.5289	−0.2059	0.5117	−0.0903	0.0055
308.15	−1.3339	−0.0698	0.7925	0.1613	0.0046
313.15	−1.0884	0.0276	0.8186	0.1259	0.00549
			$\Delta\eta$/mPa·s		
298.15	−0.7847	−0.1293	0.0065	0.0645	0.00296
303.15	−0.7328	−0.1197	0.1384	0.0587	0.00178
308.15	−0.6393	−0.1245	0.2208	0.0967	0.00359
313.15	−0.5606	−0.1118	0.2922	0.141	0.00333
			K_S^E/TPa^{-1}		
298.15	−121.777	−64.3513	−24.8534	−29.8607	0.37123
303.15	−133.145	−66.2282	−46.6216	−17.4037	0.21847
308.15	−144.714	−73.4879	−59.3712	4.1751	0.27323
313.15	−157.833	−74.2561	−62.2923	10.1465	0.51612
			ΔR		
298.15	−0.7478	0.2009	−0.1111	0.2177	0.0089
303.15	−0.6486	0.1994	0.0606	0.045	0.0056
308.15	−0.5417	0.2093	0.2408	−0.0917	0.0127
313.15	−0.4454	0.1947	0.3148	−0.1187	0.0167
			ΔG^E/J·mol^{-1}		
298.15	133.9592	209.2619	16.9998	168.234	64.9657
303.15	246.905	198.0077	349.8235	15.7865	3.63471
308.15	488.086	69.5143	603.5378	59.8815	7.55398
313.15	724.2593	−0.9946	832.3713	64.9657	8.13559

where K_S^{id} stands for isentropic compressibility for an ideal mixture calculated using the relation

$$K_S^{id} = \sum_{i=1}^{2} \Phi_i \left[K_{S,i} + \frac{TV_i\left(a_i^2\right)}{C_{p,i}} \right] - \left\{ \frac{T\left(\sum_{i=1}^{2} x_i V_i\right)\left(\sum_{i=1}^{2} \Phi_i a_i\right)^2}{\sum_{i=1}^{2} x_i C_{p,i}} \right\} \quad (9)$$

recommended by Kiyohara and Benson [4]. (1979) and Douhéret et al. [5] (1997), where a_i and $C_{p,i}$ are the volume expansion coefficient and heat capacity of the ith components.

3.4. Molar Refraction. Refractive indices have been used for the calculation of Molar refraction (R_m) obtained by using Lorenz-Lorenz [6] equation

$$R_m = \frac{\left[n_D^2 - 1\right]}{\left[n_D^2 - 2\right]} \sum \frac{x_i M_i}{\rho_m}, \qquad R_i = \frac{\left[n_D^2 - 1\right]}{\left[n_D^2 - 2\right]} \frac{M_i}{\rho_i}. \quad (10)$$

Deviation in molar refraction (ΔR) is calculated by equation

$$\Delta R = R_m - \sum \Phi_i R_i, \qquad \Phi = \frac{x_i}{\sum xjVj}, \quad (11)$$

where n_D refers to refractive index, R_m is molar refraction of the mixture, R_i is molar refraction of the ith component, and Φ is ideal state volume fraction.

All the deviations (V^E, ΔR, $\Delta\eta$, Δu, K_S^E) have been fitted to Redlich-Kister [7] polynomial regression of the type

$$\Delta Y = x_1 x_2 \sum_{i=1}^{m} A_{j-1} \left(1 - 2x_1\right)^{j-1} \quad (12)$$

to derive the constant A_j using the method of least square. Standard deviation for each case is calculated by

$$\sigma = \left[\frac{\sum \left(\Delta Y_{\text{expt}\,l} - \Delta Y_{\text{calc}\,d}\right)^2}{m - n} \right]^{0.5}, \quad (13)$$

where m is the number of data points and n is the number of coefficients. Derived parameters of the redlich-Kister (12) equation and standard deviations (13) are presented in Table 3

Table 4: Interaction parameters for the McAllister model (5) and herric correletion (6) for viscosity at different temperatures.

T/K	η_{12}	η_{21}	$\sigma(\eta)$/mPa·s
	McAllister model		
298.15	0.975300	0.909494	0.00014
303.15	0.968905	0.905236	0.00032
308.15	0.962426	0.900935	0.00044
313.15	0.958670	0.900372	0.00051
T/K	α_{12}	α'_{12}	$\sigma(\eta)$/mPa·s
	Herric correlation		
298.15	−0.02338	−0.01368	0.00014
303.15	−0.02600	−0.01677	0.00032
308.15	−0.02874	−0.01993	0.00044
313.15	−0.03047	−0.02084	0.00052

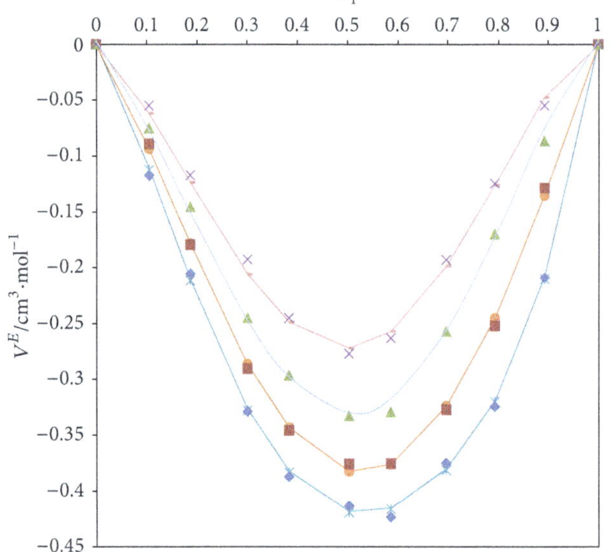

Figure 1: Experimental and calculated excess molar for the DMSO(1) + CUMENE(2) at ✦, 298.15 K; ■, 303.15 K; ▲, 308.15 K; ×, 313.15 K; symbols represent the experimental values; lines are optimised by Redlich-Kister parameters.

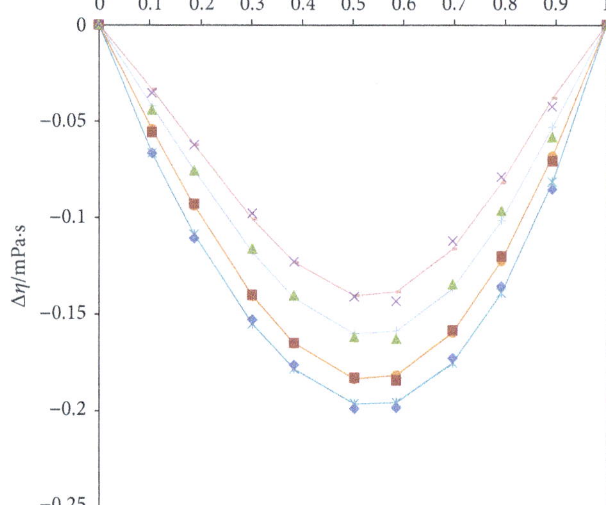

Figure 2: Experimental and calculated deviations in viscosity for DMSO(1) + CUMENE(2) at ✦, 298.15 K; ■, 303.15 K; ▲, 308.15 K; ×, 313.15 K; symbols represent the experimental values; lines are optimised by Redlich-Kister parameters.

4. Discussions

The deviations in excess molar volume at 298.15 to 313.15 K versus the mole fraction of DMSO are shown in Figure 1. The molar volume of the mixture and the viscosity data have also been used for the calculation of Gibb's free energy presented in Figure 5. The negative values of V^E indicate an interaction between the molecules which decreases with an increase of temperature. The large negative V^E values indicate a contraction of the volume and can be explained in terms of the heteroassociation in the mixture and suggest the strongest association occurs in this binary system [8].

The viscosity and deviations are presented in Table 2 and plotted in Figure 2, respectively. The negative values of $\Delta\eta$ obtained for the investigated mixture suggest that there may be reduction in the strength of H-bonds upon mixing. The viscosity data can be qualitatively explained by considering that the Cumene has a branching CH_3 group, leading to packing of molecules in the pure state and a consequent lower density and larger mobility (lower viscosity) of the structure. This may be because interaction of Cumene with DMSO is less [6].

The viscosity data is also fitted to the two- and the three-parameter model, that is, herric correlation and the McAllister model and the evaluated parameters are presented in Table 4. The deviations decrease with the increase in temperature.

The deviations in molar refraction are shown in Figure 3. The ΔR values are negative for the whole composition range which goes on decreasing as the temperature of the solution increases.

The results of derived K_S^E are also plotted in Figure 4. The deviations are negative over the whole composition range.

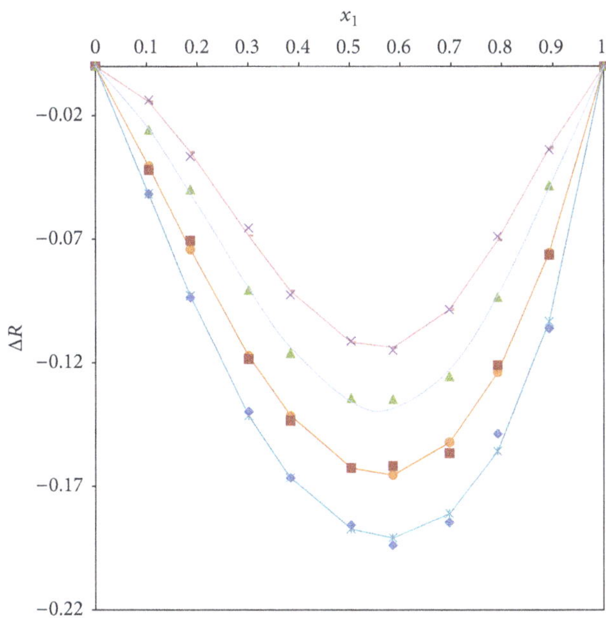

FIGURE 3: Experimental and calculated deviations in molar refraction for DMSO(1) + CUMENE(2) at ◆, 298.15 K; ■, 303.15 K; ▲, 308.15 K; ×, 313.15 K; symbols represent the experimental values; lines are optimised by Redlich-Kister parameters.

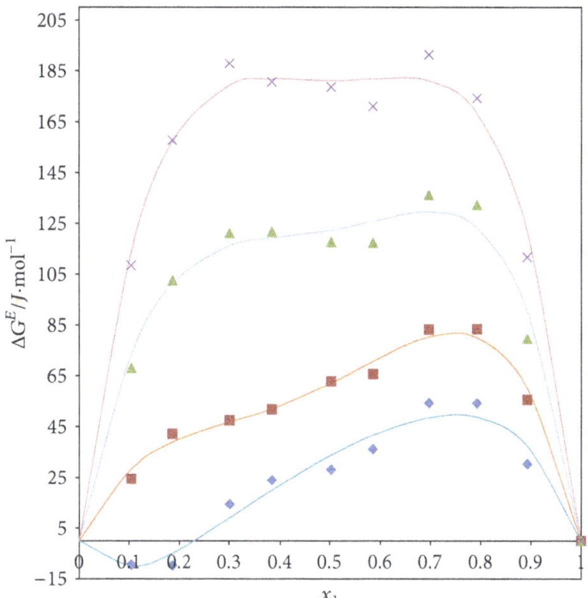

FIGURE 5: Experimental and calculated deviations in Gibbs free energy of activation for DMSO(1) + CUMENE(2) at ◆, 298.15 K; ■, 303.15 K; ▲, 308.15 K; ×, 313.15 K; symbols represent the experimental values; Lines are optimised by Redlich-Kister parameters.

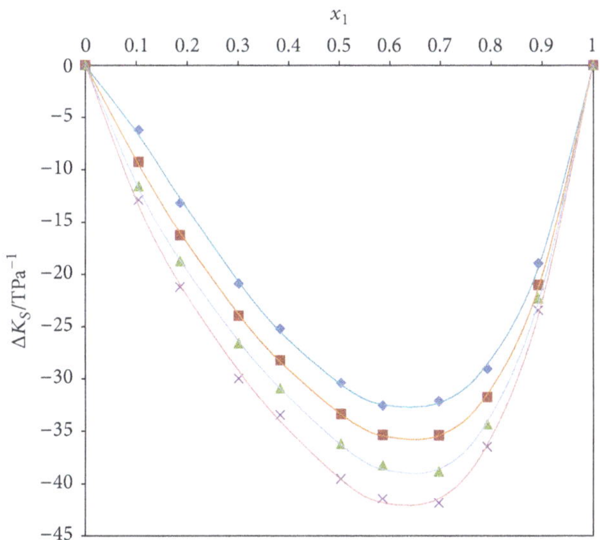

FIGURE 4: Experimental and calculated deviations in isentropic compressibility for DMSO(1) + CUMENE(2) at ◆, 298.15 K; ■, 303.15 K; ▲, 308.15 K; ×, 313.15 K; symbols represent the experimental values; lines are optimised by Redlich-Kister parameters.

The K_S^E is negative near the composition of $x_1 = 0.7$ and then abruptly goes to zero.

Symbols

A_0, A_1, A_2, A_3: Parameters of Redlich-Kister equation
A_{12}, A_{21}: Interaction coefficient of McAllister model
$\alpha_{12}, \alpha'_{12}$: Coefficients of Herric's correlation
v: Kinematic viscosity (m^2/s)
ρ: Density (g/cm^3)
σ: Standard deviation
α: Expansion coefficient
η: Viscosity (cP)
ΔG^E: Excess Gibbs free energy (J/mol)
V^E: Excess molar volume
ΔK_S^E: Excess isentropic compressibility
R: Universal gas constant (8.314 J/mol.K)
T: Absolute temperature (K)
Φ_i: Volume fraction (dimensionless).

References

[1] J. A. Riddick, W. B. Bunger, and T. K. Sakano, *Organic Solvents: Physical Properties and Methods of Purific tions. Techniques of Chemistry*, vol. 2, John Wiley & Sons, New York, NY, USA, 1986.

[2] V. K. Rattan, S. Kapoor, and K. Tochigi, "Viscosities and densities of binary mixtures of toluene with acetic acid and propionic acid at (293.15, 303.15, 313.15, and 323.15) K," *Journal of Chemical and Engineering Data*, vol. 47, no. 5, pp. 1182–1184, 2002.

[3] R. A. McAllister, "The viscosity of liquid mixtures," *AIChE Journal*, vol. 6, pp. 427–431, 1960.

[4] D. Kiyohara and G. C. Benson, "Evaluation of excess isentropic compressibilities and isochoric heat capacities," *Th Journal of Chemical The modynamics*, vol. 11, p. 861, 1979.

[5] G. Douhéret, M. I. Davis, I. J. Fjellanger, and H. Høiland, "Ultrasonic speeds and volumetric properties of binary mixtures of water with poly(ethylene glycol)s at 298.15 K," *Journal of the Chemical Society, Faraday Transactions*, vol. 93, no. 10, pp. 1943–1949, 1997.

[6] J. G. Baragi, M. I. Aralaguppi, T. M. Aminabhavi, M. Y. Kariduraganavar, and A. S. Kittur, "Density, viscosity, refractive index, and speed of sound for binary mixtures of anisole with 2-chloroethanol, 1,4-dioxane, tetrachloroethylene, tetrachloroethane, DMF, DMSO, and diethyl oxalate at (298.15, 303.15, and 308.15) K," *Journal of Chemical and Engineering Data*, vol. 50, no. 3, pp. 910–916, 2005.

[7] O. Redlich and A. T. Kister, "Algebraic representation of thermodynamic properties and the classification of solutions," *Industrial & Engineering Chemistry*, vol. 40, pp. 345–348, 1948.

[8] M. A. Saleh, M. Habibullah, M. Shamsuddin Ahmed et al., "Excess molar volumes and viscosities of some alkanols with cumene," *Physics and Chemistry of Liquids*, vol. 44, no. 1, pp. 31–43, 2006.

8

Studies on Excess Volume, Viscosity, and Speed of Sound of Binary Mixtures of Methyl Benzoate in Ethers at $T = (303.15, 308.15, \text{ and } 313.15)$ K

M. V. Rathnam,[1] Devappa R. Ambavadekar,[1] and M. Nandini[2]

[1] *Physical Chemistry Research Laboratory, B.N. Bandodkar College of Science, Th ne 400601, India*
[2] *Department of Chemistry, Dr. P.R.Ghogrey Science College, Deopur, Dhule 424005, India*

Correspondence should be addressed to M. V. Rathnam; mvrathnam58@rediffmail.com

Academic Editor: Felix Sharipov

Densities, viscosities, and speed of sound have been determined at $T = (303.15, 308.15, \text{ and } 313.15)$ K for the binary mixtures of methyl benzoate with tetrahydrofuran, 1,4-dioxane, anisole, and butyl vinyl ether over the entire range of composition. Using these measured values, excess volume V^E, deviation in viscosities $\Delta\eta$, excess Gibb's free energy of activation for viscous flow ΔG^{*E}, and deviation in isentropic compressibility Δk_s have been calculated. These calculated binary data have been fitted to Redlich-Kister equation to determine the appropriate coefficients. The values of excess volume V^E and deviation in viscosities $\Delta\eta$ are negative over the entire range of composition for all the binary systems at the studied temperatures. The behavior of these parameters with composition of the mixture has been discussed in terms of molecular interactions between the components of liquids.

1. Introduction

The excess or deviation properties of liquid-liquid mixtures are useful in several industrial applications owing to their influence upon the effectiveness of the operations. Nowadays, the determination of ultrasonic velocity and the related acoustical parameters derived from it has attracted the attention of several researchers. Much work has been done in solutions of polymers [1–3], pharmamaterials [4], electrolytes [5–7], and nonelectrolytes [8–10].

Tetrahydrofuran is a polar ($\mu = 1.75D$), aprotic solvent. It's main use is a precursor to polymers. It is also used as an industrial solvent for PVC and in varnishes. Dye carrier formulations based on methyl benzoate are useful in textile processing. Further methyl benzoate finds its primary uses as a perfume or flavouring agent and also used as a source of benzoyl radical. While ethers are excellent solvents, because they are relatively unreactive, yet they solvate a wide variety of compounds. Tetrahydrofuran, a cyclic ether which is one of the most polar simple ethers, is used as a solvent for polymers and also commercially used for the production of several compounds. Anisole, an aryl ether, is a major constituent of essential oil of anise seed.

In view of the abovementioned significances, these solvents are chosen with an aim to investigate the effect of the presence of two lone pairs of electrons on the oxygen atom of ether, when it is mixed with methyl benzoate. Moreover, it would be interesting to know the behavior of ether in an environment of aromatic ester molecules. The main objective of the present study is to determine the density ρ, viscosity η, and speed of sound u for the binary mixtures formed by the methyl benzoate with ether compounds.

In our earlier papers [11–13], we have reported some data on thermodynamic, transport, acoustical, and optical properties on mixtures of methyl benzoate with hydrocarbons and ketones and analysed the data in terms of molecular interactions. Several authors have also published density, viscosity, and speed of sound data [14–20] of binary mixtures of methyl benzoate with different types of organic solvents. In the present study, we report the results of density ρ, viscosity η, and speed of sound u for the binary mixtures of methyl benzoate with ethers, namely, tetrahydrofuran, 1,4-dioxane,

TABLE 1: Comparison of experimental density ρ and viscosity η of pure liquids with the literature values at (303.15, 308.15, and 313.15) K.

Liquid	T/K	ρ/g·cm^{-3}		η/mPa·s	
		Exptl.	Lit.	Exptl.	Lit.
Methyl benzoate	303.15	1.0785	1.0788 [18]	1.678	1.673 [28]
			1.0790 [22]		1.656 [18]
	308.15	1.0743	1.0740 [18]	1.517	1.510 [18]
			1.0741 [19]		
					1.504 [20]
			1.07399 [20]		1.510 [21]
	313.15	1.0696	1.0690 [27]	1.373	1.365 [27]
Tetrahydrofuran	303.15	0.8787	0.8771 [22]	0.439	
	308.15	0.8730	0.87214 [22]	0.429	
	313.15	0.8669	0.86719 [22]	0.390	
1,4-Dioxane	303.15	1.0227	1.02271 [23]	1.090	1.102 [23]
					1.095 [24]
	308.15	1.0178	1.0172 [24]	0.999	1.008 [24]
	313.15	1.0116	1.01132 [23]	0.946	0.946 [23]
Anisole	303.15	0.9853	0.984374 [25]	0.923	0.931 [25]
	308.15	0.9792	0.9788 [26]	0.849	0.849 [26]
	313.15	0.9728		0.764	
Butyl vinyl ether	303.15	0.7741		0.387	
	308.15	0.7682		0.365	
	313.15	0.7633		0.354	

anisole, and butyl vinyl ether measured at (303.15, 308.15, and 313.15) K over the entire mixture composition. To the best of our knowledge, such data on the abovementioned mixtures are not available in the earlier literature. Using the experimental data of ρ, η, and u, various parameters such as excess volume V^E, deviation in viscosity $\Delta\eta$, excess Gibb's free energy of activation for viscous flow ΔG^{*E}, and deviation in isentropic compressibility Δk_s were determined. These computed data are discussed to study the nature of behaviors between the components of the mixtures.

2. Experimental

2.1. Materials. Methyl benzoate (Fluka AG >0.996), tetrahydrofuran, 1,4-dioxane, anisole, and butyl vinyl ether (all Sigma-Aldrich, AR grade) of mass purity >0.997 were used without further purification. These chemicals were kept over molecular sieves for several days before use. The mass fraction purities as determined by gas chromatography (HP 8610) using FID were as follows: methyl benzoate (>0.996), tetrahydrofuran, (>0.996), 1,4-dioxane, (>0.998), anisole, (>0.998), and butyl vinyl ether (>0.997). The purity of the samples was further checked by comparing the measured density ρ and viscosity η with the literature values [18–28] as shown in Table 1.

TABLE 2: Values of density ρ, excess volume V^E, viscosity η, speeds of sound u, and isentropic compressibility k_s for the binary liquid mixtures.

x_1	ρ/g·cm^{-3}	V^E/cm^3·mol^{-1}	η/mPa·s	u/m·s^{-1}	k_s/Tpa^{-1}
Methyl benzoate (1) + tetrahydrofuran (2)					
T = 303.15 K					
0.0689	0.9020	−0.261	0.496	1260	698
0.1417	0.9246	−0.496	0.562	1272	669
0.2202	0.9468	−0.699	0.638	1284	641
0.3030	0.9680	−0.858	0.727	1296	615
0.3932	0.9885	−0.943	0.829	1308	591
0.4930	1.0084	−0.938	0.951	1320	569
0.6018	1.0272	−0.822	1.096	1332	549
0.7225	1.0452	−0.590	1.262	1344	530
0.8535	1.0623	−0.277	1.446	1356	512
T = 308.15 K					
0.0689	0.8964	−0.269	0.484	1232	735
0.1417	0.9192	−0.510	0.546	1236	712
0.2202	0.9413	−0.719	0.616	1244	686
0.3030	0.9627	−0.882	0.699	1256	659
0.3932	0.9833	−0.969	0.790	1268	633
0.4930	1.0033	−0.964	0.904	1284	605
0.6018	1.0222	−0.847	1.032	1300	679
0.7225	1.0403	−0.612	1.181	1316	555
0.8535	1.0575	−0.294	1.343	1336	530
T = 313.15 K					
0.0689	0.8905	−0.283	0.443	1216	759
0.1417	0.9134	−0.538	0.505	1220	736
0.2202	0.9358	−0.749	0.567	1228	709
0.3030	0.9572	−0.913	0.641	1240	679
0.3932	0.9779	−0.999	0.729	1252	652
0.4930	0.9980	−0.990	0.830	1268	623
0.6018	1.0171	−0.877	0.943	1284	596
0.7225	1.0354	−0.645	1.078	1300	572
0.8535	1.0527	−0.317	1.220	1320	545
Methyl benzoate (1) + 1,4-dioxane (2)					
T = 303.15 K					
0.0898	1.0307	−0.059	1.130	1328	550
0.1468	1.0355	−0.105	1.157	1336	541
0.2280	1.0421	−0.185	1.198	1344	531
0.3145	1.0485	−0.254	1.244	1352	521
0.4054	1.0543	−0.281	1.293	1360	513
0.5056	1.0599	−0.278	1.352	1368	504
0.6161	1.0653	−0.242	1.418	1372	499
0.7316	1.0701	−0.161	1.491	1372	496
0.8597	1.0749	−0.062	1.573	1372	494
T = 308.15 K					
0.0898	1.0261	−0.093	1.038	1324	556
0.1468	1.0311	−0.157	1.066	1332	547
0.2280	1.0377	−0.237	1.106	1340	537
0.3145	1.0440	−0.297	1.147	1348	530
0.4054	1.0499	−0.333	1.192	1352	521
0.5056	1.0555	−0.329	1.246	1356	515
0.6161	1.0608	−0.282	1.308	1360	510
0.7316	1.0655	−0.189	1.376	1360	507
0.8597	1.0702	−0.077	1.449	1360	505

TABLE 2: Continued.

x_1	ρ/g·cm^{-3}	V^E/cm^3·mol^{-1}	η/mPa·s	u/m·s^{-1}	k_s/Tpa^{-1}
		T = 313.15 K			
0.0898	1.0205	−0.132	0.984	1320	562
0.1468	1.0257	−0.207	1.006	1328	553
0.2280	1.0326	−0.304	1.040	1336	543
0.3145	1.0390	−0.363	1.075	1340	536
0.4054	1.0450	−0.398	1.110	1348	527
0.5056	1.0507	−0.392	1.157	1352	521
0.6161	1.0560	−0.331	1.208	1356	515
0.7316	1.0608	−0.234	1.263	1356	513
0.8597	1.0655	−0.106	1.324	1356	510
		Methyl benzoate (1) + Anisole (2)			
		T = 303.15 K			
0.0860	0.9950	−0.052	0.974	1388	522
0.1821	1.0056	−0.113	1.038	1380	522
0.2756	1.0157	−0.179	1.101	1372	523
0.3978	1.0282	−0.229	1.187	1368	520
0.4637	1.0347	−0.248	1.234	1368	516
0.5659	1.0442	−0.238	1.307	1368	512
0.6701	1.0534	−0.202	1.391	1368	507
0.7776	1.0624	−0.140	1.480	1368	503
0.8850	1.0710	−0.063	1.572	1368	499
		T = 308.15 K			
0.0860	0.9891	−0.062	0.894	1368	540
0.1821	1.0000	−0.145	0.949	1356	544
0.2756	1.0102	−0.209	1.007	1352	542
0.3978	1.0229	−0.264	1.086	1348	538
0.4637	1.0295	−0.285	1.129	1348	535
0.5659	1.0392	−0.282	1.195	1348	530
0.6701	1.0485	−0.242	1.272	1348	525
0.7776	1.0576	−0.175	1.355	1348	520
0.8850	1.0663	−0.093	1.439	1348	516
		T = 313.15 K			
0.0860	0.9830	−0.082	0.800	1356	553
0.1821	0.9941	−0.170	0.850	1344	557
0.2756	1.0045	−0.241	0.902	1340	554
0.3978	1.0174	−0.297	0.972	1336	551
0.4637	1.0241	−0.317	1.012	1336	547
0.5659	1.0340	−0.318	1.074	1336	541
0.6701	1.0435	−0.282	1.144	1336	537
0.7776	1.0528	−0.218	1.220	1336	532
0.8850	1.0616	−0.127	1.301	1336	528
		Methyl benzoate (1) + Butyl vinyl ether (2)			
		T = 303.15 K			
0.1026	0.8056	−0.108	0.437	1104	1019
0.2055	0.8375	−0.260	0.497	1128	938
0.3060	0.8690	−0.423	0.571	1152	867
0.4047	0.8999	−0.544	0.657	1176	804
0.5057	0.9313	−0.605	0.761	1204	741
0.6065	0.9623	−0.596	0.891	1232	685
0.7061	0.9926	−0.524	1.038	1260	635
0.8034	1.0218	−0.388	1.205	1292	586
0.9021	1.0511	−0.201	1.422	1328	540

TABLE 2: Continued.

x_1	ρ/g·cm^{-3}	V^E/cm^3·mol^{-1}	η/mPa·s	u/m·s^{-1}	k_s/Tpa^{-1}
		T = 308.15 K			
0.1026	0.8001	−0.141	0.421	1092	1048
0.2055	0.8322	−0.320	0.478	1116	965
0.3060	0.8637	−0.476	0.546	1136	891
0.4047	0.8947	−0.602	0.657	1160	825
0.5057	0.9261	−0.651	0.626	1188	760
0.6065	0.9572	−0.639	0.720	1216	702
0.7061	0.9875	−0.550	0.974	1248	650
0.8034	1.0168	−0.408	1.124	1280	600
0.9021	1.0462	−0.212	1.319	1316	552
		T = 313.15 K			
0.1026	0.7954	−0.182	0.400	1084	1070
0.2055	0.8275	−0.367	0.452	1108	984
0.3060	0.8591	−0.541	0.515	1132	908
0.4047	0.8901	−0.667	0.585	1156	841
0.5057	0.9215	−0.713	0.671	1184	774
0.6065	0.9526	−0.697	0.778	1212	715
0.7061	0.9829	−0.601	0.897	1244	657
0.8034	1.0121	−0.439	1.034	1276	607
0.9021	1.0414	−0.221	1.205	1308	561

2.2. Methods. Binary mixtures were prepared by mass in airtight ground stoppered bottles. Mass measurements accurate to ±0.01 mg were made on a digital electronic balance (Mettler AE 240, Switzerland). The resulting uncertainty in mole fraction was estimated to be less than ±0.0001. Each mixture was immediately used after it was well mixed by shaking. The densities of the pure and their binary mixtures were determined by using an Anton Paar density meter (DMA 4100). The uncertainty in the density measurements was found to be less than ±0.0004 g·cm^{-3}.

Viscosities were determined using an Ubbelohde viscometer with an uncertainty of ±0.008 mPa.s. The detailed method of measurement of viscosity has been reported earlier [11]. The speeds of sound were measured with a single crystal variable path interferometer (Mittal Enterprises, New Delhi, India) at an operating frequency of 2 MHz that had been calibrated with double distilled water and benzene. The uncertainty in speed of sound was estimated to be ±1 m·s^{-1}.

3. Results and Discussion

The experimental values of densities ρ, excess volumes V^E, viscosities η, speeds of sound u, and isentropic compressibility k_s of the binary mixtures of methyl benzoate with tetrahydrofuran, 1,4-dioxane, anisole, and butyl vinyl ether at temperatures 303.15, 308.15, and 313.15 K are listed in Table 2.

The excess volume V^E of the mixtures was deduced from the measured densities using the following relation:

$$V^E = \frac{(x_1M_1 + x_2M_2)}{\rho_{12}} - \left(\frac{x_1M_1}{\rho_1} + \frac{x_2M_2}{\rho_2}\right), \quad (1)$$

where x, M, and ρ are the mole fraction, molar mass, and density, respectively, of pure components 1 and 2. ρ_{12} is

Table 3: Derived parameters of excess functions by Redlich-Kister equation and standard deviations σ of binary liquid mixtures.

Function	T/K	a_0	a_1	a_2	a_3	σ
Methyl benzoate (1) + tetrahydrofuran (2)						
V^E	303.15	−3.7342	1.2905	1.1561	—	0.003
	308.15	−3.8390	1.2879	1.1173	—	0.003
	313.15	−3.9542	1.2921	1.9336	—	0.004
$\Delta\eta$	303.15	−0.3703	0.0783	−0.0054	—	0.002
	308.15	−0.2820	0.0762	−0.0017	—	0.002
	313.15	−0.2206	0.0052	−0.0013	—	0.001
ΔG^{*E}	303.15	1303.56	9.2800	−47.1207	−173.82	2.5
	308.15	1332.27	−4.3204	−21.9964	−216.76	3.3
	313.15	1442.68	−177.56	50.2305	−79.613	3.8
Δk_s	303.15	−17.457	7.2762	−0.2584	1.4106	0.008
	308.15	−11.729	−0.0066	3.5145	−0.134	0.093
	313.15	−11.726	0.5719	4.7747	1.1539	0.090
Methyl benzoate (1) + 1,4-dioxane (2)						
V^E	303.15	−1.144	0.290	0.846	—	0.002
	308.15	−1.326	0.425	0.753	—	0.002
	313.15	−1.568	0.552	0.575	—	0.003
$\Delta\eta$	303.15	−0.122	0.018	−0.011	—	0.001
	308.15	−0.096	0.007	0.009	—	0.001
	313.15	−0.062	0.019	0.045	—	0.001
ΔG^{*E}	303.15	152.594	48.243	−4.7463	−34.315	0.7
	308.15	191.571	23.232	37.394	−27.738	2.8
	313.15	201.888	29.749	100.81	−116.35	2.4
Δk_s	303.15	−8.7386	1.8216	3.7965	−1.7238	0.094
	308.15	−9.2275	2.7930	−0.9078	−0.8437	0.044
	313.15	−11.303	2.1793	−5.0991	−0.8657	0.048
Methyl benzoate (1) + anisole (2)						
V^E	303.15	−0.9866	0.2637	0.5566	—	0.003
	308.15	−1.1427	−0.0217	0.4368	—	0.003
	313.15	−1.2778	−0.0899	0.2005	—	0.002
$\Delta\eta$	303.15	−0.1440	0.0055	0.0234	—	0.002
	308.15	−0.1774	−0.0067	−0.0351	—	0.001
	313.15	−0.1602	−0.0069	−0.0270	—	0.002
ΔG^{*E}	303.15	177.724	−46.913	99.649	70.211	2.2
	308.15	116.959	−70.397	12.737	80.252	2.5
	313.15	56.636	−50.389	−50.388	63.520	1.2
Δk_s	303.15	3.5076	−3.8057	0.9701	5.8130	0.036
	308.15	4.7056	−4.2092	5.0184	4.5773	0.053
	313.15	5.3764	−4.7277	6.5667	2.5075	0.050
Methyl benzoate (1) + butyl vinyl ether (2)						
V^E	303.15	−2.4161	−0.7060	1.0922	—	0.004
	308.15	−2.6110	−0.5335	0.9973	—	0.004
	313.15	−2.8690	−0.4280	0.9846	—	0.005
$\Delta\eta$	303.15	−1.0856	−0.3096	−0.1033	—	0.002
	308.15	−0.9534	−0.2594	−0.0654	—	0.003
	313.15	−0.8261	−0.2221	−0.0264	—	0.003
ΔG^{*E}	303.15	−663.33	174.93	−102.07	−27.064	5.8
	308.15	−627.07	153.92	−43.740	−3.6493	7.2
	313.15	−596.84	106.49	131.85	−19.948	5.7
Δk_s	303.15	−20.411	5.1677	5.1277	−1.1243	0.080
	308.15	−22.018	5.1564	3.8155	−3.0159	0.085
	313.15	−25.224	3.8500	−3.2901	2.8527	0.072

the density of the liquid mixture. The Gibb's free energy of activation for viscous flow ΔG^{*E} was calculated from the viscosity data using the following relation:

$$\Delta G^{*E} = \mathrm{RT}\left[\ln(\eta v) - \{x_1 \ln(\eta_1 v_1) + x_2 \ln(\eta_2 v_2)\}\right], \quad (2)$$

where $v = (x_1 M_1/\rho_1 + x_2 M_2)/\rho_{12}$ is the molar volume of the mixture and v_1 and v_2 are the molar volumes of the pure components.

The isentropic compressibilities k_s were calculated from the densities ρ and speeds of sound u using the Newton-Laplace equation:

$$k_s = \frac{1}{u^2 \rho}. \quad (3)$$

The deviation in viscosity $\Delta\eta$ and deviation in isentropic compressibility Δk_s were calculated using the general equation:

$$\Delta Y = Y_m - x_1 Y_1 - x_2 Y_2, \quad (4)$$

where ΔY is the deviation property in question, Y_m refers to the property of the mixture, and $x_1 Y_1$ and $x_2 Y_2$ refer to the mole fraction and specific property of the pure components 1 and 2, respectively. The results of these excess or deviation properties were fitted by the method of least squares to the Redlich-Kister [29] polynomial type equation:

$$Y = x_1 x_2 \sum_{j=1}^{p} a_{j-1} (x_1 - x_2)^{j-1}, \quad (5)$$

where Y is V^E, $\Delta\eta$, ΔG^{*E}, or Δk_s and x_1 and x_2 are the mole fractions of pure components 1 and 2, respectively. a_{j-1} is the polynomial coefficient, and p is the polynomial degree. The degree of (5) was optimised by applying the F-test. The correlated results are shown in Table 3 in which the tabulated standard deviation σ was calculated using the following relation:

$$\sigma = \left[\frac{\sum\left(y^E_{\text{exp}} - y^E_{\text{cal}}\right)^2}{(n - J)}\right]^{0.5}, \quad (6)$$

where n is the number of data points and j is the number of coefficients. The subscripts exp and cal denote experimental and calculated values, respectively.

The variations of excess volume V^E with mole fraction x_1 of methyl benzoate at the studied temperatures for the binary mixtures are displayed in Figure 1. It is observed that all the studied systems exhibit negative deviations. The V^E data for different components of the mixtures vary in the sequence:

tetrahydrofuran > butyl vinyl ether > 1,4-dioxane > anisole

The effect of temperature on V^E as displayed in Figure 1 is quite significant, as the negative V^E values increase with increase in temperature. The behavior of V^E can be explained

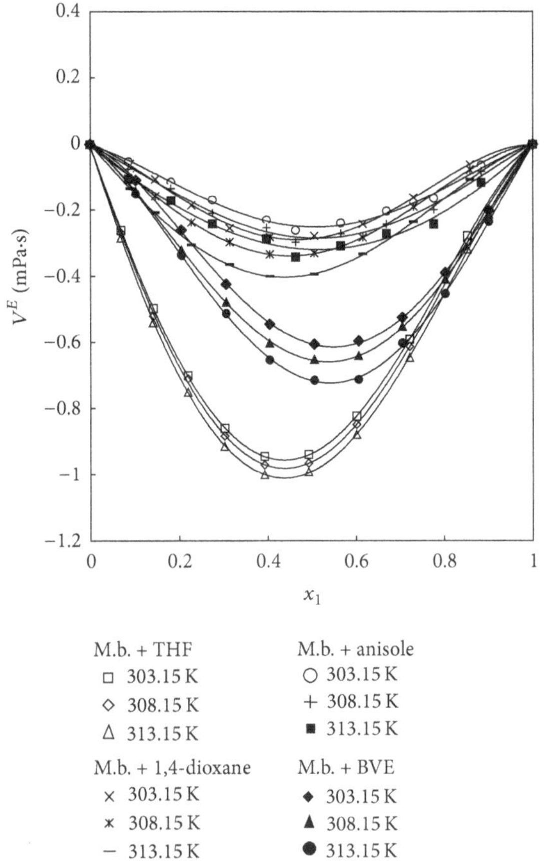

FIGURE 1: Curves of excess volumes (V^E) versus mole fraction for the binary mixtures of methyl benzoate + tetrahydrofuran at (□, 303.15; ◊, 308.15; Δ, 313.15) K, methyl benzoate + 1,4-dioxane at (×, 303.15; ∗, 308.15; —, 313.15) K, methyl benzoate + anisole at (O, 303.15; +, 308.15; ■, 313.15) K, and methyl benzoate + butyl vinyl ether at (⧫, 303.15; ▲, 308.15; •, 313.15) K.

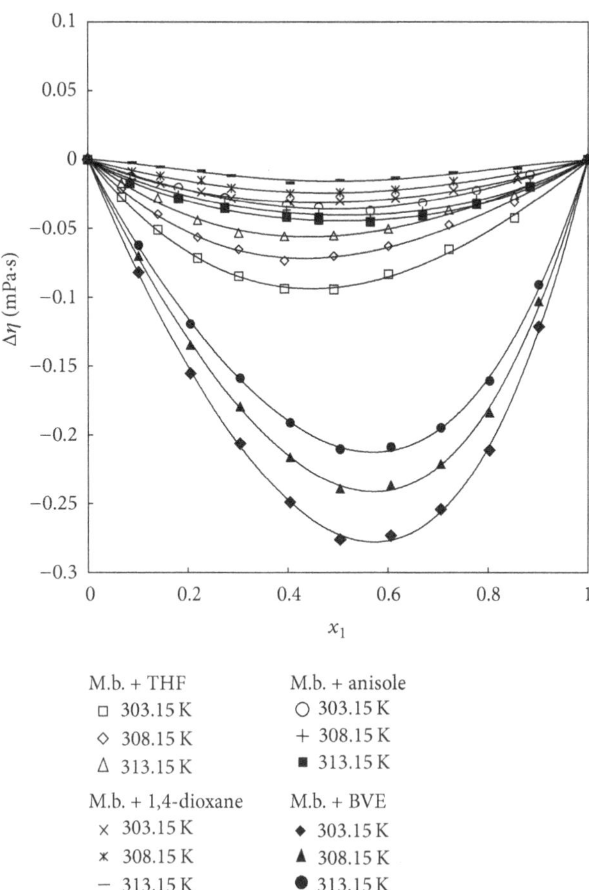

FIGURE 2: Curves of deviation in viscosity ($\Delta\eta$) versus mole fraction for the binary mixtures of methyl benzoate + tetrahydrofuran at (□, 303.15; ◊, 308.15; Δ, 313.15) K, methyl benzoate + 1,4-dioxane at (×, 303.15; ∗, 308.15; —, 313.15) K, and methyl benzoate + anisole at (O, 303.15; +, 308.15; ■, 313.15) K.

by considering the type of component molecules in pure state and in mixture. In the present case, the negative V^E values may be attributed to the differences in the dielectric constants of the components. The dielectric constants of methyl benzoate, tetrahydrofuran, 1,4-dioxane, anisole, and butyl vinyl ether are $\epsilon = 6.02, 7.261, 2.209, 4.33, 3.9$, respectively. Further, there is a possibility of electron donor-acceptor type or charge transfer interactions [30] between the highly electronegative oxygen of ether and the π-electron of ring of aromatic ester molecule resulting in negative V^E values. The less negative V^E values in case of methyl benzoate + anisole may be due to the presence of methyl ($-CH_3$) group in anisole. This observation suggests that with an increase in methyl group in mixture, the donor-acceptor interactions between the unlike molecules tend to decrease.

The results of $\Delta\eta$ as displayed in the Figure 2 are all negative over the entire range of composition and the magnitude values of $\Delta\eta$ vary according to the following sequence:

butyl vinyl ether > tetrahydrofuran > anisole > 1,4-dioxane

It is observed that negative values of $\Delta\eta$ decrease with increase in the temperature indicating the effect of temperature on $\Delta\eta$. Fort and Moore [31] observed that the negative $\Delta\eta$ values indicate the dispersion forces without involving the formation of any hetero-molecular complexes. A perusal of Figures 1 and 2 reveals that both V^E and $\Delta\eta$ exhibit the negative deviations over the entire range of composition; however, the magnitude values of the studied mixtures follow a different sequence. This type of behavior supports the observation of Rastogi et al. [32] and Kaulgud [33], suggesting that the strength of specific or dispersion forces is not the only factor influencing the $\Delta\eta$, but the molecular size and shape of the components are also equally important.

The dependence of ΔG^{*E} on the mole fraction is shown in Figure 3, where it was observed that ΔG^{*E} is positive for mixtures of methyl benzoate with tetrahydrofuran, 1,4-dioxane, and anisole, whereas for mixtures of methyl benzoate + butyl vinyl ether, the values of ΔG^{*E} are negative over the entire range of composition. The positive ΔG^{*E} values may be an indicative of specific interactions [31, 34, 35] and the negative ΔG^{*E} values indicate the presence of dispersive forces in these mixtures.

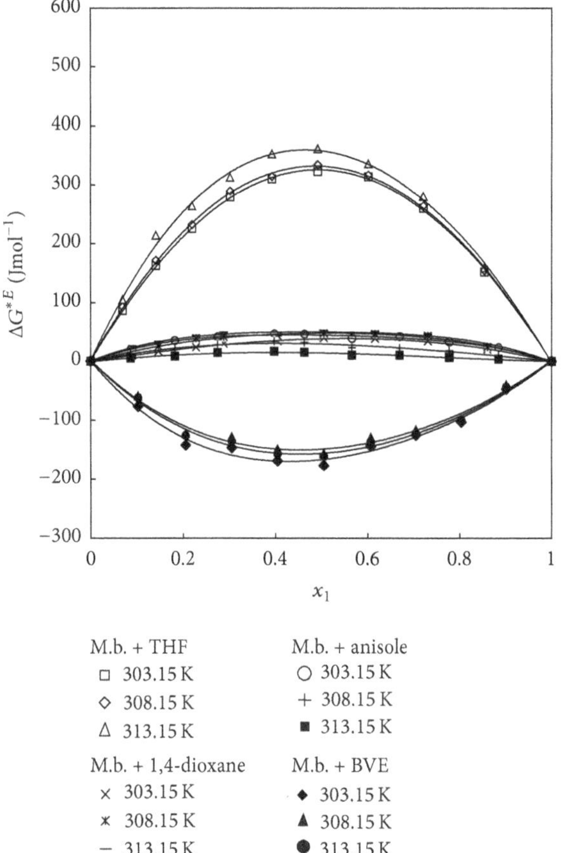

FIGURE 3: Curves of excess Gibb's free energy of activation of viscous flow (ΔG^{*E}) versus mole fraction for the binary mixtures of methyl benzoate + tetrahydrofuran at (□, 303.15; ◊, 308.15; Δ, 313.15) K, methyl benzoate + 1,4-dioxane at (×, 303.15; ∗, 308.15; —, 313.15) K, methyl benzoate + anisole at (O, 303.15; +, 308.15; ■, 313.15) K, and methyl benzoate + Butyl vinyl ether at (⧫, 303.15; ▲, 308.15; •, 313.15) K.

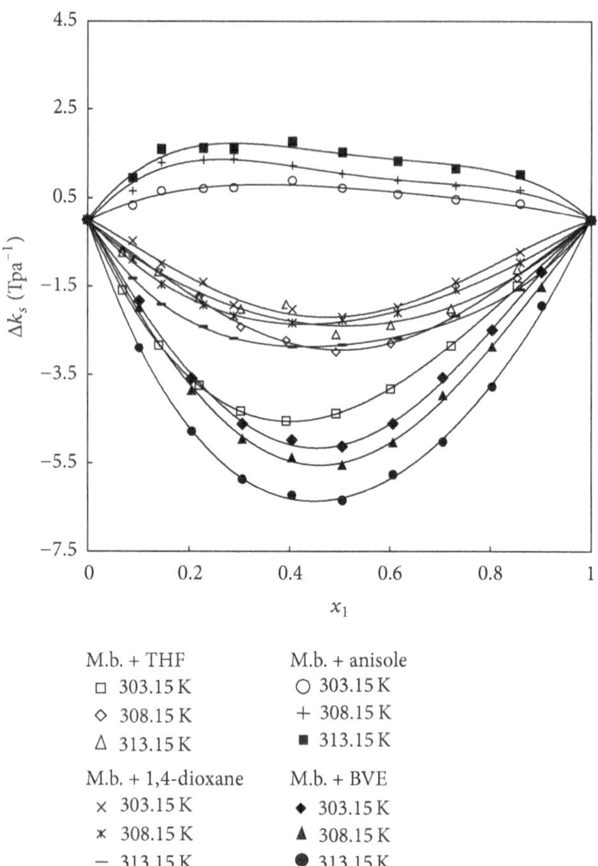

FIGURE 4: Curves of deviation in isentropic compressibility (Δk_s) versus mole fraction for the binary mixture of methyl benzoate + tetrahydrofuran at (□, 303.15; ◊, 308.15; Δ, 313.15) K, methyl benzoate + 1,4-dioxane at (×, 303.15; ∗, 308.15; —, 313.15) K, methyl benzoate + anisole at (O, 303.15; +, 308.15; ■, 313.15) K, and methyl benzoate + Butyl vinyl ether at (⧫, 303.15; ▲, 308.15; •, 313.15) K.

The plots of Δk_s, as displayed in Figure 4, show that for methyl benzoate with tetrahydrofuran, 1,4-dioxane, and butyl vinyl ether, the values of Δk_s exhibit negative deviation over the entire range of mixture composition; while for methyl benzoate + anisole, the values of Δk_s exhibit positive deviations. The positive Δk_s values are a sign of weak interaction between component molecules, which may be attributed to the mutual disruption in molecules associated with pure liquids. These positive Δk_s values are accompanied by a decrease in sound velocity over the entire range of composition of ester. The negative values of Δk_s may be attributed to the formation of weak bonds [36–38] by dipole-induced dipole interaction between unlike molecules and the geometrical fitting of component molecules in to each other's structure.

4. Conclusions

The densities ρ, excess volumes V^E, viscosities η, speeds of sound u, and data of binary mixtures methyl benzoate with tetrahydrofuran, 1,4-dioxane, anisole, and butyl vinyl ether have been reported at (303.15, 308.15 and 313.15) K. The excess molar volumes V^E, deviation in viscosity $\Delta\eta$, excess Gibb's free energies of activation for viscous flow ΔG^{*E}, and deviation in isentropic compressibility Δk_s were evaluated using the experimental data of ρ, η, and u and discussed in terms of the interactions between the components of the mixtures. Both V^E and $\Delta\eta$ exhibit negative deviations, while ΔG^{*E} and Δk_s exhibit both positive and negative deviations. It is concluded that the negative V^E may be attributed to the differences in the dielectric constants of the components and the possibility of the electron donor-acceptor type or charge transfer interactions between the electronegative oxygen of ether and the π-electrons of aromatic ester resulting into contraction in volumes.

Acknowledgments

The financial support from University Grants Commission, New Delhi, India, through Major Research Project (no. 38-24/2009 SR) to the corresponding author (M. V. Rathnam) is gratefully acknowledged. The authors also sincerely express their gratitude to the editor-in-chief and the reviewers for their valuable comments.

References

[1] J. N. Prassianakis, "Correlation of mechanical and acoustical properties of plasticized epoxy polymers," *Journal of Applied Polymer Science*, vol. 39, no. 10, pp. 2031–2041, 1990.

[2] V. Kannappan, S. Mahendran, P. Sathyamoorthy, and D. Roopsingh, "Ultrasonic method of determination of glass transition temperatures of certain thermotropic liquid crystalline copolyesters," *Journal of Polymer Materials*, vol. 18, no. 4, pp. 409–416, 2001.

[3] D. R. Godhani and P. H. Parsania, "Studies on acoustical properties of poly(4,4′-diphenylphthalidediphenyl-4,4′-disulfonate) at 30°C," *Journal of the Indian Chemical Society*, vol. 79, no. 7, pp. 620–622, 2002.

[4] M. Levina and M. H. Rubinstein, "The effect of ultrasonic vibration on the compaction characteristics of paracetamol," *Journal of Pharmaceutical Sciences*, vol. 89, no. 6, pp. 705–723, 2000.

[5] V. K. Syal, V. Bhalla, and S. Chauhan, "Ultrasonic studies of some tetraalkylammonium salts in acetonitrile+dioxane mixtures at 35°C," *Acustica*, vol. 81, no. 3, pp. 276–278, 1995.

[6] D. Geetha and P. S. Ramesh, "Ultrasonic studies on polymer blend (natural/synthetic) in strong electrolyte solutions," *Journal of Molecular Liquids*, vol. 136, no. 1-2, pp. 50–53, 2007.

[7] E. S. Balankina and A. K. Lyaschenko, "The acoustical and structural changes of aqueous solutions due to transition to molten supercooled salts," *Journal of Molecular Liquids*, vol. 101, no. 1–3, pp. 273–283, 2002.

[8] M. Gowrishankar, S. Sivarambabu, P. Venkateswarlu, and K. S. Kumar, "Excess volumes, speeds of sound, isentropic compressibilities and viscosities of binary mixtures of N-ethyl aniline with some aromatic ketones at 303.15 K," *Bulletin of the Korean Chemical Society*, vol. 33, no. 5, pp. 1686–1692, 2012.

[9] A. C. H. Chandrasekhar, S. K. RamanJaneyulu, and A. Krishnaiah, "Volumetric and ultrasonic behaviour of acetophenone and alcohol mixtures," *Physics and Chemistry of Liquids*, vol. 19, no. 3, pp. 171–179, 1989.

[10] M. Yasmin, K. P. Singh, S. Parveen, M. Gupta, and J. P. Shukla, "Thermoacoustical excess properties of binary liquid mixtures: a comparative experimental and theoretical study," *Acta Physica Polonica A*, vol. 115, no. 5, pp. 890–900, 2009.

[11] M. V. Rathnam, R. T. Sayed, K. R. Bhanushali, and M. S. S. Kumar, "Molecular interaction study of binary mixtures of methyl benzoate: viscometric and ultrasonic study," *Journal of Molecular Liquids*, vol. 166, pp. 9–16, 2012.

[12] M. V. Rathnam, S. Mankumare, and M. S. S. Kumar, "Density, viscosity, and speed of sound of (Methyl Benzoate + Cyclohexane), (Methyl Benzoate + n-Hexane), (Methyl Benzoate + Heptane), and (Methyl Benzoate + Octane) at temperatures of (303.15, 308.15, and 313.15) K," *Journal of Chemical and Engineering Data*, vol. 55, no. 3, pp. 1354–1358, 2010.

[13] M. V. Rathnam, S. Mohite, and M. S. Kumar, "Thermophysical properties of isoamyl acetate or methyl benzoate + hydrocarbon binary mixtures, at (303.15 and 313.15) K," *Journal of Chemical and Engineering Data*, vol. 54, no. 2, pp. 305–309, 2009.

[14] P. J. Raju, K. Rambabu, D. Ramchandran, M. Nagshwar Rao, and C. Rambabu, "Densities, adiabatic compressibility, freelength, viscosities and excess volumes of dimethyl sulfoxide with some aromatic esters at 303.15-318.15 K," *Physics and Chemistry of Liquids*, vol. 34, pp. 51–60, 1997.

[15] B. Garcia, R. Alcade, S. Aparicio, and J. M. Leal, "Volumetric properties, viscosities and refractive indices of binary mixed solvents containing methyl benzoate," *Physical Chemistry Chemical Physics*, vol. 4, pp. 5833–5840, 2002.

[16] Y. W. Sheu and C. H. Tu, "Densities, viscosities, refractive indices, and surface tensions for 12 flavor esters from T = 288.15 K to T = 358.15 K," *Journal of Chemical and Engineering Data*, vol. 50, no. 5, pp. 1706–1710, 2005.

[17] Y. W. Sheu and C. H. Tu, "Densities and viscosities of binary mixtures of ethyl acetoacetate, ethyl isovalerate, methyl benzoate, benzyl acetate, ethyl salicylate, and benzyl propionate with ethanol at T = (288.15, 298.15, 308.15, and 318.15) K," *Journal of Chemical and Engineering Data*, vol. 51, no. 2, pp. 545–553.

[18] T. M. Aminabhavi, S. K. Raikar, and R. H. Balundgi, "Densities, viscosities, refractive indices, and speeds of sound in methyl acetoacetate + methyl acetate, + ethyl acetate, + n-butyl acetate, + methyl benzoate, and + ethyl benzoate at 298.15, 303.15, and 308.15 K," *Journal Of Chemical And Engineering Data*, vol. 38, no. 3, pp. 441–445, 1993.

[19] A. M. Blanco, J. Ortega, B. Garcia, and J. M. Leal, "Studies on densities and viscosities of binary mixtures of alkyl benzoates in n-heptane," *Th rmochimica Acta*, vol. 222, no. 1, pp. 127–136, 1993.

[20] B. Garcia, R. Alcade, S. Aparicio, and J. M. Leal, "Thermophysical behavior of methylbenzoate + n-alkanes mixed aolvents: application of cubic equations of state and viscosity models," *Industrial and Engineering Chemistry Research*, vol. 41, no. 17, pp. 4399–4408, 2002.

[21] T. M. Aminabhavi and S. K. Raikar, "Thermodynamic interactions in binary mixtures of 2-methoxyethanol with alkyl and aryl esters at 298. 15, 303. 15 and 308. 15 K," *Collection of Czechoslovak Chemical Communications*, vol. 58, pp. 1761–1776, 1993.

[22] A. K. Nain, "Densities and volumetric properties of binary mixtures of tetrahydrofuran with some aromatic hydrocarbons at temperatures from 278.15 to 318.15 K," *Journal of Solution Chemistry*, vol. 35, no. 10, pp. 1417–1439, 2006.

[23] A. G. Oskoei, N. Sataei, and J. Ghasemi, "Densities and viscosities for binary and ternary mixtures of 1, 4-dioxane + 1-hexanol + N,N-dimethylaniline from T = (283.15 to 343.15) K," *Journal of Chemical and Engineering Data*, vol. 53, no. 2, pp. 343–349, 2008.

[24] J. G. Baragi, M. I. Aralaguppi, T. M. Aminabhavi, M. Y. Kariduraganavar, and S. S. Kulkarni, "Density, viscosity, refractive index, and speed of sound for binary mixtures of 1,4-dioxane with different organic liquids at (298.15, 303.15, and 308.15) K," *Journal of Chemical and Engineering Data*, vol. 50, no. 3, pp. 917–923, 2005.

[25] J. A. Al-Kandary, A. S. Al-Jimaz, and A.-H. M. Abdul-Latif, "Viscosities, densities, and speeds of sound of binary mixtures of benzene, toluene, o-xylene, m-xylene, p-xylene, and mesitylene with anisole at (288.15, 293.15, 298.15, and 303.15) K," *Journal of Chemical and Engineering Data*, vol. 51, no. 6, pp. 2074–2082, 2006.

[26] J. N. Nayak, M. I. Aralaguppi, and T. M. Aminabhavi, "Density, viscosity, refractive index, and speed of sound in the binary mixtures of ethyl chloroacetate with aromatic liquids at 298.15, 303.15, and 308.15 K," *Journal of Chemical and Engineering Data*, vol. 47, no. 4, pp. 964–969, 2002.

[27] T. M. Aminabhavi, H. T. S. Phayde, R. S. Khinnavar, B. Gopalakrishna, and K. C. Hansen, "Densities, refractive indices, speeds of sound, and shear viscosities of diethylene glycol dimethyl ether with ethyl acetate, methyl benzoate, ethyl benzoate, and diethyl succinate in the temperature range from 298.15 to 318.15 K," *Journal of Chemical and Engineering Data*, vol. 39, no. 2, pp. 251–260, 1994.

[28] J. A. Reddick, W. B. Bunger, and T. K. Sakano, *Organic Solvents, Physical Properties and Method of Purification, Techniques of Chemistry, II*, Wiley-Interscience, New York, NY, USA, 1986.

[29] O. Redlich and A. T. Kister, "Algebraic representation of thermodynamic properties and the classification of solutions," *Industrial and Engineering Chemistry Research*, vol. 40, no. 2, pp. 345–348.

[30] C. Yang, P. Ma, and Q. Zhouy, "Excess molar volumes and viscosities of binary mixtures of sulfolane with benzene, toluene, ethylbenzene, *p*-Xylene, *o*-Xylene, and *m*-Xylene at 303.15 and 323.15 K and atmospheric pressure," *Journal of Chemical and Engineering Data*, vol. 49, pp. 881–885, 2004.

[31] R. J. Fort and W. R. Moore, "Viscosities of binary liquid mixtures," *Transactions of the Faraday Society*, vol. 62, pp. 1112–1119, 1966.

[32] R. P. Rastogi, J. Nath, and J. Misra, "Thermodynamics of weak interactions in liquid mixtures. I. Mixtures of carbon tetrachloride, benzene, toluene, and p-xylene," *Journal of Physical Chemistry*, vol. 71, no. 5, pp. 1277–1286, 1967.

[33] M. V. Kaulgud, "Ultrasonic velocity and compressibility in Binary liquid mixtures," *Zeitschrift Für Physikalische Chemie*, vol. 36, pp. 365–370, 1963.

[34] R. PalePu, J. Oliver, and B. Mackinnon, "Viscosities and densities of binary liquid mixtures of *m*-cresol with substituted anilines," *Canadian Journal of Chemistry*, vol. 63, no. 5, pp. 1024–1030, 1985.

[35] T. M. Reed III and T. E. Taylor, "Viscosities of liquid mixtures," *Journal of Physical Chemistry*, vol. 63, no. 1, pp. 58–67, 1959.

[36] R. Mehra and M. Pancholi, "Effect of temperature variation on acoustic and thermodynamic properties in the multicomponent system of heptane + 1-dodecanol + cyclohexane at 298, 308 and 318 K and suitability of different theories of sound speed," *Journal of the Indian Chemical Society*, vol. 83, no. 11, pp. 1149–1152, 2006.

[37] A. Ali, A. K. Nain, D. Chand, and B. Lal, "Molecular interactions in binary mixtures of anisole with benzyl chloride, chlorobenzene and nitrobenzene at 303.15 K: an ultrasonic, volumetric, viscometric and refractive index study," *Indian Journal of Chemistry A*, vol. 44, no. 3, pp. 511–515, 2005.

[38] R. Palani and K. Meenakshi, "Investigation of molecular interactions in ternary liquid mixtures using ultrasonic velocity," *Indian Journal of Chemistry A*, vol. 46, no. 2, pp. 252–257, 2007.

Nonequilibrium Thermodynamics of Cell Signaling

Enrique Hernández-Lemus

Computational Genomics Department, National Institute of Genomic Medicine, Periférico Sur 4809, Col. Arenal Tepepan, Delegación Tlalpan, 14610 Mexico City, DF, Mexico

Correspondence should be addressed to Enrique Hernández-Lemus, ehernandez@inmegen.gob.mx

Academic Editor: Ali-Akbar Saboury

Signal transduction inside and across the cells, also called cellular signaling, is key to most biological functions and is ultimately related with both life and death of the organisms. The processes giving rise to the propagation of biosignals are complex and extremely cooperative and occur in a far-from thermodynamic equilibrium regime. They are also driven by activation kinetics strongly dependent on local energetics. For these reasons, a nonequilibrium thermodynamical description, taking into account not just the activation of second messengers, but also transport processes and dissipation is desirable. Here we present a proposal for such a formalism, that considers cells as small thermodynamical systems and incorporates the role of fluctuations as intrinsic to the dynamics in a spirit guided by mesoscopic nonequilibrium thermodynamics. We present also a minimal model for cellular signaling that includes contributions from activation, transport, and intrinsic fluctuations. We finally illustrate its feasibility by considering the case of FAS signaling which is a vital signal transduction pathway that determines either cell survival or death by apoptosis.

1. Introduction

Survival of living organisms is intimately linked to their ability to react quite efficiently to even extremely weak external signals. Common examples are the reaction of the human eye to single light photons [1, 2], the reaction of a male butterfly to a single pheromone molecule coming from a female at a distance that sometimes is in the order of kilometers [3], and so forth. Cellular receptors react to hormones, cytokines, or antigens at very low concentrations. This strong reaction to a weak impulse is attained by an amplification process which is performed by means of special pathways of free energy transduction. Mechanisms such as immune system response, thermal-shock inhibitions, and cardiovascular rearrangement in response to environmental changes are all mediated by signaling processes. Signal transduction (information flow) is, thus, equally important, if not more important, for the functioning of a living organisms than metabolism and energy flow.

Signal transduction or *cell signaling* is the generic name of the set of concatenated processes or stages in which a cell transforms a certain signal or stimulus—either intercellular or intracellular—into another signal or a specific response. Cell signaling affects the complex arrangement of biochemical reactions inside the cell that takes place by means of enzymes that are bounded to other molecules called *second messengers*. Each process takes place in fast times, with dynamic ranges between a few milliseconds in most cases, to a few seconds in the case of more complex signaling cascades. Intricate and very sensible molecular biology experiments have shown cell signaling to be rate processes, that is, kinetic-guided phenomena determined by previous systems settings [4].

The wide variety of physicochemical signals to which cells may respond may seem to imply on a wide range of signal transduction mechanisms. However, only a handful of event chains is able to generate proper response to every stimulus in different cell subtypes which points to generalistic strategies commonly beginning with the action of cell receptors. Many signal transduction processes are then usually started by the adhesion of a ligand protein to a membrane receptor that then activates either itself or other receptor (or series of receptors) thus converting the initial stimulus into a response that once inside the cell provokes a chain of biochemical events known as a *signaling cascade* or *second messenger pathway* which results in the amplification of the signal.

The archetypal example here is that of the *epinephrine cascade*. It is known that epinephrine (adrenaline) stimulates the liver to convert glycogen to glucose in liver cells, but epinephrine alone would not convert glycogen to glucose. In an outstanding experimental *tour de force* that granted him the 1971 Nobel Prize in Physiology or Medicine, Earl Sutherland found that epinephrine had to trigger a second messenger, cyclic AMP, for the liver to convert glycogen to glucose [5]. Secondary messenger systems can be synthesized and activated by the action of enzymes, for instance, cyclases that synthesize cyclic nucleotides. Second messengers also form by opening of ion channels to allow influx of metal ions (e.g., calcium signaling). These second messengers then bind and activate protein kinases, ion channels, and other proteins continuing the signaling cascade.

The role that activation kinetics and other energy-driven dynamic processes play in cell signaling makes evident the need for a thermodynamic description. Most studies to date are based in equilibrium thermodynamics assumptions [6, 7] or, in any case, coarse-grained approaches [8]. Specific applications of nonequilibrium thermodynamics have been studied in the past [9–13] focusing on single features such as switching, sensitivity, and controllability. Thus, a nonequilibrium thermodynamics analysis of cell signaling, describing transport processes, activation kinetics and nonlocal effects is desirable.

Some particular cases of signaling dynamics have been studied, even at the nonequilibrium statistical physics level of description, for instance, by means of information theoretical approaches [14]. In such studies, a positive correlation between the *channel capacity* (i.e., the information-carrying capacity of the signaling networks) and free-energy expenditure has been observed. For phosphorylation-dephosphorylation switches, hydrolysis-free energy is in the sustained high concentration of ATP and low concentrations of ADP, that is, away from thermochemical equilibrium. This deviation from equilibrium implies, among other things, that useful hydrolysis-free energy does not come from the phosphate bond of the ATP molecule alone but from more complex-systemic mechanisms.

Nonequilibrium thermodynamic entropy and entropy production have been studied, to gain dynamic signaling information transfer, in insulin transduction [11]. In that case, entropy production rates show a broad secondary peak in time that represents a possible evidence of the decrease of the concentration of membrane GLUT4 (a so called *backflow*), thus to the reduction of insulin efficiency. Interestingly, at least in that case entropic contributions take a leading role in controlling signaling efficiency. Pathway selectivity driven by receptor-receptor interaction has also been studied by means of thermodynamic models [8]. Ligand-induced oligomerization of cell-surface receptors is driven by cooperative behavior. Oligomerization occurs due to interaction between nearest-neighbor receptors. This type of cooperativity can exhibit a first-order phase transition, corresponding to a jump in the surface density of ligand-receptor complexes. Clustering could be described by the statistical mechanics of a simple lattice Hamiltonian. Receptor-receptor interaction may lead to a first-order phase transition with a discontinuous jump in the receptor density as a function of the receptor chemical potential and/or the ligand concentration [8].

Thermodynamical studies of biomolecular switches could be quantitatively described by a simple 3-state population-shift model, in which the equilibrium between a non-binding, nonsignaling state and the binding-competent, signaling state is shifted toward the latter upon target binding. Performance of biomolecular switches can be sensibly tuned via mutations that alter their switching thermodynamics [7]. Thermodynamic conditions in the intracellular medium hence alter sensitivity, control, and effective information transfer in signaling networks [14].

Moreover, as we may see later, typical settings in signal transduction correspond with complex nonequilibrium stages [15]. In fact, even relatively simple signaling models such as the phosphorylation-dephosphorylation switches exhibit bistability due to feedback, and the related nonequilibrium steady state even presents a phase transition [16]. Such complex behavior led some researchers to propose that non-linear deterministic biochemical behavior is dynamically *trapped* between stochastic dynamics, both at the molecular signaling level and at the cellular evolution level [16]. This proposition raised from an analysis of the so-called *chemical master equation* which is founded in the tenets of non-linear nonequilibrium thermodynamics [10, 17].

Thermodynamic models of cell signaling aim to model and describe these phenomena at the basic level, and applications of thermodynamic modeling in search of therapeutic action have been recently developed [18]. Claims have been made that fast binding kinetics was advantageous for most targets with a couple of exceptions, that targeting some protein kinases could enhance rather than attenuate the pathway, and that therapeutic doses could be sensitive to the kinetic parameters of drug binding. Thermo-kinetic rates have been shown to play an important role in the dynamics of signaling and immune response. Plasmon resonance-based thermodynamics points out to slow-signaling modified second-messenger variants have similar affinities but distinctly faster dissociation rates that compared with the original messengers and that this may be behind their lower activity. Signaling deregulation could be starting not at the biochemical (recognition) but at the thermodynamical (dissociation rates) levels [19]. In fact, thermodynamic studies are now part of the drug-design tools of pharmaceutical chemistry. In fact, thermodynamic and kinetic analyses are sources of deeper insight into specificity of molecular recognition processes and signaling [20]. Such advances had led to research efforts combining statistical thermodynamic models in combination with experimental data [21]. Preliminary results of these studies are very promising.

A proposal for a nonequilibrium thermodynamics formalism including the role of fluctuations as intrinsic to the dynamics, and the role of transport processes is made in this work. We detail a minimal model for cellular signaling that considers activation, transport, and intrinsic fluctuations.

The rest of the paper is organized as follows: in Section 2 we discuss the role that stochastic fluctuations and transport processes play in biological signal transduction, in Section 3 we develop a nonequilibrium thermodynamics formalism of cell signaling and from it we derive a minimalist model, Section 4 deals with the biology of direct FAS signaling in apoptosis that in its simplest version (here presented) is akin to our minimalist model, finally in Section 5 we discuss some potential applications, the scope and limitations of our formalism as well as some perspectives.

2. Cell Signaling and Stochasticity and Nonlocality

One source of complexity in the nonequilibrium thermodynamical characterization of cell signaling is the fact that a cell is a *small system*; that is, cellular dimensions do not permit the immediate application of the thermodynamic limit. Specifically, the role that fluctuations and stochasticity may play within such scenarios is not completely clear. A formalism to study *small systems thermodynamics* in equilibrium has been developed [22, 23], and some results were even expected to extend to local equilibrium settings within cellular sized biosystems [24]. However, one important drawback in completing such theoretical frameworks at that time was the lack of proper experimental settings to test their hypotheses. Nevertheless, with the development of modern techniques, such as microscopic manipulation by means of atomic force microscopes, optical tweezers, and cold traps, this situation has been overcome at least partially. Theories have been developed including mesoscopic thermodynamical approaches [25–29] and also studies made by means of fluctuation theorems [30–33]. Some of these theoretical results have been even experimentally tested.

Due to the low copy number of many reactants in cells, and the nonequilibrium nature of the many intracellular reactions, signal transduction may result from stochastic intracellular events. Distribution analyses of cell responses provide a means to probe the stochastic character of intracellular signaling. A goal is to determine the class of stochasticity that affects intracellular pathways [34]. Stochasticity has been measured experimentally, it has been also incorporated in molecular simulations, and it has been discovered that locality and Gaussian behavior are not always present. In fact, transient multipeak distributions have been observed in computer simulations of cell-signaling dynamics. The emergence of these complex distributions cannot be explained using either deterministic chemical kinetics or simple Gaussian noise approximation [35].

Multipeak distributions are typically transient and eventually evolve into single-peak distributions in certain cases these distributions may be stable in the limit of long times. It has also been shown that introducing positive feedback loops results in diminution of the probability distribution complexity. This effect is so strong that even stochastic resonance has been reported in signaling cascades [36] where certain *optimal reaction rates* minimize the average threshold-crossing time. A noisy signal reaches the threshold more easily when the upstream and downstream reaction time scales are related in a specific way, indicating the existence of internal resonances embedded in cellular signaling cascades [36]; that is, nonequilibrium thermodynamic couplings exist between different modes (as characterized by their corresponding relaxation times) a feature that can be accounted by certain nonequilibrium thermodynamics formalisms (see next section). This may seem to point out on how the rates of various nodes could be *collectively tuned* in protein-signaling networks in such a way that signals are optimally picked up and biological information is transmitted through the signaling cascade.

3. Irreversible Thermodynamics of Cell Signaling

As we have just pointed out, systems outside the thermodynamic limit are characterized by large fluctuations and hence stochastic effects. The classic thermodynamic theory of irreversible process (also called linear irreversible thermodynamics, LIT) [37] provides us with a *coarse-grained* description of the systems, thus ignoring the molecular nature of matter studying it as a continuum media by means of a phenomenological field theory. As such LIT is not suitable for the description of small systems because it ignores fluctuations that could become the dominant factor in the system's response. Nevertheless, in many instances, it would be desirable to have a thermodynamic theoretical framework to study small systems, most noticeably in cellular and subcellular processes like signal transduction. One way to do so is by considering the stochastic nature of the time evolution of small nonequilibrium systems. This is the approach of Mesoscopic Nonequilibrium Thermodynamics (MNETs) [26]. MNET for small systems can be understood as the extension of the equilibrium thermodynamics of small systems developed by Hill and Chamberlin [22] and Hill [23, 24].

MNET, for instance, was developed to analyze nonequilibrium small systems. Any reduction of the spatio-temporal scale description of a system implies an increase in the number of noncoarse-grained degrees of freedom. These degrees of freedom could be related with the extended variables in extended irreversible thermodynamics (EITs) [38]. In order to characterize such variables, let us say that there exist a set $\Upsilon = \{v_i\}$ of such non-equilibrated degrees of freedom. $P(\Upsilon, t)$ is the probability that the system is at a state given by Υ at time t. If one assumes [27] that the evolution of the degrees of freedom could be described as a diffusion process in Υ-space, then the corresponding Gibbs equation could be written as

$$\delta S = -\frac{1}{T} \int \mu(\Upsilon) P(\Upsilon, t) d\Upsilon. \quad (1)$$

$\mu(\Upsilon)$ is a generalized chemical potential related to the probability density, whose time-dependent expression could be explicitly be written as

$$\mu(\Upsilon, t) = k_B T \ln \frac{P(\Upsilon, t)}{P(\Upsilon)_{\text{equil}}} + \mu_{\text{equil}} \quad (2)$$

or in terms of a nonequilibrium work term ΔW as follows:

$$\mu(\Upsilon, t) = k_B T \ln P(\Upsilon, t) + \Delta W. \quad (3)$$

The time-evolution of the system could be described as a generalized diffusion process over a potential landscape in the space of mesoscopic variables Υ. This process is driven by a generalized mesoscopic-thermodynamic force $(\partial/d\Upsilon)(\mu/T)$ whose explicit stochastic origin could be tracked back by means of a Fokker-Planck-like analysis [26, 27]. MNET seems to be a good candidate theory for describing nonequilibrium thermodynamics for small systems, *provided that one has a suitable model* or microscopic means to infer the probability distribution $P(\Upsilon, t)$.

MNET and similar approaches are appropriate to deal with activated processes, like a system crossing a potential barrier. Biochemical reactions like the ones involved in signal transduction are clearly in this case. According to [27] the diffusion current in this Υ space could be written in terms of a local fugacity defined as

$$z(\Upsilon) = \exp \frac{\mu(\Upsilon)}{k_B T}, \quad (4)$$

and the expression for the associated flux it will be

$$J = -k_B L \frac{1}{z} \frac{\partial z}{\partial \Upsilon}. \quad (5)$$

L is an Onsager-like coefficient. After defining a *diffusion coefficient* D and the associated affinity $A = \mu_2 - \mu_1$, the integrated rate is given as

$$\bar{J} = J_o \left(1 - \exp \frac{A}{k_B T}\right) \quad (6)$$

with $J_o = D \exp(\mu_1 / k_B T)$.

MNET then gives rise to nonlinear kinetic laws like (6). MNET has been applied successfully to biomolecular processes at the cellular level or description [28]. Non-linear kinetics are used to express, for example, RNA unfolding rates as *diffusion currents*, modeled via transition state theory, giving rise to Arrhenius-type non-linear equations. In that case the current was proportional to the chemical potential difference, so the entropy production was quadratic in that chemical potential gradient.

Signal transduction consists of a series of biochemical reactions, and many of these have unexplored chemical kinetics, due to this fact a detailed MNET analysis such as the one described above is unattainable at the present moment. On what follows, we will explore a phenomenologically based approach that takes into account similar considerations as the MNET framework already sketched but does so in a more informal, modeling-like manner. This phenomenological approach is based on the EIT assumption of enlargement of the thermodynamical variables space [39, 40].

Assuming that a generalized entropy-like function Ψ exists, we can write down a Gibbs equation which may be written in the following form [38, 41]:

$$\frac{d\Psi}{dt} = T^{-1} \left[\frac{dU}{dt} + p \frac{dv}{dt} - \sum_i \mu_i \frac{dC_i}{dt} - \sum_j \mathcal{A}_j \frac{d\xi_j}{dt} - \sum_k \mathcal{X}_k \odot \frac{d\Phi_k}{dt} \right] \quad (7)$$

or as a differential form

$$d_t \Psi = T^{-1} \left[d_t U + p d_t v - \sum_i \mu_i d_t C_i - \sum_j \mathcal{A}_j d_t \xi_j - \sum_k \mathcal{X}_k \odot d_t \Phi_k \right]. \quad (8)$$

Quantities are defined as usual, U is the internal energy per mol, T the absolute temperature, p the pressure, v the molar volume, μ_i is the chemical potential for the i species, C_i its concentration (mole fraction), $\mathcal{A}_j$ the molar chemical affinity for the reaction producing species, j (i.e., $\mathcal{A}_j = \sum_i \nu_i^j \mu_i^j$ being ν_i^j the stoichiometric coefficient for the ith species in the jth reaction and μ_i^j the corresponding chemical potential), ξ_j the reaction coordinate for the production of species, j, $\mathcal{X}_k$, and Φ_k are extended fluxes and forces for diverse k processes, and $\odot$ is the appropriate scalar product.

Here we are considering the presence of thermal processes, but also the energetics of three different contributions due to signal transduction: the effect of *bulk* chemical potentials related to concentration changes of the signaling molecules in the cellular environment (identified with the subscript i), activation kinetics (considered as generalized chemical reactions) related to the chemical affinities between ligand proteins, membrane receptors and effector proteins in the signaling cascade (identified with the subscript j), and generalized *transport* processes (including the effects of nonlocal dynamics and delays) considered as *extended* variables or *generalized fluxes and forces* (identified with the the subscript k).

3.1. Activation Kinetics. We will introduce a simple—although general—model for signal transduction including the action of ligand proteins (LPs), membrane receptors (MRs), effector proteins (EPs), and finally response proteins (RPs). In this idealized model, LPs and MRs play the role of *pulls and triggers*, then a series of m EP steps (not necessarily, but possibly sequential) constitute the core of the signaling cascade and finally, and the RPs when activated constitute the

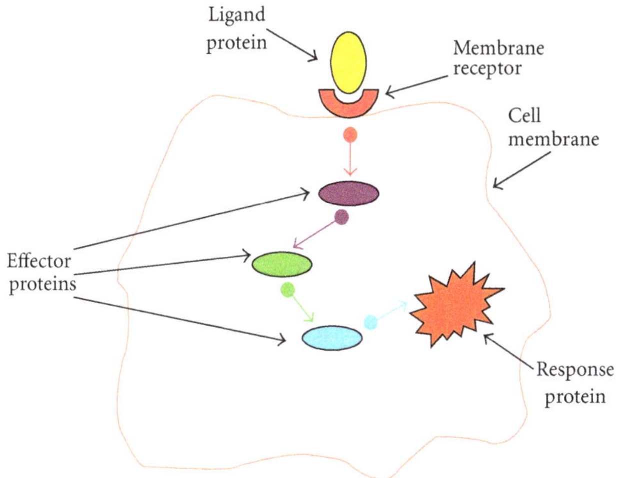

FIGURE 1: A toy model of cell signaling.

cell's response to the initial stimulus. The pseudo-chemical reactions could be written as follows:

$$LP + MR \longrightarrow MR^{\star} \quad (9)$$

$$MR^{\star} + EP_1 \longrightarrow EP_1^{\star} \quad (10)$$

$$EP_1^{\star} + EP_2 \longrightarrow EP_2^{\star}$$

$$\vdots \quad (11)$$

$$EP_i^{\star} + EP_{i+1} \longrightarrow EP_{i+1}^{\star} \quad (12)$$

$$EP_m^{\star} + RP \longrightarrow RP^{\star}. \quad (13)$$

Here the superscript $\star$ refers to the *activated* form of the molecule, that is, the form which presents the corresponding biological signaling activity. For a pictorial representation, please refer to Figure 1.

3.2. Generalized Transport Processes. Let us recall (8). If we write down explicit expression for second and third terms in the r.h.s. of (8) in the context of signal transduction, we have the following generalized Gibbs form:

$$d_t G = -\Psi d_t T + \sum_i \mu_i d_t C_i + \sum_j \mathcal{A}_j d_t \xi_j + \sum_k \mathcal{X}_k \odot d_t \Phi_k. \quad (14)$$

Since signal transduction occurs within the cell, it is possible to relate an internal *work* term with the regulation process itself, being this a *far-from equilibrium* contribution. This nonlocal contribution is given by the generalized force-flux term (last term in the r.h.s. of (14)). This is so as cell signaling often does not occur *in situ* and also since is the only way to take into account (albeit indirectly) the changes in the local chemical potentials that cause the long tails in the fluctuations distributions characteristic of nonequilibrium small systems (e.g., cells). The term relating second messenger *flows* due to transduction could be written as a product of extended fluxes Φ_k and forces $\mathcal{X}_k$. Here $k = 1, \dots m$ refers to the different second messenger species involved.

3.3. A Minimalist Model of Signal Transduction. In order to present a full detail of the different energetic contributions in (14), let us consider a *minimalist* model consisting of just four molecules under isothermal conditions: one triggering ligand protein (LP), one membrane receptor (MR), one effector protein (EP), and one response protein (RP). In such case we have the following 3 pseudoreactions:

$$LP + MR \longrightarrow MR^{\star} \quad (15)$$

$$MR^{\star} + EP \longrightarrow EP^{\star} \quad (16)$$

$$EP^{\star} + RP \longrightarrow RP^{\star}. \quad (17)$$

That will give rise to the following form of (14):

$$\begin{aligned} d_t G = {} & \mu_{LP} d_t C_{LP} + \mu_{MR} d_t C_{MR} + \mu_{EP} d_t C_{EP} + \mu_{RP} d_t C_{RP} \\ & + \mathcal{A}_{MR^{\star}} d_t \xi_{MR^{\star}} + \mathcal{A}_{EP^{\star}} d_t \xi_{EP^{\star}} + \mathcal{A}_{RP^{\star}} d_t \xi_{RP^{\star}} \\ & + \mathcal{X}_{MR^{\star}} \odot d_t \Phi_{MR^{\star}} + \mathcal{X}_{EP^{\star}} \odot d_t \Phi_{EP^{\star}} + \mathcal{X}_{RP^{\star}} \odot d_t \Phi_{RP^{\star}}. \end{aligned} \quad (18)$$

Equation (18) considers the energies of formation for four molecules (as given by the μ's), the energies of activation of three species (as given by the chemical affinities $\mathcal{A}$'s) as well as the energies related with transport of the active species, given by their respective thermodynamic forces ($\mathcal{X}$'s). If we now refer to (6) for the definition of *signaling fluxes* [25, 28], we can write down expressions for the Φ's, namely,

$$\Phi_{MR^{\star}} = D_{MR^{\star}} \exp^{\mu_{MR}/K_B T} \times \left(1 - \exp^{\mathcal{A}_{MR^{\star}}/K_B T}\right) \quad (19)$$

$$\Phi_{EP^{\star}} = D_{EP^{\star}} \exp^{\mu_{EP}/K_B T} \times \left(1 - \exp^{\mathcal{A}_{EP^{\star}}/K_B T}\right) \quad (20)$$

$$\Phi_{RP^{\star}} = D_{RP^{\star}} \exp^{\mu_{RP}/K_B T} \times \left(1 - \exp^{\mathcal{A}_{RP^{\star}}/K_B T}\right). \quad (21)$$

Hence their temporal derivatives are given by:

$$\begin{aligned} d_t \Phi_{MR^{\star}} = {} & \frac{D_{MR^{\star}}}{K_B T} \exp^{\mu_{MR}/K_B T} \\ & \times \left[\left(1 - \exp^{\mathcal{A}_{MR^{\star}}/K_B T}\right) d_t \mu_{MR} - \exp^{\mathcal{A}_{MR^{\star}}/K_B T} d_t \mathcal{A}_{MR^{\star}}\right] \end{aligned} \quad (22)$$

$$\begin{aligned} d_t \Phi_{EP^{\star}} = {} & \frac{D_{EP^{*}}}{K_B T} \exp^{\mu_{EP}/K_B T} \\ & \times \left[\left(1 - \exp^{\mathcal{A}_{EP^{*}}/K_B T}\right) d_t \mu_{EP} - \exp^{\mathcal{A}_{EP^{\star}}/K_B T} d_t \mathcal{A}_{EP^{\star}}\right] \end{aligned} \quad (23)$$

$$\begin{aligned} d_t \Phi_{RP^{\star}} = {} & \frac{D_{RP^{\star}}}{K_B T} \exp^{\mu_{RP}/K_B T} \\ & \times \left[\left(1 - \exp^{\mathcal{A}_{RP^{\star}}/K_B T}\right) d_t \mu_{RP} - \exp^{\mathcal{A}_{RP^{\star}}/K_B T} d_t \mathcal{A}_{RP^{\star}}\right]. \end{aligned} \quad (24)$$

If we consider, as it is often done in irreversible thermodynamics, that the thermodynamic forces $\mathcal{X}$'s are proportional to the fluxes Φ's, with proportionality constant $\mathcal{R}$, we have

$$\mathcal{X}_{\mathrm{MR}^\star} = \mathcal{R}_{\mathrm{MR}^\star} D_{\mathrm{MR}^\star} \exp^{\mu_{\mathrm{MR}}/K_B T} \times \left(1 - \exp^{\mathcal{A}_{\mathrm{MR}^\star}/K_B T}\right) \quad (25)$$

$$\mathcal{X}_{\mathrm{EP}^\star} = \mathcal{R}_{\mathrm{EP}^\star} D_{\mathrm{EP}^\star} \exp^{\mu_{\mathrm{EP}}/K_B T} \times \left(1 - \exp^{\mathcal{A}_{\mathrm{EP}^\star}/K_B T}\right) \quad (26)$$

$$\mathcal{X}_{\mathrm{RP}^\star} = \mathcal{R}_{\mathrm{RP}^\star} D_{\mathrm{RP}^\star} \exp^{\mu_{\mathrm{RP}}/K_B T} \times \left(1 - \exp^{\mathcal{A}_{\mathrm{RP}^\star}/K_B T}\right). \quad (27)$$

We now define the following generalized transport coefficients:

$$\Theta_{\mathrm{MR}^\star} = \frac{\mathcal{R}_{\mathrm{MR}^\star} D^2_{\mathrm{MR}^\star}}{K_B T};$$
$$\Theta_{\mathrm{EP}^\star} = \frac{\mathcal{R}_{\mathrm{EP}^\star} D^2_{\mathrm{EP}^\star}}{K_B T}; \quad (28)$$
$$\Theta_{\mathrm{RP}^\star} = \frac{\mathcal{R}_{\mathrm{RP}^\star} D^2_{\mathrm{RP}^\star}}{K_B T}.$$

By substitution of (22) to (28) in (18) we have

$$\begin{aligned} d_t G = {} & \mu_{\mathrm{LP}} d_t C_{\mathrm{LP}} + \mu_{\mathrm{MR}} d_t C_{\mathrm{MR}} + \mu_{\mathrm{EP}} d_t C_{\mathrm{EP}} + \mu_{\mathrm{RP}} d_t C_{\mathrm{RP}} \\ & + \mathcal{A}_{\mathrm{MR}^\star} d_t \xi_{\mathrm{MR}^\star} + \mathcal{A}_{\mathrm{EP}^\star} d_t \xi_{\mathrm{EP}^\star} + \mathcal{A}_{\mathrm{RP}^\star} d_t \xi_{\mathrm{RP}^\star} \\ & + \Theta_{\mathrm{MR}^\star} \exp^{2\mu_{\mathrm{MR}}/K_B T} \left[1 - \exp^{\mathcal{A}_{\mathrm{MR}^\star}/K_B T}\right]^2 d_t \mu_{\mathrm{MR}} \\ & - \Theta_{\mathrm{MR}^\star} \exp^{2\mu_{\mathrm{MR}}/K_B T} \left[\exp^{\mathcal{A}_{\mathrm{MR}^\star}/K_B T} - \exp^{2\mathcal{A}_{\mathrm{MR}^\star}/K_B T}\right] d_t \mathcal{A}_{\mathrm{MR}^\star} \\ & + \Theta_{\mathrm{EP}^\star} \exp^{2\mu_{\mathrm{EP}}/K_B T} \left[1 - \exp^{\mathcal{A}_{\mathrm{EP}^\star}/K_B T}\right]^2 d_t \mu_{\mathrm{EP}} \\ & - \Theta_{\mathrm{EP}^\star} \exp^{2\mu_{\mathrm{EP}}/K_B T} \left[\exp^{\mathcal{A}_{\mathrm{EP}^\star}/K_B T} - \exp^{2\mathcal{A}_{\mathrm{EP}^\star}/K_B T}\right] d_t \mathcal{A}_{\mathrm{EP}^\star} \\ & + \Theta_{\mathrm{RP}^\star} \exp^{2\mu_{\mathrm{RP}}/K_B T} \left[1 - \exp^{\mathcal{A}_{\mathrm{RP}^\star}/K_B T}\right]^2 d_t \mu_{\mathrm{RP}} \\ & - \Theta_{\mathrm{RP}^\star} \exp^{2\mu_{\mathrm{RP}}/K_B T} \left[\exp^{\mathcal{A}_{\mathrm{RP}^\star}/K_B T} - \exp^{2\mathcal{A}_{\mathrm{RP}^\star}/K_B T}\right] d_t \mathcal{A}_{\mathrm{RP}^\star}. \end{aligned} \quad (29)$$

Equation (29) gives a complete irreversible thermodynamical description of the minimal model given by (15) to (17). The model is then to be supplemented with the appropriate constitutive relations; in this case, the time evolution for the concentrations, chemical potentials, and chemical affinities as given by biochemical kinetics.

The free-energy coupling given by the corresponding generalized Maxwell relations (since $d_t G$ is an exact differential form, integrability conditions imply the existence of Maxwell-like relations [41]), as well as Gibbs-Duhem constrains (not all the concentrations and chemical potentials are independent) once explicit kinetics are given, constitutes the energetic core behind the complex processes of signal transduction. This is possibly the key contribution of this work, the explicit derivation from a nonequilibrium thermodynamics formalism showing that cell signaling *control* is, indeed, an energy-driven process. Of course, free-energy transduction has been known to be responsible for *the initiation* of signaling cascades. However, our model has shown that *every step* of the process is controlled and *locked* via the local chemical potentials even in the presence of stochasticity and fluctuations, provided that the assumptions of MNET hold.

4. Case Study: FAS Signaling in Apoptosis

An important family of signal transduction pathways is related with the onset and regulation of *programmed cellular death* or, apoptosis. Any functional disruption in the balance that apoptotic cells encounter may affect death signaling thus leading to diseases ranging from cancer in the case of subnormal apoptosis to degenerative disorders in supernormal apoptosis. Hence the control of the process as given by signal transduction pathways is of foremost relevance. One of the simplest example of such pathways is apoptosis regulation by FAS signaling. FAS is a cell-surface receptor protein that when triggered by an stimulus induces apoptosis in FAS-expressing cells. This process is highly linked with immune response, as the ligand for FAS, FAS-L, is mostly present on cytotoxic T cells and TH1 cells central players in innate immunity. FAS is composed of an extracellular region, one transmembrane domain, and an intracellular region. FAS activity is governed by interaction with its ligand (FAS-L). Activation of FAS through binding to its ligand or FAS antibody induces apoptosis, which has been confirmed by many experiments [42].

In order for signal transduction to occur, cross-linking of FAS with its ligand must occur. FAS trimerizes to properly bind to its ligand, which exists as a trimer. This creates a clustering of FAS that is necessary for signaling. In its intracellular region, FAS contains a conserved sequence deemed as *death domain*. An adaptor protein, FADD, interacts with the death domain on the FAS receptor. Subsequent binding to another region of FADD by procaspase 8 promotes grouping of pro-caspase 8 molecules bound to each of the clustered FADD proteins. This entire cluster is sometimes called a *death-inducing signaling complex*, or DISC [43]. Pro-caspase 8 transactivates itself once grouped, cleaving and releasing active caspase 8 molecules intracellularly.

As is clear from Figure 2, there is a correspondence between the model given by (15) to (17) and direct FAS signaling (Figure 2). In this case the ligand protein (LP) is the FAS-L molecule, the membrane receptor (MR) is FAS-R that when activated ($\mathrm{MR}^\star$) becomes FAS and then interacts with the effector protein (EP), in this case FADD, that carries the biosignal activating procaspase-8 (a response protein) that when activated ($\mathrm{RP}^\star$) becomes caspase-8, the molecule responsible for the no-return apoptotic response leading to cellular death.

FAS signaling is a well-characterized process [45], some thermodynamic parameters may be thus obtained by experiments [46, 47] or by means of molecular simulations [48]. This is the case of activation energies—especially when activation occurs by means of ATP produced by oxidative

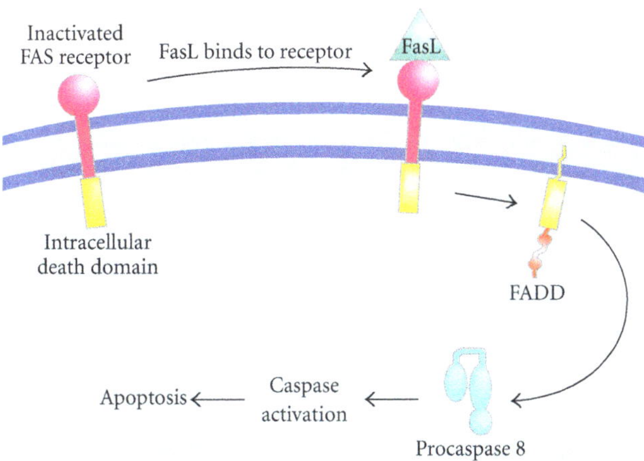

FIGURE 2: A *real-life* toy model of cell signaling: FAS signaling in apoptosis [44].

phosphorylation—and free energies of formation. However, transport processes have not been measured accurately (and in most cases have not been even measured at all). Being signaling pathways so important for the understanding of cell function, and in many instances for their biomedical importance as pharmacological targets; we hope that this situation soon will change. At the present moment, some insight on particular signaling pathways may be obtained by molecular dynamics simulations [49–54].

Experimental techniques have been refined that allow thermomolecular characterization of signaling processes. The technical challenges are, however, gigantic. Cell-signaling thermodynamic parameters must be experimentally measured by combining many different methodologies involving different scales of description: protein-protein electrostatic interactions, the electrohydrodynamic effect of the medium, cleavage and protein structure, free energies of folding/unfolding, transport processes, and so forth. Nonetheless, progress is being made in the actual realization of such experimental challenges, by using a clever combination of surface plasmon resonance (SPR), isothermal titration calorimetry (ITC), and ultracentrifugation (UC) of the thermodynamics of T-cell signaling in the MHC pathway have been unveiled [55].

SPR is extensively used to study receptor/ligand binding both qualitatively and quantitatively. However, results are commonly ambiguous, and every conclusion needs to be independently verified. SPR provides both kinetic and equilibrium data; data acquisition is fast, comparative studies are easily performed, and low affinities can be detected with relatively low amounts of protein. The accuracy of the SPR-derived kinetic constants depends crucially on other various parameters such as mass transport, sensor chip capacity, and flow rate. In conclusion, SPR experiments are useful but partial and sometimes even dubious. In contrast, equilibrium methods, such as ITC, often result in more reliable, specially when used in conjunction with SPR and analytical techniques as ultracentrifugation in which sedimentation velocity, and equilibrium experiments provide insight into the hydrodynamic and thermodynamic properties of the sample, thus enabling the inference of transport parameters and free energies via association constants; on the other hand, SPR experiments could shed some light on biochemical kinetics and their associated relaxation times [55]. ITC has also used in the experimental analysis of the interaction between TRAF and tumor necrosis factor receptors [56].

FAS signaling proceeds by typical physicochemical mechanisms. Being this the case, common ranges for the parameters in cell signaling may be used as proxy values instead. For instance, the characteristic concentration of signaling protein molecules ranges from 0.01 to 1 μ-molar, with molecule counts between 120,000 and 20,000,000 depending on molecular weights and type of cell [57]. RAS concentration in HeLa cells has been measured to be 0.4 μ-molar [58]. Thresholding signal duration times for whole processes range between 2 minutes and 24 hours. NF-κB signaling (which is related with FAS signaling) takes about 320 minutes in epithelial cells [59, 60].

In order to figure out the order of magnitude of kinetic parameters, let us consider the case of the values of fluxes and dissociation constants in the Wnt-signaling pathway [61]. Dissociation constants for several second messenger molecules range around 10–1200 nM, with protein concentrations in the 15 to 100 nM regime. The degradation flux of β-catenin via the proteasome is 25 nM/h. The characteristic time of the associated phosphorylation-dephosphorylation switch is 2.5 minutes for APC. Relaxation times for GSk3-β association/dissociation is 1 minute, and that of Axin degradation is 6 minutes [61]. Decay rates (half-life times) $t^{1/2}$, for signaling molecules in the MAPK pathway and the STAT pathway, are valued between 0.75 and 24 h [57]. More closely related with FAS signaling, duration times on switch of apoptosis and duration of apoptotic death in HeLa cells exposed to different levels of TRAIL are between 19 and 27 minutes for switching and between 140 and 660 minutes for cell death [62]. Rate constants for diffusion-limited enzymes may vary around 10^8 and $10^{10}\,M^{-1}\,s^{-1}$ [63] although there are other kinetic mechanisms for second messengers that in some cases seem to be cell-type specific [64], these figures may serve as reference to infer chemical kinetic behavior since the general behavior seems to be quite common [65]. Once the rate constants are measured, one can infer activation energies from them, following a kinetic model [66].

In relation to energetics, it is known that ATP hydrolysis releases between 28 and 33.5 kJ/mol [67] depending on cell type and condition whereas for other energy-rich compounds involved in substrate level phosphorylation range around 23.44 and 88 kJ/mol [68]. Free energy profiles may also help us to understand the role that protein-coupling plays in cell signaling. In Ras signaling, for instance, binding free energy determines the fate of Ras/Raf dynamics [69]. Diffusion coefficients measured inside the cell differ according to cell medium and molecule size. Inside the cell nucleus typical diffusion constants vary between 10 and 100 $\mu m^2/s$ [70]. In reference with signaling proteins, this is usually also the case even in cytoplasm and/or aqueous solution. The diffusion rate of phosphoglycerate kinase (around 45 kDa) has been measured as 63.8 $\mu m^2/s$ [71], while heavier molecules as 62 kDa Dextran move slowly, around 39 $\mu m^2/s$ [72]. Smaller second messenger molecules like insulin (5.808 kDa) can diffuse much faster, $D = 150\,\mu m^2/s$ [73]. The combination of different experimental/computational modeling techniques and estimated parameters just sketched may allow to construct quantitative thermodynamic models for cell signaling, following the lines of the present work, in the near future.

5. Discussion

Signaling transduction is a quite complex yet extremely important physicochemical phenomenon in cell biology. As we have seen cellular signaling is characterized by a combination of stochastic effects, activation biochemical kinetics, and multiple transport processes all setup in a far-from thermodynamic equilibrium setting. Is is known that free-energy transduction plays a key role in the process of signaling cascades. For this reason a nonequilibrium thermodynamics description at the mesoscopic level is desirable. In the present work we have presented such a formalism in the context of MNET [26].

The role of stochasticity is taken into account (albeit in an indirect way) by means of incorporating the probability distribution for the nonequilibrated degrees of freedom into a generalized chemical potential as is described in (2) to (6). In this scheme, the thermodynamic forces (equations (25) to (27))—that reflect in a coarse grained way the effect of stochasticity—are identified as the gradients in the space of mesoscopic variables of the logarithm of the ratio of the probability density to its equilibrium value. The main idea is to generalize the definition of the chemical potential to account for these additional mesoscopic variables. Thus it is possible to assume that the evolution of these degrees of freedom is described by a diffusion process and formulate the corresponding Gibbs equation.

The effect of generalized transport processes related with the distribution of relaxation times of the kinetics (it takes some (in general different) time for every biochemical reaction to activate the corresponding signaling molecule) and nonlocalities in the molecular processes (i.e., a second messenger has to travel some distance, say by diffusion, until it reaches its target molecule) is given by the last term at the r.h.s. of (14). In particular, relaxation times for the coarse-grained processes are given in terms of generalized transport coefficients (28). Our formalism is written in such a way that we can distinguish between local equilibrium effects (corresponding to the energetics of non-activated molecules at the top of the signaling cascade, as given by the first 4 terms at the r.h.s. of (29)), deterministic activation kinetics depending exclusively on the rate equations for the chemical reactions (corresponding to the 5th to 7th terms at the r.h.s. of (29)), and far-from equilibrium effects, involving both stochasticity and transport processes (terms 8th–13th at the r.h.s. of (29)) involving the dynamics of the evolution of nonconserved variables. We could think of this structure as a hierarchy in which *trains of signals* are coupled with each other via their relative relaxation times (as given by the corresponding generalized transport coefficients, (28)).

The potential application of such a formalism is wide, in particular with respect to the detailed study of the dynamics for important biological pathways. Consider, for instance, the extremely important scenario of calcium signaling. Phenomena like *calcium waves* [74, 75] and *calcium-induced calcium* release [76] could be understood more clearly (and even modeled and simulated) in the light of a nonequilibrium thermodynamic description like the one presented above. For instance, the role of energy releasing pathways in the dynamics and control of cell signaling under a system biology-like philosophy becomes almost crystal clear. In turn these free-energy triggers may be appropriate candidates for pharmacological targets for drug-therapy in cases of diseases associated with abnormal signaling. This may be the case of cardiac arrhythmias, neurological disorders [75, 77], and metabolic diseases [76].

Of course, such general modeling strategy has some drawbacks. On the one hand being a fully thermodynamical description, this framework depends entirely on experimental data for the activation kinetics and other constitutive relations or in any case in a good set of molecular simulations. On the other hand, ultrafast kinetics may be accompanied by noncompensated stochasticity that could not be handled entirely under the MNET paradigm. In conclusion, much work has still to be done in order to establish the validity and feasibility of these physical models into biological and biomedical research.

Acknowledgments

The author gratefully acknowledges support by Grant PIUTE10-92 (ICyT-DF) (Contract 281-2010) as well as federal funding from the National Institute of Genomic Medicine (México). He also acknowledges suggestions of the anonymous reviewers that, no doubt, contributed greatly to improve this work.

References

[1] D. A. Baylor, T. D. Lamb, and K. W. Yau, "Responses of retinal rods to single photons," *Journal of Physiology*, vol. 288, pp. 613–634, 1979.

[2] S. Hecht, S. Schlaer, and M. H. Pirenne, "Energy, quanta and vision," *Journal of the Optical Society of America A*, vol. 38, pp. 196–208, 1942.

[3] J. Andersson, A. K. Borg-Karlson, N. Vongvanich, and C. Wiklund, "Male sex pheromone release and female mate choice in a butterfly," *Journal of Experimental Biology*, vol. 210, no. 6, pp. 964–970, 2007.

[4] F. H. Johnson, H. Eyring, and B. J. Stover, *The Theory of Rate Processes in Biology and Medicine*, John Wiley & Sons, New York, NY, USA, 1974.

[5] T. N. Raju, "The Nobel chronicles. 1971: Earl Wilbur Sutherland, Jr. (1915–74)," *The Lancet*, vol. 354, no. 9182, article 961, 1999.

[6] E. Mertz, J. B. Beil, and S. C. Zimmerman, "Kinetics and thermodynamics of amine and diamine signaling by a

trifluoroacetyl azobenzene reporter group," *Organic Letters*, vol. 5, no. 17, pp. 3127–3130, 2003.
[7] A. Vallée-Bélisle, F. Ricci, and K. W. Plaxco, "Thermodynamic basis for the optimization of binding-induced biomolecular switches and structure-switching biosensors," *Proceedings of the National Academy of Sciences of the United States of America*, vol. 106, no. 33, pp. 13802–13807, 2009.
[8] C. Guo and H. Levine, "A thermodynamic model for receptor clustering," *Biophysical Journal*, vol. 77, no. 5, pp. 2358–2365, 1999.
[9] H. Qian, "Thermodynamic and kinetic analysis of sensitivity amplification in biological signal transduction," *Biophysical Chemistry*, vol. 105, no. 2-3, pp. 585–593, 2003.
[10] H. Qian and T. C. Reluga, "Nonequilibrium thermodynamics and nonlinear kinetics in a cellular signaling switch," *Physical Review Letters*, vol. 94, no. 2, Article ID 028101, 4 pages, 2005.
[11] E. Liu and J. M. Yuan, "Dynamic sensitivity and control analyses of metabolic insulin signalling pathways," *IET Systems Biology*, vol. 4, no. 1, pp. 64–81, 2010.
[12] D. Hu and J. M. Yuan, "Time-dependent sensitivity analysis of biological networks: coupled MAPK and PI3K signal transduction pathways," *Journal of Physical Chemistry A*, vol. 110, no. 16, pp. 5361–5370, 2006.
[13] N. R. Nené, J. Garca-Ojalvo, and A. Zaikin, "Speed-dependent cellular decision making in nonequilibrium genetic circuits," *PLoS ONE*, vol. 7, no. 3, Article ID e32779, 2012.
[14] H. Qian and S. Roy, "An information theoretical analysis of Kinase activated phosphorylation dephosphorylation cycle," *IEEE Transactions on NanoBioscience*, vol. 99, no. 1, pp. 1–17, 2012.
[15] M. Kurzynski, *The Thermodynamic Machinery of Life*, Springer, Berlin, Germany, 2006.
[16] H. Ge and H. Qian, "Non-equilibrium phase transition in mesoscopic biochemical systems: from stochastic to nonlinear dynamics and beyond," *Journal of the Royal Society Interface*, vol. 8, no. 54, pp. 107–116, 2011.
[17] T. G. Kurtz, "The relationship between stochastic and deterministic models for chemical reactions," *Journal of Chemical Physics*, vol. 57, no. 7, pp. 2976–2978, 1972.
[18] M. Goyal, M. Rizzo, F. Schumacher, and C. F. Wong, "Beyond thermodynamics: drug binding kinetics could influence epidermal growth factor signaling," *Journal of Medicinal Chemistry*, vol. 52, no. 18, pp. 5582–5585, 2009.
[19] K. Matsui, J. J. Boniface, P. Steffner, P. A. Reay, and M. M. Davis, "Kinetics of T-cell receptor binding to peptide/I-E(k) complexes: correlation of the dissociation rate with T-cell responsiveness," *Proceedings of the National Academy of Sciences of the United States of America*, vol. 91, no. 26, pp. 12862–12866, 1994.
[20] N. J. De Mol, F. J. Dekker, I. Broutin, M. J. E. Fischer, and R. M. J. Liskamp, "Surface plasmon resonance thermodynamic and kinetic analysis as a strategic tool in drug design. Distinct ways for phosphopeptides to plug into Src- and Grb2 SH2 domains," *Journal of Medicinal Chemistry*, vol. 48, no. 3, pp. 753–763, 2005.
[21] C. C. Mello and D. Barrick, "An experimentally determined energy landscape for protein folding," *Proceedings of the National Academy of Sciences of the United States of America*, vol. 101, pp. 169–178, 2004.
[22] T. L. Hill and R. V. Chamberlin, "Extension of the thermodynamics of small systems to open metastable states: an example," *Proceedings of the National Academy of Sciences of the United States of America*, vol. 95, no. 22, pp. 12779–12782, 1998.
[23] T. L. Hill, *Thermodynamics of Small Systems*, Dover, New York, NY, USA, 2002.
[24] T. L. Hill, *Free Energy Transduction and Biochemical Cycle Kinetics*, Dover, New York, NY, USA, 2004.
[25] J. M. Rubí, "Non-equilibrium thermodynamics of small-scale systems," *Energy*, vol. 32, no. 4, pp. 297–300, 2007.
[26] D. Reguera, J. M. Rubí, and J. M. G. Vilar, "The mesoscopic dynamics of thermodynamic systems," *Journal of Physical Chemistry B*, vol. 109, no. 46, pp. 21502–21515, 2005.
[27] J. M. Rubí, "Mesoscopic non-equilibrium thermodynamics, atti dell'accademia peloritana dei pericolanti, classes di scienze fisiche," *Matematiche E Naturali*, vol. 86, C1S081020, supplement 1, Article ID 081020, 2008.
[28] J. M. Rubí, D. Bedeaux, and S. Kjelstrup, "Unifying thermodynamic and kinetic descriptions of single-molecule processes: RNA unfolding under tension," *Journal of Physical Chemistry B*, vol. 111, no. 32, pp. 9598–9602, 2007.
[29] F. Ritort, "The nonequilibrium thermodynamics of small systems," *Comptes Rendus Physique*, vol. 8, no. 5-6, pp. 528–539, 2007.
[30] D. J. Evans and D. J. Searles, "Fluctuation theorem for stochastic systems," *Physical Review E*, vol. 50, no. 2, pp. 1645—1648, 1994.
[31] G. Gallavotti and E. G. D. Cohen, "Dynamical ensembles in nonequilibrium statistical mechanics," *Physical Review Letters*, vol. 74, no. 14, pp. 2694–2697, 1995.
[32] C. Jarzynski, "Nonequilibrium equality for free energy differences," *Physical Review Letters*, vol. 78, no. 14, pp. 2690–2693, 1997.
[33] G. E. Crooks, "Nonequilibrium measurements of free energy differences for microscopically reversible Markovian systems," *Journal of Statistical Physics*, vol. 90, no. 5-6, pp. 1481–1487, 1998.
[34] M. N. Artyomov, J. Das, M. Kardar, and A. K. Chakraborty, "Purely stochastic binary decisions in cell signaling models without underlying deterministic bistabilities," *Proceedings of the National Academy of Sciences of the United States of America*, vol. 104, no. 48, pp. 18958–18963, 2007.
[35] Y. Lan and G. A. Papoian, "Evolution of complex probability distributions in enzyme cascades," *Journal of Theoretical Biology*, vol. 248, no. 3, pp. 537–545, 2007.
[36] Y. Lan and G. A. Papoian, "Stochastic resonant signaling in enzyme cascades," *Physical Review Letters*, vol. 98, no. 22, Article ID 228301, 4 pages, 2007.
[37] S. R. de Groot and P. Mazur, *Non-Equilibrium Thermodynamics*, Dover, New York, NY, USA, 1984.
[38] D. Jou, C. Pérez-García, L. S. García-Colín, M. Lapez De Haro, and R. F. Rodríguez, "Generalized hydrodynamics and extended irreversible thermodynamics," *Physical Review A*, vol. 31, no. 4, pp. 2502–2508, 1985.
[39] D. Jou, J. Casas-Vazquez, and G. Lebon, *Extended Irreversible Thermodynamics*, Springer, New York, NY, USA, 1998.
[40] D. Jou, J. Casas-Vazquez, and G. Lebon, "Extended irreversible thermodynamics," *Reports on Progress in Physics*, vol. 51, no. 8, article 02, pp. 1105–1179, 1988.
[41] M. Chen and B. C. Eu, "On the integrability of differential forms related to nonequilibrium entropy and irreversible thermodynamics," *Journal of Mathematical Physics*, vol. 34, no. 7, pp. 3012–3029, 1993.
[42] N. Itoh, S. Yonehara, A. Ishii et al., "The polypeptide encoded by the cDNA for human cell surface antigen fas can mediate apoptosis," *Cell*, vol. 66, no. 2, pp. 233–243, 1991.
[43] A. O. Hueber, M. Zörnig, D. Lyon, T. Suda, S. Nagata, and G. I. Evan, "Requirement for the CD95 receptor-ligand pathway

in c-myc-induced apoptosis," *Science*, vol. 278, no. 5341, pp. 1305–1309, 1997.

[44] http://upload.wikimedia.org/wikipedia/commons/f/f5/Fas-signalling.png.

[45] Y.-Y. Mo and W. T. Beck, "DNA damage signals induction of Fas ligand in tumor cells," *Molecular Pharmacology*, vol. 55, no. 2, pp. 216–222, 1999.

[46] J. Albanese, S. Meterissian, M. Kontogiannea et al., "Biologically active fas antigen and its cognate ligand are expressed on plasma membrane-derived extracellular vesicles," *Blood*, vol. 91, no. 10, pp. 3862–3874, 1998.

[47] T. Nguyen and J. Russell, "The regulation of FasL expression during activation-induced cell death (AICD)," *Immunology*, vol. 103, no. 4, pp. 426–434, 2001.

[48] J. D. Suever, Y. Chen, J. M. McDonald, and Y. Song, "Conformation and free energy analyses of the complex of calcium-bound calmodulin and the Fas death domain," *Biophysical Journal*, vol. 95, no. 12, pp. 5913–5921, 2008.

[49] N. Ota and D. A. Agard, "Intramolecular signaling pathways revealed by modeling anisotropic thermal diffusion," *Journal of Molecular Biology*, vol. 351, no. 2, pp. 345–354, 2005.

[50] H. Park, W. Im, and C. Seok, "Transmembrane signaling of chemotaxis receptor tar: insights from molecular dynamics simulation studies," *Biophysical Journal*, vol. 100, no. 12, pp. 2955–2963, 2011.

[51] J. E. Pessin and A. L. Frattali, "Molecular dynamics of insulin/IGF-I receptor transmembrane signaling," *Molecular Reproduction and Development*, vol. 35, no. 4, pp. 339–345, 1993.

[52] K. Watanabe, K. Saito, M. Kinjo et al., "Molecular dynamics of STAT3 on IL-6 signaling pathway in living cells," *Biochemical and Biophysical Research Communications*, vol. 324, no. 4, pp. 1264–1273, 2004.

[53] Y. Kong and M. Karplus, "Signaling pathways of PDZ2 domain: a molecular dynamics interaction correlation analysis," *Proteins*, vol. 74, no. 1, pp. 145–154, 2009.

[54] E. P. G. Arêas, P. G. Pascutti, S. Schreier, K. C. Mundim, and P. M. Bisch, "Molecular dynamics simulations of signal sequences at a membrane/water interface," *Journal of Physical Chemistry*, vol. 99, no. 40, pp. 14885–14892, 1995.

[55] M. G. Rudolph, J. G. Luz, and I. A. Wilson, "Structural and thermodynamic correlates of T cell signalling," *Annual Review of Biophysics and Biomolecular Structure*, vol. 31, pp. 121–149, 2002.

[56] H. Ye and H. Wu, "Thermodynamic characterization of the interaction between TRAF2 and tumor necrosis factor receptor peptides by isothermal titration calorimetry," *Proceedings of the National Academy of Sciences of the United States of America*, vol. 97, no. 16, pp. 8961–8966, 2000.

[57] S. Legewie, H. Herzel, H. V. Westerhoff, and N. Blüthgen, "Recurrent design patterns in the feedback regulation of the mammalian signalling network," *Molecular Systems Biology*, vol. 4, article 190, 2008.

[58] A. Fujioka, K. Terai, R. E. Itoh et al., "Dynamics of the Ras/ERK MAPK cascade as monitored by fluorescent probes," *The Journal of Biological Chemistry*, vol. 281, no. 13, pp. 8917–8926, 2006.

[59] D. D. Hershko, B. W. Robb, C. J. Wray, G. J. Luo, and P. O. Hasselgren, "Superinduction of IL-6 by cycloheximide is associated with mRNA stabilization and sustained activation of p38 map kinase and NF-κB in cultured Caco-2 cells," *Journal of Cellular Biochemistry*, vol. 91, no. 5, pp. 951–961, 2004.

[60] A. Hoffmann, A. Levchenko, M. L. Scott, and D. Baltimore, "The IκB-NF-κB signaling module: temporal control and selective gene activation," *Science*, vol. 298, no. 5596, pp. 1241–1245, 2002.

[61] E. Lee, A. Salic, R. Krüger, R. Heinrich, and M. W. Kirschner, "The roles of APC and axin derived from experimental and theoretical analysis of the Wnt pathway," *PLoS Biology*, vol. 1, no. 1, article e10, 2003.

[62] J. G. Albeck, J. M. Burke, S. L. Spencer, D. A. Lauffenburger, and P. K. Sorger, "Modeling a snap-action, variable-delay switch controlling extrinsic cell death," *PLoS Biology*, vol. 6, no. 12, article e299, pp. 2831–2852, 2008.

[63] M. E. Stroppolo, M. Falconi, A. M. Caccuri, and A. Desideri, "Superefficient enzymes," *Cellular and Molecular Life Sciences*, vol. 58, no. 10, pp. 1451–1460, 2001.

[64] C. Kiel and L. Serrano, "Cell type-specific importance of Ras-c-Raf complex association rate constants for mapk signaling," *Science Signaling*, vol. 2, no. 81, article ra38, 2009.

[65] C. Kiel, D. Aydin, and L. Serrano, "Association rate constants of ras-effector interactions are evolutionarily conserved," *PLoS Computational Biology*, vol. 4, no. 12, Article ID e1000245, 2008.

[66] C. Kiel and L. Serrano, "Affinity can have many faces: thermodynamic and kinetic properties of Ras effector complex formation," *Current Chemical Biology*, vol. 1, pp. 215–225, 2007.

[67] J. Rosing and E. C. Slater, "The value of G degrees for the hydrolysis of ATP," *Biochim Biophys Acta*, vol. 267, no. 2, pp. 275–290, 1972.

[68] R. K. Thauer, K. Jungermann, and K. Decker, "Energy conservation in chemotrophic anaerobic bacteria," *Bacteriological Reviews*, vol. 41, no. 1, pp. 100–180, 1977.

[69] H. Gohlke, C. Kiel, and D. A. Case, "Insights into protein-protein binding by binding free energy calculation and free energy decomposition for the Ras-Raf and Ras-RalGDS complexes," *Journal of Molecular Biology*, vol. 330, no. 4, pp. 891–913, 2003.

[70] T. Misteli, "Physiological importance of RNA and protein mobility in the cell nucleus," *Histochemistry and Cell Biology*, vol. 129, no. 1, pp. 5–11, 2008.

[71] P. G. Squire and M. E. Himmel, "Hydrodynamics and protein hydration," *Biochem Biophys*, vol. 6, no. 1, pp. 165–177, 1979.

[72] R. Peters, "Nucleo-cytoplasmic flux and intracellular mobility in single hepatocytes measured by fluorescence microphotolysis," *The EMBO Journal*, vol. 3, no. 8, pp. 1831–1836, 1984.

[73] S. Vogel, *Life's Devices: The Physical World of Animals and Plants*, Princeton University Press, Princeton, NJ, USA, 1988.

[74] J. Amundson and D. Clapham, "Calcium waves," *Current Opinion in Neurobiology*, vol. 3, no. 3, pp. 375–382, 1993.

[75] W. N. Ross, "Understanding calcium waves and sparks in central neurons," *Nature Reviews Neuroscience*, vol. 13, no. 3, pp. 157–168, 2012.

[76] M. Shahidul Islam, P. Rorsman, and P. O. Berggren, "Ca^{2+}-induced Ca^{2+} release in insulin-secreting cells," *FEBS Letters*, vol. 296, no. 3, pp. 287–291, 1992.

[77] C. Trueta, S. Sánchez-Armass, M. A. Morales, and F. F. De-Miguel, "Calcium-induced calcium release contributes to somatic secretion of serotonin in leech Retzius neurons," *Journal of Neurobiology*, vol. 61, no. 3, pp. 309–316, 2004.

10

Excess Molar Volumes and Viscosities for the Binary Mixtures of n-Octane, n-Decane, n-Dodecane, and n-Tetradecane with Octan-2-ol at 298.15 K

Arvind R. Mahajan and Sunil R. Mirgane

P. G. Department of Chemistry, Jalna Education Society's R. G. Bagdia Arts, R. Benzonji Science College, S. B. Lakhotia Commerce, Jalna, Maharashtra 431 203, India

Correspondence should be addressed to Arvind R. Mahajan; marvind22@yahoo.co.in

Academic Editor: M. A. Rosen

Experimental values of densities (ρ) and viscosities (η) in the binary mixtures of n-octane, n-decane, n-dodecane, and n-tetradecane with octan-2-ol are presented over the whole range of mixture composition at $T = 298.15$ K. From these data, excess molar volume (V_m^E), deviations in viscosity ($\Delta\eta$), and excess Gibbs free energy of activation ΔG^{*E} have been calculated. These results were fitted to Redlich-Kister polynomial equations to estimate the binary coefficients and standard errors. Jouyban-Acree model is used to correlate the experimental values of density and viscosity at $T = 298.15$ K. The values of V_m^E have been analyzed using Prigogine-Flory-Patterson (PFP) theory. The results of the viscosity composition are discussed in the light of various viscosity equations suggested by Grunberg-Nissan, Tamara and Kurata, Hind et al., Katti and Chaudhri, Heric, Heric and Brewer, and McAllister multibody model. The values of $\Delta \ln \eta$ have also been analyzed using Bloomfield and Dewan model. The experiments on the constituted binaries are analyzed to discuss the nature and strength of intermolecular interactions in these mixtures.

1. Introduction

Density and viscosity data for liquid mixtures are important from practical and theoretical points of view. Experimental measurements of these properties for binary mixtures have gained much importance in many chemical industries and engineering disciplines [1]. Experimental liquid viscosities of pure hydrocarbons and their mixtures are needed for the design of chemical processes where heat and mass transfer and fluid mechanics are important. Prediction of the liquid behavior of hydrocarbon mixture viscosities is not yet possible within the experimental uncertainty. Therefore, experimental measurements are needed to understand the fundamental behavior of this property and then to develop new models [2]. Alkanes are important series of homologous, nonpolar, and organic solvents. They have often been used in the study of solute dynamics because their physicochemical properties as a function of chain length are well-known [3]. They are also employed in a large range of chemical processes [4]. The physicochemical properties play an important role in the understanding of several industrial processes. Properties such as viscosity or surface tension are required in many empirical equations for different operations such as mass and heat transfer processes. For example, it is necessary to know the mass transfer coefficient to design gas-liquid contactors. To determine the equations that modelize the mass transfer process requires knowledge of the density, viscosity, and surface tension of the liquid phase [5].

Viscosities and excess molar volumes of binary mixtures of methylbenzene [6] and methylcyclohexane [7] with octan-2-ol at $T = 298.15$ K are only reported. To the best of our knowledge, the properties of the binary mixtures of n-octane, n-decane, n-dodecane, and n-tetradecane with octan-2-ol have not been reported earlier.

In the present paper, we report density and viscosity data for the binary mixtures of n-octane, n-decane, n-dodecane, and n-tetradecane with octan-2-ol $T = 298.15$ K. This work will also provide a test of various semiempirical equations to correlate viscosity of binary mixtures. The types of used relations are Tamara and Kurata, Heric, Heric and Brewer, Hind et al., Katti and Chaudhri, and McAllister multibody model.

2. Experimental

Chemicals used in the present study were of analytical grade and supplied by S.D. Fine Chemicals Pvt Ltd. Mumbai with quoted mass fraction purities: n-octane (>0.99), n-decane (>99.6), n-dodecane (>99.8), and n-tetradecane (>99.7). Octan-2-ol (purity >99.3) was supplied by E-Merck. Prior to use, all liquids were stored over 0.4 nm molecular sieves to reduce the water content and were degassed. The binary mixtures of varying composition were prepared by mass in special air-tight bottles. The masses were recorded on a Mettler balance to an accuracy of $\pm 1 \times 10^{-5}$ g. Care was taken to avoid evaporation and contamination during mixing. The estimated uncertainty in mole fraction was $< 1 \times 10^{-4}$.

The densities of the solutions were measured using a single capillary pycnometer made up of borosil glass with a bulb of 8 cm^3, and capillary with internal diameter of 0.1 cm was chosen for the present work. The detailed pertaining to calibration, experimental set up, and operational procedure has been previously described [6–10]. An average of triplicate measurement was taken in to account. The reproducibility of density measurement was $\pm 3 \times 10^{-5}$ g/cm^3.

The dynamic viscosities were measured using an Ubbelohde suspended level viscometer [6–10] calibrated with conductivity water. An electronic digital stop watch with readability of ±0.01 s was used for the flow time measurements. At least three repetitions of each data reproducible to ±0.05 s were obtained, and the results were averaged. Since all flow times were greater than 300 s and capillary radius (0.1 mm) was far less than its length (50 to 60) mm, the kinetic energy and end corrections, respectively, were found to be negligible. The uncertainties in dynamic viscosities are of the order of ±0.003 mPa·s.

The purity of the samples and accuracy of data were checked by comparing the measured densities and viscosities of the pure compounds with the literature values, which are given in Table 1. Thus, our results are in good agreement with those listed in the literature.

Table 1: Comparison of experimental densities (ρ), viscosities (η), and speeds of sound (u) of pure liquids with the literature values at $T = 298.15$ K.

Pure liquids	$\rho \times 10^{-3}$ (Kg m^{-3})		η (mPa·s)	
	Experimental	Literature	Experimental	Literature
n-Octane	0.69867	0.69866[a]	0.512	0.515[b]
n-Decane	0.72635	0.7267[c]	0.845	0.861[d]
n-Dodecane	0.74518	0.74172[e]	1.336	1.332[f]
n-Tetradecane	0.75913	0.75911[g]	2.081	2.077[g]
Octan-2-ol	0.81705	0.81708[h]	6.429	6.435[h]

[a][39], [b][37], [c][48], [d][42], [e][41], [f][38], [g][38], [h][6].

3. Results and Discussion

Experimental values of densities (ρ) and viscosities (η) of mixtures at 298.15 K are listed as a function of mole fraction in Table 2. The density values have been used to calculate excess molar volumes (V^E) using the following equation:

$$V_m^E/\text{m}^3 \cdot \text{mol}^{-1} = \frac{(x_1 M_1 + x_2 M_2)}{\rho_{12}} - \left(\frac{x_1 M_1}{\rho_1}\right) - \left(\frac{x_2 M_2}{\rho_2}\right), \quad (1)$$

where ρ_{12} is the densities of the mixture and x_1, M_1, ρ_1, and x_2, M_2, ρ_2 are the mole fractions, the molecular weights, and the densities of pure components 1 and 2, respectively.

The variation in excess molar volumes, V_m^E with mole fraction of the binary mixtures of n-octane, n-decane, n-dodecane, and n-tetradecane with octan-2-ol at $T = 298.15$ K, is displayed in Figure 1. The V_m^E curve for the mixture of n-octane and octan-2-ol is sigmoidal and tends to change to positive values at higher mole fractions ($x_1 \geq 0.65$) of n-octane while V_m^E values for n-decane, n-dodecane, and n-tetradecane with octan-2-ol mixtures show positive deviation over the entire composition range.

Generally, V_m^E can be considered as arising from three types of interactions between component molecules of liquid mixtures [11–13]: (1) physical interactions consisting of mainly of dispersion forces or weak dipole-dipole interaction making a +ve contribution, hereby the contraction volume and compressibility of the mixtures, (2) chemical or specific interactions, which include charge transfer, forming of H-bonds and other complex forming interactions, resulting in a −ve contribution, and (3) structural contribution due to differences in size and shape of the component molecules of the mixtures, due to fitting of component molecules into each other's structure, hereby reducing the volume and compressibility of the mixtures, resulting in a −ve contribution.

The large positive V_m^E values (Figure 1) for n-decane, n-dodecane, and n-tetradecane with octan-2-ol mixture are attributed to the breaking up of three-dimensional H-bonded network of octan-2-ol due to the addition of solute, which is not compensated by the weak interactions between unlike molecules. The V_m^E exhibits an inversion in sign in the mixtures f (n-octane + 2-octanol). The values of V_m^E are negative in the lower region of x_1 due to interstitial accommodation is more as compared to the de-clustering of 1-octanol molecules and beyond ($x_1 \geq 0.65$) as the amount of n-octane increases in the mixture, due to dispersion forces, thereby making a positive contribution to V_m^E.

The viscosity deviations ($\Delta\eta$) were calculated using

$$\Delta\eta/\text{mPa} \cdot \text{s} = \eta_{12} - x_1\eta_1 - x_2\eta_2, \quad (2)$$

where η_{12} is the viscosities of the mixture and x_1, x_2 and η_1, η_2 are the mole fraction and the viscosities of pure components 1 and 2, respectively.

Figure 2 depicts the variation of $\Delta\eta$ with mole fraction of the binary mixtures of n-octane, n-decane, n-dodecane, and n-tetradecane with octan-2-ol at $T = 298.15$ K.

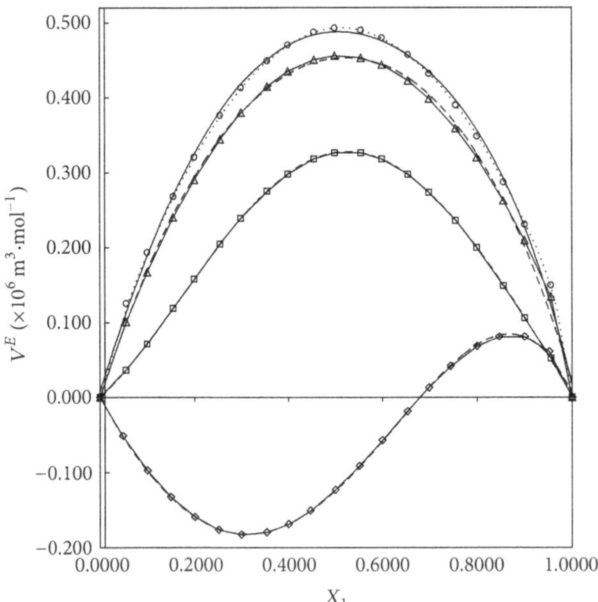

FIGURE 1: Plot of excess molar volumes (V_m^E) against mole fraction of octan-2-ol with (◊) n-octane; (□) n-decane; (△) n-dodecane, and n-tetradecane (○) at $T = 298.15$ K. The corresponding dotted (- - -) curves have been derived from PFP theory.

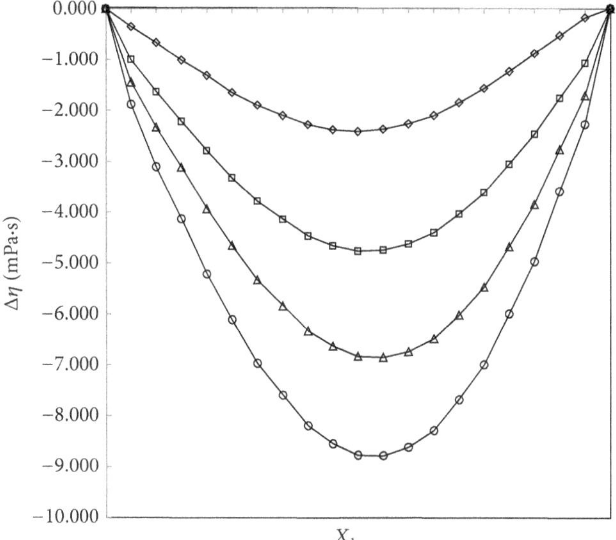

FIGURE 2: Plot of viscosity deviations (Δη) against mole fraction of octan-2-ol with (◊) n-octane; (□) n-decane; (△) n-dodecane and n-tetradecane (○) at $T = 298.15$ K.

The deviations viscosity may be generally explained by considering the following factors [14]. (1) The difference in size and shape of the component molecules and the loss of dipolar association to a decrease in viscosity; (2) specific interactions between unlike molecules such as H-bond formation and charge transfer complexes may cause increase in viscosity in mixtures rather than in pure component. The former effect produces negative in excess viscosity, and latter effect produces positive in excess viscosity. Positive values of $\Delta\eta$ are indicative of strong interactions whereas negative values indicate weaker interactions [15].

The negative deviations in viscosity support the main factor of breaking of the self-associated alcohols and weak interactions between unlike molecules. The negative values viscosity deviation decreases in the following sequence: n-octane > n-decane > n-dodecane > n-tetradecane.

Excess Gibbs free energies of activation of viscous flow ΔG^{*E} for binary mixtures can be calculated as

$$\Delta G^{*E} = RT\left[\ln\left(\frac{\eta v}{\eta_2 v_2}\right) - x_1 \ln\left(\frac{\eta_1 v_1}{\eta_2 v_2}\right)\right], \quad (3)$$

where v is the molar volume of the mixture, v_i is the molar volume of the pure component, R is the gas constant, T is the absolute temperature, and η is the dynamic viscosity of the mixture, respectively. η_i is the dynamic viscosity of the pure component i and x_1 the mole fraction in component. The ΔG^{*E} values of all binary systems are shown in Table 2. The values of ΔG^{*E} for all binary mixtures are negative over entire mole fraction. According to Meyer et al. [16], negative values of ΔG^{*E} correspond to the existence of solute-solute association.

The excess molar volumes and deviations in viscosity were fitted to Redlich and Kister [17] equation of the type

$$Y = x_1 x_2 \sum_i^n a_i (x_1 - x_2)^i, \quad (4)$$

where Y is either V^E or $\Delta\eta$ and n is the degree of polynomial. Coefficient a_i was obtained by fitting (5) to experimental results using a least-squares regression method. In each case, the optimum number of coefficients is ascertained from an examination of the variation in standard deviation (σ).

σ was calculated using the relation

$$\sigma(Y) = \left[\frac{\sum\left(Y_{\text{expt}} - Y_{\text{calc}}\right)^2}{N-n}\right]^{1/2}, \quad (5)$$

where N is the number of data points and n is the number of coefficients. The calculated values of the coefficients a_i along with the standard deviations (σ) are given in Table 3.

4. Theoretical Analysis

4.1. Semiempirical Models for Analyzing Viscosity of Liquid Mixtures. Several empirical and semiempirical relations have been used to represent the dependence of viscosity on concentration of components in binary liquid mixtures, and these are classified according to the number of adjustable parameters used to account for the deviation from some average [18, 19]. We will consider here some of the most commonly used semiempirical models for analyzing viscosity of liquid mixtures based on one, two, and three parameters. An attempt has been made to check the suitability of equations for experimental data fits by taking into account the number of empirical adjustment coefficients.

TABLE 2: Densities (ρ), viscosities (η), excess molar volumes V^E, viscosities deviations $\Delta\eta$, and excess Gibbs free energy ΔG^{*E} of binary mixtures at T = 298.15 K.

x_1	$\rho \times 10^{-3}$ kg·m^{-3}	$V_m^E \times 10^6$ m^3·mol^{-1}	η mPa·s	$\Delta\eta$ mPa·s	ΔG^{*E} KJmol^{-1}
		n-Octane(1) + octan-2-ol(2)			
0	0.81705	0	6.429	0	0
0.0487	0.8114	−0.051	5.956	−0.185	115
0.1002	0.8054	−0.097	5.301	−0.535	149
0.1498	0.79959	−0.133	4.654	−0.888	137
0.2006	0.79361	−0.16	3.998	−1.244	78
0.2516	0.78756	−0.177	3.364	−1.576	−30
0.3003	0.78176	−0.183	2.796	−1.856	−183
0.3538	0.77534	−0.18	2.228	−2.108	−410
0.4003	0.76975	−0.169	1.79	−2.27	−661
0.4466	0.76415	−0.151	1.411	−2.375	−960
0.5001	0.75767	−0.123	1.049	−2.421	−1359
0.5517	0.75142	−0.091	0.778	−2.387	−1776
0.6002	0.74554	−0.057	0.59	−2.287	−2157
0.6539	0.73905	−0.018	0.454	−2.106	−2469
0.6994	0.73357	0.013	0.39	−1.901	−2560
0.7449	0.72811	0.042	0.366	−1.656	−2431
0.8	0.72157	0.068	0.376	−1.32	−2019
0.8465	0.7161	0.081	0.402	−1.018	−1571
0.9003	0.70985	0.081	0.434	−0.668	−1034
0.9521	0.70394	0.061	0.441	−0.355	−670
1	0.69867	0	0.512	0	0
		n-Decane(1) + octan-2-ol(2)			
0	0.81705	0	6.429	0	0
0.0554	0.81077	0.036	5.222	−0.898	−233
0.0998	0.80583	0.071	4.638	−1.234	−301
0.1554	0.79974	0.119	3.972	−1.589	−403
0.1999	0.79499	0.158	3.491	−1.822	−497
0.2554	0.7892	0.205	2.951	−2.052	−632
0.2999	0.78467	0.239	2.564	−2.19	−755
0.3555	0.77916	0.275	2.136	−2.308	−927
0.3998	0.7749	0.298	1.836	−2.361	−1078
0.4554	0.76969	0.318	1.509	−2.377	−1285
0.4999	0.76564	0.326	1.284	−2.354	−1461
0.5555	0.76074	0.326	1.046	−2.281	−1690
0.5998	0.75694	0.318	0.889	−2.191	−1871
0.6555	0.7523	0.298	0.73	−2.039	−2080
0.7	0.7487	0.274	0.63	−1.89	−2223
0.7554	0.74432	0.236	0.538	−1.673	−2337
0.7998	0.74091	0.2	0.487	−1.476	−2362
0.8554	0.73674	0.149	0.45	−1.203	−2281
0.8999	0.73347	0.106	0.438	−0.966	−2126
0.9554	0.72948	0.052	0.445	−0.649	−1811
1	0.72635	0	0.845	0	0
		n-Dodecane(1) + octan-2-ol(2)			
0	0.81705	0	6.429	0	0
0.0554	0.81098	0.101	5.513	−0.634	−159
0.0999	0.80637	0.167	4.916	−1.004	−296
0.1554	0.80091	0.24	4.254	−1.383	−404

TABLE 2: Continued.

x_1	$\rho \times 10^{-3}$ kg·m^{-3}	$V_m^E \times 10^6$ m^3·mol^{-1}	η mPa·s	$\Delta\eta$ mPa·s	ΔG^{*E} KJmol^{-1}
0.1998	0.79676	0.291	3.786	−1.625	−517
0.2555	0.7918	0.345	3.272	−1.856	−659
0.2998	0.78805	0.38	2.916	−1.986	−770
0.3554	0.78355	0.415	2.532	−2.087	−902
0.3998	0.78012	0.435	2.27	−2.123	−999
0.4554	0.77603	0.451	1.995	−2.115	−1102
0.4999	0.7729	0.456	1.813	−2.071	−1167
0.5555	0.76916	0.453	1.627	−1.973	−1219
0.5998	0.76631	0.444	1.508	−1.866	−1236
0.6555	0.76288	0.422	1.391	−1.699	−1221
0.6998	0.76027	0.398	1.32	−1.545	−1180
0.7554	0.75713	0.359	1.253	−1.328	−1095
0.7998	0.75473	0.32	1.213	−1.142	−1005
0.8555	0.75185	0.263	1.176	−0.896	−869
0.8999	0.74965	0.21	1.152	−0.694	−750
0.9555	0.747	0.134	1.124	−0.439	−599
0.9997	0.74519	0	1.337	0	0
		n-Tetradecane(1) + octan-2-ol(2)			
0	0.81705	0	6.429	0	0
0.0555	0.81134	0.126	5.619	−0.568	−144
0.0999	0.80721	0.193	5.163	−0.831	−230
0.1555	0.80239	0.268	4.632	−1.121	−330
0.1998	0.7988	0.321	4.238	−1.321	−417
0.2554	0.79459	0.376	3.786	−1.533	−531
0.2998	0.79144	0.413	3.455	−1.67	−628
0.3555	0.78774	0.449	3.081	−1.802	−751
0.3999	0.78496	0.47	2.813	−1.876	−850
0.4554	0.78171	0.487	2.519	−1.93	−968
0.4999	0.77925	0.492	2.314	−1.941	−1054
0.5555	0.77637	0.49	2.098	−1.915	−1144
0.5998	0.7742	0.48	1.957	−1.863	−1196
0.6554	0.77163	0.457	1.82	−1.759	−1226
0.7	0.76969	0.432	1.741	−1.644	−1217
0.7555	0.76741	0.39	1.683	−1.46	−1154
0.8	0.76569	0.348	1.668	−1.282	−1060
0.8555	0.76365	0.287	1.688	−1.02	−887
0.9	0.76211	0.23	1.736	−0.779	−703
0.9553	0.76029	0.15	1.835	−0.439	−424
0.9997	0.75914	0	2.081	0	0

The equation of Grunberg-Nissan, Tamara and Kurata Hind et al., and Katti and Chaudhri has one adjustable parameter.

Gruenberg-Nissan provided the following empirical equation containing one adjustable parameter [20]. The equation is

$$\ln \eta_{12} = x_1 \ln \eta_1 + x_2 \ln \eta_2 + x_1 x_2 G_{12}, \quad (6)$$

where G_{12} may be regarded as a parameter proportional to the interchange energy also an approximate measure of the strength of the interaction between the components.

The one-parameter equation due to Tamura and Kurata [21] gave the equation of the form

$$\eta_m = x_1 \eta_1 \Phi + x_2 \eta_2 \Phi_2 + 2(x_1 x_2 \Phi \Phi_2)^{1/2} T_{12}, \quad (7)$$

where Φ and Φ_2 are the volume fractions of components 1 and 2, respectively; T_{12} is Tamura and Kurata constant.

TABLE 3: Coefficients a_i of (4) and corresponding standard deviation (σ) of (5) at $T = 298.15$ K.

System	a_0	a_1	a_2	a_3	σ
n-Octane + octan-2-ol					
$V_m^E \times 10^6/(\text{m}^3\cdot\text{mol}^{-1})$	0.125	0.1577	−0.41	0.1523	0.943
$\Delta\eta/(\text{mPa}\cdot\text{s})$	−9.6807	0.4921	4.661	−2.7989	0.024
n-Decane + octan-2-ol					
$V_m^E \times 10^6/(\text{m}^3\cdot\text{mol}^{-1})$	0.129	0.167	−0.462	0.1437	0.0023
$\Delta\eta/(\text{mPa}\cdot\text{s})$	−8.4704	2.213	−6.8274	−1.0134	0.1273
n-Dodecane + octan-2-ol					
$V_m^E \times 10^6/(\text{m}^3\cdot\text{mol}^{-1})$	0.363	−1.276	0.289	0.812	0.529
$\Delta\eta/(\text{mPa}\cdot\text{s})$	−7.989	3.365	−2.905	8.812	0.060
n-Tetradecane + octan-2-ol					
$V_m^E \times 10^6/(\text{m}^3\cdot\text{mol}^{-1})$	0.880	−0.116	0.009	0.708	0.020
$\Delta\eta/(\text{mPa}\cdot\text{s})$	−6.322	−4.730	−10.25	15.66	0.377

Hind et al. [22] proposed the following equation:

$$\eta_m = x_1^2\eta_1 + x_2^2\eta_2 + 2x_1x_2H_{12}, \tag{8}$$

where H_{12} is attributed to unlike pair interactions.

Katti and Chaudhri [23] derived the following equation

$$\ln \eta V = x_1 \ln \eta V_1^0 + x_2 \ln \eta V_2^0 + \frac{x_1x_2W_{\text{vis}}}{RT}, \tag{9}$$

where W_{vis} is an interaction term and v_i is the molar volume of pure component i.

Heric [24] expression is

$$\begin{aligned}\ln \eta_m = {} & x_1 \ln \eta_1 + x_2 \ln \eta_2 + x_1 \ln \eta_1 \\ & + x_2 \ln \eta_2 + \ln (x_1\eta_1 + x_2\eta_2) + \delta_{12},\end{aligned} \tag{10}$$

where $\delta_{12} = \alpha_{12}x_1x_2$ is a term representing departure from a noninteracting system and $\alpha_{12} = \alpha_{21}$ is the interaction parameter. Either α_{12} or α_{21} can be expressed as a linear function of composition:

$$\alpha_{12} = \beta'_{12} + \beta''_{12}\,(x_1 - x_2). \tag{11}$$

From an initial guess of the values of coefficients β'_{12} and β''_{12}, the values of α_{12} are computed.

Heric and Brewer [25] equation is

$$\begin{aligned}\ln \nu = {} & x_1 \ln \nu_1 + x_2 \ln \nu_2 + x_1 \ln M_1 + x_2 \ln M_2 \\ & - \ln [x_1M_1 + x_2M_2] + x_1x_2 [\alpha_{12} + \alpha_{21}(x_1 - x_2)].\end{aligned} \tag{12}$$

M_1 and M_2 are molecular weights of components 1 and 2, and α_{12} and α_{21} are interaction parameters, and other terms involved have their usual meaning. α_{12} and α_{21} are parameters, which can be calculated from the least-squares method.

McAllister's multibody interaction model [26] was widely used to correlate kinematic viscosity (ν) data. The two-parameter McAllister equation based on Eyring's [27] theory of absolute reaction rates has taken into account interaction of both like and unlike molecules by two-dimensional three-body model. The three-body interaction model is

$$\begin{aligned}\ln \nu_m = {} & x_1^3 \ln \nu_1 + 3x_1^2x_2 \ln Z_{12} + 3x_1x_2^2 \ln Z_{21} \\ & + x_2^3 \ln \nu_2 - \ln \left[x_1 + \frac{(x_2M_2)}{M_1}\right] \\ & + 3x_1^2x_2 \ln \left[\frac{2}{3} + \frac{M_2}{(3M_1)}\right] \\ & + 3x_1x_2^2 \ln \left[\frac{1}{3} + \frac{2M_2}{(3M_1)}\right] x_2^3 \ln \left(\frac{M_2}{M_1}\right).\end{aligned} \tag{13}$$

And four-body model was given by

$$\begin{aligned}\ln \nu_m = {} & x_1^4 \ln \nu_1 + 4x_1^3x_2 \ln Z_{1112} \\ & + 6x_1^2x_2^2 \ln Z_{1122} + 4x_1x_2^3 \ln Z_{2221} + x_2^4 \ln \nu_2 \\ & - \ln \left[x_1 + x_2\left(\frac{M_2}{M_1}\right)\right] \\ & + 4x_1^3x_2 \ln \left[3 + \frac{(M_2/M_1)}{4}\right] \\ & + 6x_1^2x_2^2 \ln \left[1 + \frac{(M_2/M_1)}{2}\right] \\ & + 4x_1x_2^3 \ln \left[1 + \frac{(3M_2/M_1)}{4}\right] + x_2^4 \ln \left(\frac{M_2}{M_1}\right),\end{aligned} \tag{14}$$

where Z_{12}, Z_{21}, Z_{1112}, Z_{1122}, and Z_{2221} are interaction parameters and M_i and ν_i are the molecular mass and kinematic viscosity of pure component i, respectively.

The correlating ability of each of (6)–(14) was tested as well as their adjustable parameters and standard deviations (σ):

$$\sigma\,(\%) = \left[\left(\frac{1}{(n-k)}\sum \frac{100\left(\eta_{\text{exptl}} - \eta_{\text{calcd}}\right)}{\eta_{\text{exptl}}}\right)^2\right]^{1/2}, \tag{15}$$

where n represents the number of data points and k is the number of numerical coefficients given in Table 4. The interaction parameter G_{12} is negative for binary systems. Nigam and Mahl [28] concluded from the study of binary mixtures that (1) if $\Delta\eta > 0$, $G_{12} > 0$ and magnitude of both are large then strong specific interaction would be present; (2) if $\Delta\eta < 0$, $G_{12} > 0$ then weak specific interaction would be present; (3) if $\Delta\eta < 0$, $G_{12} < 0$ magnitude of both are large then the dispersion force would be dominant. According to Fort and Moore [29] and Ramamoorty and Varadachari [30], system exhibits strong interaction if the G_{12} is positive; if it is negative they show weak interaction. On this basis, we can say that there is a weak interaction in the system studied.

Interaction parameter W_{vis} shows almost the same trend as that of G_{12}. In fact, one could say that the parameters

TABLE 4: Adjustable parameters of (6)–(13) and standard deviations of binary mixture viscosities for x_1 n-alkanes + $(1 - x_1)$ octan-2-ol at $T = 298.15$ K.

Equation	System including n-alkanes + octan-2-ol			
	n-Octane	n-Decane	n-Dodecane	n-Tetradecane
		Grunberg-Nissan		
G_{12}	−2.398	−3.226	−1.997	−2.002
σ	3.759	3.904	0.858	1.075
		Tamura and Kurata		
T_{12}	0.872	−1.117	0.184	0.675
σ	1.737	1.884	0.812	0.782
		Hind et al.		
H_{12}	−0.91	−1.409	−0.401	0.079
σ	1.758	2.145	1.172	0.711
		Katti and Chaudhri		
W_{vis}	−2.402	−3.198	−1.96	−1.871
σ	3.755	3.905	0.857	1.081
		Heric and Brewer		
α_{12}	−2.381	−3.936	−2.109	−2.104
α_{21}	−4.123	−5.529	−1.506	−0.827
σ	−0.926	2.449	−0.559	0.866
		McAllister's three-body		
Z_{12}	0.185	0.094	0.879	1.439
Z_{21}	6.363	7.068	3.929	3.544
σ	0.926	2.449	0.559	0.866
		McAllister's four-body		
Z_{1112}	0.269	0.51	0.817	1.140
Z_{1122}	0.936	19.584	3.639	7.334
Z_{2221}	6.926	1.998	3.598	2.871
σ	5.149	5.865	0.622	1.622

G_{12} and W_{vis} exhibit almost similar behaviour, which is not unlikely in view of logarithmic nature of both equations.

Tamara and Kurata and Hind et al. represent the binary mixture satisfactory as compared to Gruenberg-Nissan and Katti and Chaudhri. Use of three parameters equation reduces the σ values significantly below that of two parameters equation. From this study, it can be concluded that the correlating ability significantly improves for these nonideal systems as number of adjustable parameters is increased. From Table 4, it is clear that McAllister's three-body interaction model is suitable for correlating the kinematic viscosities of the binary mixtures studied.

4.2. Prigogine-Flory-Patterson (PFP) Theory. The Prigogine-Flory-Patterson (PFP) theory [31–34] has been commonly employed to estimate and analyze excess thermodynamic functions theoretically. This theory has been described in details by Patterson and coworkers [35, 36]. According to PFP theory, V_m^E can be separated into three factors: (1) an interactional contribution, V_m^E (int.) (2) a free volume contribution, V_m^E(fv), and (3) an internal pressure contribution, $V_m^E(P^*)$. The expression for these three contributions are given as V_m^E:

$$V_m^E(\text{int}) = \left[\frac{\left(v^{1/3}-1\right)v^{2/3}\psi_1\theta_2}{\left(4/3v^{-1/3}-1\right)P_1^*\chi_{12}}\right],$$

$$V_m^E(\text{fv}) = \frac{\left[\left(v_1-v_2\right)^2\left(14/9v^{-1/3}-1\right)\psi_1\psi_2\right]}{\left[\left(14/9v^{-1/3}-1\right)v\right]}, \quad (16)$$

$$V_m^E(P^*) = \frac{\left[\left(v_1-v_2\right)\left(P_1^*+P_2^*\right)\psi_1\psi_2\right]}{\left(P_1^*\psi_1+P_2^*\psi_2\right)}.$$

Thus, the excess molar volume V_m^E is given as

$$\frac{V_m^E}{\left(x_1V_1+x_2V_2\right)} = V_m^E(\text{int}) - V_m^E(\text{fv}) + V_m^E(P^*), \quad (17)$$

where ψ, θ, and P^* represent the contact energy fraction, surface site fraction, and characteristic pressure, respectively, and are calculated as

$$\psi_1 = \left(1-\psi_2\right) = \frac{\Phi_1P_1^*}{\left(\Phi_1P_1^*+\Phi_2P_2^*\right)}, \quad (18)$$

$$\theta_2 = \left(1-\theta_1\right) = \frac{\Phi_2}{\left[\Phi_1\left(V_2^*/V_1^*\right)\right]}, \quad (19)$$

$$P^* = \frac{Tv^2\alpha}{\kappa_T}. \quad (20)$$

The details of the notations and terms used in (16)–(19) may be obtained from the literature [31–34, 37, 38]. The other parameters pertaining to pure liquids and the mixtures are obtained from the Flory theory [7, 31, 38] while α and κ_T values are taken from the literature [39–44]. The interaction parameter χ_{12} required for the calculation of V_m^E using PFP theory has been derived by fitting the V_m^E expression to the experimental equimolar value of V_m^E for each system under study.

The values of χ_{12}, θ_2, three PFP contributions interactional, free volume, P^* effect, and experimental and calculated (using PFP theory) V_m^E values at near equimolar composition are presented in Table 5. Study of the data presented in Table 6 reveals that the interactional and free volume contributions are positive, whereas P^* contributions are negative for all the three systems under investigation. For these binary mixtures, it is only the interactional contribution which dominates over the remaining two contributions. The P^* contribution, which depends both on the differences of internal pressures and differences of reduced volumes of the components, has little significance for the studied binary mixtures as compared to the other two.

Furthermore, in order to check whether χ_{12}, derived from nearly equimolar V_m^E values, can predict the correct composition dependence, V_m^E has been calculated theoretically using χ_{12} over the entire composition range. The theoretically calculated values are plotted in Figure 1 for comparison with the experimental results. Figure 1 show that the PFP theory is

Table 5: Flory parameters of the pure compounds.

Components	$10^6 V^*/(\text{m}^3\cdot\text{mol}^{-1})$	$10^6 P^*$ (J·m^{-3})	T^*/K
Octan-2-ol	129.9790	535	5563
n-Octane	127.4844	444	4826
n-Decane	155.6091	453	5091
n-Dodecane	183.7700	455	5290
n-Tetradecane	212.1200	460	5479

quite successful in predicting the trend of the dependence of V_m^E on composition for the present systems.

In order to perform a numerical comparison of the estimation capability of the PFP theory, we calculated the standard percentage deviations (σ%) using the relation

$$\sigma\% = \left[\sum\left\{\frac{100\,(\text{expt} - \text{theor.})}{\text{expt}}\right\}^2 (n-1)\right]^{1/2}, \quad (21)$$

where n represents the number of experimental data points.

4.3. Bloomfi ld and Dewan Model. There are different expressions available in the literature to calculate η. Here, Bloomfield and Dewan [45] model have been applied to compare calculated $\Delta\ln\eta$ values using experimental data for each binary mixture by the following relation:

$$\Delta\ln\eta = \ln\eta - (x_1\ln\eta_1 + x_2\ln\eta_2). \quad (22)$$

Bloomfield and Dewan [45] developed the expression from the combination of the theories of free volumes and absolute reaction rate

$$\Delta\ln\eta = f(v) - \frac{\Delta G^R}{RT}, \quad (23)$$

where $f(v)$ is the characteristic function of the free volume defined by

$$f(v) = \frac{1/(v-1)\,x_1}{(v_1-1)} - \frac{x_2}{v_2-1} \quad (24)$$

and ΔG^R is the residual energy of mixing, calculated with the following expression [46]:

$$\Delta G^R = \Delta G^E + RT\left\{x_1\ln\left(\frac{x_1}{\Phi_1}\right) + x_2\ln\left(\frac{x_2}{\Phi_2}\right)\right\}, \quad (25)$$

where Φ_1 and Φ_2 are segment fractions defined by

$$\Phi_2 = 1 - \Phi_1 = \frac{x_2}{[x_2 + x_1(r_1/r_2)]}, \quad (26)$$

where r_1 and r_2 are in the ratio of respective molar core volumes V_1^* and V_2^*.

The excess energy can be obtained from the statistical theory of liquid mixtures proposed by Flory and coworkers [31, 32] and is given by

$$\begin{aligned}\Delta G^E = {} & x_1P_1^*V_1^*\left[\frac{1}{(v_1)} - \frac{1}{(v)}\right] \\ & + 3\tilde{T}_1\ln\left\{\frac{(v_1^{1/3}-1)}{(v^{1/3}-1)}\right\} \\ & + x_2P_2^*V_2^*\left[\frac{1}{(v_2)} - \frac{1}{(v)}\right] \\ & + \left[3\tilde{T}_2\ln\left\{\frac{(v_2^{1/3}-1)}{(v^{1/3}-1)}\right\}\right] \\ & + \frac{(x_1\theta_2V_1^*\chi_{12})}{v},\end{aligned} \quad (27)$$

where the various symbols used have their usual meanings.

Using the χ_{12} values from fitting values of V_m^E and the values of the parameters for the pure liquid components, we have calculated ΔG^R and $f(v)$ and finally $\Delta\ln\eta$, according to the Bloomfield and Dewan relationship, which is compared with the experimental data. Figure 3 shows that the good agreement between the estimated and experimental curves occurs for given binary systems.

4.4. Jouyban and Acree Model. Hasan et al. [6, 47, 48] has used this model for various binary system proposed a model for correlating the density and viscosity of liquid mixtures at various temperatures. The proposed equation is

$$\begin{aligned}\ln y_{m,T} = {} & f_1\ln y_{1,T} \\ & + f_2\ln y_{2,T} + f_1f_2\sum\left[\frac{A_j(f_1-f_2)^j}{T}\right],\end{aligned} \quad (28)$$

where $y_{m,T}$, $y_{1,T}$, and $y_{2,T}$ is density or viscosity of the mixture and solvents 1 and 2 at temperature T, respectively, f_1 and f_2 are the volume fractions of solvents in case of density and mole fraction in case of viscosity, and A_j are the model constants.

The correlating ability of the Jouyban-Acree model was tested by calculating the average percentage deviation (APD) between the experimental and calculated density and viscosity as

$$\text{APD} = \left(\frac{100}{N}\right)\sum\left[\frac{(|y_{\text{exptl}} - y_{\text{calcd}}|)}{y_{\text{exptl}}}\right], \quad (29)$$

where N is the number of data points in each set. The optimum numbers of constants A_j, in each case, were determined from the examination of the average percentage deviation value.

TABLE 6: Calculated values of the three contributions to the excess molar volume from the PFP theory with interaction parameter $T = 298.15$ K.

Component	$\chi_{12} \times 10^6$ (J·m^{-3})	$V^E \times 10^6$ (m^3·mol^{-1}) at $x = 0.5$		Calculated contribution		
		Experimental	PEP	$V^E \times 10^6$ (int)	$V^E \times 10^6$ (fv)	$V^E \times 10^6$ (P^*)
n-Octane + octan-2-ol	11.59	0.123	0.124	0.0331	0.1087	−0.4194
n-Decane + octan-2-ol	12.57	0.326	0.328	0.5999	0.0644	−0.2283
n-Dodecane + octan-2-ol	16.57	0.456	0.458	0.5524	0.0175	−0.0666
n-Ttetradecane + octan-2-ol	15.08	0.492	0.494	0.5157	0.0116	−0.0187

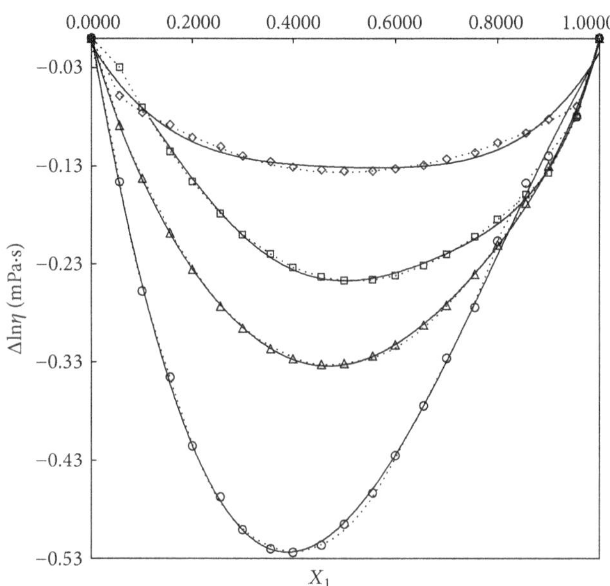

FIGURE 3: Plot of $\Delta \ln \eta$ against mole fraction of octan-2-ol with (◊) n-octane; (□) n-decane; (△) n-dodecane, and n-tetradecane (○) at $T = 298.15$ K. The corresponding dotted (- - -) curves have been derived from Bloomfield and Dewan model.

The constants A_j calculated from the least square analysis are presented in Table 7 along with the average percentage deviation (APD). The proposed model provides reasonably accurate calculations for the density, viscosity, and ultrasonic velocity of binary liquid mixtures at 298.15 K, and the model could be used in data modeling.

5. Conclusions

The present paper is a continuing effort towards the understanding of the mixing behavior of binary liquid mixtures comprising of n-alkanes + octan-2-ol. Excess molar volumes and deviations in viscosity were calculated and fitted to the Redlich-Kister equation to test the quality of the experimental values. The negative deviations in viscosity support the main factor of breaking of the self-associated alcohols and weak interactions between unlike molecules. Several empirical and semiempirical relations have been used to represent the dependence of viscosity on concentration of components in binary liquid mixtures. An attempt has been made to check the suitability of empirical and semiempirical relations for experimental viscosities data of n-alkanes + octan-2-ol fits by taking into account the number of empirical adjustment coefficients. Bloomfield and Dewan model and McAllister's three-body interaction model are suitable for the above binary system. PFP theory is also quite successful in predicting the trend of the dependence of V_m^E on composition for the present systems.

TABLE 7: Parameters of Jouyban-Acree model and average percentage deviation for densities, viscosities, and ultrasonic velocities at of binary mixtures at $T = 298.15$ K.

System	A_0	A_1	A_2	APD
Density				
n-Octane + octan-2-ol	3.7446	−1.8027	0.616	0.0293
n-Decane + octan-2-ol	−9.0926	1.0266	0.6448	0.1183
n-Dodecane + octan-2-ol	−11.8504	0.0401	0.4771	0.0816
n-Tetradecane + octan-2-ol	−12.6833	3.5123	−4.3389	0.0559
Viscosity				
n-Octane + octan-2-ol	−7.1847	−6.9259	3.3215	0.1423
n-Decane + octan-2-ol	−4.2926	11.2857	−2.7614	0.7668
n-Dodecane + octan-2-ol	−7.5692	−10.9008	4.4397	0.1031
n-Tetradecane + octan-2-ol	−6.4032	−26.9478	−7.7829	0.4296

Acknowledgments

Authors are thankful to Prof B. R. Arbad, Dr. B. A. M. university for their valuable suggestions and discussion. Authors are also thankful to Principal Dr. R. S. Agarwal, J. E. S. College, Jalna for the facilities provided.

References

[1] K. Lal, N. Tripathi, and G. P. Dubey, "Densities, viscosities, and refractive indices of binary liquid mixtures of hexane, decane, hexadecane, and squalane with benzene at 298.15 K," *Journal of Chemical and Engineering Data*, vol. 45, no. 5, pp. 961–964, 2000.

[2] J. A. Estrada-Baltazar, F. J. Juan, and A. Gustavo, "Experimental liquid viscosities of decane and octane + decane from 298.15 K to 373.15 K and up to 25 MPa," *Journal of Chemical and Engineering Data*, vol. 43, no. 3, pp. 441–446, 1998.

[3] Y. Zhang, R. M. Venable, and R. W. Pastor, "Molecular dynamics simulations of neat alkanes: the viscosity dependence of rotational relaxation," *Journal of Physical Chemistry*, vol. 100, no. 7, pp. 2652–2660, 1996.

[4] T. M. Aminabhavi, M. I. Aralaguppi, B. Gopalakrishna, and R. S. Khinnavar, "Densities, shear viscosities, refractive indices, and speeds of sound of bis(2-methoxyethyl) ether with hexane, heptane, octane, and 2,2,4-trimethylpentane in the temperature

interval 298.15-318.15 K," *Journal of Chemical and Engineering Data*, vol. 39, no. 3, pp. 522–528, 1994.

[5] D. Gómez-Díz, J. C. Mejuto, J. M. Navaza, and A. Rodríguez-Álvarez, "Viscosities, densities, surface tensions, and refractive indexes of 2,2,4-trimethylpentane + cyclohexane + decane ternary liquid systems at 298.15 K," *Journal of Chemical and Engineering Data*, vol. 47, no. 4, pp. 872–875, 2002.

[6] M. Hasan, D. F. Shirude, A. P. Hiray, A. B. Sawant, and U. B. Kadam, "Densities, viscosities and ultrasonic velocities of binary mixtures of methylbenzene with hexan-2-ol, heptan-2-ol and octan-2-ol at T = 298.15 and 308.15 K," *Fluid Phase Equilibria*, vol. 252, no. 1-2, pp. 88–95, 2007.

[7] H. Iloukhani, B. Samiey, and M. A. Moghaddasi, "Speeds of sound, isentropic compressibilities, viscosities and excess molar volumes of binary mixtures of methylcyclohexane + 2-alkanols or ethanol at T = 298.15 K," *Journal of Chemical Th rmodynamics*, vol. 38, no. 2, pp. 190–200, 2006.

[8] A. Pal and R. K. Bhardwaj, "Excess molar volumes and viscosities of binary mixtures of diethylene glycol dibutyl ether with chloroalkanes at 298.15 K," *Indian Journal of Chemistry A*, vol. 41, no. 4, pp. 706–711, 2002.

[9] A. R. Mahajan, S. R. Mirgane, and S. B. Deshmukh, "Volumetric, Viscometric and Ultrasonic studies of some amino acids in aqueous 0. 2M. $LiClO_4$. $3H_20$ solutions at 298.15 K," *Material Science Research India*, vol. 4, no. 2, pp. 345–352, 2007.

[10] A. R. Mahajan, S. R. Mirgane, and S. B. Deshmukh, "Ultrasonic studies of some amino acids in aqueous 0.02M. $LiCl0_4$. $3H_20$ solutions at 298.15 K," *Material Science Research India*, vol. 4, no. 2, pp. 373–378, 2007.

[11] A. J. Treszczanowicz and G. C. Benson, "Excess volumes for *n*-alkanols + *n*-alkanes II. Binary mixtures of n-pentanol, *n*-hexanol, *n*-octanol, and *n*-decanol + *n*-heptane," *Th Journal of Chemical The modynamics*, vol. 10, no. 10, pp. 967–974, 1978.

[12] A. Ali, A. K. Nain, V. K. Sharma, and S. Ahmad, "Molecular interactions in binary mixtures of tetrahydrofuran with alkanols (C6,C8,c10): an ultrasonic and volumetric study," *Indian Journal of Pure and Applied Physics*, vol. 42, no. 9, pp. 666–673, 2004.

[13] H. Iloukhani, M. Rezaei-Sameti, and J. Basiri-Parsa, "Excess molar volumes and dynamic viscosities for binary mixtures of toluene + n-alkanes (C_5 − C_{10}) at T = 298.15 K—Comparison with Prigogine-Flory-Patterson theory," *Journal of Chemical Th rmodynamics*, vol. 38, no. 8, pp. 975–982, 2006.

[14] R. Mehra and M. Pancholi, "Temperature-dependent studies of thermo-acoustic parameters in hexane + 1-dodecanol and application of various theories of sound speed," *Indian Journal of Physics*, vol. 80, no. 3, pp. 253–263, 2006.

[15] C. Yang, W. Xu, and P. Ma, "Thermodynamic properties of binary mixtures of *p*-xylene with cyclohexane, heptane, octane, and *N*-methyl-2-pyrrolidone at several temperatures," *Journal of Chemical and Engineering Data*, vol. 49, no. 6, pp. 1794–1801, 2004.

[16] R. Meyer, M. Meyer, J. Metzger, and A. Peneloux, "Thermodynamic and physicochemical properties of binary solvent," *Journal de Chimie Physique et de Physico-Chimie Biologique*, vol. 68, pp. 406–412, 1971.

[17] O. Redlich and A. T. Kister, "Thermodynamics of nonelectrolyte solutions. Algebraic representation of thermodynamic properties and the classification of solutions," *Industrial & Engineering Chemistry*, vol. 40, pp. 345–348, 1948.

[18] J. B. Irving, "Viscosity of binary liquid mixtures, a survey of mixture equations," NEL Report 630, National Eng Lab, East Kilbride, UK, 1977.

[19] J. B. Irving, "The effectiveness of mixture equations," NEL Report 631, National Eng Lab, East Kilbride, UK, 1977.

[20] L. Grunberg and A. H. Nissan, "Vaporisation, viscosity, cohesion and structure of the liquids," *Nature*, vol. 164, pp. 799–800, 1949.

[21] M. Tamura and M. Kurata, "Viscosity of binary mixture of liquids," *Bulletin of the Chemical Society of Japan*, vol. 25, pp. 32–37, 1952.

[22] R. K. Hind, E. McLaughlin, and A. R. Ubbelohde, "Structure and viscosity of liquids camphor+pyrene mixtures," *Transactions of the Faraday Society*, vol. 56, pp. 328–330, 1960.

[23] P. K. Katti and M. M. Chaudhri, "Viscosities of binary mixtures of benzyl acetate with dioxane, aniline, and m-cresol," *Journal of Chemical and Engineering Data*, vol. 9, no. 3, pp. 442–443, 1964.

[24] E. L. Heric, "On the viscosity of ternary mixtures," *Journal of Chemical and Engineering Data*, vol. 11, no. 1, pp. 66–68, 1966.

[25] E. L. Heric and J. G. Brewer, "Viscosity of some binary liquid nonelectrolyte mixtures," *Journal of Chemical and Engineering*, vol. 12, no. 4, pp. 574–583, 1967.

[26] R. A. McAllister, "The viscosities of lquid mixtures," *AIChE Journal*, vol. 6, pp. 427–431, 1960.

[27] S. Glasstone, K. J. Laidler, and H. Eyring, *Th Theo y of Rate Process*, McGraw-Hill, New York, NY, USA, 1941.

[28] R. K. Nigam and B. S. Mahl, "Molecular interaction in binary liquid mixtures of dimethylslphoxide with chlroehanes & chlroehenes," *Indian Journal of Chemistry*, vol. 9, p. 1255, 1971.

[29] R. J. Fort and W. R. Moore, "Viscosities of binary liquid mixtures," *Transactions of the Faraday Society*, vol. 62, pp. 1112–1119, 1966.

[30] K. Ramamoorty and P. S. Varadachari, "Study of some binary liquid mixtures," *Indian Journal of Pure and Applied Physics*, vol. 11, p. 238, 1973.

[31] P. J. Flory, "Statistical thermodynamics of liquid mixtures," *Journal of the American Chemical Society*, vol. 87, no. 9, pp. 1833–1838, 1965.

[32] A. Abe and P. J. Flory, "The thermodynamic properties of mixtures of small, nonpolar molecules," *Journal of the American Chemical Society*, vol. 87, no. 9, pp. 1838–1846, 1965.

[33] I. Prigogine, *Molecular Theories of Solutions*, North-Holland, Amsterdam, The Netherlands, 1957.

[34] D. Patterson and G. Delmas, "Corresponding states theories and liquid models," *Discussions of the Faraday Society*, vol. 49, pp. 98–105, 1970.

[35] P. Tancrède, P. Bothorel, P. De St. Romain, and D. Patterson, "Interactions in alkane systems by depolarized Rayleigh scattering and calorimetry. Part 1.—Orientational order and condensation effects in n-hexadecane + hexane and nonane isomers," *Journal of the Chemical Society, Faraday Transactions 2*, vol. 73, no. 1, pp. 15–28, 1977.

[36] P. De St. Romain, H. T. Van, and D. Patterson, "Effects of molecular flexibility and shape on the excess enthalpies and heat capacities of alkane systems," *Journal of the Chemical Society, Faraday Transactions 1*, vol. 75, pp. 1700–1707, 1979.

[37] A. T. Rodriguez and D. Patterson, "Excess thermodynamic functions of n-alkane mixtures. Prediction and interpretation through the corresponding states principle," *Journal of the Chemical Society, Faraday Transactions 2*, vol. 78, no. 3, pp. 501–523, 1982.

[38] T. M. Aminabhavi, K. Banerjee, and R. H. Balundgi, "Thermodynamic interactions in binary mixtures of 1-chloronaphthalene

and monocyclic aromatics," *Indian Journal of Chemistry A*, vol. 38, no. 8, pp. 768–777, 1999.

[39] J. A. Riddick, W. B. Bunger, and T. K. Sakano, *Techniques of Chemistry, Organic Solvents. Physical Properties and Methods of Purific tions*, vol. 2, John Wiley & Sons, New York, NY, USA, 1986.

[40] A. Krishnaiah and P. R. Naidu, "Excess thermodynamic properties of binary liquid mixtures of 1,2-dichloroethane with normal alkanes," *Journal of Chemical and Engineering Data*, vol. 25, no. 2, pp. 135–137, 1980.

[41] E. Aicart, G. Tardajos, and M. Díaz Peña, "Isothermal compressibility of cyclohexane + n-hexane, cyclohexane + n-heptane, cyclohexane + n-octane, and cyclohexane + n-nonane," *Journal of Chemical and Engineering Data*, vol. 25, no. 2, pp. 140–145, 1980.

[42] J. D. Pandey and N. Pant, "Surface tension of ternary polymeric solution," *Journal of the American Chemical Society*, vol. 104, no. 12, pp. 3299–3302, 1982.

[43] A. S. Al-Jimaz, J. A. Al-Kandary, and A.-H. M. Abdul-Latif, "Acoustical and excess properties of {chlorobenzene + 1-hexanol, or 1-heptanol, or 1-octanol, or 1-nonanol, or 1-decanol} at (298.15, 303.15, 308.15, and 313.15) K," *Journal of Chemical and Engineering Data*, vol. 52, no. 1, pp. 206–214, 2007.

[44] E. Aicart, G. Tardajos, and M. Diaz Peña, "Isothermal compressibility of cyclohexane-n-decane, cyclohexane-n-dodecane, and cyclohexane-n-tetradecane," *Journal of Chemical and Engineering Data*, vol. 26, no. 1, pp. 22–26, 1981.

[45] V. A. Bloomfield and R. K. Dewan, "Viscosity of liquid mixtures," *Journal of Physical Chemistry*, vol. 75, no. 20, pp. 3113–3119, 1971.

[46] A. Jouyban, M. Khoubnasabjafari, Z. Vaez-Gharamaleki, Z. Fekari, and W. E. Acree Jr., "Calculation of the viscosity of binary liquids at various temperatures using Jouyban-Acree model," *Chemical and Pharmaceutical Bulletin*, vol. 53, no. 5, pp. 519–523, 2005.

[47] A. Jouyban, A. Fathi-Azarbayjani, M. Khoubnasabjafari, and W. E. Acree Jr., "Mathematical representation of the density of liquid mixtures at various temperatures using Jouyban-Acree model," *Indian Journal of Chemistry A*, vol. 44, no. 8, pp. 1553–1560, 2005.

[48] A. Krishnaiah and P. R. Naidu, "Excess thermodynamic properties of binary liquid mixtures of 1,2-dichloroethane with normal alkanes," *Journal of Chemical and Engineering Data*, vol. 25, no. 2, pp. 135–137, 1980.

11

Numerical Analysis of Flow Field and Heat Transfer of 2D Wavy Ducts and Optimization by Entropy Generation Minimization Method

Ouldouz Nourani Zonouz and Mehdi Salmanpour

Department of Mechanical Engineering, Marvdasht Branch, Islamic Azad University, Marvdasht 73711-13119, Iran

Correspondence should be addressed to Ouldouz Nourani Zonouz, o_nourani@yahoo.com

Academic Editor: Felix Sharipov

This article provided a research for the trend of heat transfer and flow field through a 2-dimensional wavy duct. To construct a grid mesh, the physical domain was transferred to the computational domain and finite volume scheme was used for discretizing the governing equations. Through the simulation, the flow regime stayed in laminar mode. Constant temperature boundary condition has been used for solid walls. Air was used as a working fluid. Existence of waves makes some phenomenon like flow separation. Effect of Reynolds number, wave width, and wave number has been analyzed and velocity distribution, heat transfer coefficient, and tangential stress were computed for different cases. The final results were compared with the same straight duct. The entropy generation minimization method has been used for better comparison between final results.

1. Introduction

For many industrial thermal systems, there is much interest in reducing fuel consumption and/or increasing of system efficiency. For gas turbine recuperators, an efficient heat exchanger is required to reduce the system size and increase the cycle efficiency. The heat exchangers generally contain flow channels with various cross-sectional shapes which are corrugated and wavy curved in the mainstream to enhance heat/mass transfer rates by generating secondary flow. The heat transfer and flow characteristics in the channel with wavy plates have been widely studied previously. Sunden and Trollheden [1] studied on the heat transfer and pressure drop in the corrugated channels and the smooth tubes under constant heat flux conditions. Nishimura and Matsune [2] simulated and visualized the dynamical behavior of vortices flow in channels. Fabbri and Rossi [3, 4] studied the laminar convective heat transfer in the smooth and corrugated channels. Cheng and Wang [5] studied on the forced convection of micropolar fluid flow over the wavy surfaces. Ergin et al. [6] numerically studied periodic flow through a corrugated duct. Vasudevaiah and Balamurugan [7] studied the heat transfer in a corrugated microchannel under constant heat flux conditions. Niceno and Nobile [8] on the two-dimensional fluid flow and heat transfer in the periodic and wavy channels. Ali and Hanaoka [9] considered effects of the operating parameters on laminar flow forced-convection heat transfer of air flowing in a channel having a V-corrugated upper plate. Zimmerer et al. [10] studied effects of the inclination angle, the wave length, the amplitude, and the shape of the corrugation on the heat and mass transfer of the heat exchanger. Guzman and Amon [11–13] performed numerical investigations for high transitional Reynolds numbers in converging-diverging (symmetric wavy wall) channels. Wang and Chen [14] determined the heat transfer rates flowing through a sinusoidally curved converging-diverging channel. Savino et al. [15] studied effect of aspect ratio on convection heat transfer enhancement in the wavy channels. Hossain and Islam [16] numerically studied on the fully unsteady fluid flow and heat transfer in sine shaped wavy channels. Metwally and Manglik [17] simulated the laminar periodically developed forced convection in sinusoidal corrugated- plate channels by using the control volume finite-difference method. Naphon

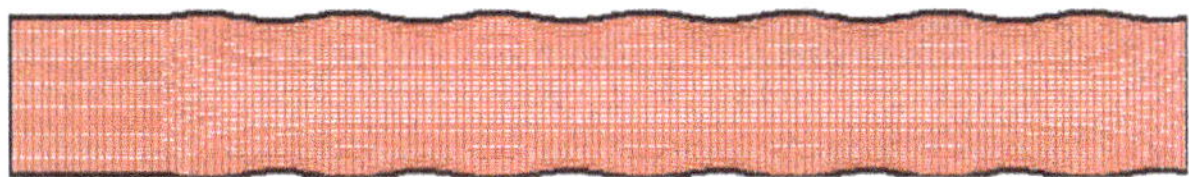

FIGURE 1: Geometry of the duct ($A = 0.001$, $b = 0.125$).

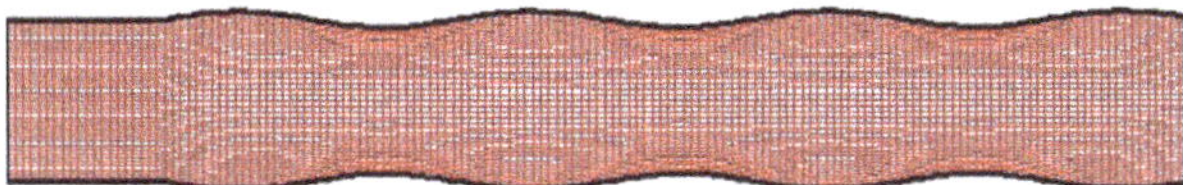

FIGURE 2: Geometry of the duct ($A = 0.002$, $b = 0.25$).

[18–20] experimentally and numerically studied on the heat transfer characteristics and pressure drop in the corrugated channel with different wavy angles and channel heights. Gradeck et al. [21] observed flow patterns of a gas-liquid flow and analyzed heat transfer in horizontal corrugated channels by using nitrogen and water as working fluids. Islamoglu et al. [22–24] numerically and experimentally studied the effect of the channel height on the enhancement of the heat-transfer characteristics in a corrugated heat-exchanger channel. In addition, artificial networks were employed to analyze the heat transfer in corrugated channels. Kuhn and Rohr [25] employed digital particle image velocimetry and planar laser-induced fluorescence techniques to examine the spatial variation of the streamwise and normal velocity components at the heated wavy surface. Chang et al. [26, 27] studied the heat transfer and pressure drop in furrowed channels with transverse sinusoidal wavy walls. Pham et al. [28] have been using large eddy simulation with dynamic modeling to investigate three-dimensional turbulent flow in wavy channels. Sui et al. [29] did some works in wavy microchannels. They simulated numerically laminar liquid-water flow and heat transfer in three-dimensional microchannels with rectangular cross section.

2. Geometry

The principle aim of this article is to improve the rate of the heat transfer thorough 2-dimensional ducts. A fully developed flow enters the duct and passes a short length with straight walls; this part reduces the effect of the entrance region. Then, the fluid flow enters the second region with wavy geometry. This sudden change in geometry made some variations for rate of heat transfer.

Figures 1 and 2 illustrate the complete geometry of problem schematically. The length of the duct is 30 cm and the width of the duct is 4 cm. A length of 4 cm from beginning of the duct is flat and a function like $y = A\sin(x/(L * b))$ is defined to the rest of the duct. In this function, A represents the height of the wave, L is the length of the duct and by changing the value of b, number of waves will be changed.

Reynolds number was defined as $\mathrm{Re} = \rho U_{in} D_h/\mu$. The hydraulic diameter is twice the entrance diameter.

3. Governing Equations

The combination of continuity, Navier-Stokes and energy equations in 2-dimensional and steady form was used as the essential governing equations. These equations and their boundary conditions are listed through numbers of 1 to 6.

Continuity.

$$\frac{\partial}{\partial x}(\rho u) + \frac{\partial}{\partial y}(\rho v) = 0. \tag{1}$$

Momentum Equation in x-Direction.

$$\left[\frac{\partial}{\partial x}(\rho u u)+\frac{\partial}{\partial y}(\rho v u)\right] = -\frac{\partial P}{\partial x}+\left[\frac{\partial}{\partial x}\left(\mu\frac{\partial u}{\partial x}\right)+\frac{\partial}{\partial y}\left(\mu\frac{\partial u}{\partial y}\right)\right]. \tag{2}$$

Momentum Equation in y-Direction.

$$\left[\frac{\partial}{\partial x}(\rho u v)+\frac{\partial}{\partial y}(\rho v v)\right] = -\frac{\partial P}{\partial y}+\left[\frac{\partial}{\partial x}\left(\mu\frac{\partial v}{\partial x}\right)+\frac{\partial}{\partial y}\left(\mu\frac{\partial v}{\partial y}\right)\right]. \tag{3}$$

Energy Equation.

$$\left[\frac{\partial}{\partial x}(\rho C_p u T)+\frac{\partial}{\partial y}(\rho C_p v T)\right] = \left[\frac{\partial}{\partial x}\left(k\frac{\partial T}{\partial x}\right)+\frac{\partial}{\partial y}\left(k\frac{\partial T}{\partial y}\right)\right] + \mu\left[\left(\frac{\partial u}{\partial y}\right)^2+\left(\frac{\partial v}{\partial x}\right)^2\right]. \tag{4}$$

Inlet Boundary Conditions.

$$\begin{gathered} T_{\text{inlet}} = 300\,\text{K}, \\ u = -6\,u_{\text{inlet}}\left[\left(\frac{y}{h_{\text{ave}}}\right)^2-\left(\frac{y}{h_{\text{ave}}}\right)\right], \\ v = 0, \\ \Pr = 0.7. \end{gathered} \tag{5}$$

Wall Boundary Conditions.

$$\begin{gathered} T_{\text{up}} = 400\,\text{K}, \\ T_{\text{down}} = 400\,\text{K}, \\ u = v = 0. \end{gathered} \tag{6}$$

In above equations h_{ave} is the distance between the up and down plates and equal to the entrance diameter. To descritize the above equations finite volume method was used. Reynolds number, number and the width of waves were changed. For each case some graphs like Nusselt number, rate of heat transfer and stress distribution were sketched.

FIGURE 3: Velocity distribution (Re = 300).

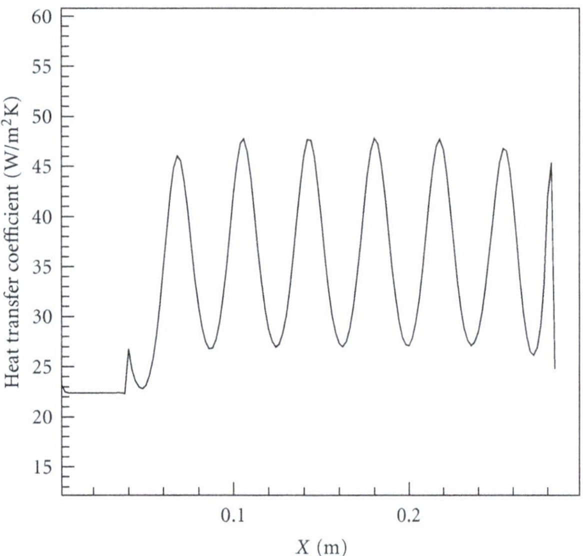

FIGURE 4: Heat transfer coefficient distribution along the wall (Re = 300).

4. Results

Three characteristic parameters like Reynolds number, number of waves, and the width of the waves have been analyzed in this research. Figure 3 illustrates the velocity distribution along the duct for Reynolds number 300. After passing the flat part of the duct, the flow enters the wavy region and senses a sudden expansion which produces wakes in concave part of waves. Before releasing these wakes from the solid walls, they act as barriers between the wall and the flow stream; hence, the fluid could not sense the hot walls and the heat transfer mechanism will be so weak. In convex part of waves, the flow velocity increases; hence the rate of heat transfer will be raised in these parts (Figure 4). This graph is divided into 2 regions. In the first region, the shape of the duct is straight and the flow is fully developed; hence the trend of graph is approximately constant. In the second region, the heat transfer coefficient finds a wavy trend because of the shape of the duct. From this graph, we can judge that the average heat transfer coefficient in second region is nearly 1.5 times of first region which means better and more powerful heat transfer mechanism.

4.1. Effect of Reynolds Number. Reynolds number is one of the most important parameter through the flow field. By increasing the value of this parameter, the velocity gradient near the solid wall will be increased and causes the separation phenomenon happened with delay. The size of the produced wakes in concave parts of the waves will be much smaller

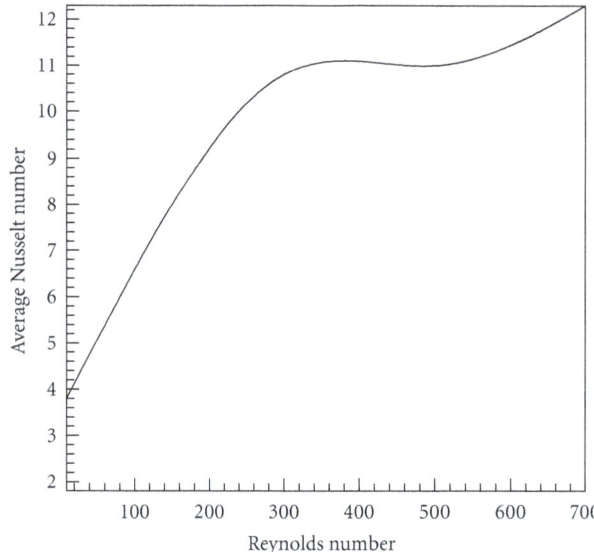

FIGURE 5: Average Nusselt number distribution versus Reynolds number.

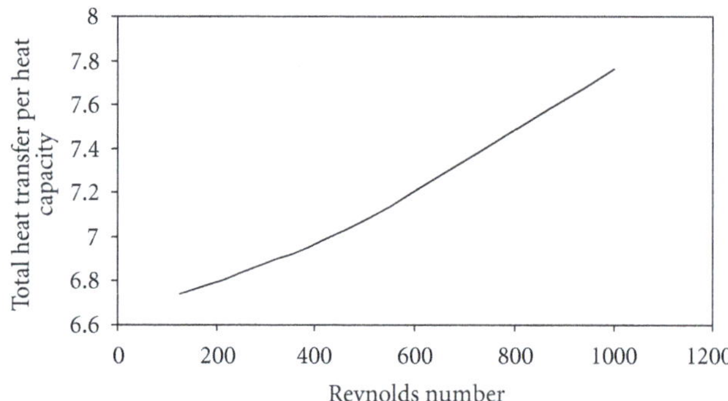

FIGURE 6: Total heat transferred by flow per heat capacity $\dot{Q}_{total}/\dot{m}c_P$ versus Reynolds number.

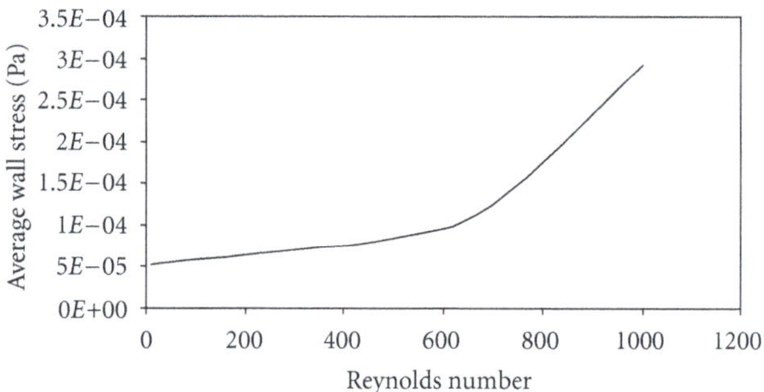

FIGURE 7: Average friction stress along the solid boundary versus Reynolds number.

by increasing the magnitude of the Reynolds number. This reduction in wake's size increases the rate of heat transfer between the fluid stream and the walls of the duct. In this study, the flow field has been studied steadily and the vortexes were not allowed to enter the flow; hence the vortexes act as barriers near the wall. With increasing the Reynolds number, the size of these barriers will be reduced.

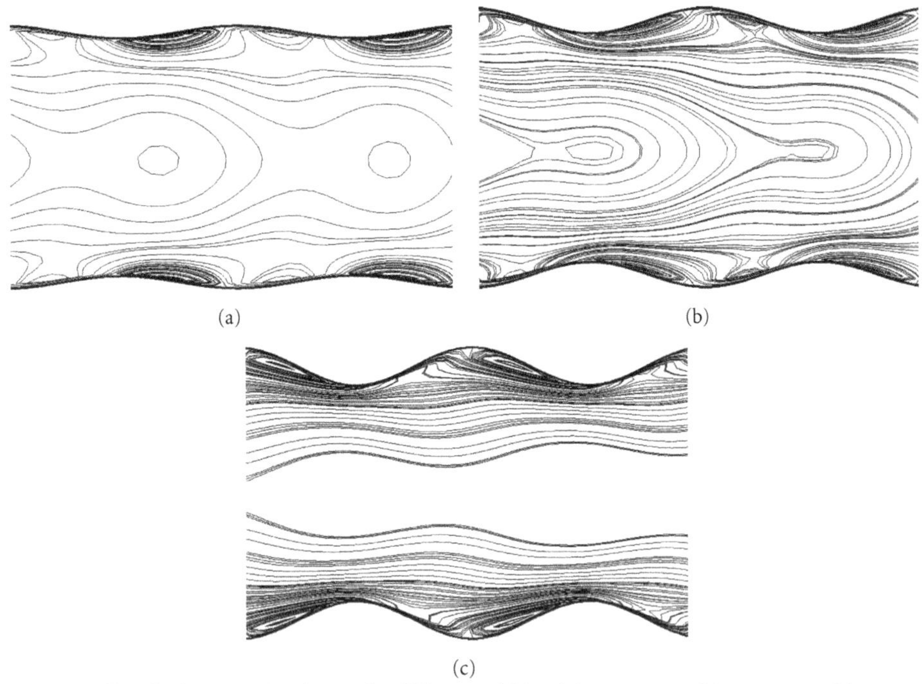

FIGURE 8: Profile of velocity in third wave for different widths: (a) $A = 0.001$, (b) $A = 0.002$, (c) $A = 0.003$.

In Figure 5, the distribution of average Nusselt number along the wall of duct has been sketched for different Reynolds numbers. By increasing the Reynolds number, the value of dimensionless heat transfer coefficient will be raised which means better heat transfer mechanism. To complete this reason, the quantity of heat which transferred by fluid flow through the same duct has been illustrated for different Reynolds numbers in Figure 6. This figure has the same trend.

Distribution of tangential stress along the wall has been sketched in Figure 7. This graph has a raising trend which means by increasing the Reynolds number, the gradient of velocity near the wall will be increased so; the magnitude of stress will be also raised.

4.2. Effect of Wave Width. Analysis of effect of wave width is one of the most important parameters for improving the quality of heat transfer. Assume a sinus correlation like $y = A\sin(y/(L * b))$ for duct geometry. By increasing the value of A which controls the width of the wave, the edges of wavy region find more sharp shape which causes the diameter of the duct in convex parts reduced. This reduction causes the velocity of the flow and also the rate of heat transfer increases considerably. Figure 8 illustrates the change of velocity profile in same place and same Reynolds number for different widths.

Figure 9 illustrates the trend of heat transfer coefficient along the wall of duct for different widths. This graph predicts that by increasing the width of the wave the magnitude of the heat transfer maximum points raised considerably.

Average Nusselt number distribution for different A coefficient is sketched in Figure 10. The value of average Nusselt number increases with increase of this parameter.

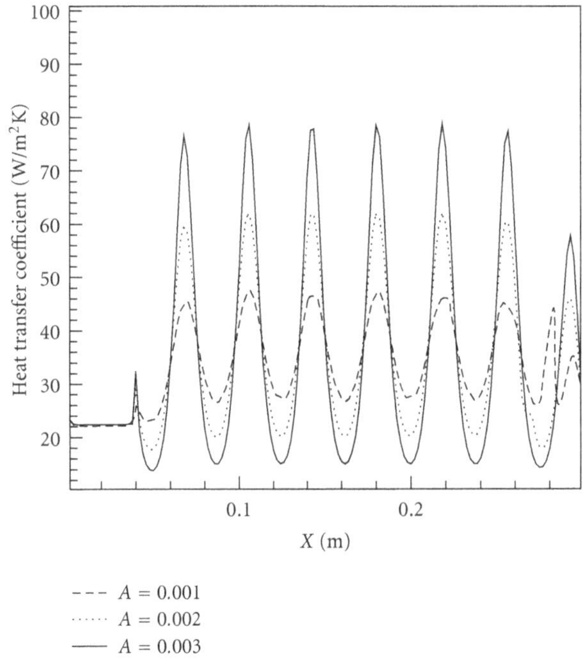

FIGURE 9: Heat transfer coefficient distributions along the length of the duct for different widths (Re = 300).

4.3. Effect of Wave Number. Number of produced waves along the duct has been studied. Figure 11 illustrates the distribution of heat transfer coefficient for different wave number. This figure shows that if the flow senses more waves along its path, the value of heat transfer coefficient will be more significant.

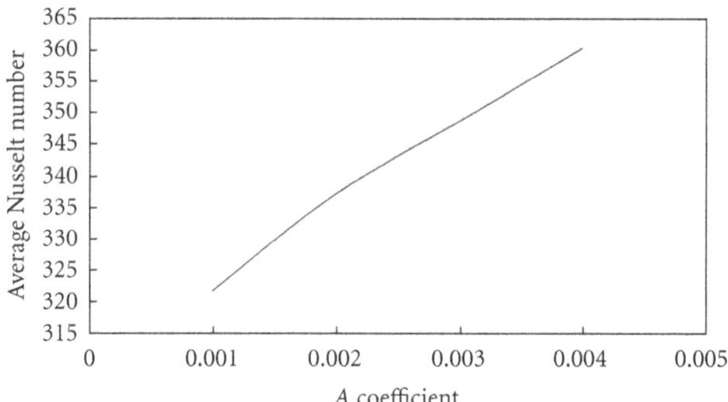

Figure 10: Average Nusselt number distribution versus A coefficient (Re = 300).

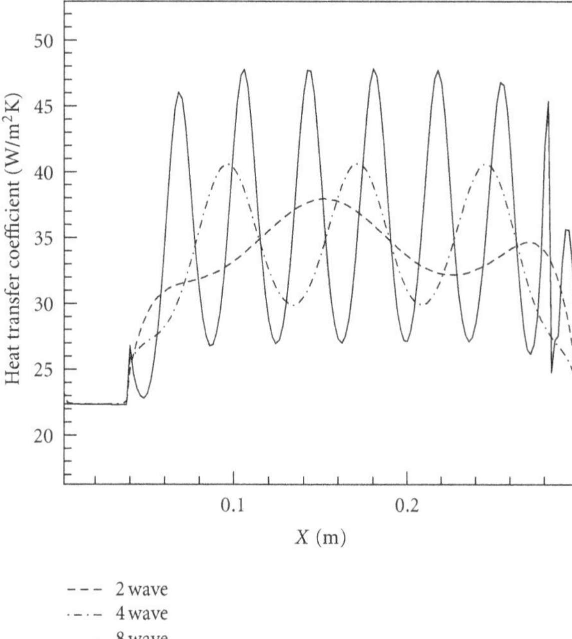

Figure 11: Heat transfer coefficient distribution along the duct for three cases: 2, 4, and 8 waves (Re = 300).

5. Entropy Generation Minimization Method

One of the newest methods for comparing and optimizing the results is entropy generation method. This method based on the rate of entropy generation. By minimizing the value of this property, the apparatus will work better and more optimize. Term of entropy generation is a summation of conductive heat transfer irreversibility and viscous dissipation which is produced from motion of fluid. By using classic thermodynamic relations, this term can be found in dimensional form. The mass conservation principle, the first and second laws are used to derive the entropy generation term:

$$\frac{\partial \rho}{\partial t} = -\rho \nabla \cdot V,$$
$$\rho \frac{\partial i}{\partial t} = -\nabla \cdot q - P\nabla \cdot V - w''', \tag{7}$$
$$\dot{s}_{\text{gen}} = \rho \frac{\partial s}{\partial t} + \nabla \cdot \left(\frac{q}{T}\right) \geq 0.$$

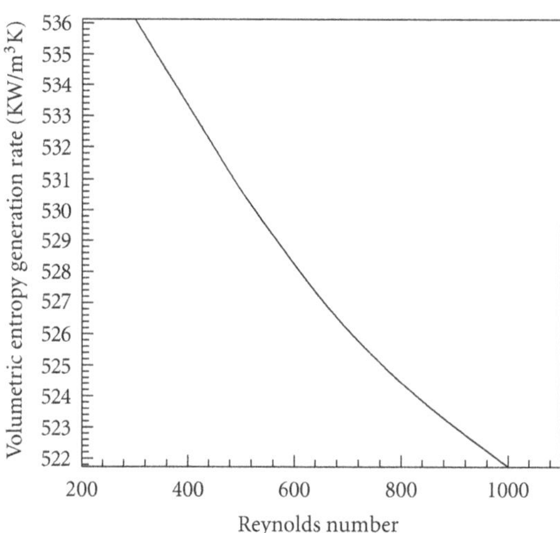

Figure 12: Total volumetric entropy generation rate distribution versus Reynolds number.

Eliminating i and s between these laws and using the Gibbs equation

$$\frac{\partial i}{\partial t} = T\frac{\partial s}{\partial t} + \frac{P}{\rho^2}\frac{\partial \rho}{\partial t} \tag{8}$$

yield the general two-term expression for the volumetric rate of entropy generation:

$$\dot{s}_{\text{gen}} = -\frac{1}{T^2} q \cdot \nabla T - \frac{w'''}{T} \geq 0. \tag{9}$$

By substituting $(-w''')$ with $\mu\phi$ and using Fourier law:

$$q = -k\nabla T, \tag{10}$$

the entropy generation equation will be derived:

$$\dot{s}_{\text{gen}} = \frac{k}{T^2}(\nabla T)^2 + \frac{\mu}{T}\phi \geq 0 \tag{11}$$

or

$$\dot{s}_{\text{gen}} = \frac{k}{T^2}\left[\left(\frac{\partial T}{\partial x}\right)^2 + \left(\frac{\partial T}{\partial y}\right)^2\right] + \frac{\mu}{T}\left[2\left(\frac{\partial u}{\partial x}\right)^2 + 2\left(\frac{\partial v}{\partial y}\right)^2 + \left(\frac{\partial u}{\partial y} + \frac{\partial v}{\partial x}\right)^2\right]. \tag{12}$$

For measuring the effect of heat transfer irreversibility and viscous dissipation in magnitude of entropy production, the Bejan number can be defined as

$$\text{Be} = \frac{(k/T^2)\left[(\partial T/\partial x)^2 + (\partial T/\partial y)^2\right]}{\dot{s}_{\text{gen}}}. \tag{13}$$

Effect of Reynolds number, wave width, and wave number on the rate of entropy generation and Bejan number has been studied.

By increasing Reynolds number, the velocity and temperature gradients will be increased and the profiles of these

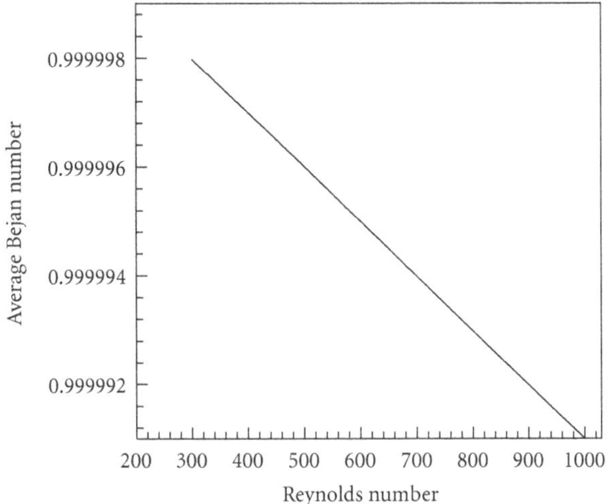

FIGURE 13: Average Bejan number distribution versus Reynolds number.

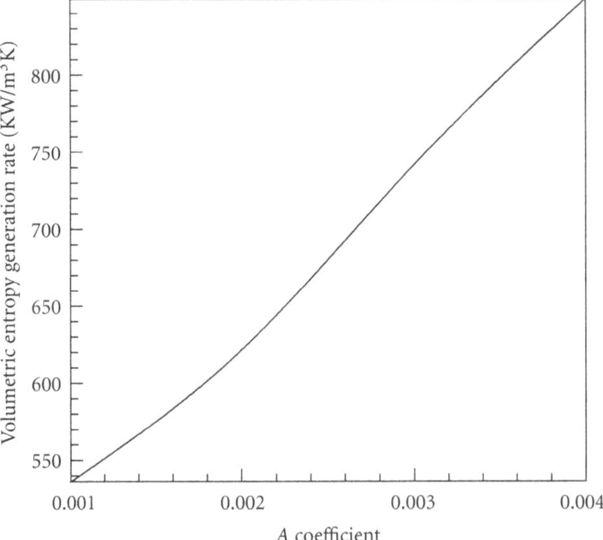

FIGURE 14: Total Volumetric entropy generation rate distribution versus A coefficient (Re = 300).

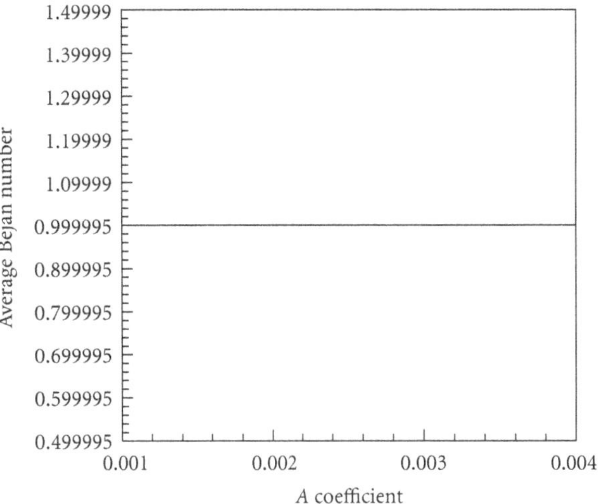

FIGURE 15: Average Bejan number distribution versus A coefficient (Re = 300).

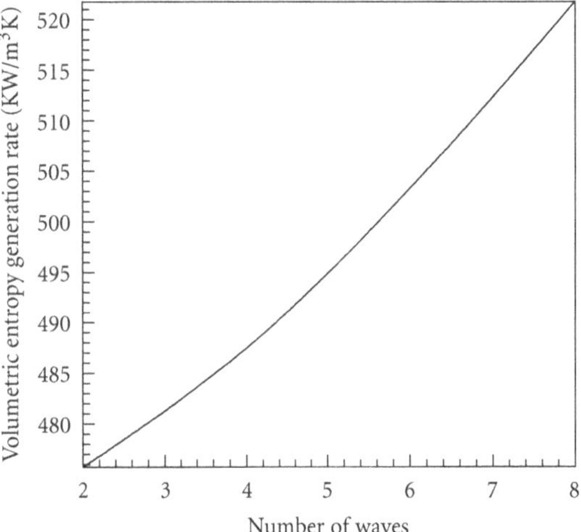

FIGURE 16: Total volumetric entropy generation rate distribution versus number of waves (Re = 1000).

properties will be more flat; hence, the rate of entropy generation will be decreased. This trend has been illustrated in Figure 12.

Figure 13 shows the trend of average Bejan number versus Reynolds number. By increasing the Reynolds number, the value of Bejan number will be decreased.

Wave width is another parameter which affects the rate of entropy generation. Figure 14 illustrates the trend of total entropy generation versus wave width. By increasing the width of the wave, the rate of entropy generation or dissipation is increased.

By changing the A coefficient, variation of Bejan number is negligible (Figure 15).

By increasing number of waves, the fluid flow senses more viscous dissipation and so the magnitude of entropy generation is raised. This trend is illustrated in Figure 16. Changing of wave's number has a very small effect on the variation of Bejan number; hence, the distribution of this parameter has a constant trend (Figure 17).

6. Comparison with Flat Duct

For comparison, a flat duct with same length and width was selected. Same thermal boundary conditions were considered. Figure 18 shows the trend of Nusselt number for two ducts. In case of wavy duct, the value of averaged Nusselt number is more than the flat duct and this parameter increases with Reynolds number; hence, with producing waves in a duct the rate of heat transfer will be increased.

7. Conclusion

Numerical analysis of flow field and heat transfer of 2D wavy ducts and optimization by entropy generation method

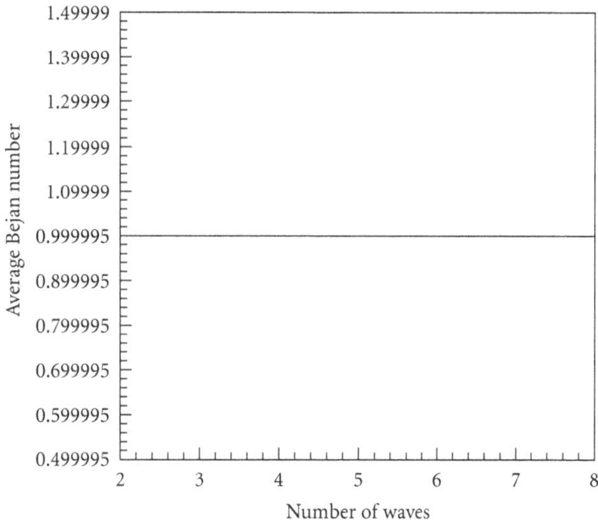

FIGURE 17: Average Bejan number distribution versus number of waves (Re = 1000).

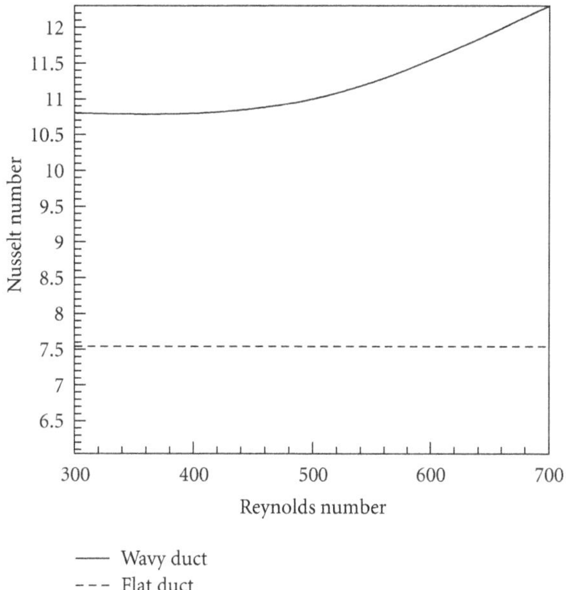

FIGURE 18: Average Nusselt number distributions versus Reynolds number for flat and wavy channels.

has been studied in this research. The governing equations with their boundary conditions were discretized with finite-volume method. The effect of Reynolds number, wave width, and the number of produced waves were obtained. The results show that by increasing Reynolds number, the value of Nusselt number will be increased. Also, by making the waves to be sharper, the heat transfer mechanism will be more powerful and the average Nusselt number will be increased along the length of the duct. The third parameter which was studied is number of waves. With increasing the number of waves, the flow will be passed more concave and convex parts; hence, the rate of heat transfer will be increased. The obtained results were analyzed by entropy generation minimization method. The volumetric entropy generation rate was reduced with increasing the Reynolds number, but in the case of wave width and the number of waves, the value of this property will be increased. In all cases, the value of Bejan number is near one, which means through the laminar flow, the term of the viscous dissipation has a negligible effect.

Nomenclature

Be: Bejan number
C_p: Constant pressure specific heat (J/Kg K)
D_h: Hydraulic diameter (m)
i: Internal energy (J/Kg)
K: Conductivity (W/m K)
L: Length of the duct (m)
Nu: Nusselt number
P: Pressure (Pa)
Pr: Prandtl number
q: Heat flux (W/m^2)
Re: Reynolds number
s: Entropy (J/Kg K)
$\dot{s}_{gen}$: Volumetric entropy generation rate (W/m^3 K)
T: Temperature (K)
t: Time (s)
u: Velocity component along the wall (m/s)
V: Velocity vector (m/s)
v: Velocity component normal to the wall (m/s)
w''': Volume power density (W/m^3)
x: Coordinate component along the wall
y: Coordinate component normal to the wall.

Greek Letters

μ: Absolute viscosity (Kg/m s)
ρ: Density (Kg/m^3)
ϕ: Viscous dissipation (J/Kg m^2).

Subscripts

down: Bottom wall
inlet: Entrance of the wall
up: Top wall.

References

[1] B. Sunden and S. Trollheden, "Periodic laminar flow and heat transfer in a corrugated two-dimensional channel," *International Communications in Heat and Mass Transfer*, vol. 16, no. 2, pp. 215–225, 1989.
[2] T. Nishimura and S. Matsune, "Vortices and wall shear stresses in asymmetric and symmetric channels with sinusoidal wavy walls for pulsatile flow at low Reynolds numbers," *International Journal of Heat and Fluid Flow*, vol. 19, no. 6, pp. 583–593, 1998.
[3] G. Fabbri, "Heat transfer optimization in corrugated wall channels," *International Journal of Heat and Mass Transfer*, vol. 43, no. 23, pp. 4299–4310, 2000.
[4] G. Fabbri and R. Rossi, "Analysis of the heat transfer in the entrance region of optimised corrugated wall channel,"

International Communications in Heat and Mass Transfer, vol. 32, no. 7, pp. 902–912, 2005.

[5] C. Y. Cheng and C. C. Wang, "Forced convection in micropolar fluid flow over a wavy surface," *Numerical Heat Transfer A*, vol. 37, no. 3, pp. 271–287, 2000.

[6] S. Ergin, M. Ota, and H. Yamaguchi, "Numerical study of periodic turbulent flow through a corrugated duct," *Numerical Heat Transfer A*, vol. 40, no. 2, pp. 139–156, 2001.

[7] M. Vasudevaiah and K. Balamurugan, "Heat transfer of rarefied gases in a corrugated microchannel," *International Journal of Thermal Sciences*, vol. 40, no. 5, pp. 454–468, 2001.

[8] B. Niceno and E. Nobile, "Numerical analysis of fluid flow and heat transfer in periodic wavy channels," *International Journal of Heat and Fluid Flow*, vol. 22, no. 2, pp. 156–167, 2001.

[9] A. Ali and Y. Hanaoka, "Experimental study on laminar flow forced-convection in a channel with upper V-corrugated plate heated by radiation," *International Journal of Heat and Mass Transfer*, vol. 45, no. 10, pp. 2107–2117, 2002.

[10] C. Zimmerer, P. Gschwind, G. Gaiser, and V. Kottke, "Comparison of heat and mass transfer in different heat exchanger geometries with corrugated walls," *Experimental Thermal and Fluid Science*, vol. 26, no. 2-4, pp. 269–273, 2002.

[11] A. M. Guzman and C. H. Amon, "Transition to chaos in converging-diverging channel flows: ruelle-takens-newhouse scenario," *Physics of Fluids A*, vol. 6, no. 6, pp. 1994–2002, 1994.

[12] A. M. Guzman and C. H. Amon, "Dynamical flow characterization of transitional and chaotic regimes in converging-diverging channels," *Journal of Fluid Mechanics*, vol. 321, pp. 25–57, 1996.

[13] A. M. Guzman and C. H. Amon, "Convective heat transfer and flow mixing in converging-diverging channel flows," in *Proceedings of the ASME International Mechanical Engineering Congress and Exposition*, pp. 61–68, November 1998, HTD-Vol. 361-1.

[14] C. C. Wang and C. K. Chen, "Forced convection in a wavy-wall channel," *International Journal of Heat and Mass Transfer*, vol. 45, no. 12, pp. 2587–2595, 2002.

[15] S. Savino, G. Comini, and C. Nonino, "Three-dimensional analysis of convection in two-dimensional wavy channels," *Numerical Heat Transfer A*, vol. 46, no. 9, pp. 869–890, 2004.

[16] M. Z. Hossain and A. K. M. Islam, "Fully developed flow structures and heat transfer in sine-shaped wavy channels," *International Communications in Heat and Mass Transfer*, vol. 31, no. 6, pp. 887–896, 2004.

[17] H. M. Metwally and R. M. Manglik, "Enhanced heat transfer due to curvature-induced lateral vortices in laminar flows in sinusoidal corrugated-plate channels," *International Journal of Heat and Mass Transfer*, vol. 47, no. 10-11, pp. 2283–2292, 2004.

[18] P. Naphon, "Laminar convective heat transfer and pressure drop in the corrugated channels," *International Communications in Heat and Mass Transfer*, vol. 34, no. 1, pp. 62–71, 2007.

[19] P. Naphon, "Heat transfer characteristics and pressure drop in channel with V corrugated upper and lower plates," *Energy Conversion and Management*, vol. 48, no. 5, pp. 1516–1524, 2007.

[20] P. Naphon, "Effect of corrugated plates in an in-phase arrangement on the heat transfer and flow developments," *International Journal of Heat and Mass Transfer*, vol. 51, no. 15-16, pp. 3963–3971, 2008.

[21] M. Gradeck, B. Hoareau, and M. Lebouche, "Local analysis of heat transfer inside corrugated channel," *International Journal of Heat and Mass Transfer*, vol. 48, no. 10, pp. 1909–1915, 2005.

[22] Y. Islamoglu and C. Parmaksizoglu, "The effect of channel height on the enhanced heat transfer characteristics in a corrugated heat exchanger channel," *Applied Thermal Engineering*, vol. 23, no. 8, pp. 979–987, 2003.

[23] Y. Islamoglu and A. Kurt, "Heat transfer analysis using ANNs with experimental data for air flowing in corrugated channels," *International Journal of Heat and Mass Transfer*, vol. 47, no. 6-7, pp. 1361–1365, 2004.

[24] Y. Islamoglu and C. Parmaksizoglu, "Numerical investigation of convective heat transfer and pressure drop in a corrugated heat exchanger channel," *Applied Thermal Engineering*, vol. 24, no. 1, pp. 141–147, 2004.

[25] S. Kuhn and P. R. V. Rohr, "Experimental investigation of mixed convective flow over a wavy wall," *International Journal of Heat and Fluid Flow*, vol. 29, no. 1, pp. 94–106, 2008.

[26] S. W. Chang, A. W. Lees, and T. C. Chou, "Heat transfer and pressure drop in furrowed channels with transverse and skewed sinusoidal wavy walls," *International Journal of Heat and Mass Transfer*, vol. 52, no. 19-20, pp. 4592–4603, 2009.

[27] S. W. Chang, A. W. Lees, T. M. Liou, and G. F. Hong, "Heat transfer of a radially rotating furrowed channel with two opposite skewed sinusoidal wavy walls," *International Journal of Thermal Sciences*, vol. 49, no. 5, pp. 769–785, 2010.

[28] M. V. Pham, F. Plourde, and S. K. Doan, "Turbulent heat and mass transfer in sinusoidal wavy channels," *International Journal of Heat and Fluid Flow*, vol. 29, no. 5, pp. 1240–1257, 2008.

[29] Y. Sui, C. J. Teo, P. S. Lee, Y. T. Chew, and C. Shu, "Fluid flow and heat transfer in wavy microchannels," *International Journal of Heat and Mass Transfer*, vol. 53, no. 13-14, pp. 2760–2772, 2010.

12

Glass Transition Behavior of the Quaternary Ammonium-Type Ionic Liquid: N,N-Diethyl-N-methyl-N-(2-methoxyethyl)ammonium Bromide-H_2O Mixtures

Yukihiro Yoshimura,[1] Naohiro Hatano,[1] Yusuke Imai,[2] Hiroshi Abe,[2] Osamu Shimada,[3] and Tomonori Hanasaki[3]

[1] *Department of Applied Chemistry, National Defense Academy, Yokosuka, Kanagawa 239-8686, Japan*
[2] *Department of Materials Science and Engineering, National Defense Academy, Yokosuka, Kanagawa 239-8686, Japan*
[3] *Department of Applied Chemistry, Ritsumeikan University, Kusatsu, Shiga 525-8577, Japan*

Correspondence should be addressed to Yukihiro Yoshimura, muki@nda.ac.jp

Academic Editor: Ramesh Gardas

By a simple differential thermal analysis (DTA) system, the concentration dependence of the glass transition temperatures (T_gs) for the quaternary ammonium-type ionic liquid, *N,N*-diethyl-*N*-methyl-*N*-(2-methoxyethyl)ammonium bromide [DEME][Br] and H_2O mixtures, after quick precooling was measured as a function of water concentration x (mol% H2O). We compared the results with the previous results of [DEME][I]-H_2O and [DEME][BF_4]-H_2O mixtures in which a double-glass transition behavior was observed. Remarkably, the [DEME][Br]-H_2O mixtures basically show one-T_g behavior and the T_g decreases monotonically with increasing H_2O content up to around $x = 91.5$. But it suddenly jumps to higher T_g value at a specific $x =\sim 92$. At this very limited point, two T_gs (T_{g1}, T_{g2}) which we might consider as a transition state from the structure belonging to the T_{g1} group to another one due to the T_{g2} group were observed. These results clearly reflect the difference in the anionic effects among Br^-, I^-, and BF_4^-. The end of the glass-formation region of [DEME][Br]-H_2O mixtures is around $x = 98.9$ and moves to more water-rich region as compared to those of [DEME][BF_4]-H_2O ($x = 96.0$) and [DEME][I]-H_2O ($x = 95.0$) mixtures.

1. Introduction

Room temperature ionic liquids (RTILs) are molten salts with melting temperature below (<373 K) and well known to have many attractive properties as solvents, for example, almost zero vapor pressure, wide electrochemical window, high recyclability, nonflammability, and so forth [1, 2], so that RTILs stimulate the interests of a wide range of applications [3, 4]. Many of the attractive features require a thorough knowledge of their thermophysical properties. Interactions in RTILs are characterized by the presence of Coulomb interactions among the constituent ions. Hydrogen-bonding, *van der Waals*, and π-π interactions also take place in RTILs and affect their nature and the liquid structures. Thus, the liquid structures of RTILs are determined by a balance between long-range electrostatic forces and local geometric factors. As a result, many RTILs can be easily supercooled and form the glassy state [5–7]. Changes in the cation and anion combinations allow the physical and chemical properties of ionic liquids to be effectively tuned, for example, for manipulating the solvent properties of RTILs in order to achieve different purposes [7–9].

Although thermodynamic properties including T_g have been so far investigated mainly on the imidazolium-based series of pure RTILs [10–12], we point out that investigations of glass transition behavior of the RTILs-H_2O mixtures are still scarce. In previous work, we reported the glass transition behavior of the quaternary ammonium-type ionic liquid, *N,N*-diethyl-*N*-methyl-*N*-(2-methoxyethyl)ammonium tetrafluoroborate [DEME][BF_4]-H_2O [13]

and iodide [DEME][I]-H_2O [14] mixtures, after quick precooling. Remarkably, double-glass transitions were observed in both mixtures. But concentrations of the two-T_gs region are completely different from each other; the double-glass transitions for the [DEME][BF_4]-H_2O mixtures were observed in the RTIL-rich region of x = 16.5 ~ 30.0 (mol% H_2O), whereas the region moves to a water-rich side of x = 77.5 ~ 85.0 (mol% H_2O) for the [DEME][I]-H_2O mixtures. These clearly reflect the difference in the anionic effect between BF_4^- and I^- on the water structure. It is interesting to quote that the ionic radius of BF_4^- anion (229 pm [15, 16]) is slightly larger than that of I^- (216 pm [17]) anion. We suspect that the subtle difference in the ionic radii between BF_4^- and I^- anions together with the anionic nature plays an important role in determining the regions where double-glass transitions occur.

Then, what happens to the phase behavior of other quaternary ammonium-type ionic liquid, [DEME][Br]-H_2O mixtures, in which the anionic radius of Br^- (195 pm [18]) is smaller than that of I^- anion? Does the double-glass transition phenomenon occur or not? The aim of the present paper is to show the glass transition behavior of [DEME][Br]-H_2O mixtures and to compare the results with those of [DEME][BF_4]-H_2O [13] and [DEME][I]-H_2O [14] mixtures. Here we have measured the T_g variation of [DEME][Br]-H_2O mixtures as a function of H_2O concentration x (mol% H_2O). A limit of the glass-formation region is also determined.

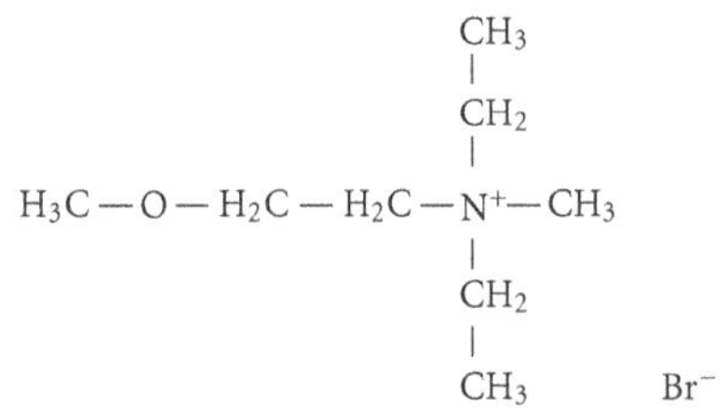

2. Experiments

2.1. Material. As an RTIL in this study, we used *N,N*-diethyl-*N*-methyl-*N*-(2-methoxyethyl)ammonium bromide, [DEME][Br] which was synthesized in our laboratory. [DEME][Br] was synthesized following procedures reported in the literature [19]. The synthetic route of this material is shown in Scheme 1. The schematic chemical structures of [DEME][Br] are shown in Figure 1.

5.37 g (61.6 mmol) of *N,N*-diethylmethylamine and 8.03 g (57.8 mmol) of 1-bromo-2-methoxyethane were dissolved in about 7 mL of anhydrous acetone. The mixture was stirred overnight at T = 373 K in an autoclave. After cooling, the reaction mixture was poured into ice-cold diethyl ether. The precipitated crude product was filtered off and was repeatedly recrystallized from anhydrous acetone. The purified product, [DEME][Br] were a colorless crystal. The molecular structures of [DEME][Br] was identified by ^1H NMR spectra (JEOL JEM-ECS400). The spectral data (400 Hz, $CDCl_3$) of [DEME][Br] are as follows: δ 3.91 (t, 2H, J = 4.3 Hz, $-CH_2-OCH_3$), δ 3.87 (br s, 2H, $-CH_2CH_2-OCH_3$), 3.66 (m, 4H, $-CH_2-CH_3$), 3.37 (s, 3H, $-OCH_3$), 3.32 (s, 3H, $-CH_3$), 1.38 (t, 6H, J = 7.2 Hz, $-CH_2-CH_3$). No other peaks were observed except for the identified peaks from the objective compound mentioned above. This result suggests that organic impurities, for example, the starting materials, solvents, and byproducts, were fully removed. [DEME][Br] is a crystalline state at room temperature as described above. The melting temperature was reported to be 358.2 K in the available literature [19].

CH_3
|
CH_2
|
$H_3C-O-H_2C-H_2C-N^+-CH_3$
|
CH_2
|
CH_3 Br^-

(a) (b)

Figure 1: Schematic chemical structure of (a) $[DEME]^+$ cation and (b) bromide anion.

We prepared the sample solutions of [DEME][Br]-H_2O mixtures at different concentrations, x = 60.0 ~ 99.8 ± 0.05 (mol% H_2O). There was a limitation of the solubility of [DEME][Br] in water (x < 60). Special care was taken with the sample preparations in a dry box to avoid atmospheric H_2O and CO_2.

2.2. DTA Measurements. To detect the glass transition temperature, T_g, a simple differential thermal analysis (DTA) system designed for quick cooling experiments was used. As a reference material for the measurements, we used benzene (Wako Pure Chemical Co.). A sample cell (about 35 mm long and 2 mm i.d. glass tube with one side sealed) was filled with a mixture, and then a thermocouple junction was placed 25~30 mm from the mouth of the sample cell. As a precooling procedure, a vitrification was done by putting the whole sample solution directly into liquid nitrogen where the quenching rate was estimated to be about 500 K/min. After taking out the sample from the liquid nitrogen, the DTA traces were recorded. The detailed procedure is basically the same as the previously reported ones [13, 14]. T_g values were found to be reproducible in this study to within 0.5 K, and the accuracy of temperature reading is estimated to be 1 K from a determination of melting temperatures of several guaranteed grade reagents (absolute ethanol, acetone and chloroform).

2.3. Raman Spectral Measurements. Raman spectra were typically measured by a JASCO NRS-1000DT Raman spectrophotometer equipped with a single monochromator and a CCD detector at room temperature (298 K). The 533 nm from green laser (Showa Optronics Co., Ltd.) with a power of 100 mW was used as an excitation source.

3. Results and Discussion

3.1. Glass Transition Behavior of [DEME][Br]-H_2O Mixed Solutions. Schematic DTA thermograms of [DEME][Br]-H_2O mixed solutions at three typical concentrations (x = 65.0, 92.0, 95.0) are shown in Figure 2. At x = 65.0, the T_g was observed at 182 K. After the glass transition, we could not see the cold crystallization [20] with a large exothermal peak, which was observed in the case of [DEME][I]-H_2O mixed solutions [14]. Just at x = 92.0, we captured the intriguing result: the DTA trace gives a first glass transition followed by an exothermic crystallization-like peak at around

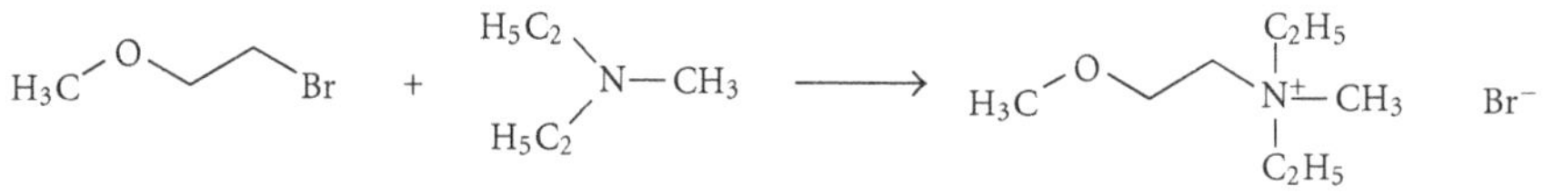

Scheme 1: The synthetic route of [DEME][Br]. Reagents and conditions: anhydrous acetone, at 373 K, overnight, in autoclave.

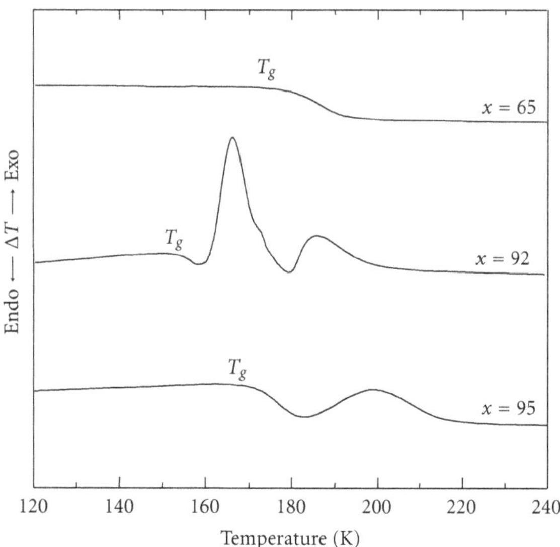

Figure 2: Schematic DTA warm-up traces of 65.0, 92.0, and 95.0 mol% H_2O mixed solutions are shown.

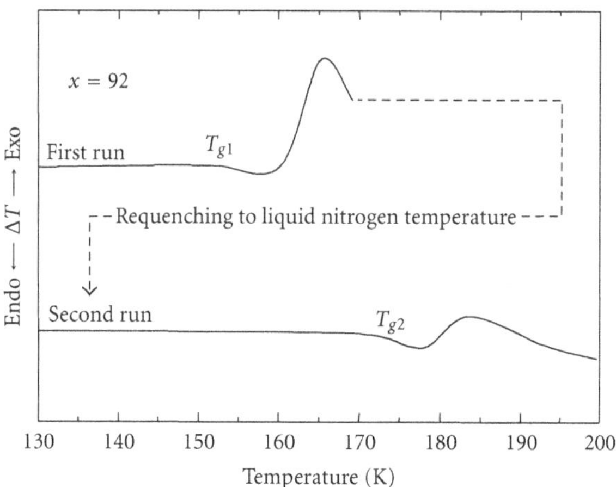

Figure 3: The requenching procedure for detecting the T_{g2} at x = 92.0 is shown.

161 K, then another set of an endothermic inflection at 172 K and an exothermic peak at 177 K in the DTA trace, respectively. To our knowledge, this is a typical DTA trace showing two glass transitions [21–23]. To ensure the second glass transition, we performed supplementary experiments using a requenching method in the following way: just after recording the first T_g (T_{g1}) and the exothermic inflection as usual in the first run, we requenched the sample to the liquid nitrogen at 77 K quickly and recorded a DTA trace repeatedly. If the exothermic peak just after the T_{g1} is due to the crystallization, we should have a clear downward C_p shift as the second T_g (T_{g2}) in the second run. The schematic DTA traces obtained in this way are shown in Figure 3. On the other hand, the DTA trace at $x = 95.0$ in Figure 2 shows one-T_g behavior, but interestingly the DTA trace keeps to show a cold crystallization peak. Taken together, glass transition behavior of the [DEME][Br]-H_2O mixed solutions basically shows one glass transition against the water content. But at only limited concentration region of $x = 91.5 \sim 92.5$, two glass transitions (T_{g1}, T_{g2}) were observed which we will describe in detail below.

Summarized results of T_g variations (●) as a function of x are shown in Figure 4. For comparisons, T_g data of [DEME][BF_4]-H_2O [13] and [DEME][I]-H_2O [14] are also shown. The T_g value monotonically decreases with increasing x up to $\sim$ 91.5 mol%. But it suddenly jumps to higher T_g value at $x \geq 91.5$. At this very limited point ($x = 91.5 \sim 92.5$), two T_gs (□: T_{g1}, ○: T_{g2}) were observed. As to an explanation for the mechanism of a double-glass transition phenomenon, the following was proposed [21–23]. When we lower a temperature of the sample, the homogeneous solution splits into two phases due to its thermodynamic instability where a metastable liquid-liquid immiscibility occurs. Based on the idea, we consider that the [DEME][Br]-H_2O mixed solution may separate into two phases from the original solution at $x = 91.5 \sim 92.5$. It is important to quote that the T_g of a glass-forming liquid has a correlation with its viscosity [24]. Thus, the two-T_gs behavior implies that the viscosities (and thus the structures) of two glassy phases are very different from each other. Looking into the results more closely, we may consider this as a transition state from the structure belonging to the T_{g1} group ($x < 91.5$) to another one due to the T_{g2} group ($x > 92.5$). At $x \geq 92.5$ some inclusions of H_2O ice crystals in the quenched samples were visually confirmed. However, the solutions keep showing the clear but small T_g value up to around 98.9 mol% meaning that the system still holds the glassy state. We find that the edge of a glass-forming composition range extends to more water-rich region as compared to those of the [DEME][I]-H_2O (x = 95.0) and/or [DEME][BF_4]-H_2O (x = 96.0) mixtures.

3.2. Comparison with the Glass Transition Behaviors of [DEME][I]-H_2O and [DEME][BF_4]-H_2O Mixed Solutions. Firstly, we compare the T_g behavior with the results of [DEME][I]-H_2O solutions. The double-glass transition phenomenon was obsevbed in the [DEME][I]-H_2O mixtures, but there is a major difference in the region where the double glass transitions appeared. One is that the double-T_gs range is relatively wide (x = 77.5 to 85.0). Another one is that the T_{g2} value changes little with x in the double-glass transition

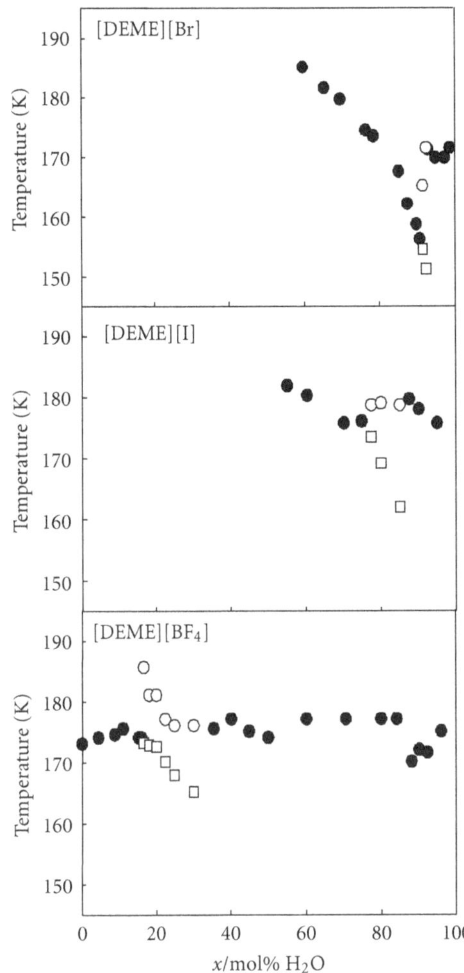

FIGURE 4: T_g variations as a function of x are shown. For comparison, T_g data of [DEME][I]-H_2O and [DEME][BF_4]-H_2O mixed solutions are also shown. □: T_{g1}, ○: T_{g2}, ●: T_g. (only one glass transition was observed except for the region $x = 91.5 \sim 92.5$).

region, though the T_{g1} value decreases with increasing x. This behavior is similar to that in the aqueous solutions of "normal salt" such as symmetrical tetraalkyl ammonium halide (R_4NX; R = Ethyl (Et), n-Propyl (n-Pr), and X = Cl^-, Br^-) [21–23].

On the other hand, the two-T_gs region moves to a very water-poor side in the case of [DEME][BF_4]-H_2O solution and the concentration range of the double glass transitions is much wider (13.5 mol%). These clearly reflect the difference in the anionic effects on the water structure. Previously, we explained that the differences in the results between the [DEME][BF_4] mixtures and the [DEME][I] mixtures come from the variation in the solvation abilities and also the relative positions of the respective anions (BF_4^- and I^-) and $[DEME]^+$ cation in the RTILs [14]. The interactions between water molecules and the anions are significantly different in the two RTILs. Considering the smaller ionic radius of the I^- anion rather than BF_4^- anion, the anion-water interaction in the [DEME][I] solution should be stronger than that in the [DEME][BF_4] solution. This was partly evidenced by the existence of nearly free hydrogen-bonded Raman band (NFHB) of water molecules in the [DEME][BF_4]-H_2O system, which will be described in the next section. NFHB is assigned to the water molecules which exist as single molecules (not self-associated state) without forming the hydrogen-bonding network among themselves as in pure liquid H_2O [14]. Unfortunately nor the detailed liquid structure of [DEME][BF_4] or [DEME][I] has apparently not been repeted as yet. It is interesting to point out that the double-glass transition behavior is systematically changing from [DEME][BF_4]-H_2O (two-T_gs behavior in water-poor region, $x = 16.5 \sim 30.0$), [DEME][I]-H_2O (two-T_gs behavior in water-rich region, $x = 77.5 \sim 85.0$), and finally to [DEME][Br]-H_2O (two T_gs at very limited concentrations, $x = 91.5 \sim 92.5$) with decreasing the anionic radii. Clearly the solvation abilities of the respective anions provide significant effect on the structures of the RTILs-H_2O mixtures which was reflected in the glass transition behavior.

3.3. Raman Spectra. To look further into states of the water structure in the mixtures, we measured Raman spectra. A vibrational spectroscopy is highly sensitive to the intermolecular interactions including hydrogen bonding among water molecules. Firstly, Figure 5(a) shows a comparison of the Raman spectra in the CH and OH stretching regions among [DEME][BF_4]-H_2O, [DEME][I]-H_2O, and [DEME][Br]-H_2O at $x = 65$ in the liquid state. The signals arising from the CH stretching modes of the [DEME] cation appear ranging from 2800 to 3100 cm^{-1}, though unfortunately the precise assignments of the respective peaks are not available at present. The overall CH frequency of the solutions shifts to a lower frequency side on going from [DEME][BF_4]-H_2O to [DEME][Br]-H_2O with slight changes in the spectral shapes. This indicates that the environment around the alkyl chains of [DEME] cation is perturbed with the change of anion.

On the contrary, peaks in the range from 3200 to 3800 cm^{-1} belong to the OH stretching vibrational mode of water molecules. Figure 5(b) shows the enlarged spectra in this region. In viewing the results, the difference in the Raman spectra are very distinct. As pointed out in a previous paper [25], the spectrum of [DEME][BF_4]-H_2O typically displays a small peak at around 3565 cm^{-1} with a shoulder of 3620 cm^{-1}. This peak is assigned to the nearly free hydrogen-bonded band (NFHB) of water molecules [26], as mentioned in the previous section. Here water molecules of the NFHB are probably very weakly interacting *via* H-bonding with the BF_4^- anions. There is no water molecules due from NFHB in the case where the anion is iodide. Instead, an intenisty of the Raman peak centered at ~3440 cm^{-1} is appreciable. This band is mainly due to the OH stretching vibrations of water molecules weakly hydrogen bonded to halide ions [27], suggesting that the interactions between water molecules and the anions are significantly different in the two RTILs, [DEME][BF_4] and [DEME][I] solutions; the anion-water interaction in the [DEME][I] solution is much stronger than that in the [DEME][BF_4] solution.

In viewing the results, the results of [DEME][Br]-H_2O show basically similar behavior to that of [DEME][I]-H_2O. The important result obtained from the comparison

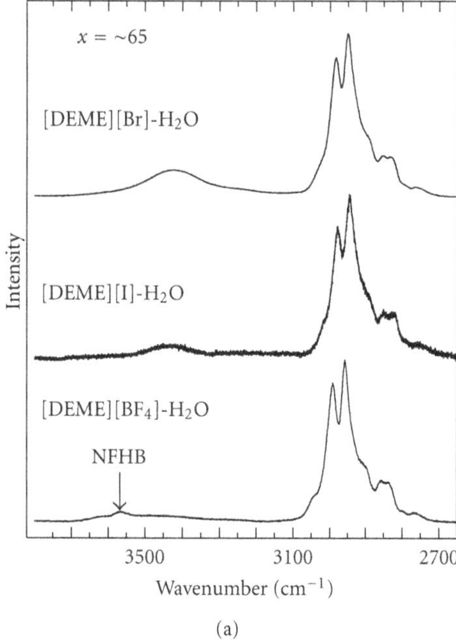

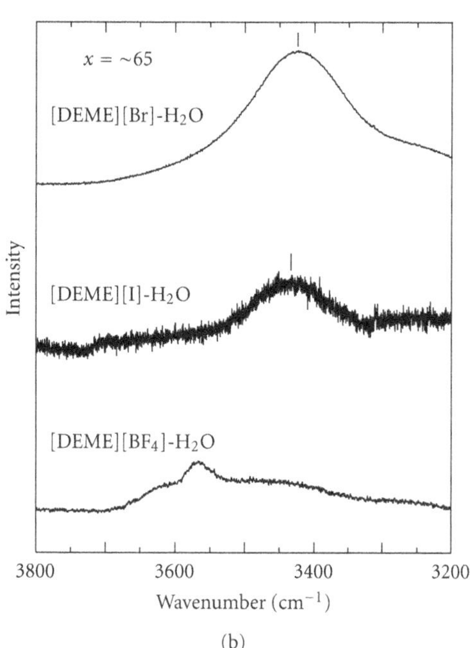

Figure 5: Comparison of the Raman spectra among $[DEME][BF_4]$-H_2O, [DEME][I]-H_2O, and [DEME][Br]-H_2O at x =~ 65 in the liquiud state. Peak intensities are normalized by the strongest peak of CH stretching vibrational mode of $[DEME]^+$ cation. (a) The CH and OH stretching vibrational modes of the solutions and (b) the OH stretching vibrational region.

of the two spectra is that the OH stretching spectrum of the [DEME][Br]-H_2O solution lies as a whole in a lower-frequency region than that of the [DEME][I]-H_2O solution, although the shift is only a marginal one at room temperature. This shows that average strength of hydrogen bonds in [DEME][Br]-H_2O solution is slightly stronger than that in [DEME][I]-H_2O solution.

In summary, we have investigated the glass transition behavior of the quaternary ammonium-type ionic liquid, [DEME][Br]-H_2O mixtures by a simple DTA method. The T_g of [DEME][Br]-H_2O mixtures decreases with increasing x up to around x = 91.5 and basically show one-T_g behavior. But at a specific concentration of around x = 92 two T_gs were observed. Additionally, we have compared the anionic effect on the glass transition behavior of the series of [DEME][X]-H_2O (X = Br^-, I^-, BF_4^-) mixtures. Similar studies on other different RTILs with, for example, imidazolium-based cation-H_2O systems, will deepen our understanding of the property of mixed solutions. Actually, we reported the glass transition temperatures of 1-butyl-3-methylimidazolium tetrafluoroborate-H_2O mixed solutions as a function of H_2O concentration [28]. Interestingly, in contrary to the results of quaternary ammonium-type ionic liquids, the multiple glass transition behavior was not observed and the system shows only one T_g throughout the whole concentration range.

Acknowledgments

The authors appreciate Dr. M. Aono and Dr. T. Takekiyo of National Defense Academy and Professor M. Kato and Mr. R. Wada of Ritsumeikan University for experimental supports and the fruitful discussion.

References

[1] T. Sato, G. Masuda, and K. Takagi, "Electrochemical properties of novel ionic liquids for electric double layer capacitor applications," *Electrochimica Acta*, vol. 49, pp. 3603–3611, 2004.

[2] M. J. Earle, P. B. McCormac, and K. R. Seddon, "Diels-Alder reactions in ionic liquids: a safe recyclable alternative to lithium perchlorate-diethyl ether mixtures," *Green Chemistry*, vol. 1, no. 1, pp. 23–25, 1999.

[3] D. Kerle, R. Ludwig, A. Geiger, and D. Paschek, "Temperature dependence of the solubility of carbon dioxide in imidazolium-based ionic liquids," *Journal of Physical Chemistry B*, vol. 113, no. 38, pp. 12727–12735, 2009.

[4] M. Galiński, A. Lewandowski, and I. Stepniak, "Ionic liquids as electrolytes," *Electrochimica Acta*, vol. 51, no. 26, pp. 5567–5580, 2006.

[5] Y. Imai, H. Abe, T. Goto, Y. Yoshimura, Y. Michishita, and H. Matsumoto, "Structure and thermal property of N,N-diethyl-N-methyl-2-methoxyethyl ammonium tetrafluoroborate-H_2O mixtures," *Chemical Physics*, vol. 352, pp. 224–230, 2008.

[6] W. Xu, L. M. Wang, R. A. Nieman, and C. A. Angell, "Ionic liquids of chelated orthoborates as model ionic glassformers," *Journal of Physical Chemistry B*, vol. 107, no. 42, pp. 11749–11756, 2003.

[7] J. G. Huddleston, A. E. Visser, W. M. Reichert, H. D. Willauer, G. A. Broker, and R. D. Rogers, "Characterization and comparison of hydrophilic and hydrophobic room temperature ionic liquids incorporating the imidazolium cation," *Green Chemistry*, vol. 3, no. 4, pp. 156–164, 2001.

[8] J. L. Anderson, J. Ding, T. Welton, and D. W. Armstrong, "Characterizing ionic liquids on the basis of multiple solvation interactions," *Journal of the American Chemical Society*, vol. 124, no. 47, pp. 14247–14254, 2002.

[9] K. R. Seddon, A. Stark, and M. J. Torres, "Influence of chloride, water, and organic solvents on the physical properties of ionic liquids," *Pure and Applied Chemistry*, vol. 72, no. 12, pp. 2275–2287, 2000.

[10] K. Nishikawa and K. I. Tozaki, "Intermittent crystallization of an ionic liquid: 1-isopropyl-3-methylimidazolium bromide," *Chemical Physics Letters*, vol. 463, no. 4–6, pp. 369–372, 2008.

[11] G. J. Kabo, A. V. Blokhin, Y. U. Paulechka, A. G. Kabo, M. P. Shymanovich, and J. W. Magee, "Thermodynamic properties of 1-butyl-3-methylimidazolium hexafluorophosphate in the condensed state," *Journal of Chemical and Engineering Data*, vol. 49, no. 3, pp. 453–461, 2004.

[12] O. Yamamuro, Y. Minamimoto, Y. Inamura, S. Hayashi, and H. O. Hamaguchi, "Heat capacity and glass transition of an ionic liquid 1-butyl-3-methylimidazolium chloride," *Chemical Physics Letters*, vol. 423, no. 4-6, pp. 371–375, 2006.

[13] Y. Imai, H. Abe, T. Miyashita, T. Goto, H. Matsumoto, and Y. Yoshimura, "Two glass transitions in *N, N*- diethyl-*N*-methyl-*N*-(2-methoxyethyl) ammonium tetrafluoroborate-H_2O mixed solutions," *Chemical Physics Letters*, vol. 486, no. 1–3, pp. 37–39, 2010.

[14] Y. Imai, H. Abe, H. Matsumoto, O. Shimada, T. Hanasaki, and Y. Yoshimura, "Glass transition behaviour of the quaternary ammonium type ionic liquid, [DEME][I]-H_2O mixtures," *Journal of Chemical Thermodynamics*, vol. 43, no. 3, pp. 319–322, 2011.

[15] R. Kiefer, S. Y. Chu, P. A. Kilmartin, G. A. Bowmaker, R. P. Cooney, and J. Travas-Sejdic, "Mixed-ion linear actuation behaviour of polypyrrole," *Electrochimica Acta*, vol. 52, no. 7, pp. 2386–2391, 2007.

[16] H. D. B. Jenkins, H. K. Roobottom, J. Passmore, and L. Glasser, "Relationships among ionic lattice energies, molecular (formula unit) volumes, and thermochemical radii," *Inorganic Chemistry*, vol. 38, no. 16, pp. 3609–3620, 1999.

[17] M. S. Sethi and M. Satake, *Periodic Tables and Periodic Properties*, Discovery Publishing House, Delhi, India, 2003.

[18] E. R. Nightingale Jr., "Phenomenological theory of ion solvation. Effective radii of hydrated ions," *Journal of Physical Chemistry*, vol. 63, no. 9, pp. 1381–1387, 1959.

[19] Z. B. Zhou, H. Matsumoto, and K. Tatsumi, "Low-melting, low-viscous, hydrophobic ionic liquids: aliphatic quaternary ammonium salts with perfluoroalkyltrifluoroborates," *Chemistry—A European Journal*, vol. 11, no. 2, pp. 752–766, 2005.

[20] Y. Imai, H. Abe, T. Goto, Y. Yoshimura, Y. Michishita, and H. Matsumoto, "Structure and thermal property of *N, N*- diethyl-*N*-methyl-*N*-2-methoxyethyl ammonium tetrafluoroborate-H_2O mixtures," *Chemical Physics*, vol. 352, no. 1–3, pp. 224–230, 2008.

[21] H. Kanno, K. Shimada, and K. Katoh, "Two glass transitions in the tetraethylammonium chloride-water system: evidence for a metastable liquid-liquid immiscibility at low temperatures," *Chemical Physics Letters*, vol. 103, no. 3, pp. 219–221, 1983.

[22] H. Kanno, K. Shimada, K. Yoshino, and T. Iwamoto, "Anomalous concentration dependence of the glass transition temperature for aqueous tetrapropylammonium chloride solution," *Chemical Physics Letters*, vol. 112, no. 3, pp. 242–245, 1984.

[23] H. Kanno, K. Shimada, and T. Katoh, "A glass formation study of aqueous tetraalkylammonium halide solutions," *Journal of Physical Chemistry*, vol. 93, no. 12, pp. 4981–4985, 1989.

[24] G. S. Fulcher, "Analysis of recent measurements of the viscosity of glasses," *Journal of the American Ceramic Society*, vol. 6, pp. 339–355, 1925.

[25] Y. Yoshimura, T. Goto, H. Abe, and Y. Imai, "Existence of nearly-free hydrogen bonds in an ionic liquid, *N, N*- diethyl-*N*-methyl-*N*-(2-methoxyethyl) ammonium tetrafluoroborate-water at 77 K," *Journal of Physical Chemistry B*, vol. 113, no. 23, pp. 8091–8095, 2009.

[26] L. Cammarata, S. G. Kazarian, P. A. Salter, and T. Welton, "Molecular states of water in room temperature ionic liquids," *Physical Chemistry Chemical Physics*, vol. 3, no. 23, pp. 5192–5200, 2001.

[27] H. Kanno and J. Hiraishi, "Raman spectroscopic study of aqueous LiX and CaX_2 solutions (X = Cl, Br, and I) in the glassy state," *Journal of Physical Chemistry*, vol. 87, no. 19, pp. 3664–3670, 1983.

[28] Y. Yoshimura, H. Kimura, C. Okamoto, T. Miyashita, Y. Imai, and H. Abe, "Glass transition behaviour of ionic liquid, 1-butyl-3-methylimidazolium tetrafluoroborate-H_2O mixed solutions," *Journal of Chemical Thermodynamics*, vol. 43, no. 3, pp. 410–412, 2011.

13

CFD Analysis for Heat Transfer Enhancement inside a Circular Tube with Half-Length Upstream and Half-Length Downstream Twisted Tape

R. J. Yadav[1] and A. S. Padalkar[2]

[1] *Department of Mechanical Engineering, MIT College of Engineering, Pune-411028, India*
[2] *Department of Mechanical Engineering, Flora Institute of Technology, Pune-412205, India*

Correspondence should be addressed to R. J. Yadav, rupesh_yadava@rediffmail.com

Academic Editor: Ahmet Z. Sahin

CFD investigation was carried out to study the heat transfer enhancement characteristics of air flow inside a circular tube with a partially decaying and partly swirl flow. Four combinations of tube with twisted-tape inserts, the half-length upstream twisted-tape condition (HLUTT), the half-length downstream twisted-tape condition (HLDTT), the full-length twisted tape (FLTT), and the plain tube (PT) with three different twist parameters ($\lambda = 0.14, 0.27$, and 0.38) have been investigated. 3D numerical simulation was performed for an analysis of heat transfer enhancement and fluid flow for turbulent regime. The results of CFD investigations of heat transfer and friction characteristics are presented for the FLTT, HLUTT, and the HLDTT in comparison with the PT case.

1. Introduction

The heat transfer enhancement technology (HTET) has been developed and widely applied to heat exchanger applications over last decade, such as refrigeration, automotives, process industry, nuclear reactors, and solar water heaters. Till date, there have been many attempts to reduce the sizes and the costs of the heat exchangers and their energy consumption with the most influential factors being heat transfer coefficients and pressure drops, which generally lead to the incurring of less capital costs.

HTET can offer significant economic benefits in various industrial processes. By "augmentation" we mean an enhancement in heat transfer, over that which is existent on the reference surface for similar operating conditions.

Bergles and Webb [1, 2] have reported comprehensive reviews on techniques for heat transfer enhancement. For a single-phase heat transfer, the enhancement has been brought using roughened surfaces and other augmentation techniques, such as swirl/vortex flow devices and modifications to duct cross sections and surfaces. These are the passive augmentation techniques, which can increase the convective heat transfer coefficient on the tube side. Many techniques for the enhancement of heat transfer in tubes have been proposed over the years.

Siddique et al. [3] reported the following heat transfer enhancers in his review paper: (a) extended surfaces including fins and microfins, (b) porous media, (c) large particles suspensions, (d) nanofluids, (e) phase-change devices, (f) flexible seals, (g) flexible complex seals, (h) vortex generators, (i) protrusions, and (j) ultrahigh thermal conductivity composite materials. Many methods that assist in heat transfer enhancement effects have been extracted from the literature.

Among of these methods discussed in the literature are using joint fins, fin roots, fin networks, biconvections, permeable fins, porous fins, and helical microfins and using complicated designs of twisted-tapes. The authors concluded that more attention should be made towards single phase heat transfer augmented with microfins in order to alleviate the disagreements between the works of the different authors.

Also, it was noted that additional attention should be made towards uncovering the main mechanisms of heat

Table 1: The highest recorded heat transfer enhancement level estimated due to each enhancer, (Siddique et al. [3]).

Heat transfer enhancer type	(Heat transfer due presence of the enhancer/heat transfer in absence of the enhancer)
Fins inside tubes	2.0, [4]
Microfins inside tubes	4.0 for laminar, [5]
Porous media	≈12.0 (keff/kf), [6]
Nanofluids	3.5, [7]
Flexible seals	2.0, [8]
Flexible complex seals	3.0, [9]
Vortex generators	2.5, [10]
Protrusions	2.0, [11, 12]
Ultra high thermal conductivity composite materials	6, [13]

transfer enhancements due to the presence of nanofluids. Further, it can be concluded that perhaps the successful modeling of flow and heat transfer inside porous media as seen in the works of Kim and Kuznetsov [14], which is a well-recognized passive enhancement method, could help in well discovering the mechanism of heat transfer enhancements due to nanofluids. This is due to some similarities between both media. Additionally, many recent works related to passive augmentations of heat transfer using vortex generators, protrusions, and ultra high thermal conductivity composite material have been reported by the authors. Whereas, Nield and Kuznetsov [15] reported an analysis of laminar forced convection in a helical pipe of circular cross section and filled by a porous medium saturated with a fluid, for the case when the curvature and torsion of the pipe are both small. Later Cheng and Kuznetsov [16, 17] carried out an investigation of laminar flow in a helical pipe filled with a fluid saturated porous medium. The maximum levels of the heat transfer enhancement estimated due to each enhancer were presented in Table 1.

Inside the round tubes, a wide range of inserts, such as tapered spiral inserts, wire coil, twisted-tape with different geometries, rings, disks, streamlined shapes, mesh inserts, spiral brush inserts, conical-nozzles, and V-nozzles, have been used Promvonge and Eiamsa-ard [18] and Promvonge [19].

Smithberg and Landis [20] have estimated the tape-fin effect assuming a uniform heat transfer coefficient on the tape wall, equal to that on the tube wall. Authors reported that the fin effect increases the heat transfer but in practice, the tape-fin effect will not attain such a high value due to the poor contact between the tape and the tube. In order to estimate the tape fin effect, Lopina and Bergles [21] conducted experiments using insulated tapes. Assuming zero contact resistance between tube and tape with equal and uniform heat transfer coefficients on tube and tape walls authors predicted from 8% to 17% of the heat is transferred through the tape.

Date [22] reported the prediction of fully developed, laminar and turbulent, and uniform-property flow in a tube containing a twisted-tape. The predictions have shown that significant augmentation in heat transfer can be obtained at high Reynolds and Prandtl numbers, low twist ratios, and high fin parameters.

Manglik and Bergles [23] presented experimental correlations for pressure drop and heat transfer coefficient for laminar, transition, and turbulent flow in isothennal-wall tubes with twisted-tape inserts. Unlike previous correlations, they included the tape thickness in order to properly account for the helical twisting of the streamlines.

Al-Fahed et al. [24] reported an experimental work to study the heat transfer and friction characteristics in a microfin tube fitted with twisted-tape inserts for three different twist/width ratios under laminar flow region.

Saha and Dutta [25] reported investigation on the experimental data on swirl flow due to twisted-tape in laminar region for friction factor and Nusselt number for a large Prandtl number range (from 205 to 518). The author observed that, on the basis of a constant pumping power, short-length twisted-tape is a good choice because in this case swirl generated by the twisted-tape decays slowly downstream which increases the heat transfer coefficient with minimum pressure drop, as compared with a full-length twisted-tape.

Whereas, the concluding remarks from earlier studies on numerical and experimental work are as follows.

Rahimi et al. [26] carried out experimental and CFD studies on heat transfer and friction factor characteristics of a tube equipped with modified twisted-tape inserts. The investigations are with the classic and three modified twisted-tape inserts. The authors observed that the Nusselt number and performance of the jagged insert were higher than other ones.

Eiamsa-ard et al. [27] carried out the numerical analysis of heat and fluid flows through a round tube fitted with twisted-tape. The author investigated the effect of tape clearance ratio on the flow, heat transfer, and friction factor.

Chiu and Jang [28] studied numerically and experimentally three-dimensional gas-fluid flow and heat transfer inside tubes with longitudinal strip inserts (both with/without holes) and twisted-tape inserts twisted at three different angles ($\alpha = 34.3^\circ$, 24.4°, and 15.3°).

From above studies it could be concluded that the tube-tape inserts in a full length of tube provide one of the most attractive heat transfer augmentative techniques for flow inside the tube on account of its simplicity and the effectiveness. The short-length twisted-tape s have been considered by many researchers due to reduction of pressure drop while augmenting the heat transfer simultaneously.

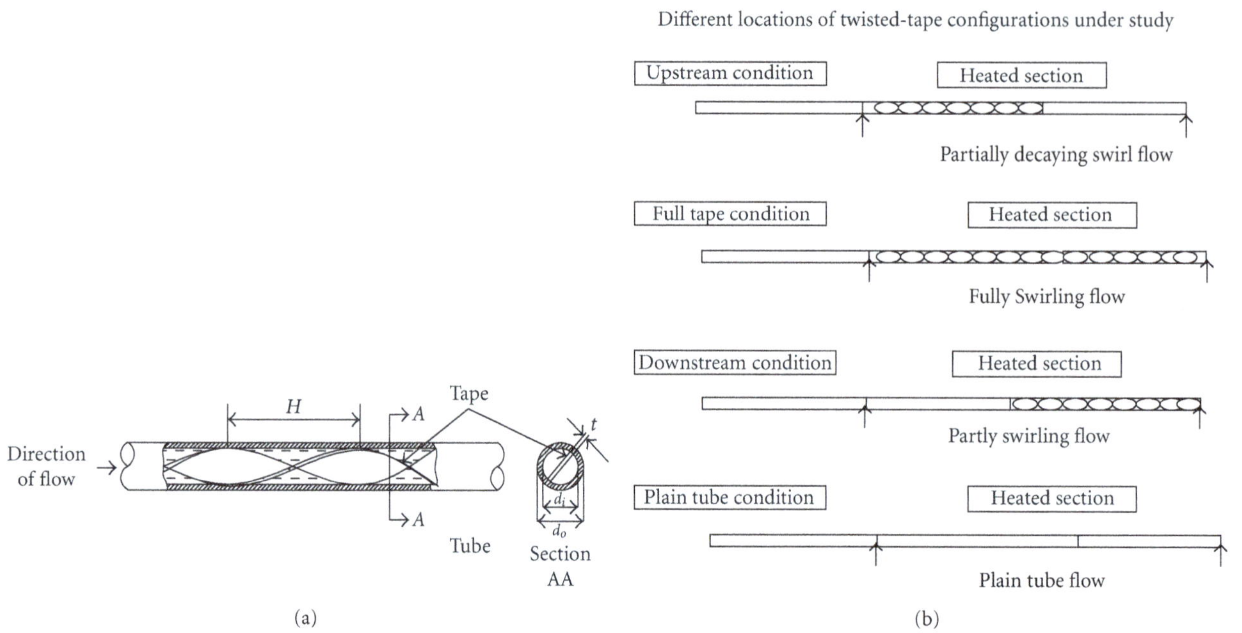

Figure 1: Problem definition geometrical configuration (a), physical model of the problem defined (b). Different configurations under study.

A CFD prediction of the heat transfer and friction characteristics of the partially decaying swirl flows in the turbulent flow regime has been taken up to study the structure of velocity and temperature fields. The effects of the twisted-tape location on pressure drop and heat transfer characteristics due to creation of swirl in the turbulent flow within the tubes were also studied.

2. Numerical Simulations

2.1. Physical Model. The numerical simulations were carried out using the CFD software package FLUENT-6.2.16 that uses the finite-volume method to solve the governing equations.

Geometry was created for air flowing in an electrically heated stainless steel tube of 22 mm diameter (D) and length (L) 90 times the diameter as in the experimental setup as shown in Figure 1. A computational model has been created in GAMBIT-2.2.30 as shown in Figure 3.

In this study, the effects of the twist parameter ($\lambda = d/H = 0.14$, 0.27, and 0.38) and the two heat flux inputs ($Q = 2300$ and $6200\,\text{W/m}^2$) on heat transfer rate (Nu), friction factor (f), and thermal performance factor (η) are examined under uniform heat flux conditions with air as the testing fluid and with different inlet frontal velocities with Reynolds number, Re between 25000 and 110000.

Twisted-tape inserts under the following locations of the twisted-tape configurations were used.

(a) Upstream condition (HLUTT)—tube with twisted-tapes located in the first half of 50 diameters of the heated section. (Partially decaying swirl flow.)

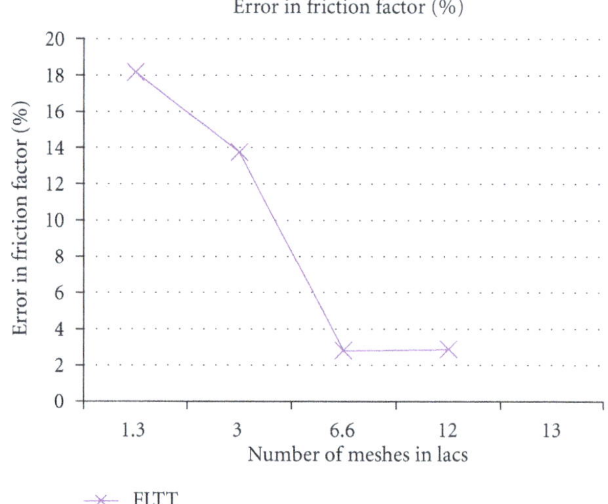

Figure 2: Grid independence test.

(b) Downstream condition (HLDTT)—tube with twisted-tapes located in the second half of 50 diameters of the heated section. (Partly swirl flow.)

(c) Full-length condition (FLTT)—tube with twisted-tapes located in the full length of the heated section. (Fully swirl flow.)

(d) Plain tube condition (PT)—tube without twisted-tapes in the full length of the heated section. (Smooth tube flow.)

2.2. Numerical Method. The available finite-difference procedures for swirling flows and boundary layer are employed

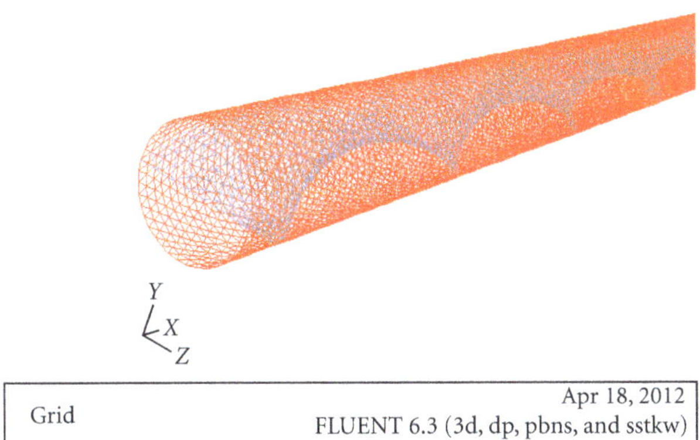

FIGURE 3: Partial view of tube with twisted-tape inside.

to solve the governing partial differential equations. Some simplifying assumptions are required for applying of the conventional flow equations and energy equations to model the heat transfer process in tube with twisted-tape.

For turbulent, steady, and incompressible air flow with constant properties, while we neglect the natural convection and radiation, we follow the three-dimensional equations of continuity, momentum, and energy, in the fluid region. These equations are as below.

Continuity equation:

$$\frac{\partial \rho}{\partial t} + \nabla \cdot (\overline{\rho v}) = 0. \quad (1)$$

Momentum equation:

$$\frac{\partial (\overline{\rho v})}{\partial t} + \nabla . (\overline{\rho v v}) = -\nabla p + \nabla . (\overline{\tau}) + \rho \overline{g}. \quad (2)$$

Energy equation:

$$\frac{\partial (\rho E)}{\partial t} + \nabla . (v(\rho E + p)) = \nabla \cdot (k_{\text{eff}} \nabla T + (\overline{\tau_{\text{eff}}} . \overline{v})). \quad (3)$$

In the Reynolds-averaged approach for turbulence modeling, the Reynolds stresses in (2) are appropriately modeled by a method that employs the Boussinesq hypothesis to relate the Reynolds stresses to the mean velocity gradients as shown below:

$$-\rho \overline{u i' u j'} = \mu_t \left(\frac{\partial u i}{\partial x j} + \frac{\partial u j}{\partial x i} \right) - \frac{2}{3} \left(\rho \kappa + \mu_t \frac{\partial u \kappa}{\partial x \kappa} \right) \delta i j. \quad (4)$$

An appropriate turbulence model is used to compute the turbulent viscosity term μ_t. The turbulent viscosity is given as

$$\mu_t = \rho C \mu \frac{k^2}{\varepsilon}. \quad (5)$$

The second-order upwind scheme was used to discretize the convective term. The linkage between the velocity and pressure was computed using the SIMPLE algorithm. The standard wall treatment model was chosen for the near-wall modeling method.

For validating the accuracy of numerical solutions, the grid independent test has been performed for the physical model. The grid is highly concentrated near the wall and in the vicinity of the twisted-tape. Four grid systems with about 130000, 300000,660000, and 1200000 cells are adopted to calculate grid independence. We compared the friction factors for these four mesh configurations as shown in Figure 2. After checking the grid independence test, the simulation grid in this study was meshed using about 6, 60,000 cells that consisted of tetrahedral grid.

Figure 3 shows an example of the partial-meshed configuration of the round tube equipped with a twisted-tape. It consists of a tube of diameter 22 mm containing twisted-tape insert, test section 2000 mm, and calming section of 1200 mm dimensions just like those in experimental setup with twist angle 0.14. To capture wall gradient effects, mesh has been finer toward the walls. There are a total of 6, 60,000 nodes in the domain simulation.

In addition, a convergence criterion of 10^{-6} was used for energy and 10^{-3} for the mass conservation of the calculated parameters.

The air inlet temperature was specified as 300 K, and three assumptions were made in the model: (1) the uniform heat flux was along the length of test section, (2) the wall of the inlet calming section was adiabatic, and (3) the physical properties of air were constant and were evaluated at the bulk mean temperature. The velocity inlet boundary condition was adopted at the inlet and outflow at the outlet of the domain shown in Figure 1.

2.3. Data Reduction. In order to express the experimental results in a more efficient way, the measured data was reduced using the following procedure. Three important parameters considered were the friction factor, the Nusselt number, and the thermal performance factor, which were used for determining the friction loss, heat transfer rate, and the effectiveness of heat transfer enhancement in the tube, respectively. The friction factor (f) is computed from pressure drop, ΔP across the length of the tube (L) using the following equation:

$$f = \frac{\Delta P}{(L/D) \cdot (\rho U^2/2)}. \quad (6)$$

The Nusselt number is defined as

$$\text{Nu} = \frac{h \cdot D}{k}. \quad (7)$$

The average Nusselt number can be obtained by

$$\text{Nu}_{\text{avg}} = \int \text{Nu}(x) \frac{\partial x}{L}. \quad (8)$$

The various characteristics of the flow, the Nusselt number, and the Reynolds number were based on the average of the tube wall and the outlet air temperatures. The local wall temperature, inlet and outlet air temperatures, the pressure drop across the test section, and the air flow velocity were measured for heat transfer of the heated tube with

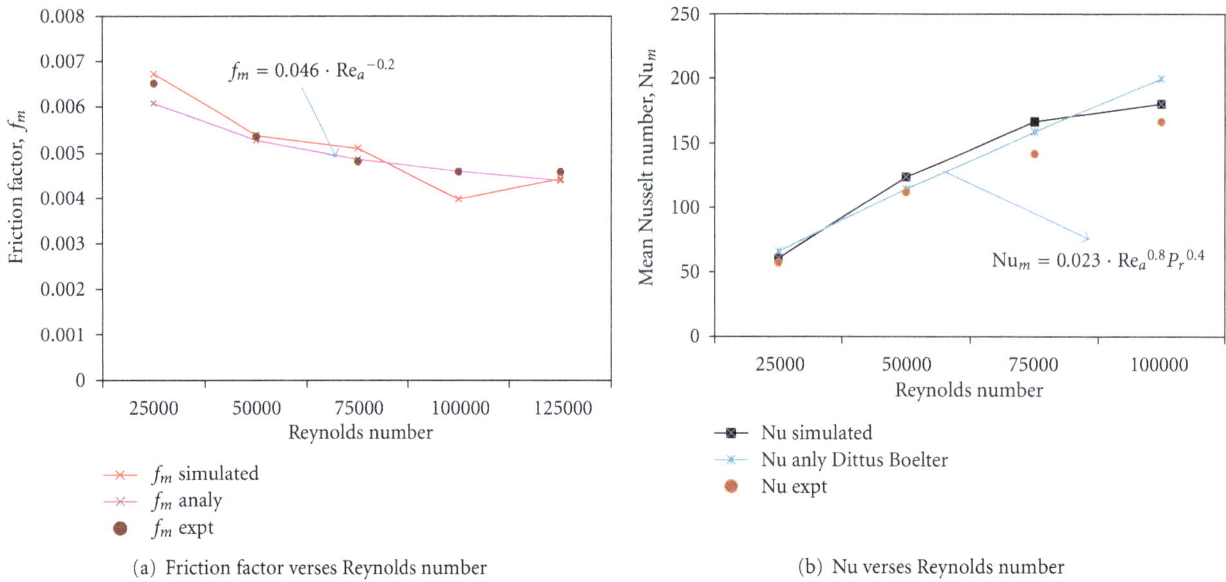

(a) Friction factor verses Reynolds number

(b) Nu verses Reynolds number

Figure 4: Comparison of CFD results with experimental data and correlations.

different tube inserts. The average Nusselt numbers and friction factors were calculated, and all fluid properties were determined at the overall bulk mean temperature.

Thermal performance factor was given by

$$\eta = \frac{\mathrm{Nu}/\mathrm{Nu}_o}{(f/f_o)^{1/3}}, \tag{9}$$

where in Nu_o, Nu, f_o, and f were the Nusselt numbers and friction factors for the plain tube and the tube with twisted-tape swirl generator, respectively.

3. Results and Discussion

3.1. Validation of Setup. The CFD simulation result of the plain tube (PT) without a twisted-tape insert has been validated with the experimental data as shown in Figures 4(a) and 4(b). The Dittus-Boelter equation for the heat transfer and the Blasius equation for the friction factor are the correlations used for the comparison. These results are within ±15% deviation for the heat transfer (Nu) and ±6% for the friction factor (f). Similarly, the CFD results for the plain tube are compared with analytical correlations. The CFD results are within ±9% deviation for the heat transfer (Nu) and ±6% for the friction factor (f) with slightly higher deviation of ±17% for Re higher than 75000.

3.2. Heat Transfer. Effect of the FLTT twisted-tapes and HLUTT twisted-tapes on the heat transfer rate is presented in Figure 5. The results for the tube fitted with HLUTT and HLDTT have been compared with those for a plain tube and the FLTT under similar operating conditions for $\lambda = 0.14$.

It was seen that the effect of different inserts on the heat transfer rate was significant for all the Reynolds numbers used due to the induction of high reverse flows and disruption of boundary layers. This technique has resulted in an improvement of the heat transfer rate over the plain tube.

It is clearly seen that as the Reynolds number goes on increasing, the heat transfer coefficient also goes on increasing.

It was found that the heat transfer coefficients in the tubes with the FLTT were 29–86% greater than those in the case of the plain tubes without inserts.

When the twisted-tape inserts with the HLUTT condition were used, the heat transfer coefficients were 8–37% higher than those of the plain tubes. However, the twisted-tape inserts with the HLUTT and HLDTT conditions had 15–95% reduction in values for the heat transfer coefficient when compared to the FLTT condition. Whereas when the twisted-tape inserts with the HLDTT condition were used, the heat transfer coefficients were 9–47% higher than those of the plain tubes.

3.3. Friction Factor. The variation of the pressure drop is presented in terms of the friction factor as shown in Figure 6. It shows the friction factor versus the Reynolds numbers for different combinations of inserts.

It is seen that the friction factors obtained from three different inserts follow a similar trend and this decreases with an increase in the Reynolds number. The increase in friction factor with swirl flow is much higher than that with an axial flow.

It was found that the pressure drop for the FLTT inserts was 203–623% higher than that for the plain tubes. For the HLUTT inserts, it was 36–170% higher than that for the plain tubes. However, the pressure drop for the HLUTT was 82–168% less than that for the FLLT inserts. For the HLDTT inserts, it was 31–144% higher than that for the plain tubes. It was seen that the highest pressure drop occurred when the tape inserts with a twist ratio $\lambda = 0.38$ were used.

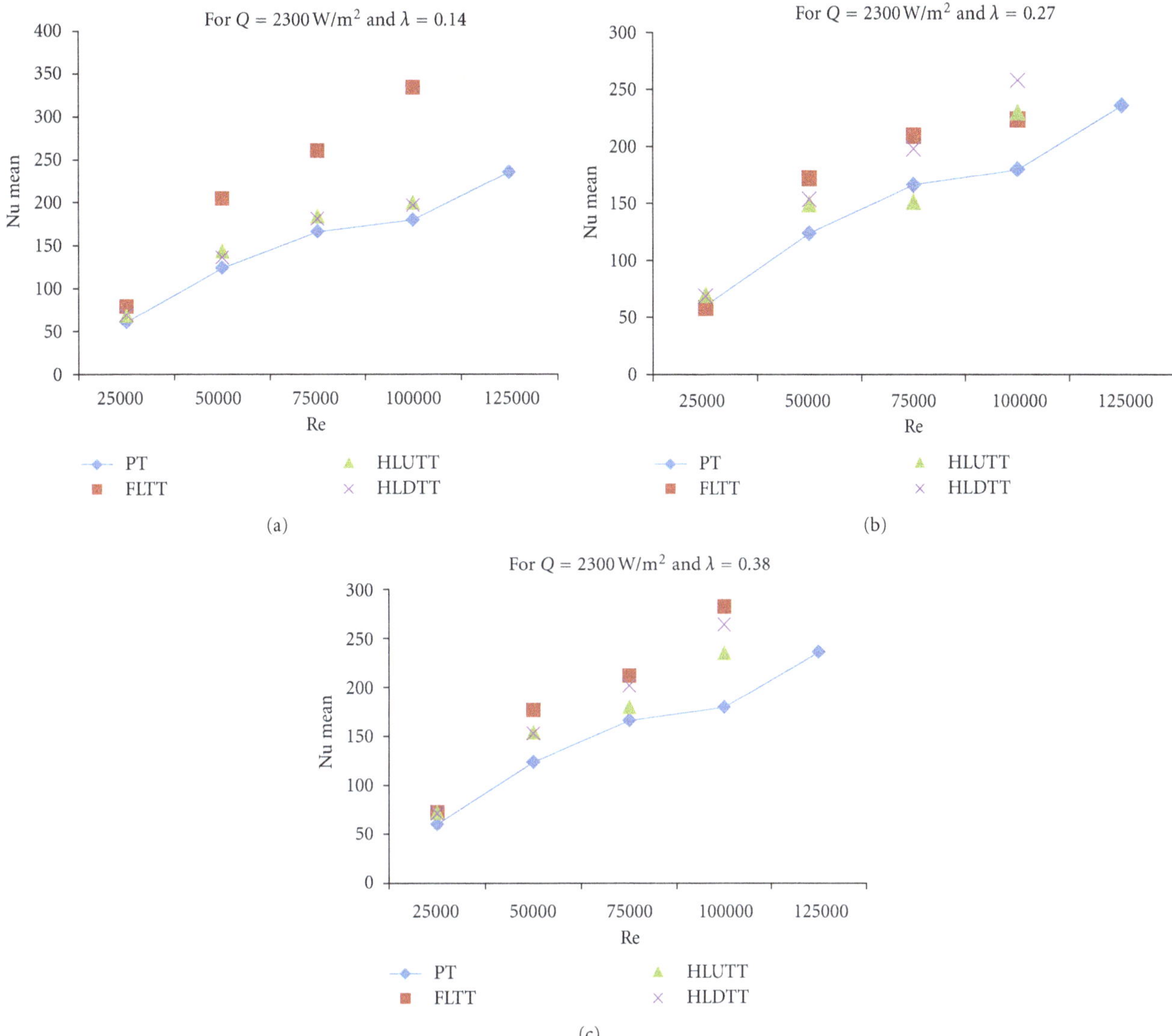

FIGURE 5: Variation of Nussle number (Nu) versus Re for different inserts: (a) for $\lambda = 0.14$, (b) for $\lambda = 0.27$ and, (c) for $\lambda = 0.38$.

3.4. Thermal Performance Factor. From Figure 7, it has been observed that the thermal performance factor tends to decrease with an increasing twist parameter and with an increase in the Reynolds number for HLUTT and HLDTT twisted-tapes. Whereas for FLTT, it was found that thermal performance factor tends to decrease with an increasing twist ratio and increase with an increase in the Reynolds number. For all the twist ratios, the HLUTT and HLDTT configurations have been seen to give the thermal performance factors in the range of 1.02–1.16, which is comparable with those provided by the FLTT (1.03–1.24).

3.5. Streamline and Pathline. Plots of pathlines through the tube with twisted-tape inserts have been shown in Figure 8. It is evident that the insertion of the tape induces the swirling flow, and the twisted-tapes generate two types of flows which are (1) a swirling flow and (2) an axial or straight flow near the tube wall. It is noteworthy that the FLTT gives higher velocity of the fluid flow through the test section compared to those with partially extending tapes where decaying of swirl flow along the length of tube takes place.

3.6. Velocity Vector Plots. Vector plots of velocity predicted for the tubes with FLTT and PT configuration are depicted in Figure 9 and Figure 10. As seen in the figures, two longitudinal vortices are generated around tapes in the core flow area. These longitudinal vortices play a critical role of disturbing the boundary layer and making the temperature uniform in the core flow. And at the same time, it has been found that a vortex tends to decay along the length in case of HLUTT and HLDTT cases which are partially decaying due to absence of twisted-tape in the latter part of the test section as shown in Figure 11. Whereas for the HLUDTT case a vortex tends to grow along the length after the initial half part of the test section as shown in Figure 12. The tangential

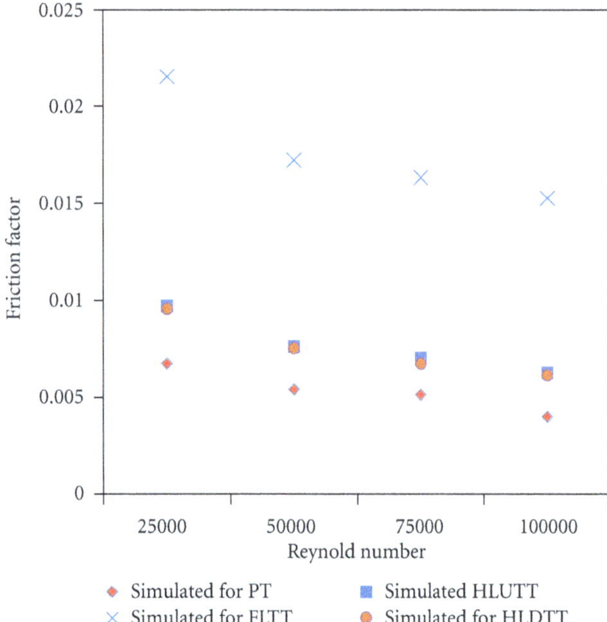

FIGURE 6: Variation of friction factor versus Re for different inserts for $\lambda = 0.14$ at $Q = 2300\,\mathrm{W/m^2}$.

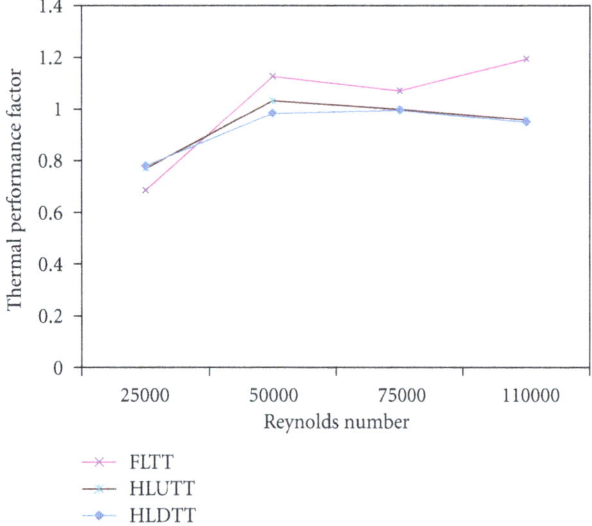

FIGURE 7: Variation of thermal performance factor versus Re for different inserts for $\lambda = 0.14$ at $Q = 2300\,\mathrm{W/m^2}$.

velocity is almost zero for the plain tube at all the Reynolds numbers. However, it is seen that this velocity component increases when any of the above mentioned inserts are placed inside the tube.

3.7. Temperature Profiles Analysis

(a) Plain Tube Data. It is observed from the smooth tube temperature profile that the maximum wall temperature in the test section, $T_{wx,max}$, and the maximum temperature difference between the tube wall and the fluid, $(T_{wx} - T_{bx})_{max}$,

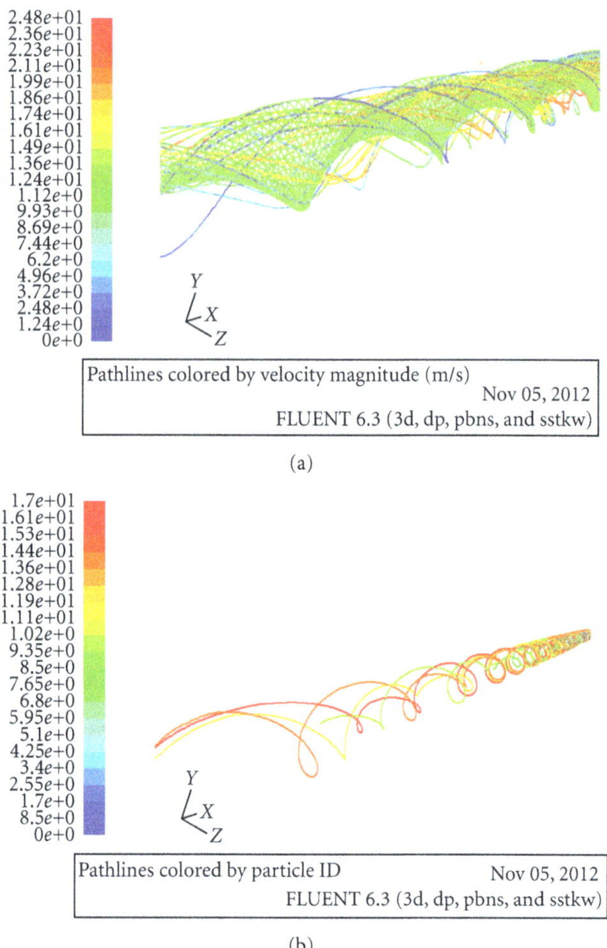

FIGURE 8: (a) Pathline through twisted-tape; (b) particle track through the tube with twisted-tape inserts.

both occur at $X/D = 86$. $T_{wx,max}$ varies from 48°C to 215°C, $(T_{wx} - T_{bx})_{max}$ varies from 10°C to 56°C, and the wall temperature profile shows fully developed characteristics at roughly 17 diameters from entrance. At any location wall temperature decreases with Reynolds number and increases with heat flux.

(b) Twisted-Tape Data. Figures 13 and 14 show wall temperature profiles, for all the cases investigated. The data presented reveal the following trend.

(i) *Effect of Reynolds Number on T_{wx}.* At any axial location the local wall temperature decreases with increasing Reynolds number. This is quite expected since heat transfer coefficients increase with Reynolds number bringing down both the wall to fluid temperature difference and the absolute wall temperature.

(ii) *Effect of Heat Flux on T_{wx}.* In all the cases, an increase in heat flux results in an increase in the local wall temperature. For the upstream condition at $\lambda = 0.14$, an increase in heat flux caused, in addition to an increase

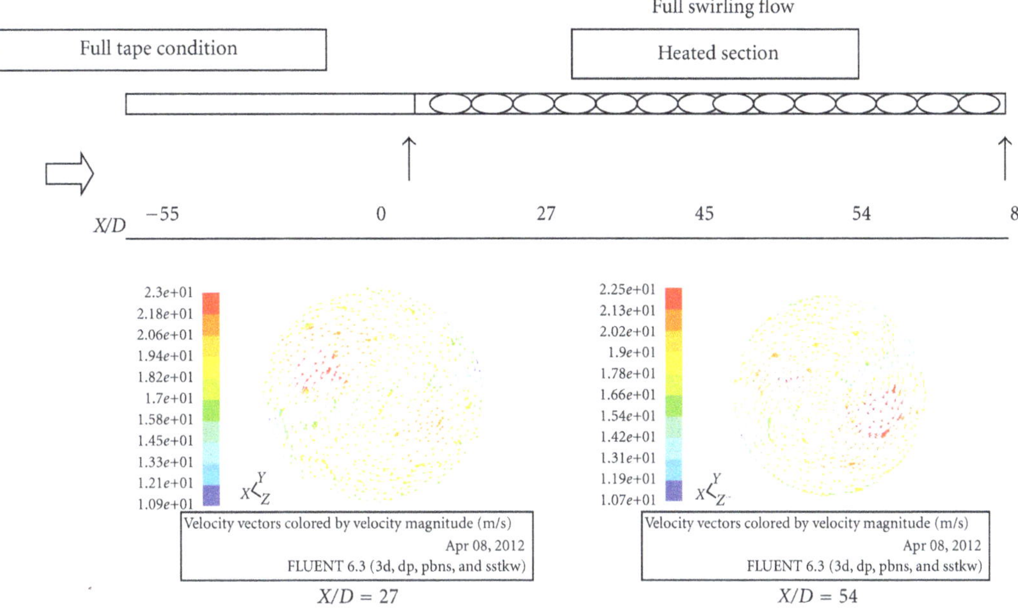

FIGURE 9: Vector plots across tube cross section along the axial length for FLTT with $\lambda = 0.14$, Re = 25000, and $q = 2300\,\mathrm{W/m^2}$.

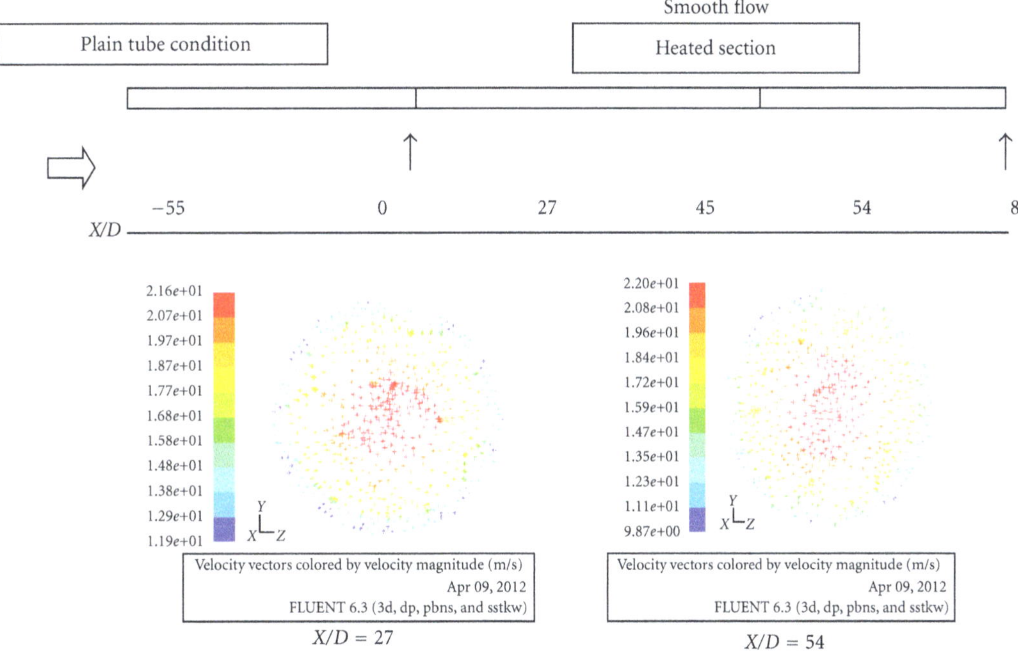

FIGURE 10: Vector plots across tube cross section along the axial length for PT with $\lambda = 0.14$, Re = 25000, and $q = 2300\,\mathrm{W/m^2}$.

in the local wall temperature, a shift in the location of $(T_{wx} - T_{bx})_{min}$ from $X/D = 36$ to $X/D = 45$.

(iii) *Effect of Twist Parameter on* T_{wx}. It is observed that the twist parameter λ has a significant effect on both the magnitude of T_{wx} and its variation along the test section. Effect of λ on T_{wx} will be discussed separately for the upstream and downstream conditions.

Upstream Condition. A dip in wall temperature is observed at the end of tape section for all values of λ. In the swirl decay section following the tape, the local wall temperature increases with an increase in λ.

Downstream Condition. The maximum wall temperature, $T_{wx,max}$ is located at $X/D = 86$, for all values of λ. A steep temperature drop is noticed rear the entrance to tape section,

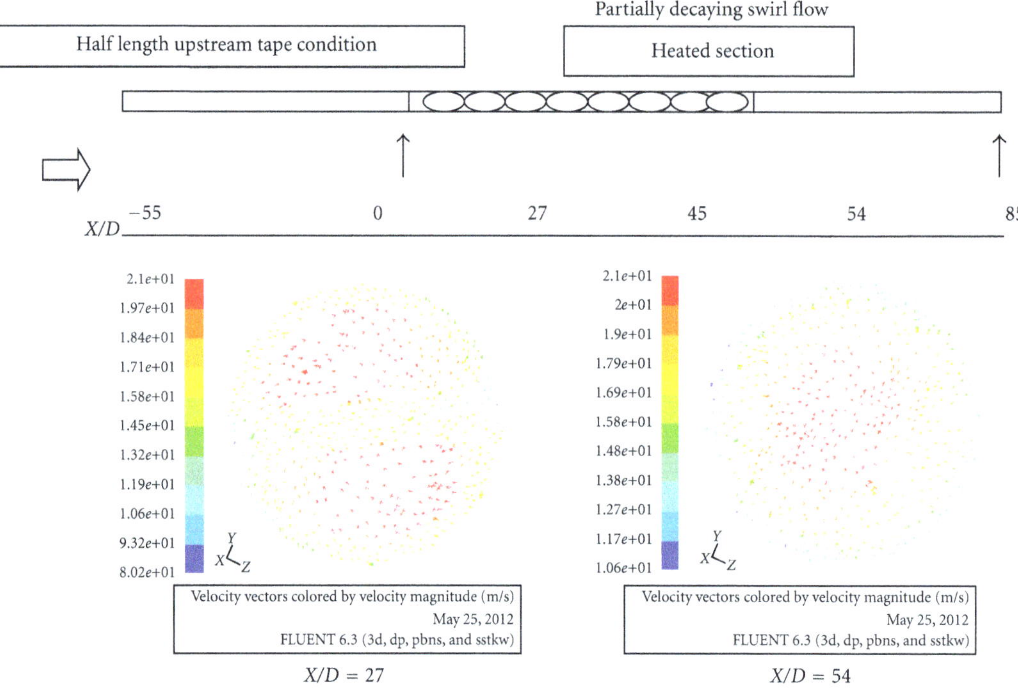

FIGURE 11: Vector plots across tube cross section along the axial length for HLUTT with $\lambda = 0.14$, Re = 25000, and $q = 2300\ \mathrm{W/m^2}$.

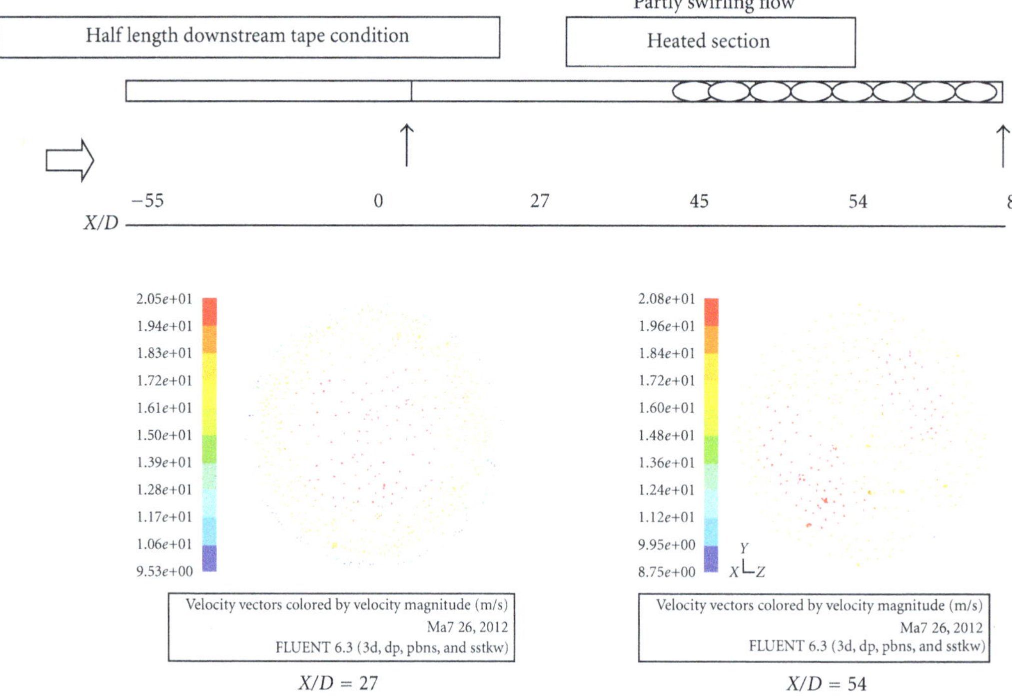

FIGURE 12: Vector plots across tube cross section along the axial length for HLDTT with $\lambda = 0.14$, Re = 25000, and $q = 2300\ \mathrm{W/m^2}$.

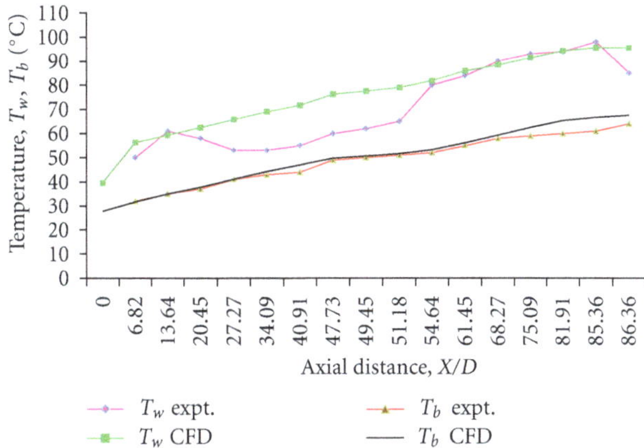

FIGURE 13: Experimental and CFD wall temp and surface temp for upstream location for $q = 2300\,\text{W/m}^2$, Re = 25000, and $\lambda = 0.14$.

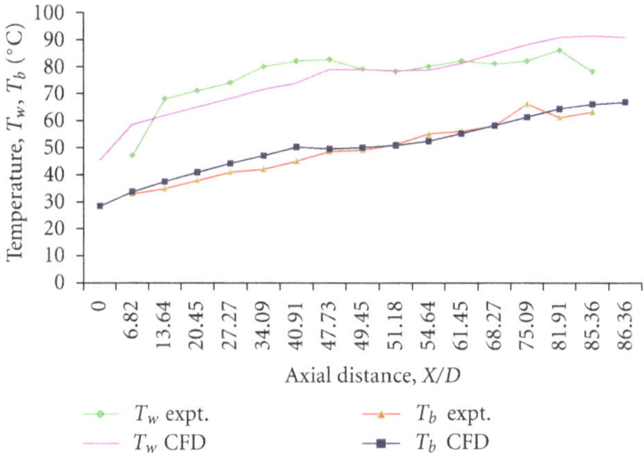

FIGURE 14: Expt. and CFD wall temp and surface temp for Downstream location for $q = 2300\,\text{W/m}^2$, Re = 25000, and $\lambda = 0.14$.

for $\lambda = 0.38$ and 0.27, except for $\lambda = 0.14$. For all values of λ, the tape section records the lowest wall temperature and the smooth section, the highest.

(iv) *Effect of Tape Location on* T_{wx}. It is observed that the local wall temperatures are least for downstream condition. The effect of tape location on the maximum test section temperature, $T_{wx,max}$, is given below. Values are given as a percentage decrease from corresponding smooth tube values. See Table 2.

It is seen that downstream location of tape is most effective in bringing down the maximum test section wall temperature.

3.8. Local Nusselt Number Analysis. An examination of the local Nusselt number profiles for the upstream and downstream conditions as shown in Figure 15 shows that the Nusselt number attains local peaks, a characteristic which was noticed by Klepper [29] also in his experiments on partially extending tapes. This unusual behavior of the local Nusselt number has not been reported by any other investigator. That the occurrence of these peaks is real and appears beyond doubt when it is observed that they occur at all Re, λ and, q and for both the downstream and upstream tape locations.

TABLE 2: The effect of tape location on the maximum test section temperature, Twx, max.

Tape location	Min	Max
HLDTT	5.2%	19%
HLUTT	0.0%	6%

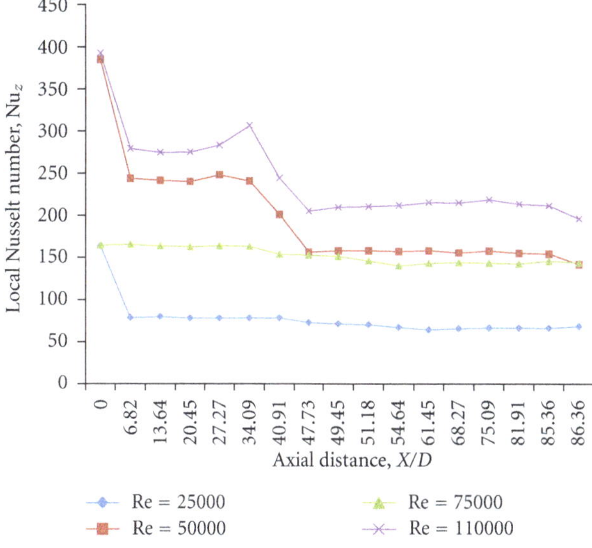

FIGURE 15: Local Nusselt number along axial distance for HLUTT for $\lambda = 0.14$, $q = 2300\,\text{W/m}^2$.

While, the characteristic of local peaks observed in the investigation can be used in avoiding the local hot spots in heat exchanger application, with possible application in such diverse areas as the cooling of an overheated rocket nozzle throat, prevention of burnout in space and earth power plants, and reduction of wall temperature in circulating fuel reactors and in the heat exchange equipment used in process industries. In most of the above applications temperatures critical to material life are likely to be reached, and as such any reduction in wall temperature would imply an improvement in performance.

4. Conclusion

The important issue in the present work can be expressed as the understanding of heat transfer and temperature analysis for fully, partially decaying, and partly swirl flow using the FLTT, HLUTT, and HLDTT twisted-tape insert.

The performance of this insert was compared with those of the FLTT twisted-tape inserts and the PT.

It was found that the heat transfer coefficient and the pressure drop in the tubes with the FLTT were 29–86% and 203–623% greater than those in the case of the plain tubes without inserts.

When the twisted-tape inserts with the HLUTT condition were used, the heat transfer coefficient and the pressure drop were estimated at 8–37% and 36–170% higher than those in the case of the plain tubes.

When the twisted-tape inserts with the HLDTT condition were used, the heat transfer coefficient and the pressure drop were estimated at 9–47% and 31–144% higher than those in the plain tube case.

It was found that thermal performance and local peak in heat transfer could be increased by using a combination of inserts with different geometries in the plain tubes while reducing the pressure drop. The characteristic of local peaks observed in the investigation can be used in avoiding the local hot spots in heat exchanger application.

Since the Nusselt number peaks were observed for both the downstream and upstream tape locations, the choice of tape location would be governed by the actual location of hot spots.

Nomenclature

E: Total energy, J
f: Friction factor
h: Enthalpy, J or convective heat transfer coefficient, $Wm^{-2} K^{-1}$
k: Thermal conductivity, $Wm^{-1} K^{-1}$
k_{eff}: Effective thermal conductivity, $Wm^{-1} K^{-1}$,
L: Test section length, m
Nu: Nusselt number
p: Static pressure, Pa; Δp pressure drop, Pa
Re: Reynolds number, Re_a= axial Reynolds number
u: Mean velocity, ui' fluctuation velocity components, ms^{-1}
w: Tape width, m
y: Twist ratio = H/D, Hpitch for 180° rotation of the twisted-tape (mm)
λ: Twist parameter = D/H or $1/y$
T_{wm}: Mean inside wall temperature
T_{bm}: Mean bulk fluid temperature.

Greek Symbols

μ_t: Eddy viscosity, $kg\ s^{-1}\ m^{-1}$
η: Thermal performance factor
δ: Thickness of the twisted-tape (mm)
ε: Turbulent dissipation rate, m^2s^{-3}.

Acknowledgments

The authors would like to acknowledge the keen interest taken by Late Dr. M.S. Lonath to start this research work. The moral support given to this investigation by Professor Dr. M.T. Karad is also appreciated and deeply recognized.

References

[1] A. E. Bergles, "Techniques to augment heat transfer," in *Handbook of Heat Transfer Applications*, J. P. Hartnett, W. M. Rohsenow, and E. N. Ganic, Eds., chapter 1, McGraw-Hill, New York, NY, USA, 2nd edition, 1985.

[2] R. L. Webb, *Principle of Enhanced Heat Transfer*, John Wiley, New York, NY, USA, 1994.

[3] A. R. A. Khaled, M. Siddique, N. I. Abdulhafiz, and A. Y. Boukhary, "Recent advances in heat transfer enhancements: a review report," *International Journal of Chemical Engineering*, vol. 2010, Article ID 106461, 28 pages, 2010.

[4] W. E. Hilding and C. H. Coogan, "Heat transfer and pressure loss measurements in internally finned tubes," in *Proceedings of the ASME Symposium on Air Cooled Heat Exchangers*, pp. 57–85, 1964.

[5] P. Bharadwaj, A. D. Khondge, and A. W. Date, "Heat transfer and pressure drop in a spirally grooved tube with twisted tape insert," *International Journal of Heat and Mass Transfer*, vol. 52, no. 7-8, pp. 1938–1944, 2009.

[6] M. A. Al-Nimr and M. K. Alkam, "Unsteady non-Darcian forced convection analysis in an annulus partially filled with a porous material," *Journal of Heat Transfer*, vol. 119, no. 4, pp. 799–804, 1997.

[7] Y. Ding, H. Alias, D. Wen, and R. A. Williams, "Heat transfer of aqueous suspensions of carbon nanotubes (CNT nanofluids)," *International Journal of Heat and Mass Transfer*, vol. 49, no. 1-2, pp. 240–250, 2006.

[8] K. Vafai and A. R. A. Khaled, "Analysis of flexible microchannel heat sink systems," *International Journal of Heat and Mass Transfer*, vol. 48, no. 9, pp. 1739–1746, 2005.

[9] A. R. A. Khaled and K. Vafai, "Analysis of thermally expandable flexible fluidic thin-film channels," *Journal of Heat Transfer*, vol. 129, no. 7, pp. 813–818, 2007.

[10] S. Tiwari, P. L. N. Prasad, and G. Biswas, "A numerical study of heat transfer in fin-tube heat exchangers using winglet-type vortex generators in common-flow down configuration," *Progress in Computational Fluid Dynamics*, vol. 3, no. 1, pp. 32–41, 2003.

[11] E. M. Sparrow, J. E. Niethammer, and A. Chaboki, "Heat transfer and pressure drop characteristics of arrays of rectangular modules encountered in electronic equipment," *International Journal of Heat and Mass Transfer*, vol. 25, no. 7, pp. 961–973, 1982.

[12] E. M. Sparrow, A. A. Yanezmoreno, and D. R. Otis Jr., "Convective heat transfer response to height differences in an array of block-like electronic components," *International Journal of Heat and Mass Transfer*, vol. 27, no. 3, pp. 469–473, 1984.

[13] Y. M. Chen and J. M. Ting, "Ultra high thermal conductivity polymer composites," *Carbon*, vol. 40, no. 3, pp. 359–362, 2002.

[14] S. Y. Kim, J. M. Koo, and A. V. Kuznetsov, "Effect of anisotropy in permeability and effective thermal conductivity on thermal performance of an aluminum foam heat sink," *Numerical Heat Transfer; Part A*, vol. 40, no. 1, pp. 21–36, 2001.

[15] D. A. Nield and A. V. Kuznetsov, "Forced convection in a helical pipe filled with a saturated porous medium," *International Journal of Heat and Mass Transfer*, vol. 47, no. 24, pp. 5175–5180, 2004.

[16] L. Cheng and A. V. Kuznetsov, "Heat transfer in a laminar flow in a helical pipe filled with a fluid saturated porous medium," *International Journal of Thermal Sciences*, vol. 44, no. 8, pp. 787–798, 2005.

[17] L. Cheng and A. V. Kuznetsov, "Investigation of laminar flow in a helical pipe filled with a fluid saturated porous medium," *European Journal of Mechanics, B/Fluids*, vol. 24, no. 3, pp. 338–352, 2005.
[18] P. Promvonge and S. Eiamsa-ard, "Heat transfer enhancement in a tube with combined conical-nozzle inserts and swirl generator," *Energy Conversion and Management*, vol. 47, no. 18-19, pp. 2867–2882, 2006.
[19] P. Promvonge, "Heat transfer behaviors in round tube with conical ring inserts," *Energy Conversion and Management*, vol. 49, no. 1, pp. 8–15, 2008.
[20] E. Smithberg and F. Landis, "Friction and forced convection heat transfer characteristics in tubes with twisted tape swirl generators," *Journal of Heat Transfer*, vol. 86, no. 1, pp. 39–49, 1964.
[21] R. F. Lopina and A. E. Bergles, "Heat transfer and pressure drop in tape-generated swirl flow of single-phase water," *Journal of Heat Transfer*, vol. 91, pp. 434–442, 1968.
[22] A. W. Date, "Prediction of fully-developed flow in a tube containing a twisted-tape," *International Journal of Heat and Mass Transfer*, vol. 17, no. 8, pp. 845–859, 1974.
[23] R. M. Manglik and A. E. Bergles, "Heat transfer and pressure drop correlations for twisted-tape inserts in isothermal tubes: part I—laminar flows," *Journal of Heat Transfer*, vol. 115, no. 4, pp. 881–889, 1993.
[24] S. Al-Fahed, L. M. Chamra, and W. Chakroun, "Pressure drop and heat transfer comparison for both microfin tube and twisted-tape inserts in laminar flow," *Experimental Thermal and Fluid Science*, vol. 18, no. 4, pp. 323–333, 1998.
[25] S. K. Saha and A. Dutta, "Thermohydraulic study of laminar swirl flow through a circular tube fitted with twisted tapes," *Journal of Heat Transfer*, vol. 123, no. 3, pp. 417–427, 2001.
[26] M. Rahimi, S. R. Shabanian, and A. A. Alsairafi, "Experimental and CFD studies on heat transfer and friction factor characteristics of a tube equipped with modified twisted tape inserts," *Chemical Engineering and Processing*, vol. 48, no. 3, pp. 762–770, 2009.
[27] S. Eiamsa-ard, C. Thianpong, P. Eiamsa-ard, and P. Promvonge, "Convective heat transfer in a circular tube with short-length twisted tape insert," *International Communications in Heat and Mass Transfer*, vol. 36, no. 4, pp. 365–371, 2009.
[28] Y. W. Chiu and J. Y. Jang, "3D numerical and experimental analysis for thermal-hydraulic characteristics of air flow inside a circular tube with different tube inserts," *Applied Thermal Engineering*, vol. 29, no. 2-3, pp. 250–258, 2009.
[29] O. H. Klepper, *Experimental tudy of Heat Transfer and pressure drop for gas flowing in tubes containing a short twisted tape [M.S. thesis]*, University of Tennessee, 1971.

14

Computation of Isobaric Vapor-Liquid Equilibrium Data for Binary and Ternary Mixtures of Methanol, Water, and Ethanoic Acid from T, p, x, and H_m^E Measurements

Daming Gao,[1,2] Hui Zhang,[1,2] Peter Lücking,[2,3] Hong Sun,[1,2] Jingyu Si,[1] Dechun Zhu,[1,2] Hong Chen,[1] and Jianjun Shi[1,2]

[1] *Department of Chemistry and Materials Engineering, Hefei University, Anhui, Hefei 230022, China*
[2] *Sino-German Research Center for Process Engineering and Energy Technology, Anhui, Hefei 230022, China*
[3] *Department of Engineering, Jade University of Applied Science, 26389 Wilhelmshaven, Germany*

Correspondence should be addressed to Daming Gao, dmgao@hfuu.edu.cn

Academic Editor: Ahmet Z. Sahin

Vapor-liquid equilibrium (VLE) data for the strongly associated ternary system methanol + water + ethanoic acid and the three constituent binary systems have been determined by the total pressure-temperature-liquid-phase composition-molar excess enthalpy of mixing of the liquid phase (p, T, x, H_m^E) for the binary systems using a novel pump ebulliometer at 101.325 kPa. The vapor-phase compositions of these binary systems had been calculated from Tpx and H_m^E based on the Q function of molar excess Gibbs energy through an indirect method. Moreover, the experimental T, x data are used to estimate nonrandom two-liquid (NRTL), Wilson, Margules, and van Laar model parameters, and these parameters in turn are used to calculate vapor-phase compositions. The activity coefficients of the solution were correlated with NRTL, Wilson, Margules, and van Laar models through fitting by least-squares method. The VLE data of the ternary system were well predicted from these binary interaction parameters of NRTL, Wilson, Margules, and van Laar model parameters without any additional adjustment to build the thermodynamic model of VLE for the ternary system and obtain the vapor-phase compositions and the calculated bubble points.

1. Introduction

New strategies for the correlation and accurate prediction of the vapor-liquid equilibrium (VLE) data play a vital role in the distillation and separation process in chemical industry. The common technique for obtaining VLE data is by direct measurement on the system. That is to say, when the VLE is established, and phases are sampled and analyzed. Consequently, the experimental technique must be rather highly accurate to ensure meaningful results in the operation of equilibrium stills. Actually, when the vapor-phase components are sampled and analyzed, the whole compositions of components in solution and vapor have been changed. Accordingly, the behavior of the systems has been changed with the amount of compositions. Moreover, it has been long realized that the analysis of vapor-liquid composition for the infinite dilute solution is very difficult. In addition, for VLE measurements of mixtures containing a highly volatile compound, the accurate measurement of the vapor-phase composition can be difficult. This fact, coupled with the necessity for much analytical work, tends to enhance interest in exploring new methods for the determination of equilibrium data that do not involve sampling and analysis of the vapor-phase components.

Several methods have been explored for the calculation of component behavior from gross solution behavior. Van Ness and coworkers have suggested the classification of these methods into two categories, direct and indirect methods [1]. The direct methods involve calculation of vapor compositions by integration of the coexistence equation, a first-order differential equation derived from the Gibbs-Duhem equation relating phase compositions at equilibrium. Hala and coworkers have given a detailed discussion of the basic direct method [2], and Van Ness and coworkers have discussed techniques for handling nonconstant temperature or pressure conditions as well as nonideal vapor phase behavior [1, 3]. Moreover, in the total pressure method one can calculate y from T, p, x measurements using an indirect method discussed by Mixon et al. [4].

Of all the methods, the indirect methods involve first the measurement, by some appropriate means, the temperature and total pressure of the system, the liquid-phase compositions, and subsequent calculation of vapor-phase compositions there from. These methods usually involve ascertaining which of selected solution equations to the Gibbs-Duhem equation lead to the best fit to the experimental data, and of the determination of the parametric values producing the best fit. For example, Barker has developed a procedure based on the assumption that the excess free energy can be represented as a polynomial function of composition [5].

There is a basic difference in the degree of rigor associated with the direct method and the indirect method of Barker. In the former, one makes no assumptions about the solution behavior for the nature of molecular interactions. Solution behavior is determined directly from the experimental data. The Barker method necessitates the assumption of a particular model and the estimation of its parameters, this deficiency in the method of Barker has been recognized by Tao, who has presented another indirect method in which the necessity for the a priori assumption of a particular functional form for the excess free energy has been removed. Tao's procedure involves calculation of the activity coefficient essentially by integration of an equation resembling the coexistence equation. His procedure, though indirect, retains the rigor usually associated with the direct method [6]. The method of Tao appears specific to binary systems and does not seem to be easily generalized. However, an indirect method such as that of Barker is readily and easily generalized to ternary and higher-ordered systems, but this method retains the disadvantage of lacking rigor as compared with the direct method.

This paper presents the vapor-liquid compositions calculated based on the measurements of VLE data for temperature-total pressure-liquid-phase composition-molar excess enthalpy energy of mixing of the liquid phase (T, p, x, H^E_m) at 101.325 kPa according to the Q function of molar excess Gibbs energy through an indirect method. We know that the reaction of methanol carboxylated with carbon monoxide is the most common and important technology for synthesis of ethanoic acid in the chemical industrial process. Modeling the thermodynamic properties and correlating and predicting the phase equilibria of a mixture involving associating components forming hydrogen bonding such as carboxylic acid remain a challenging problem, since such systems show extremely nonideal behavior and the formation of monomer, dimer, and even trimer in vapor and liquid phase. In addition, for the VLE measurements of vapor-phase components containing a highly volatile compound such as carboxylic acid, the accurate measurement of the vapor-phase composition can be difficult. Many attempts have been made to describe the vapor-liquid equilibria of carboxylic acid containing mixtures using the concept of multiscale association [7]. Arlt reported the isothermal VLE data of a new apparatus for phase equilibria in reaction mixtures containing water with ethanoic acid and propanoic acid at (333 to 363) K [8]. Xu and Chuang have developed a new correlation for the prediction of the vapor-liquid equilibrium of methyl acetate-methanol-water-ethanoic acid mixtures [9]. Sawistowski and Pilavakis explored the vapor-liquid equilibrium behavior of the quaternary system methyl acetate-methanol-water-ethanoic acid modeled by using the Margules equation in combination with Marek's method for the association of ethanoic acid [10]. Moreover, Guan et al. investigated that the isobaric vapor-liquid equilibria for water + ethanoic acid + n-pentyl acetate, isopropyl acetate, N-methyl pyrrolidone, or N-methyl acetamide were correlated and predicted by both the nonrandom two-liquid (NRTL) and universal quasichemical activity coefficient (UNIQUAC) models used in combination with the Hayden-O'Connell (HOC) method [11–14]. In our recent work, we have concluded that the VLE data for the associating systems containing the carboxylic acid system can be correlated and predicted [15, 16]. Although the VLE data of the mixture containing the associating systems were previously reported by the different research groups [17–21], respectively, the VLE data for the associating binary system containing the carboxylic acid have been still extensively studied because of the extensive association effects occurring in them and the difficulty of properly calculating the activity coefficients [22–25]. Nominally, the system is binary but in practice, it is multicomponent. Because the carboxylic acid monomer undergoes partial dimerization and even higher polymerization. This association is attributed to the formation of hydrogen bonds and occurs in both the vapor and liquid phase. Therefore, the challenge for the VLE data of the associating systems has evoked more and more researchers to focus on new strategies for exploring them. The VLE data for methanol + water + ethanoic acid ternary system and the constituent binary systems are indispensable in distillation separation process to the product of methanol carboxylation through the correlation and prediction by the new method, while some of the isobaric VLE data for these binary and ternary systems are correlated and predicted earlier [9–25]. To provide the new correlation and prediction for some necessary basic thermodynamic data on the separation process of methanol carboxylation, therefore, it is very indispensable for these systems studied on the VLE data of the constituent binary and ternary systems using the new method. This paper reports a novel correlation and prediction for the VLE data for these systems containing the associating component that has been developed. The VLE data for methanol + water + ethanoic acid system and constituent binary systems were measured by the total pressure-temperature-liquid-phase composition-molar excess enthalpy of mixing of the liquid phase (p, T, x, H^E_m) for the binary systems using the novel pump-ebulliometer at 101.325 kPa. Owing to the association of ethanoic acid molecules, Marek's method in combination with Hayden-O'Connell (HOC) model was used to deal with the associating properties of the liquid and vapor phase and the nonideality of vapor phase, respectively [26–28]. However, the nonideality of liquid phase was corrected by the calculation of its activity coefficient, which was obtained based on NRTL, Wilson, Margules, and van Laar models as the function of T, x through the nonlinear fit of the least-squares method. NRTL, Wilson, Margules, and van Laar models were applied to correlate the VLE data for the three constituent binary systems, and the model parameters together with the deviations of temperature and vapor-phase molar fractions calculated from T, p, x, H_m^E according to the Q function of molar excess Gibbs energy by the indirect method were obtained by the least-squares method. The VLE data of the ternary system were well predicted from these binary interaction parameters of NRTL, Wilson, Margules, and van Laar models without any additional adjustment to build the thermodynamic model of VLE for the ternary system and obtain the vapor-phase compositions and the calculated bubble points. The excess Gibbs free energies for these binary systems as the function of liquid-phase composition and activity coefficient were calculated through the activity coefficient correlation to NRTL model parameters with the experimental data.

2. Modeling Section

There is an added complexity when working with carboxylic acids because they associate in the vapor and liquid phases. This association can be represented by assuming that the organic acid exists as monomer and dimer according to the Marek's method [26, 27]. This fact, coupled with the necessity for much analytical work, tends to enhance interest in exploring the new strategies for correlation and prediction of the VLE data for the systems containing the associating carboxyl acid.

According to the Marek's chemical theory, there are monomer and dimer carboxylic acid molecules in both liquid and vapor phases. The equilibrium constant of vapor-phase dimerization, C_E^V, is calculated by the expression

$$C_E^V = \frac{y_D^*}{(y_M^*)^2 p} \tag{1}$$

and the equilibrium constant of liquid-phase dimerization, C_E^L, is calculated by the expression

$$C_E^L = \frac{x_D^*}{(x_M^*)^2}. \tag{2}$$

In equations mentioned above, y_D^* and x_D^* are the mole fractions of dimers of ethanoic acid molecules in both vapor and liquid phases, respectively, and C_E^V can be defined as the function of temperature by the expression obtained from the literature [27]

$$C_E^V = \exp\left(\frac{7290}{T} - 21.980\right), \tag{3}$$

where C_E^V was presented in kPa^{-1} and T in K.

When the dimers of ethanoic acid molecules mainly exist, the binary system for the methanol or water (1) + ethanoic acid (2) is nominally binary; however, it is actually ternary for the methanol or water (1) + monomer ethanoic acid (M) + dimer ethanoic acid (D) system. In this system, the mole fractions of vapor-liquid equilibrium phases are y_1^*, y_M^*, y_D^*, x_1^*, x_M^*, and x_D^*, respectively. Therefore, the VLE relations of the nonassociating and associating components are individually calculated by the expressions

$$\begin{aligned} p y_1^* \Phi_1 &= p_1^s x_1^* \gamma_1, \\ p y_M^* \Phi_M &= p_M^s x_M^* \gamma_M, \\ p y_D^* \Phi_D &= p_D^s x_D^* \gamma_D. \end{aligned} \tag{4}$$

In (4), the actual mole fractions can be denoted by the apparent mole fractions of easily determined components (y_1, y_2, x_1, and x_2 multiplied by a modified coefficient. Likewise, the measured apparent vapor pressure saturated of ethanoic acid (p_2^s) multiplied by a modified coefficient can also denote the actual vapor pressures saturated of monomer and dimer ethanoic acid (p_M^s, p_D^s). Consequently, the apparent compositions and vapor pressures saturated substitute for the actual ones, and the VLE relation of the nonassociating component, methanol or water (1), is expressed by

$$p y_1 \alpha_1 \Phi_1 = p_1^s x_1 \beta_1 \gamma_1, \tag{5}$$

where

$$\alpha_1 = \frac{2}{(2 - y_2)} \frac{1 - y_2 - \sqrt{1 + 4C_E^V p y_2 (2 - y_2)}}{1 + \sqrt{1 + 4C_E^V p y_2 (2 - y_2)}}, \tag{6}$$

$$\beta_1 = \frac{2}{(2 - x_2)} \frac{1 - x_2 - \sqrt{1 + 4C_E^L p x_2 (2 - x_2)}}{1 + \sqrt{1 + 4C_E^L p x_2 (2 - x_2)}}, \tag{7}$$

and for ethanoic acid (2), which is the associating component, its relation is expressed by

$$p y_2 \alpha_2 \Phi_2 = p_2^s \theta_2^s x_2 \beta_2 \gamma_2, \tag{8}$$

where

$$\alpha_2 = \frac{2}{1 + \sqrt{1 + 4C_E^V p y_2 (2 - y_2)}}, \tag{9}$$

$$\beta_2 = \frac{2}{1 + \sqrt{1 + 4C_E^L x_2 (2 - x_2)}}, \tag{10}$$

$$\theta_2^s = \frac{-1 + \sqrt{1 + 4C_E^V p_2^s}}{2C_E^V p_2^s}. \tag{11}$$

In (5) to (11), α_1 and α_2 can be regarded as modified coefficients for the deviation from ideality in vapor phase on account of association. The fugacity of coefficients Φ_1 and Φ_2 is not negligible, and their values were obtained through the HOC model [28]. From another point of view, θ_2^s can be viewed as a modified coefficient for the vapor pressure of the associating component. Moreover, β_1 and β_2 denote modified coefficients of the deviation from ideality in the liquid phase by reason of the existence of association, and γ_1 and γ_2 express the deviation from ideality in liquid

TABLE 1: Physical properties of the pure compounds: densities ρ, refractive indexes n_D at 298.15 K and normal boiling points T_b.

Compound	ρ/kg·m^{-3}		n_D		T_b/K	
	expt	lit[a]	expt	lit[a]	expt	lit[a]
Methanol	786.44	786.37	1.3264	1.3265	337.71	337.69
Water	997.01	997.05	1.3324	1.3325	373.16	373.15
Ethanoic acid	1043.90	1043.92	1.3716	1.3718	391.53	391.15

[a] Riddick et al. [29].

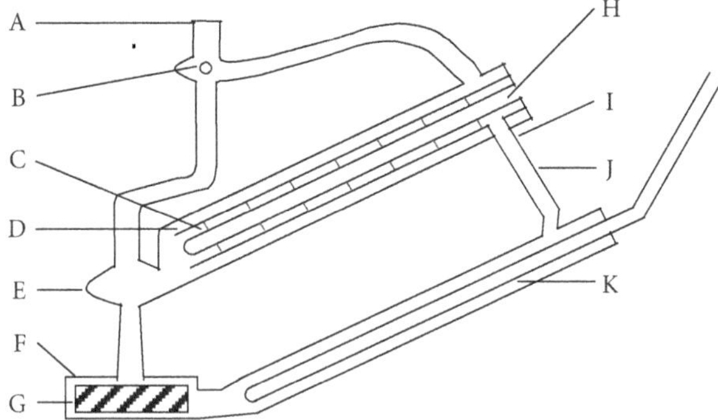

FIGURE 1: Schematic illustration for the structure of pump ebulliometer. (A) Normal distilling adaptor; (B) feed inlet; (C) coiled glass thread; (D) constant temperature introducing tube; (E) mixed vial section; (F) magnetic recirculating bump; (G) magnetic stirrer; (H) measured thermowell (filled with conducting oil); (I) tube window to observe in VLE status; (J) internal heating tube; (K) part-heating tube (resistance heating wire inserted).

phase because of other factors. Herein, in liquid phase, the modified coefficients of deviation from ideality, β_1 and β_2, can be incorporated to the activity coefficients, γ_1 and γ_2, respectively. So (5) and (8) are obtained by the expressions

$$py_1\alpha_1\Phi_1 = p_1^s x_1\gamma_1, \quad (12)$$

$$py_2\alpha_2\Phi_2 = p_2^s\theta_2^s x_2\gamma_2. \quad (13)$$

3. Experimental Section

3.1. Materials. Methanol (99.8 mass %) and ethanoic acid (99.8 mass %) were purchased from Sigma. The purities of the chemicals are provided by the manufacture's specifications. Ultra sound was used to dispel the solvent air in the materials, which were dried on molecular sieve (pore diameter 30 nm from Fluka) to remove all possible traces of moisture before use, but no other treatments were applied. The densities and refractive indices at 298.15 K and normal boiling points at 101.325 kPa of the pure component were compared with the literature values of Riddick et al. [29]. The results show that the measured values are approximately in agreement with those of the literature, as presented in Table 1. The measurement method of the composition dependence of densities and refractive indices has been previously reported [30].

3.2. Apparatus and Procedure. A new type of magnetic pump ebulliometer described in detail by Qiu et al. [31] was used to measure the boiling points with different liquid phase compositions at the 101.325 kPa. The experimental main apparatus for pump-ebulliometer is schematically shown in Figure 1. The recirculation still is entirely constructed from borosilicate glass. The main parts are a magnetic recirculating bump (F), a magnetic stirrer (G), a thermo-well (filled with a conducting oil) provided to enable good measurement of equilibrium temperature (H), a part-heating tube (resistance heating wire inserted) (K), and a tube window to observe in VLE status (I). The apparatus is an all-glass dynamic recirculation still with total volume of about 1.00×10^{-4} m^3. During the run, to avoid the over heating, the still was submerged in a constant temperature bath at about 3°C below the equilibrium boiling point, which was obtained by the Nichrome wire in the part-heating tube (K) to partially heat the known mass of the material. The equilibrium pressure was measured using a Fischer digital manometer with a precision of ±0.01 kPa. The pressure measurement for the manometer had an uncertainty of ±0.07 kPa, as provided by the manufacturer's specifications. The total uncertainty of calibration and pressure measurement is estimated to be ±0.15 kPa because of the uncertainty of the calibration and measurement errors. Since the barometric pressure changed slightly, the experimental temperatures of the systems were automatically calibrated to that at 101.325 kPa with self-adjusted pressure system. The temperature was measured using a Heraeus QuaT100 quartz thermometer with a thermosensor, with an accuracy of ±0.01 K. The calibration of the thermometers was carried out at the accredited calibration laboratory (Quality and Technique Bureau, Anhui). The total uncertainty of calibration and temperature measurement is evaluated to be ±0.085 K because of the uncertainty of the calibration, the probe's position, and the pressure fluctuations. The equilibrium temperature, T, was measured by means of the thermosensor inserted into the thermowell (filled with conducting oil) (H). In each experiment, a known mass of the material was introduced from the injector into the still from feed inlet (B) and heated to equilibrium boiling point of the system at a fixed pressure of 101.325 kPa by an automatic pressure regulation system. The liquid mixtures of required composition were prepared gravimetrically, with the use of a Sartorius electronic analytic balance (model ER-182A) with an accuracy of ±0.0001 g. The values of mole fraction were reproducible to ±0.0001 and have uncertainty of 0.01%. The ebulliometer was charged with the mixture of desired composition, and the boiler was then heated by nichrome wire wound around the boiler. After the liquid mixture started boiling, the bubbles along with the drops of liquid slowly spurted on the thermowell one by one through the tube window observed in VLE status (I). When the VLE state was attained by adjusting the pressure to 101.325 kPa, it remained constant for 20 min to ensure the stationary state, and then the boiling temperature was measured.

A flow microcalorimeter (model 2107, LKB produkter, Bromma, Sweden) was applied to measure the molar excess enthalpies, H_m^E, of the mixtures. The electrical-calibration apparatus and its operating procedure have been described elsewhere by Francesconi and Comelli [34]. Two automatic burets (ABU, Radiometer, Copenhagen, Denmark) were used to pump liquids through the mixing cell of the calorimeter. The performance of the calorimeter was checked by measuring the H_m^E of the test mixture hexane + cyclohexane system reported by Gmehling [35]. Agreement with literature is better than 0.5% over the central range of mole fraction of hexane.

The uncertainties in mole fraction and H_m^E are estimated to be 0.0005 and 0.5%, respectively. The liquid-phase mole fraction of component i, x_i, could be calculated from the known mass of the material added to the still. The vapor-phase mole fraction of component i, y_i, was calculated from the experimental T, p, x, and H_m^E data based on the Q function (the function of molar excess Gibbs energy) by the indirect method.

4. Results and Discussion

4.1. Calculation of Vapor-Phase Mole Fraction y for the Binary Systems from Tpx and H_m^E. The vapor-phase mole fraction y_i of the components was calculated from

$$y_i \Phi_i p = x_i \gamma_i p_i^s, \tag{14}$$

and the fugacity of coefficients Φ_i of the components was obtained by the expression

$$\Phi_i = \frac{\hat{\varphi}_i^V}{\varphi_i^s} \exp\left[-\frac{V_i^L (p - p_i^s)}{RT}\right], \tag{15}$$

where $\hat{\varphi}_i^V$ is the fugacity coefficient of component i in the vapor mixture, φ_i^s is the fugacity coefficient of component i at saturation, V_i^L is the molar volume of component i in the liquid phase, R is the universal gas constant, T is the experimental temperature, p is the total pressure (101.325 kPa), and p_i^s is the vapor pressure of pure component i. The Antoine equations were applied to calculate the values of these vapor pressures. The Antoine constants A_i, B_i, C_i, and the values of T_c, p_c, V_c, Z_c and ω were obtained from Shi et al. [36], as shown in Table 2. The Poynting correction factor was also included in the calculation of fugacity at the reference state. The liquid molar volumes were evaluated from the modified Rackett equation [37].

According to the Gibbs-Duhem equation, any extensive molar thermodynamic property of a given phase, such as the Gibbs and Helmholtz energies, the enthalpy, and molar volume, must satisfy the following differential relation:

$$\left(\frac{\partial M}{\partial T}\right)_{p, x_{j \neq i}} dT + \left(\frac{\partial M}{\partial p}\right)_{T, x_{j \neq i}} dp - \sum_{i=1}^{C} x_i d\overline{M}_i = 0. \tag{16}$$

In (16) M is a general molar property, x_i the molar fraction of component i in the phase under consideration, $x_{j \neq i}$ the pertinent set of compositions, C the number of components, and $\overline{M}_i$ the partial contribution of component i to M.

When considering VLE, the molar excess Gibbs energy, G^E, can be evaluated from measurable (T, p, x) data using activity coefficient relations. Replacing M by G^E/RT in the following equation yields the well-known relation

$$-\frac{H_m^E}{RT^2} dT + \frac{V_m^E}{RT} dp - \sum_{i=1}^{C} x_i d \ln \gamma_i = 0, \tag{17}$$

where H_m^E and V_m^E are the molar excess enthalpy and volume of mixing of the liquid phase. According to the thermodynamic principles, the activity coefficients γ_i of the components were calculated from the expression as follows:

$$Q = \frac{G^E}{RT} = \sum_{i=1}^{C} x_i \ln \gamma_i, \tag{18}$$

where Q is the function of molar excess Gibbs energy and combining (17) and (18) yields

$$d\left(\frac{G^E}{RT}\right) - \sum_{i=1}^{C} \ln \gamma_i dx_i = \sum_{i=1}^{C} x_i d \ln \gamma_i = -\frac{H_m^E}{RT^2} dT + \frac{V_m^E}{RT} dp. \tag{19}$$

Application to a binary system gives

$$d\left(\frac{G^E}{RT}\right) - \ln \frac{\gamma_1}{\gamma_2} dx_1 = -\frac{H_m^E}{RT^2} dT + \frac{V_m^E}{RT} dp. \tag{20}$$

Simultaneous solution of (18) and (20) yields

$$\begin{aligned} \ln \gamma_1 &= \frac{G^E}{RT} + x_2 \left[\frac{d}{dx_1}\left(\frac{G^E}{RT}\right) + \frac{H_m^E}{RT^2}\frac{dT}{dx_1} - \frac{V_m^E}{RT}\frac{dp}{dx_1}\right], \\ \ln \gamma_2 &= \frac{G^E}{RT} - x_1 \left[\frac{d}{dx_1}\left(\frac{G^E}{RT}\right) + \frac{H_m^E}{RT^2}\frac{dT}{dx_1} - \frac{V_m^E}{RT}\frac{dp}{dx_1}\right]. \end{aligned} \tag{21}$$

Equation (18) substituted into (21) reduced it to

$$\begin{aligned} \gamma_1 &= \exp\left[Q + (1 - x_1)\left(\frac{dQ}{dx_1} + \frac{H_m^E}{RT^2}\frac{dT}{dx_1} - \frac{V_m^E}{RT}\frac{dp}{dx_1}\right)\right], \\ \gamma_2 &= \exp\left[Q - x_1\left(\frac{dQ}{dx_1} + \frac{H_m^E}{RT^2}\frac{dT}{dx_1} - \frac{V_m^E}{RT}\frac{dp}{dx_1}\right)\right]. \end{aligned} \tag{22}$$

For a binary system comprised of species 1 and 2 at VLE state, at constant pressure, $(dp/dx_1) = 0$, substitution into (22) reduced it to

$$\gamma_i = \exp\left[Q + (1 - x_i)\left[\left(\frac{dQ}{dx_i}\right) + \frac{H_m^E}{RT^2}\left(\frac{dT}{dx_i}\right)\right]\right], \quad (i = 1, 2). \tag{23}$$

The right-hand side, $(H_m^E/RT^2)(dT/dx_i)$, of (23) cannot be neglected. Proper use of (23) requires the availability

Table 2: Antoine coefficients of the compounds and published parameters [36] used to calculate fugacity coefficients: critical temperature T_c, critical pressure p_c, critical volume V_c, critical compression Z_c, and acentric factor ω of pure compounds.

Compound	A_i	B_i	C_i	T_c/K	p_c/MPa	$V_c/\mathrm{m^3 \cdot kmol^{-1}}$	Z_c	ω
Methanol	7.19736	1574.99	−34.29	512.6	8.096	0.118	0.224	0.559
Water	7.07404	1657.46	−46.13	647.3	22.048	0.056	0.229	0.344
Ethanoic acid	6.42452	1479.02	−56.34	594.4	5.786	0.171	0.200	0.454

of heats of mixing as a function of composition and temperature. The activity coefficients γ_i of the components as functions of the excess Gibbs energy are as follows:

$$\gamma_1 = \exp\left\{Q + (1 - x_1)\left[\left(\frac{dQ}{dx_1}\right) + \frac{H^E_m}{RT^2}\left(\frac{dT}{dx_1}\right)\right]\right\},$$
$$\gamma_2 = \exp\left\{Q - x_1\left[\left(\frac{dQ}{dx_1}\right) + \frac{H^E_m}{RT^2}\left(\frac{dT}{dx_1}\right)\right]\right\}. \tag{24}$$

From (14), this equation is rearranged to obtain

$$y_i = \frac{x_i \gamma_i p_i^s}{\Phi_i p}. \tag{25}$$

Because $\sum y_i = 1$, for the binary system, the equation may be summed to give

$$\frac{x_1\gamma_1\varphi_1^s p_1^s \exp\left[V_1^L(p - p_1^s)/RT\right]}{\hat{\varphi}_1^V p} + \frac{x_2\gamma_2\varphi_2^s p_2^s \exp\left[V_2^L(p - p_2^s)/RT\right]}{\hat{\varphi}_2^V p} = 1. \tag{26}$$

In (26), solved for y_i by difference method. Suppose that $[0, 1]$ is subdivided into n subintervals $[x_k, x_{k+1}]$ of equal step size $h = 1/n$ by using $x_k = kh$ for $k = 0, 1, \ldots, n$. In k difference point, we obtain

$$F_k = 1 - \left\{\frac{x_1\gamma_1\varphi_1^s p_1^s \exp\left[V_1^L(p - p_1^s)/RT\right]}{\hat{\varphi}_1^V p}\right\}_k - \left\{\frac{x_2\gamma_2\varphi_2^s p_2^s \exp\left[V_2^L(p - p_2^s)/RT\right]}{\hat{\varphi}_2^V p}\right\}_k = 0. \tag{27}$$

Meanwhile, (23) may be shown as follows:

$$\gamma_i|_k = \exp\left\{Q_k + (1 - x_i|_k)\left[\frac{Q|_{k+1} - Q|_{k-1}}{2/h} + \frac{H^E_m}{RT^2}\left(\frac{dT}{dx_1}\right)\Big|_k\right]\right\}. \tag{28}$$

And (27) is linearized to obtain:

$$-F_k = \Delta Q|_{k-1}\left(\frac{\delta F_k}{\delta Q|_{k-1}}\right) + \Delta Q|_k\left(\frac{\delta F_k}{\delta Q|_k}\right) + \Delta Q|_{k+1}\left(\frac{\delta F_k}{\delta Q|_{k+1}}\right). \tag{29}$$

The number of n linear equation from (28) is solved for $\Delta Q|_k$ by chasing method

$$Q_k^{j+1} = Q_k^j + t\Delta Q_k^j, \tag{30}$$

where t is relaxation factor, finally, γ_{ik} ($k = 1 \sim n$) is obtained by difference method. The block diagram for calculation procedure was detailedly shown in Figure 2.

4.2. Calculation of Vapor-Phase Mole Fraction y for the Binary Systems from Model. There are many methods concerning the correlation and prediction of VLE data. The model-free approach data treatment of vapor-liquid equilibrium is also one of the best strategies for the correlation and prediction of VLE data. Wisniak's group has developed that the novel model-free computation techniques and limiting conditions have been applied to VLE data for azeotropic systems [42]. Moreover, Segura and coworkers reported that a model-free approach dealt with VLE data in application of ternary systems [38, 43]. For the three binary systems of this study, the activity coefficients were correlated with the NRTL [39], Wilson [40], Margules [41], and van Laar [44] equations, respectively. The interaction parameters optimized were achieved by the objective function (OF) minimized using the least-squares fitting as follows:

$$\mathrm{OF} = \sum_{i=1}^{N}\left(x_{i,\mathrm{cal}} - x_{i,\mathrm{exp}}\right)^2, \tag{31}$$

where N is the number of experimental points $x_{i,\mathrm{cal}}$ and $x_{i,\mathrm{exp}}$ are the liquid-phase mole fraction of component i calculated and experimental values from the (12) or (13) and from measured data, respectively.

Because carboxylic acids are always present in an associated form, like a dimer or trimer, in both the vapor and liquid phases even at low pressures, a significant deviation in fugacity coefficient may exist using the ideal gas assumption. To illustrate the deviation from ideal behavior, Marek's chemical theory and HOC model were applied to deal with the associating component and modify the deviation from idealities of vapor phase [26–28], respectively. The Poynting correction factor was also included in the calculation of fugacity at the reference state. The liquid molar volumes were evaluated from the modified Rackett equation [37]. However, under isobaric conditions, the most volatile component cannot exist in the liquid state, only as superheated vapor. Hence, there is no way to calculate or measure this property for the molar volumes of the pure liquids. Therefore, the correct procedure for isobaric measurements is to calculate the overall values of A_{ij} and A_{ij}, as adjustable parameters,

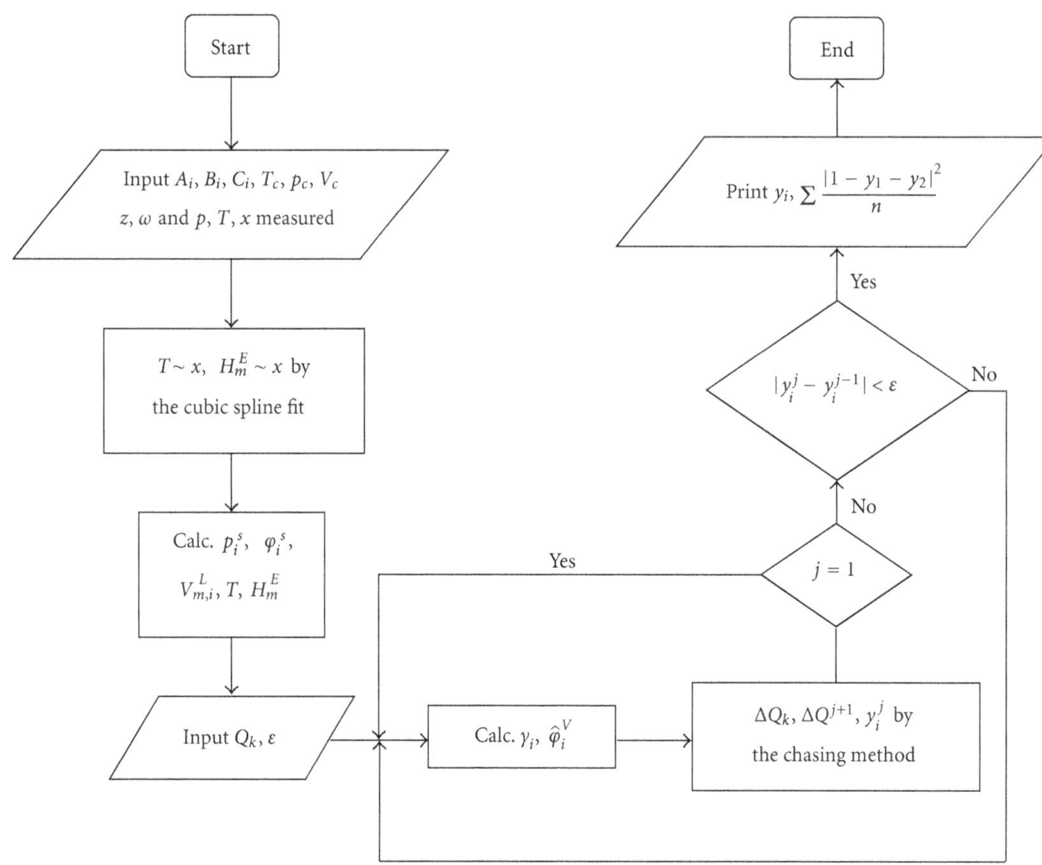

FIGURE 2: Block diagram for the calculation y by T, p, x, and H_m^E.

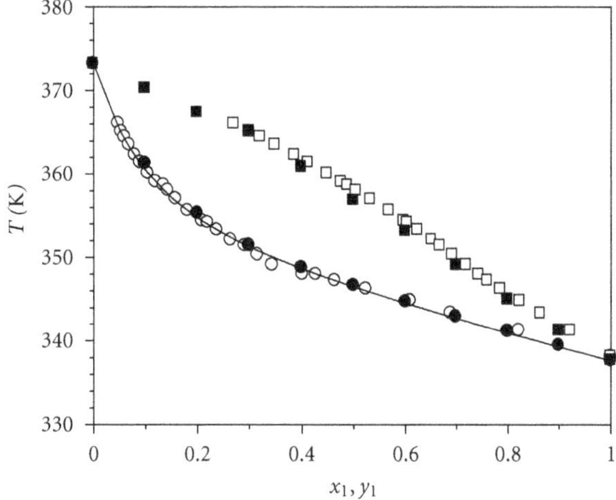

FIGURE 3: T-x_1-y_1 diagram for methanol (1) + water (2) at 101.325 kPa: □, vapor-phase mole fraction y_1 from Tpx and H_m^E; ■, vapor-phase mole fraction y_1 from the literature [32, 33]; ∘, liquid-phase experimental temperature; •, liquid-phase mole fraction x_1 from the literature [32, 33]; —, NRTL correlation temperature.

and not the values of the interaction excess energy. Hence, for Wilson model, A_{ij} and A_{ij} were reported as adjustable parameters.

The activity coefficients computed on the basis of NRTL model were used to evaluated dimensionless excess Gibbs function at 101.325 kPa for three binary systems over the overall range of composition. Liquid-phase mole fraction x_1, experimental boiling point temperature T_{exp}, calculated bubble point temperature T_{cal}, vapor-phase mole fraction y_1 from the model, activity coefficients γ_1 and γ_2 using NRTL equation correlation, fugacity coefficients $\hat{\varphi}_1^V$ and $\hat{\varphi}_2^V$, and the molar excess enthalpies of mixing of the liquid phase H_m^E are included in Table 3. The T-x_1-y_1 diagrams for the methanol (1) + water (2), methanol (1) + ethanoic acid (2), and water (1) + ethanoic acid (2) three binary systems at 101.325 kPa are shown in Figures 3, 4, and 5, respectively. The plot of excess Gibbs energy function G^E/RT versus liquid-phase mole fraction x_1 for the three binary systems is shown in Figure 6. All the mixtures exhibit deviations from ideality with a range that may be attributed to interactions leading to the formation of various associated aggregates. Observed nonideal behavior is indicative of the magnitude of experimental activity coefficients γ_i, as well as of the variation of excess Gibbs energy function, G^E/RT, with composition, as depicted in Figure 6. The obtained absolute maximum values of G^E/RT for the methanol (1) + water (2), methanol (1) + ethanoic acid (2), and water (1) + ethanoic acid (2) three binary systems are 0.1703, 0.0180, and 0.0892, respectively. The values of excess Gibbs energy

TABLE 3: VLE data for the methanol (1) + water (2), methanol (1) + ethanoic acid (2), and water (1) + ethanoic acid (2). Three binary systems at 101.325 kPa: liquid-phase mole fraction x_1, experimental boiling point temperature T_{exp}, calculated bubble point temperature T_{cal}, vapor-phase mole fraction y_1, activity coefficients γ_1 and γ_2 using NRTL equation correlation, fugacity coefficients $\hat{\varphi}_1^V$ and $\hat{\varphi}_2^V$, and molar excess enthalpies of mixing of the liquid phase H_m^E.

x_1	T_{exp}/K	T_{cal}/K	y_1	γ_1	$\hat{\varphi}_1^V$	γ_2	$\hat{\varphi}_2^V$	H_m^E/J·mol^{-1}
				Methanol (1) + water (2)				
0.0000	373.15	373.15	0.0000			1.0000	0.9882	0.00
0.0484	366.00	365.54	0.2704	2.1436	0.9852	1.0034	0.9875	225.74
0.0539	365.05	364.88	0.3009	2.1147	0.9849	1.0042	0.9875	248.74
0.0607	364.40	364.12	0.3213	2.0799	0.9846	1.0053	0.9874	276.88
0.0696	363.50	363.19	0.3492	2.0358	0.9843	1.0069	0.9874	312.53
0.0806	362.25	362.13	0.3866	1.9835	0.9839	1.0092	0.9873	354.36
0.0911	361.35	361.21	0.4126	1.9360	0.9836	1.0117	0.9872	392.94
0.1056	360.05	360.05	0.4492	1.8739	0.9832	1.0155	0.9871	443.59
0.1197	359.05	359.04	0.4762	1.8172	0.9829	1.0197	0.9870	489.29
0.1354	358.60	358.02	0.4885	1.7584	0.9826	1.0248	0.9869	538.32
0.1436	358.00	357.53	0.5047	1.7293	0.9824	1.0277	0.9869	562.20
0.1586	356.95	356.70	0.5323	1.6788	0.9822	1.0334	0.9868	602.83
0.1815	355.55	355.56	0.5677	1.6083	0.9819	1.0428	0.9867	659.52
0.2092	354.35	354.35	0.5969	1.5325	0.9815	1.0555	0.9865	719.43
0.2202	354.10	353.91	0.6031	1.5049	0.9814	1.0609	0.9865	741.25
0.2380	353.25	353.24	0.6237	1.4633	0.9812	1.0701	0.9864	772.62
0.2648	352.05	352.32	0.6519	1.4066	0.9809	1.0847	0.9862	812.77
0.2918	351.35	351.48	0.6681	1.3561	0.9807	1.1005	0.9861	846.82
0.3166	350.30	350.76	0.6923	1.3148	0.9805	1.1157	0.9860	870.15
0.3442	349.05	350.02	0.7197	1.2740	0.9803	1.1336	0.9859	889.06
0.4015	347.95	348.61	0.7431	1.2038	0.9800	1.1730	0.9856	909.80
0.4282	347.90	348.00	0.7443	1.1767	0.9798	1.1924	0.9854	911.28
0.4644	347.20	347.21	0.7613	1.1449	0.9796	1.2194	0.9853	903.04
0.5240	346.20	345.98	0.7860	1.1026	0.9793	1.2656	0.9850	869.55
0.6091	344.80	344.33	0.8231	1.0595	0.9789	1.3338	0.9845	783.83
0.6899	343.31	342.85	0.8633	1.0325	0.9785	1.3995	0.9840	666.92
0.8213	341.20	340.56	0.9221	1.0086	0.9779	1.5038	0.9832	419.31
1.0000	338.10	337.66	1.0000	1.0000	0.9771			0.00
				Methanol (1) + ethanol acid (2)				
0.0000	390.15	391.04	0.0000			1.0000	0.9703	0.00
0.0359	387.32	386.23	0.1574	0.9388	0.9765	0.9999	0.9685	77.28
0.0410	386.21	385.56	0.1664	0.9393	0.9768	0.9999	0.9682	86.70
0.0648	383.89	382.68	0.2719	0.9419	0.9779	0.9997	0.9666	134.05
0.0741	382.73	381.90	0.3133	0.9428	0.9782	0.9996	0.9660	149.55
0.0859	381.20	380.60	0.3333	0.9441	0.9787	0.9995	0.9652	171.14
0.1040	380.03	379.06	0.3830	0.9460	0.9791	0.9992	0.9639	205.37
0.1280	378.46	377.35	0.4602	0.9485	0.9797	0.9988	0.9622	245.43
0.1629	375.62	374.70	0.5213	0.9521	0.9802	0.9981	0.9595	299.80
0.1919	373.21	372.43	0.5873	0.9551	0.9805	0.9973	0.9572	341.32
0.2510	369.31	368.44	0.6495	0.9609	0.9808	0.9954	0.9528	411.44
0.3391	364.35	363.56	0.7538	0.9692	0.9809	0.9916	0.9461	490.73
0.4401	358.54	357.71	0.8232	0.9776	0.9805	0.9857	0.9388	536.56
0.5829	351.77	350.75	0.8999	0.9874	0.9797	0.9748	0.9292	526.43
0.7050	346.47	346.39	0.9389	0.9936	0.9790	0.9633	0.9217	449.37
0.7509	344.90	344.06	0.9533	0.9955	0.9787	0.9583	0.9189	401.45
0.8109	342.81	342.35	0.9693	0.9974	0.9783	0.9515	0.9155	330.61
0.8711	341.60	340.75	0.9766	0.9988	0.9779	0.9442	0.9122	241.41
0.9161	339.65	340.00	0.9859	0.9995	0.9776	0.9384	0.9097	163.78
1.0000	338.15	337.66	1.0000	1.0000	0.9771			0.00

Table 3: Continued.

x_1	T_{exp}/K	T_{cal}/K	y_1	γ_1	$\hat{\varphi}_1^V$	γ_2	$\hat{\varphi}_2^V$	H_m^E/J·mol^{-1}
Water (1) + ethanoic acid (2)								
0.0000	390.15	391.04	0.0000			1.0000	0.9703	0.00
0.0665	387.22	387.67	0.1140	1.4161	0.9899	1.0025	0.9693	245.31
0.0749	387.00	387.31	0.1288	1.4046	0.9898	1.0031	0.9692	273.17
0.0855	386.48	386.88	0.1603	1.3903	0.9897	1.0040	0.9691	307.50
0.0995	385.86	386.34	0.1938	1.3722	0.9896	1.0054	0.9689	350.32
0.1507	384.16	384.60	0.2802	1.3117	0.9892	1.0121	0.9683	489.63
0.1722	383.44	383.96	0.3155	1.2888	0.9891	1.0156	0.9681	540.35
0.1926	382.99	383.39	0.3375	1.2684	0.9890	1.0193	0.9679	585.57
0.2168	382.39	382.77	0.3676	1.2458	0.9890	1.0242	0.9677	632.34
0.2502	381.53	381.98	0.4101	1.2171	0.9889	1.0316	0.9674	689.92
0.2936	380.94	381.06	0.4407	1.1838	0.9887	1.0425	0.9670	750.61
0.3303	380.20	380.36	0.4815	1.1589	0.9887	1.0528	0.9667	788.98
0.3756	379.40	379.57	0.5253	1.1318	0.9886	1.0666	0.9663	821.39
0.4465	378.27	378.48	0.5890	1.0965	0.9885	1.0908	0.9658	841.59
0.4946	377.61	377.82	0.6274	1.0768	0.9885	1.1087	0.9654	836.35
0.5615	376.80	376.99	0.6777	1.0543	0.9884	1.1355	0.9649	802.60
0.6036	376.24	376.51	0.7137	1.0427	0.9884	1.1533	0.9645	767.63
0.6517	375.59	376.00	0.7546	1.0315	0.9884	1.1746	0.9642	714.78
0.7067	375.11	375.46	0.7849	1.0213	0.9884	1.1998	0.9637	639.01
0.7045	375.11	375.48	0.7848	1.0217	0.9884	1.1988	0.9638	642.76
0.7507	374.71	375.05	0.8173	1.0148	0.9883	1.2207	0.9634	568.24
0.8008	374.31	374.62	0.8501	1.0091	0.9883	1.2451	0.9630	475.20
0.8723	373.80	374.05	0.9011	1.0035	0.9883	1.2809	0.9624	323.52
0.9116	373.54	373.76	0.9315	1.0016	0.9883	1.3009	0.9620	230.13
0.9546	373.24	373.45	0.9630	1.0004	0.9883	1.3231	0.9616	121.95
1.0000	373.15	373.15	1.0000	1.0000	0.9882			0.00

Table 4: Correlation parameters for activity coefficients, average deviation for studied systems and vapor-phase composition mean absolute deviation from the literatures.

Equation	Parameters or deviations	Methanol (1) + water (2)	Methanol (1) + ethanoic acid (2)	Water (1) + ethanoic acid (2)
NRTL[a]	$(g_{12} - g_{11})$/J·mol^{-1}	−436.50	−85.50	−352.42
	$(g_{21} - g_{22})$/J·mol^{-1}	1159.55	0.98	715.43
	α_{12}	0.241	0.30	0.23
	dT/K	0.28	0.51	0.33
	dy	0.0072	0.0054	0.0096
Wilson[a]	Λ_{12}/J·mol^{-1}	389.37	50.46	644.71
	Λ_{21}/J·mol^{-1}	368.45	95.58	−87.59
	dT/K	0.43	0.47	0.64
	dy	0.0036	0.0059	0.0053
Margules[b]	A_{12}	0.95	0.12	0.29
	A_{21}	0.65	−0.46	0.57
	dT/K	0.37	0.32	0.56
	dy	0.0049 (0.0057)[c]	0.0075 (0.0234)[c]	0.0086 (0.0123)[c]
van Laar[b]	A_{12}	0.86	0.24	0.32
	A_{21}	0.56	0.03	0.55
	dT/K	0.27	0.58	0.47
	dy	0.0069	0.0078	0.0098

[a]Wilson's interaction parameters (J·mol^{-1}), NRTL's interaction parameters (J·mol^{-1}). [b]Margules and van Laar interaction parameters (dimensionless). $dT = \sum |T_{exp} - T_{cal}|/N$; N: number of data points; T_{cal}: calculated bubble point from model, K; T_{exp}: experimental boiling point temperature, K. $dy = \sum |y_{cal} - y_{mod}|/N$; N: number of data points; y_{cal}: calculated vapor-phase mole fraction from Tpx; y_{mol}: calculated vapor-phase mole fraction from model. [c]The values in parentheses from the literatures [38–41].

function G^E/RT are positive for methanol (1) + water (2) and water (1) + ethanoic acid (2) binary systems. However, for methanol (1) + ethanoic acid (2) system, the values of those are negative in the overall range of mole fraction. G^E/RT values follow the order methanol (1) + water (2) > water (1) + ethanoic acid (2) > methanol (1) + ethanoic acid (2). The absolute maximum value of G^E/RT is approximate at an equimolar fraction in three binary systems. Figures 3–5 show that the comparison of the predicted vapor-phase and experimental liquid-phase compositions with those of the literature [10, 11, 32, 33]. Comparing with the values of vapor-phase and liquid-phase component from the literatures, the values of those from the paper are in good agreement with the literature, as shown in Figures 3–5. The results have demonstrated that the methods for Tpx and H_m^E are appropriate for representing the experimental data of the three binary systems. The optimum model interaction parameter of liquid activity coefficient and the absolute average deviations for the different models, and from Tpx and H_m^E are listed in Table 4. Herein, we obtained the results by the four different types of correlations for the prediction of activity coefficients in these systems, which reveal that the deviations of NRTL, Wilson, Margules, and van Laar equations are reasonably small in Table 4. For comparison, the mean deviations obtained by Gmehling and Onken [32, 33, 45, 46] are also shown in the Table 4. It can seem that two sets of deviation values are comparable. Since the superiority of one method over the others is not always obvious, practice must rely on experience and analogy. The comprehensive comparisons of four of the methods (NRTL, Wilson, van Laar, and Margules) were made in Table 4. From the data analysis, the temperature deviations between the experimental and calculated values of four different types of model are very similar in the three binary systems, and the vapor-phase mole fraction deviations between calculated values from Tpx and H_m^E, and from the model are very similar. Therefore, the activity coefficient models are appropriate for representing the experimental data of the three binary systems. In Table 4, the absolute average deviations dT of the difference between boiling point temperature from experiment and bubbling point temperature from calculation by NRLT model parameters for the three binary systems are 0.28°C, 0.51°C, and 0.33°C, respectively. And the absolute average deviations dy of the difference between vapor-phase mole fraction from Tpx and H_m^E calculation and from NRTL model calculation are 0.0072, 0.0054, and 0.0096, respectively.

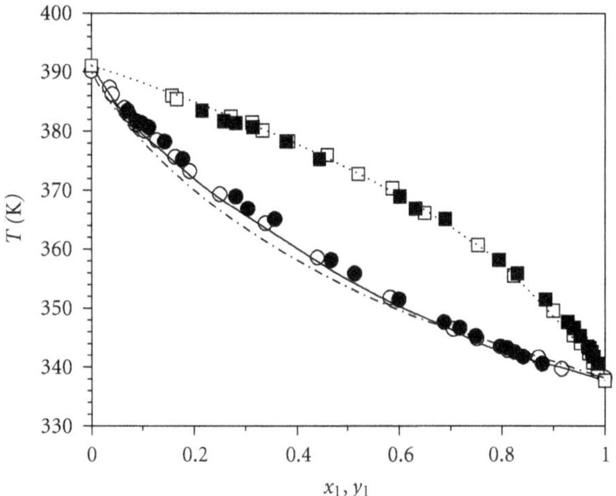

FIGURE 4: T-x_1-y_1 diagram for methanol (1) + ethanoic acid (2) at 101.325 kPa: □, vapor-phase mole fraction y_1 from Tpx and H_m^E; ■, vapor-phase mole fraction y_1 from literature [10]; —, dot line, vapor-phase mole fraction y_1 with Block diagram for the calculation y by T, p, x, and H_m^E neglected from our previous work [15]; ∘, liquid-phase experimental temperature; •, liquid-phase experimental temperature from literature [10]; -·-·-, dash dot, liquid-phase experimental temperature from our previous work [15]; —, NRTL correlation temperature.

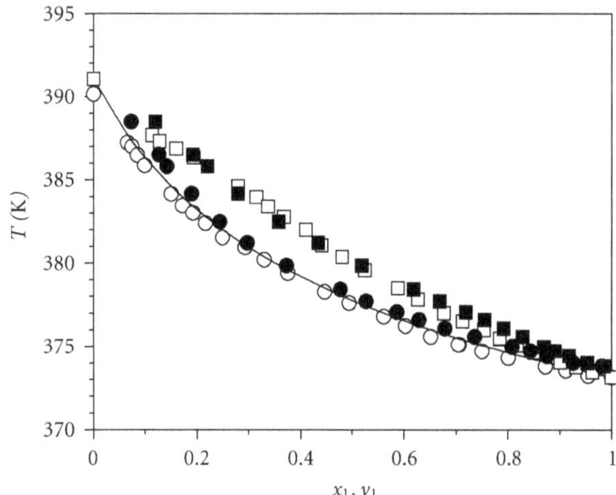

FIGURE 5: T-x_1-y_1 diagram for water (1) + ethanoic acid (2) at 101.325 kPa: □, vapor-phase mole fraction y_1 from Tpx and H_m^E; ■, vapor-phase mole fraction y_1 from literature [11]; •, liquid-phase experimental temperature from the literature; ∘, liquid-phase experimental temperature [11]; —, NRTL correlation temperature.

4.3. Correlation and Prediction of VLE of Ternary System. The binary interaction parameters of the NRTL, Wilson, Margules, and van Laar model given in Table 4 were used to correlate and predict the VLE data of the ternary system. VLE data for methanol (1) + water (2) + ethanoic acid (3) at 101.325 kPa included liquid-phase mole fraction x_1, x_2, and x_3, experimental boiling point temperature T_{exp}, calculated bubble point temperature T_{cal}, calculated vapor-phase mole fraction y_1, y_2, and y_3, activity coefficients γ_1, γ_2, and γ_3, the average deviation in the bubble temperatures of the ternary system using NRTL equation correlation listed in Table 5. The absolute average and maximum deviation between the boiling point from experimental data and the bubble point from NRTL model calculation are 0.77°C and 1.94°C, respectively. The average and maximum deviations using Wilson, Margules, and van Laar equation individually are 0.79°C, 1.90°C, 0.85°C, 1.89°C, and 0.96°C, 1.91°C. Diagram of VLE for the ternary system methanol (1) + water (2) + ethanoic acid (3) at 101.325 kPa is shown in Figure 7.

TABLE 5: VLE data for the methanol (1) + water (2) + ethanoic acid (3) Ternary system at 101.325 kPa: liquid-phase mole fraction x_1, x_2, and x_3, experimental boiling point temperature T_{exp}, calculated bubble point temperature T_{cal}, vapor-phase mole fraction y_1, y_2, and y_3, activity coefficients γ_1, γ_2, and γ_3 using NRTL equation correlation.

x_1	x_2	x_3	T_{exp}/K	T_{cal}/K	y_1	y_2	y_3	γ_1	γ_2	γ_3
0.7919	0.2080	0.0001	341.08	340.94	0.9108	0.0870	0.0022	1.0152	1.5155	0.9163
0.7502	0.1967	0.0531	342.79	342.12	0.9024	0.0882	0.0094	1.0156	1.5068	0.9294
0.6923	0.1816	0.1261	344.60	343.90	0.8886	0.0872	0.0242	1.0147	1.4953	0.9444
0.6623	0.1841	0.1536	345.45	344.77	0.8783	0.0910	0.0307	1.0152	1.4838	0.9493
0.5917	0.1645	0.2438	348.36	347.27	0.8561	0.0895	0.0544	1.0114	1.4719	0.9632
0.4619	0.4554	0.0827	349.12	348.30	0.7566	0.2242	0.0191	1.1037	1.2749	0.9612
0.4122	0.3196	0.2682	350.38	352.17	0.7339	0.1929	0.0732	1.0461	1.3352	0.9808
0.5333	0.1483	0.3184	350.67	349.55	0.8338	0.0881	0.0781	1.0072	1.4623	0.9722
0.4263	0.4203	0.1534	351.53	350.19	0.7359	0.2254	0.0387	1.0874	1.2853	0.9763
0.4946	0.1375	0.3679	351.93	351.19	0.8167	0.0869	0.0964	1.0040	1.4558	0.9772
0.3166	0.4774	0.2060	352.84	354.29	0.6481	0.2893	0.0627	1.1178	1.2310	1.0116
0.3959	0.3903	0.2138	353.19	351.89	0.7162	0.2258	0.0580	1.0733	1.2939	0.9847
0.2537	0.5812	0.1651	354.90	355.93	0.5880	0.3565	0.0555	1.1972	1.1675	1.0536
0.2148	0.6454	0.1398	356.16	357.08	0.5479	0.4013	0.0508	1.2669	1.1308	1.0932
0.1750	0.4020	0.4230	363.36	363.66	0.4673	0.3513	0.1814	1.0664	1.2364	1.0246
0.0585	0.9055	0.0360	364.18	364.83	0.2965	0.6797	0.0238	1.9547	1.0150	1.5148
0.1563	0.3589	0.4848	365.42	365.67	0.4367	0.3420	0.2213	1.0459	1.2516	1.0186
0.0478	0.9228	0.0294	366.09	365.86	0.2627	0.7164	0.0209	2.0514	1.0104	1.5759
0.1413	0.3246	0.5341	366.99	367.35	0.4109	0.3321	0.2570	1.0318	1.2630	1.0145
0.0404	0.9348	0.0248	367.10	366.66	0.2359	0.7454	0.0187	2.1259	1.0077	1.6231
0.0350	0.9435	0.0215	367.41	367.30	0.2142	0.7689	0.0169	2.1843	1.0059	1.6602
0.1290	0.2963	0.5747	368.56	368.80	0.3888	0.3218	0.2895	1.0214	1.2720	1.0116
0.0241	0.9592	0.0167	368.82	368.78	0.1626	0.8230	0.0144	2.2986	1.0032	1.7357
0.0203	0.9657	0.0140	369.63	369.35	0.1425	0.8449	0.0126	2.3510	1.0023	1.7687
0.0175	0.9704	0.0121	370.14	369.79	0.1266	0.8622	0.0112	2.3906	1.0017	1.7938
0.0154	0.9740	0.0106	370.54	370.13	0.1141	0.8759	0.0100	2.4220	1.0013	1.8136
0.0884	0.2313	0.6803	371.87	373.49	0.3017	0.3004	0.3979	1.0004	1.2864	1.0068
0.0779	0.2038	0.7183	373.44	375.08	0.2771	0.2814	0.4415	0.9933	1.2933	1.0050
0.0690	0.1805	0.7505	375.20	376.50	0.2548	0.2629	0.4822	0.9878	1.2987	1.0037
0.0625	0.1635	0.7740	375.65	377.59	0.2376	0.2480	0.5144	0.9841	1.3022	1.0030
0.0413	0.1802	0.7785	379.37	379.33	0.1656	0.2871	0.5473	0.9859	1.2885	1.0044
0.0359	0.1567	0.8074	379.87	380.59	0.1486	0.2617	0.5897	0.9812	1.2942	1.0032
0.0314	0.1369	0.8317	381.03	381.69	0.1338	0.2380	0.6282	0.9775	1.2986	1.0023
0.0280	0.1222	0.8498	381.64	382.55	0.1220	0.2192	0.6588	0.9749	1.3016	1.0018
Deviations			$dT = 0.77$							

5. Conclusions

VLE data for the ternary system methanol + water + ethanoic acid and three constituent binary systems at 101.325 kPa: methanol + water, methanol + ethanoic acid, and water + ethanoic acid were determined at different liquid-phase compositions using a novel pump ebulliometer. The equilibrium composition of the vapor phase was calculated from T, p, x, and H_m^E based on the Q function of excess Gibbs free energy by the indirect method. The experimental data were correlated using the NRTL, Wilson, Margules, and van Laar equations. It was shown that the deviations of NRTL, Wilson, Margules, and van Laar equations are reasonably small. The VLE data of ternary system were predicted by NRTL, Wilson, Margules, and van Laar equation. The calculated bubble points accorded well with the experimental data. The results show that the calculated bubble point is fitted by the models, which satisfy the need for the design and operation of separation process in chemical industry. Moreover, the method will provide theoretical guidance for the research of VLE data of strongly associating system of vapor and liquid phase in nonideal behavior.

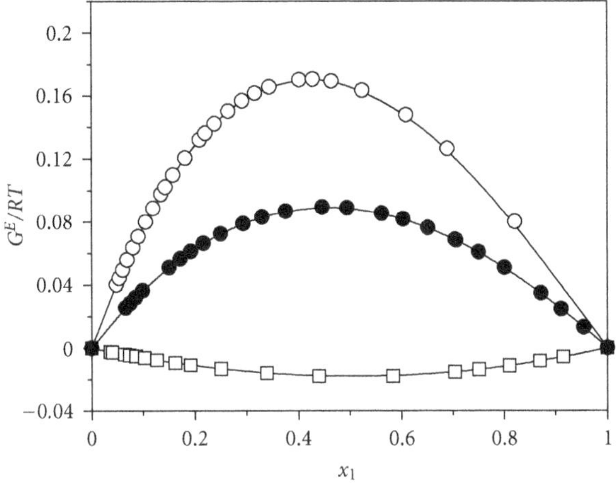

FIGURE 6: Excess Gibbs energy functions (G^E/RT) versus liquid-phase mole fraction of component 1 (x_1) diagram: ∘, methanol (1) + water (2); •, water (1) + ethanoic acid (2); □, methanol (1) + ethanoic acid (2).

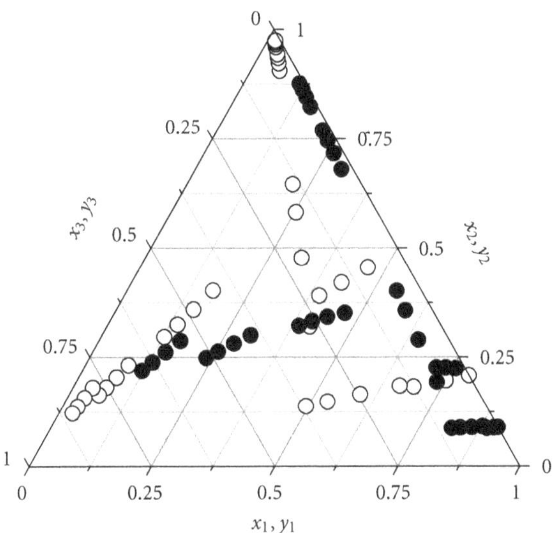

FIGURE 7: Diagram of VLE for the ternary system methanol (1) + water (2) + ethanoic acid (3) at 101.325 kPa: ∘, liquid-phase mole fraction; •, vapor-phase mole fraction.

Acknowledgments

This work was supported by the Deutscher Akademischer Austausch Dienst (DAAD) (Ref. Code: A/11/06441), the National Natural Science Foundation of China (Grant no. 21075026), and the Natural Science Foundation of the Higher Education Institutions of Anhui Province (Grant no. ZD200902).

References

[1] J. J. Ljunglin and H. C. van Ness, "Calculation of vapour-liquid equilibria from vapour pressure data," *Chemical Engineering Science*, vol. 17, no. 7, pp. 531–539, 1962.

[2] E. Hala, J. Pick, V. Fried, and O. Vilim, *Vapor-Liquid Equilibrium*, Pergamon, New York, NY, USA, 1958.

[3] H. C. Van Ness, "On the integration of the coexistence equation for binary vapor-liquid equilibrium," *AIChE Journal*, vol. 16, no. 1, pp. 18–22, 1970.

[4] F. O. Mixon, B. Gumowski, and B. H. Carpenter, "Computation of vapor-liquid equilibrium data from solution vapor pressure measurements," *Industrial and Engineering Chemistry Fundamentals*, vol. 4, no. 4, pp. 455–459, 1965.

[5] J. A. Barker, "Determination of activity coefficients from total pressure measurements," *Australian Journal of Chemistry*, vol. 6, no. 3, pp. 207–210, 1953.

[6] L. C. Tao, "How to compute binary vapor-liquid equilibrium compositions from experimental *P-x* or *t-x* data," *Industrial and Engineering Chemistry*, vol. 53, no. 4, pp. 307–309, 1961.

[7] J. M. Prausnitz, T. Anderson, E. Grens, C. Eckert, R. Hsieh, and J. P. O. 'Connell, *Computer Calculations For Multicomponent Vapor-Liquid and Liquid-Liquid Equilibria*, Prentice-Hall, Englewood Cliffs, NJ, USA, 1980.

[8] W. Arlt, "A new apparatus for phase equilibria in reacting mixtures," *Fluid Phase Equilibria*, vol. 158–160, pp. 973–977, 1999.

[9] Z. P. Xu and K. T. Chuang, "Correlation of vapor-liquid equilibrium data for methyl acetate methanol-water-acetic acid mixtures," *Industrial and Engineering Chemistry Research*, vol. 36, no. 7, pp. 2866–2870, 1997.

[10] H. Sawistowski and P. A. Pilavakis, "Vapor-liquid equilibrium with association in both phases. Multicomponent systems containing acetic acid," *Journal of Chemical and Engineering Data*, vol. 27, no. 1, pp. 64–71, 1982.

[11] W. Chang, G. Guan, X. Li, and H. Yao, "Isobaric vapor-liquid equilibria for water + acetic acid + (n-pentyl acetate or isopropyl acetate)," *Journal of Chemical and Engineering Data*, vol. 50, no. 4, pp. 1129–1133, 2005.

[12] C. Zhang, H. Wan, L. Xue, and G. Guan, "Investigation on isobaric vapor liquid equilibrium for acetic acid+water+(n-propyl acetate or iso-butyl acetate)," *Fluid Phase Equilibria*, vol. 305, no. 1, pp. 68–75, 2011.

[13] Q. Xie, H. Wan, M. Han, and G. Guan, "Investigation on isobaric vapor-liquid equilibrium for acetic acid + water + methyl ethyl ketone + isopropyl acetate," *Fluid Phase Equilibria*, vol. 280, no. 1-2, pp. 120–128, 2009.

[14] W. Chang, H. Wan, G. Guan, and H. Yao, "Isobaric vapor-liquid equilibria for water + acetic acid + (N-methyl pyrrolidone or N-methyl acetamide)," *Fluid Phase Equilibria*, vol. 242, no. 2, pp. 204–209, 2006.

[15] D. Gao, D. Zhu, H. Sun, L. Zhang, H. Chen, and J. Si, "Isobaric vapor-liquid equilibria for binary and ternary mixtures of methanol, ethanoic acid, and propanoic acid," *Journal of Chemical and Engineering Data*, vol. 55, no. 9, pp. 4002–4009, 2010.

[16] D. Gao, D. Zhu, L. Zhang et al., "Isobaric vapor—liquid equilibria for binary and ternary mixtures of propanal, propanol, and propanoic acid," *Journal of Chemical and Engineering Data*, vol. 55, no. 12, pp. 5887–5895, 2010.

[17] A. Tamir and J. Wisniak, "Vapour-liquid equilibria in associating solutions," *Chemical Engineering Science*, vol. 30, no. 3, pp. 335–342, 1975.

[18] A. Apelblat and F. Kohler, "Excess Gibbs energy of methanol + propionic acid and of methanol + butyric acid," *The Journal of Chemical Thermodynamics*, vol. 8, no. 8, pp. 749–756, 1976.

[19] A. Rius, J. L. Otero, and A. Macarron, "Equilibres liquide-vapeur de mélanges binaires donnantune réaction chimique: systèmes méthanol-acide acétique; éthanol-acide acétique; n-propanol-acide acétique; n-butanol-acide acétique," *Chemical Engineering Science*, vol. 10, no. 1-2, pp. 105–111, 1959.

[20] Z. H. Wang, B. Wu, J. W. Zhu, K. Chen, and W. Q. Wang, "Isobaric vapor-liquid-liquid equilibrium for isopropenyl acetate plus water plus acetic acid at 101 kPa," *Fluid Phase Equilibria*, vol. 314, no. 2, pp. 152–155, 2012.

[21] S. Bernatová, K. Aim, and I. Wichterle, "Isothermal vapour-liquid equilibrium with chemical reaction in the quaternary water + methanol + acetic acid + methyl acetate system, and in five binary subsystems," *Fluid Phase Equilibria*, vol. 247, no. 1-2, pp. 96–101, 2006.

[22] A. Tamir and J. Wisniak, "Association effects in ternary vapour-liquid equilibria," *Chemical Engineering Science*, vol. 31, no. 8, pp. 625–630, 1976.

[23] A. Tamir and J. Wisniak, "Vapor equilibrium in associating systems (water-formic acid-propionic acid)," *Industrial and Engineering Chemistry Fundamentals*, vol. 15, no. 4, pp. 274–280, 1976.

[24] A. Tamir and J. Wisniak, "Activity coefficient calculations in multicomponent associating systems," *Chemical Engineering Science*, vol. 33, no. 6, pp. 651–656, 1978.

[25] J. Wisniak and A. Tamir, "Vapor-liquid equilibria in the ternary systems water-formic acid-acetic acid and water-acetic acid-propionic acid," *Journal of Chemical and Engineering Data*, vol. 22, no. 3, pp. 253–260, 1977.

[26] J. Marek and G. Standart, "Vapor-liquid equilibria in mixtures containing an associating substance. I. Equilibrium relationships for systems with an associating component," *Collection of Czechoslovak Chemical Communications*, vol. 19, pp. 1074–1084, 1954.

[27] J. Marek, "Vapor-liquid equilibria in mixtures containing an associating substance. II. Binary mixtures of acetic acid at atmospheric pressure," *Collection of Czechoslovak Chemical Communications*, vol. 20, pp. 1490–1502, 1955.

[28] J. G. Hayden and J. P. O'Conell, "A generalized method for predicting second virial coefficient," *Industrial & Engineering Chemistry Process Design and Development*, vol. 14, no. 3, pp. 209–215, 1975.

[29] J. A. Riddick, W. B. Bunger, and T. K. Sakano, *Organic Solvents: Physical Properties and Methods of Purification*, Wiley-Interscience, New York, NY, USA, 4th edition, 1986.

[30] M. V. P. Rao and P. R. Naidu, "Excess volumes of binary mixtures of alcohols in methylcohexane," *Canadian Journal of Chemistry*, vol. 52, no. 5, pp. 788–790, 1974.

[31] Z. M. Qiu, Z. C. Luo, and Y. Hu, "The pump ebullionmeter," *Journal of Chemical Engineering of Chinese Universities*, vol. 11, no. 1, pp. 74–77, 1997.

[32] J. Gmehling and U. Onken, *Vapor-Liquid Equilibrium Data Collection-Aqueous Organic Systems*, vol. 1, part 1 of *DECHEMA Chemistry Data Series*, DECHEMA, Frankfurt, Germany, 1977.

[33] J. Gmehling and U. Onken, *Vapor-Liquid Equilibrium Data Collection-Organic Hydroxy Compounds: Alcohols*, vol. 1, part 2 of *DECHEMA Chemistry Data Series*, DECHEMA, Frankfurt, Germany, 1977.

[34] R. Francesconi and F. Comelli, "Liquid-phase enthalpy of mixing for the system 1,3-dioxolane-chlorobenzene in the temperature range 288.15–313.15 K," *Journal of Chemical and Engineering Data*, vol. 31, no. 2, pp. 250–252, 1986.

[35] J. Gmehling, "Excess enthalpies for 1,1,1-trichloroethane with alkanes, ketones, and esters," *Journal of Chemical and Engineering Data*, vol. 38, no. 1, pp. 143–146, 1993.

[36] J. Shi, J. D. Wang, and G. Z. Yu, *Handbook of Chemical Engineering*, Chemical Industry Press, Beijing, China, 2nd edition, 1996.

[37] C. Tsonopoulos, "An empirical correlation of second virial coefficients," *AIChE Journal*, vol. 20, no. 2, pp. 263–272, 1974.

[38] E. Lam, A. Mejía, H. Segura, J. Wisniak, and S. Loras, "A model-free approach data treatment of vapor-liquid equilibrium data in ternary systems. 1. Theory and numerical procedures," *Industrial and Engineering Chemistry Research*, vol. 40, no. 9, pp. 2134–2148, 2001.

[39] H. Renon and J. M. Prausnitz, "Estimation of parameters for the NRTL equation for excess gibbs energies of strongly nonideal liquid mixtures," *Industrial & Engineering Chemistry Process Design and Development*, vol. 8, no. 3, pp. 413–419, 1969.

[40] G. M. Wilson, "Vapor-liquid equilibrium. XI. A new expression for the excess free energy of mixing," *Journal of the American Chemical Society*, vol. 86, no. 2, pp. 127–130, 1964.

[41] M. S. Margules, "On the composition of saturated vapors of mixtures," *Akademie der Wissenschaften in Wien, Mathematisch-Naturwissenschaftliche Klasse Abteilung II*, vol. 104, pp. 1234–1239, 1895.

[42] J. Wisniak, A. Apelblat, and H. Segura, "Application of model-free computation techniques and limiting conditions for azeotropy. An additional assessment of experimental data," *Chemical Engineering Science*, vol. 52, no. 23, pp. 4393–4402, 1997.

[43] E. Lam, A. Mejía, H. Segura, J. Wisniak, and S. Loras, "A model-free approach data treatment of vapor-liquid equilibrium data in ternary systems. 2. Applications," *Industrial and Engineering Chemistry Research*, vol. 40, no. 9, pp. 2149–2159, 2001.

[44] J. J. van Laar, "The vapor pressure of binary mixtures," *Zeitschrift für Physikalische Chemie*, vol. 72, pp. 723–751, 1910.

[45] J. Gmehling and U. Onken, *Vapor-Liquid Equilibrium Data Collection-Carboxylic Acids, Anhydrides, Esters*, vol. 1, part 5 of *DECHEMA Chemistry Data Series*, DECHEMA, Frankfurt, Germany, 1977.

[46] J. Gmehling and U. Onken, *Vapor-Liquid Equilibrium Data Collection-Organic Hydroxy Compounds: Alcohols (Supplement 3)*, vol. 1, part 2 of *DECHEMA Chemistry Data Series*, DECHEMA, Frankfurt, Germany, 1977.

15

Effects of Exothermic/Endothermic Chemical Reactions with Arrhenius Activation Energy on MHD Free Convection and Mass Transfer Flow in Presence of Thermal Radiation

Kh. Abdul Maleque

Department of Mathematics, American International University-Bangladesh, House-23, 17, Kamal Ataturk Avenue, Banani, Dhaka 1213,Bangladesh

Correspondence should be addressed to Kh. Abdul Maleque; maleque@aiub.edu

Academic Editor: Felix Sharipov

A local similarity solution of unsteady MHD natural convection heat and mass transfer boundary layer flow past a flat porous plate within the presence of thermal radiation is investigated. The effects of exothermic and endothermic chemical reactions with Arrhenius activation energy on the velocity, temperature, and concentration are also studied in this paper. The governing partial differential equations are reduced to ordinary differential equations by introducing locally similarity transformation (Maleque (2010)). Numerical solutions to the reduced nonlinear similarity equations are then obtained by adopting Runge-Kutta and shooting methods using the Nachtsheim-Swigert iteration technique. The results of the numerical solution are obtained for both steady and unsteady cases then presented graphically in the form of velocity, temperature, and concentration profiles. Comparison has been made for steady flow ($A = 0$) and shows excellent agreement with Bestman (1990), hence encouragement for the use of the present computations.

1. Introduction

In free convection boundary layer flows with simultaneous heat mass transfer, one important criteria that is generally not encountered is the species chemical reactions with finite Arrhenius activation energy. The modified Arrhenius law (IUPAC Goldbook definition of modified Arrhenius equation) is usually of the form (Tencer et al. [1])

$$K = B\left(\frac{T}{T_\infty}\right)^n \exp\left[\frac{-E_a}{kT}\right], \quad (1)$$

where K is the rate constant of chemical reaction and B that is the preexponential factor simply prefactor (constant) is based on the fact that increasing the temperature frequently causes a marked increase in the rate of reactions. E_a is the activation energy, and $k = 8.61 \times 10^{-5}$ eV/K is the Boltzmann constant which is the physical constant relating energy at the individual particle level with temperature observed at the collective or bulk level.

In areas such as geothermal or oil reservoir engineering, the prvious phenomenon is usually applicable. Apart from experimental works in these areas, it is also important to make some theoretical efforts to predict the effects of the activation energy in flows mentioned above. But in this regard very few theoretical works are available in the literature. The reason is that the chemical reaction processes involved in the system are quite complex and generally the mass transfer equation that is required for all the reactions involved also becomes complex. Theoretically, such an equation is rather impossible to tackle. Form chemical kinetic viewpoint this is a very difficult problem, but if the reaction is restricted to binary type, a lot of progress can be made. The thermomechanical balance equations for a mixture of general materials were first formulated by Truesdell [2]. Thereafter Mills [3] and Beevers and Craine [4] have obtained some exact solutions for the boundary layer flow of a binary mixture of incompressible Newtonian fluids. Several problems relating to the mechanics of oil and water emulsions, particularly with regard to applications in lubrication practice, have been considered within the context of a binary mixture theory by Al-Sharif et al. [5] and Wang et al. [6].

A simple model involving binary reaction was studied by Bestman [7]. He considered the motion through the plate to be large which enabled him to obtain analytical solutions (subject to same restrictions) for various values of activation energy by employing the perturbation technique proposed by Singh and Dikshit [8]. Bestman [9] and Alabraba et al. [10] took into account the effect of the Arrhenius activation energy under the different physical conditions. Recently Kandasamy et al. [11] studied the combined effects of chemical reaction, heat and mass transfer along a wedge with heat source and concentration in the presence of suction or injection. Their result shows that the flow field is influenced appreciably by chemical reaction, heat source, and suction or injection at the wall of the wedge. Recently Makinde et al. [12, 13] studied the problems of unsteady convection with chemical reaction and radiative heat transfer past a flat porous plate moving through a binary mixture in an optically thin environment. More recently Abdul Maleque [14, 15] investigated the similarity solution on unsteady incompressible fluid flow with binary chemical reactions and activation energy. In the present paper, we investigate a numerical solution of unsteady Mhd natural convection heat and mass transfer boundary layer flow past a flat porous plate taking into account the effect of Arrhenius activation energy with exothermic/endothermic chemical reactions and thermal radiation. This problem is an extension work studied by Abdul Maleque [16].

2. Governing Equations

We consider the boundary wall to be of infinite extend, so that all quantities are homogeneous in x, and hence all derivatives with respect to x are neglected. The x-axis is taken along the plate, and y-axis is perpendicular to the plate. Assume that a uniform magnetic field B_0 is applied perpendicular to the plate and the plate is moving with uniform velocity U_o in its own plane. In presence of exothermic/endothermic binary chemical reaction with Arrhenius activation energy, a uniform magnetic field and thermal radiation thus the governing equations are

$$\frac{\partial v}{\partial y} = 0, \tag{2}$$

$$\frac{\partial u}{\partial t} + v\frac{\partial u}{\partial y} = \upsilon\frac{\partial^2 u}{\partial y^2} + g\beta_1 (T - T_\infty) + g\beta_2 (C - C_\infty) - \frac{\sigma B_0^2}{\rho}u, \tag{3}$$

$$\rho c_p\left(\frac{\partial T}{\partial t} + v\frac{\partial T}{\partial y}\right) = \kappa\frac{\partial^2 T}{\partial y^2} + \beta k_r^2\left(\frac{T}{T_\infty}\right)^n \times \exp\left[\frac{-E_a}{kT}\right](C - C_\infty) - \frac{\partial q_r}{\partial y}, \tag{4}$$

$$\frac{\partial C}{\partial t} + v\frac{\partial C}{\partial y} = D_m\frac{\partial^2 C}{\partial y^2} - k_r^2\left(\frac{T}{T_\infty}\right)^n \exp\left[\frac{-E_a}{kT}\right](C - C_\infty). \tag{5}$$

The boundary conditions of previous system are

$$u = U_0, \quad v = v(t), \quad T = T_w, \quad C = C_w \quad \text{at } y = 0,\ t > 0, \tag{6}$$
$$u = 0, \quad T \longrightarrow T_\infty, \quad C \longrightarrow C_\infty \quad \text{at } y = \infty,\ t > 0,$$

where (u, v) is the velocity vector, $T$is the temperature, $C$is the concentration of the fluid, μ is the fluid viscosity, σ is the electrical conductivity, υ is the kinematic coefficient of viscosity, β_1 and β_2 are the coefficients of volume expansions for temperature and concentration, respectively, κ is the heat diffusivity coefficient, c_p is the specific heat at constant pressure, D_m is the coefficient of mass diffusivity, k_r^2 is the chemical reaction rate constant, $\beta(= \pm 1)$ is exothermic/endothermic parameter, and $(T/T_\infty)^n \exp[-E_a/k\, T]$ is the Arrhenius function where n is a unit less constant exponent fitted rate constants typically lie in the range $-1 < n < 1$.

The radiative heat flux q_r is described by Roseland approximation such that

$$q_r = -\frac{4\sigma_1}{3k_1}\frac{\partial T^4}{\partial y}, \tag{7}$$

where σ_1 and k_1 are the Stefan Boltzmann constant and mean absorption coefficient, respectively. Following Makinde and Olanrewaju [13], we assume that the temperature differences within the flow are sufficiently small, so that the T^n can be expressed as a linear function. By using Taylor's series, we expand T^n about the free stream temperature T_∞ and neglecting higher order terms. This result of the following approximation:

$$T^n \approx (1 - n)\,T_\infty^n + nT_\infty^{n-1}T. \tag{8}$$

Using (9), we have

$$\frac{\partial q_r}{\partial y} = -\frac{4\sigma_1}{3k_1}\frac{\partial^2 T^4}{\partial y^2} \approx -\frac{16T_\infty^3\sigma_1}{3k_1}\frac{\partial^2 T}{\partial y^2}, \tag{9}$$

$$\left(\frac{T}{T_\infty}\right)^n \approx (1 - n) + n\frac{T}{T_\infty}. \tag{10}$$

3. Mathematical Formulations

In order to solve the governing equations (2) to (5) under the boundary conditions (6), we adopt the well-defined similarity technique to obtain the similarity solutions. For this purpose the following nondimensional variables are now introduced:

$$\eta = \frac{y}{\delta(t)}, \qquad \frac{u}{U_0} = f(\eta), \qquad \frac{T - T_\infty}{T_w - T_\infty} = \theta(\eta), \qquad \frac{C - C_\infty}{C_w - C_\infty} = \phi(\eta). \tag{11}$$

From the equation of continuity (3), we have

$$v(t) = -\frac{v_0 \upsilon}{\delta(t)}, \tag{12}$$

where v_0 is the dimensionless suction/injection velocity at the plate, $v_0 > 0$ corresponds to suction, and $v_0 < 0$ corresponds to injection.

Introducing (9), (10), the dimensionless quantities from (11) and v from (12) in (4), (5), and (6), we finally obtain the nonlinear ordinary differential equations as

$$f'' + \left(\eta \frac{\delta\delta'}{\upsilon} + v_0\right) f' - Mf + G_0\gamma\theta + G_m\phi = 0, \tag{13}$$

$$\theta'' + \left(\frac{P_r}{1+N}\right)\left[\left(\eta\frac{\delta\delta'}{\upsilon} + v_0\right)\theta' + \beta\lambda^2(1+n\gamma\theta)\exp\left(-\frac{E}{1+\gamma\theta}\right)\phi\right] = 0, \tag{14}$$

$$S_c^{-1}\phi'' + \left(\eta\frac{\delta\delta'}{\upsilon} + v_0\right)\phi' - \lambda^2(1+n\gamma\theta)\exp\left(-\frac{E}{1+\gamma\theta}\right)\phi = 0. \tag{15}$$

Here, Grashof number $G_r = G_0\gamma = (g\beta T_\infty \delta^2/\upsilon U_0)\gamma$; the temperature relative parameter $\gamma = (T_w - T_\infty)/T_\infty$.

Modified (Solutal) Grashof number $G_m = g\beta^*(C_w - C_\infty)\delta^2/\upsilon U_0$; Prandtl number $P_r = \rho\upsilon c_p/\kappa$; magnetic interaction parameter $M = \sigma\delta^2 B_0^2/\mu$; schmidt number $S_c = \upsilon/D_M$; the nondimensional activation energy $E = E_a/k(T_w - T_\infty)$; the dimensionless chemical reaction rate constant $\lambda^2 = k_r^2\delta^2/\upsilon$; radiation parameter $N = 16\sigma_1 T_\infty^3/3\kappa k_1$.

Equations from (13) to (15) are similar except for the term $\delta\delta'/\upsilon$, where time t appears explicitly. Thus the similarity condition requires that $\delta\delta'/\upsilon$ must be a constant quantity. Hence following Abdul Maleque [17], one can try a class of solutions of (13) to (15) by assuming that

$$\frac{\delta\delta'}{\upsilon} = A \ (\text{constant}). \tag{16}$$

From (16), we have

$$\delta(t) = \sqrt{2A\upsilon t + L^2}, \tag{17}$$

where the constant of integration L is determined through the condition that $\delta = L$ when $t = 0$. Here $A = 0$ implies that $\delta = L$ represents the length scale for steady flow and $A \neq 0$, that is, δ represents the length scale for unsteady flow. Since δ is a scaling factor as well as similarity parameter, any other values of A in (13) would not change the nature of the solution except that the scale would be different. Finally introducing (16) in (13) to (15), respectively, we have the following dimensionless nonlinear ordinary differential equations:

$$f'' + (A\eta + v_0) f' - Mf + G_0\gamma\theta + G_m\phi = 0 \tag{18}$$

$$\theta'' + \left(\frac{P_r}{1+N}\right)\left[(A\eta + v_0)\theta' + \beta\lambda^2(1+n\gamma\theta)\exp\left(-\frac{E}{1+\gamma\theta}\right)\phi\right] = 0, \tag{19}$$

$$S_c^{-1}\phi'' + (A\eta + v_0)\phi' - \lambda^2(1+n\gamma\theta)\exp\left(-\frac{E}{1+\gamma\theta}\right)\phi = 0. \tag{20}$$

The boundary conditions equation (4) becomes

$$\begin{gathered} f(0) = 1, \quad \theta(0) = 1, \quad \phi(0) = 1, \\ f(\infty) = 0, \quad \theta(\infty) = 0, \quad \phi(\infty) = 0. \end{gathered} \tag{21}$$

In all over equations primes denote the differentiation with respect to η. Equations from (18)–(20) are solved numerically under the boundary conditions (21) using Nachtsheim-Swigert iteration technique.

4. Numerical Solutions

Equations (18)–(20) are solved numerically under the boundary conditions (21) using Nachtsheim-Swigert [18] iteration technique. In (21), there are three asymptotic boundary conditions and hence follow three unknown surface conditions $f'(0)$, $\theta'(0)$, and $\phi'(0)$. Within the context of the initial-value method and the Nachtsheim-Swigert iteration technique the outer boundary conditions may be functionally represented by the first-order Taylor's series as

$$\begin{aligned} f(\eta_{\max}) &= f(X,Y,Z) \\ &= f_c(\eta_{\max}) + \Delta X f_X + \Delta Y f_Y + \Delta Z f_Z = \delta_1, \end{aligned} \tag{22}$$

$$\begin{aligned} \theta(\eta_{\max}) &= \theta(X,Y,Z) \\ &= \theta_c(\eta_{\max}) + \Delta X \theta_X + \Delta Y \theta_Y + \Delta Z \theta_Z = \delta_2, \end{aligned} \tag{23}$$

$$\begin{aligned} \phi(\eta_{\max}) &= \phi(X,Y,Z) \\ &= \phi_c(\eta_{\max}) + \Delta X \phi_X + \Delta Y \phi_Y + \Delta Z \phi_Z = \delta_3, \end{aligned} \tag{24}$$

with the asymptotic convergence criteria given by

$$\begin{aligned} f'(\eta_{\max}) &= f'(X,Y,Z) \\ &= f'_c(\eta_{\max}) + \Delta X f'_X + \Delta Y f'_Y + \Delta Z f'_Z = \delta_4, \end{aligned} \tag{25}$$

$$\begin{aligned} \theta'(\eta_{\max}) &= \theta'(X,Y,Z) \\ &= \theta'_c(\eta_{\max}) + \Delta X \theta'_X + \Delta Y \theta'_Y + \Delta Z \theta'_Z = \delta_5, \end{aligned} \tag{26}$$

$$\begin{aligned} \phi'(\eta_{\max}) &= \phi'(X,Y,Z) \\ &= \phi'_c(\eta_{\max}) + \Delta X \phi'_X + \Delta Y \phi'_Y + \Delta Z \phi'_Z = \delta_6, \end{aligned} \tag{27}$$

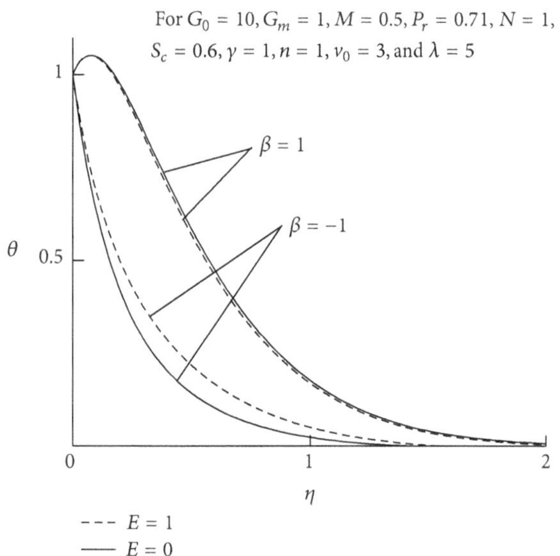

FIGURE 1: Effects of β and E on temperature profiles.

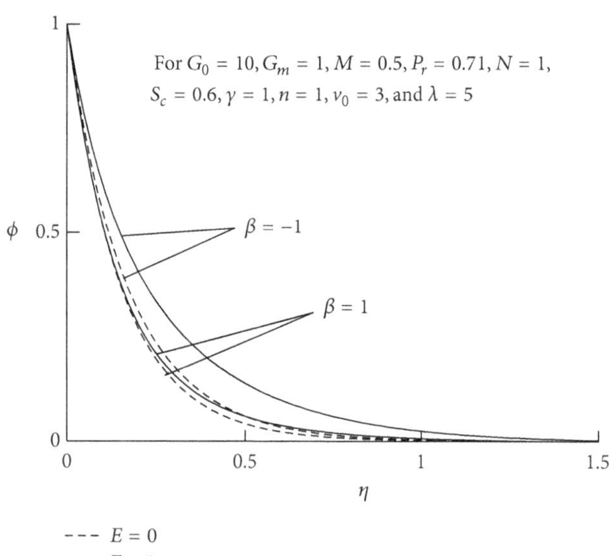

FIGURE 2: Effects of β and E on concentration profiles.

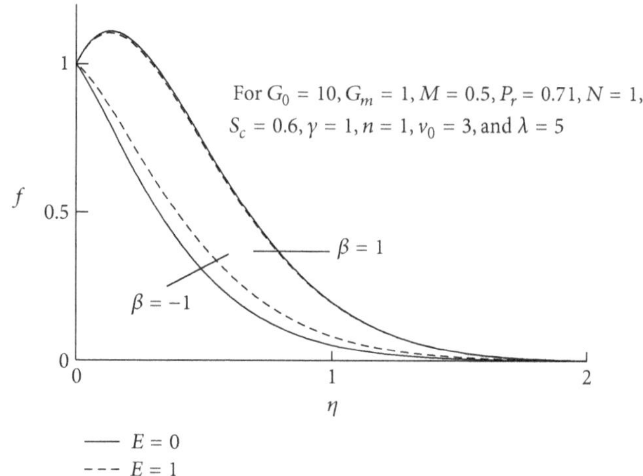

FIGURE 3: Effects of β and E on velocity profiles.

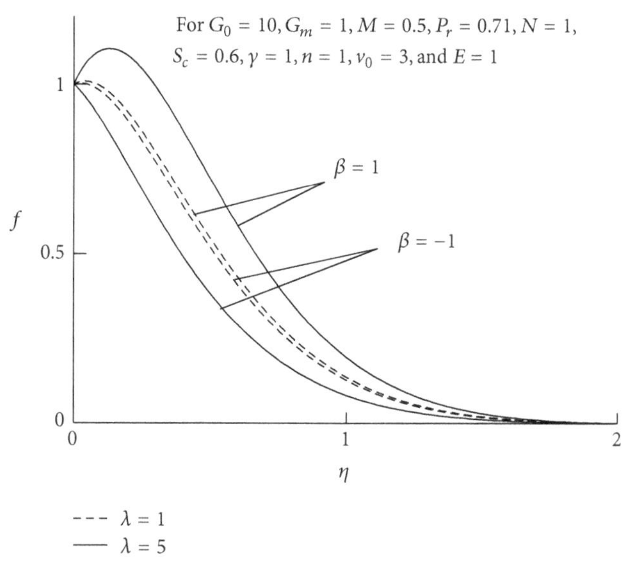

FIGURE 4: Effects of β and λ on velocity profiles.

where $X = f'(0)$, $Y = \theta'(0)$, $Z = \phi'(0)$, and X, Y, Z subscripts indicate partial differentiation, for example, $f'_Y = \partial f'(\eta_{\max})/\partial\theta'(0)$. The subscript c indicates the value of the function at $\eta_{\max}$ to be determined from the trial integration.

Solutions of these equations in a least square sense require determining the minimum value of $E = \delta_1^2 + \delta_2^2 + \delta_3^2 + \delta_4^2 + \delta_5^2 + \delta_6^2$ with respect to X, Y, and Z. To solve ΔX, ΔY, and ΔZ, we require to differentiate E with respect to X, Y and Z, respectively. Thus adopting this numerical technique, a computer program was set up for the solutions of the basic nonlinear differential equations of our problem where the integration technique was adopted as the six-ordered Runge-Kutta method of integration. The results of this integration are then displayed graphically in the form of velocity, temperature, and concentration profiles in Figures 1–15. In the process of integration, the local skin-friction coefficient, the local rates of heat, and mass transfer to the surface, which are of chief physical interest, are also calculated out. The equation defining the wall skin friction is

$$\tau = -\mu\left(\frac{\partial u}{\partial y}\right)_{y=0} = -\frac{\mu U_0}{\delta} f'(0). \tag{28}$$

Hence the skin-friction coefficient is given by

$$\frac{1}{2} C_f = \frac{\tau}{\rho U_0^2} \Longrightarrow \frac{1}{2} R_e C_f = -f'(0), \tag{29}$$

here, the Reynolds number $R_e = \rho U_0 \delta/\mu$.

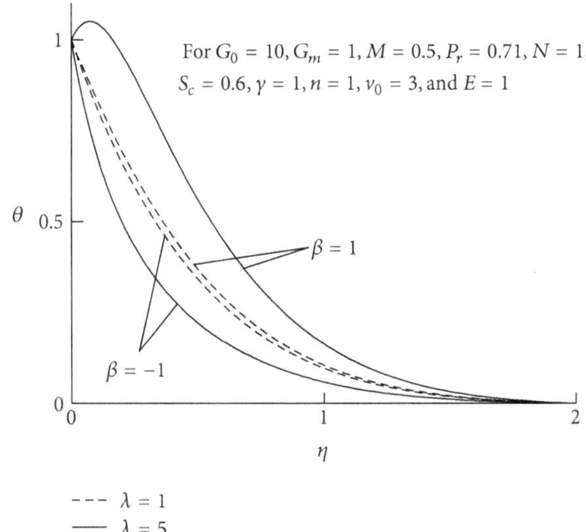

Figure 5: Effects of β and λ on temperature profiles.

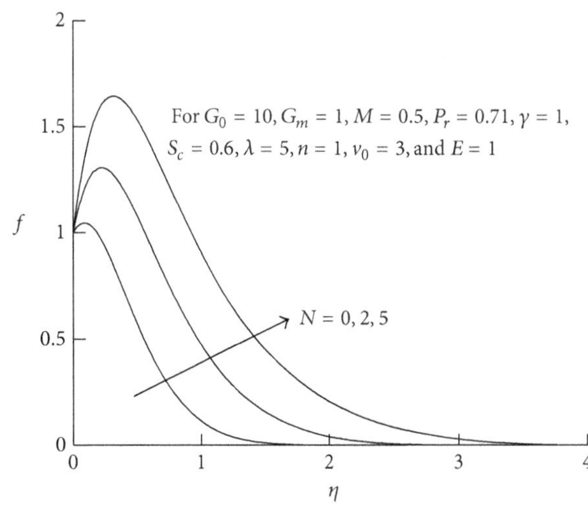

Figure 8: Effects of N on velocity profiles.

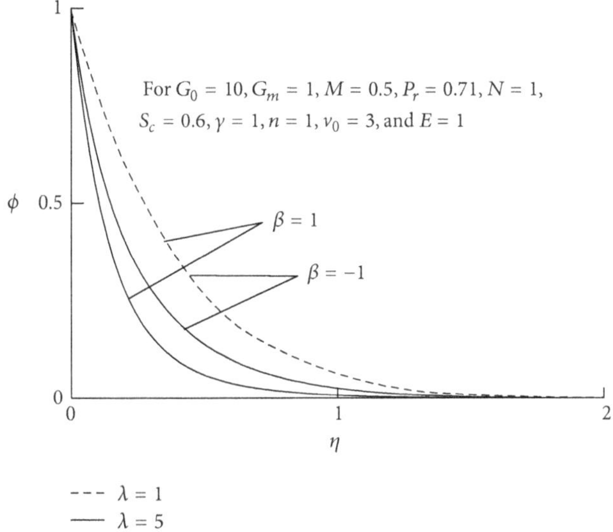

Figure 6: Effects of β and λ on concentration profiles.

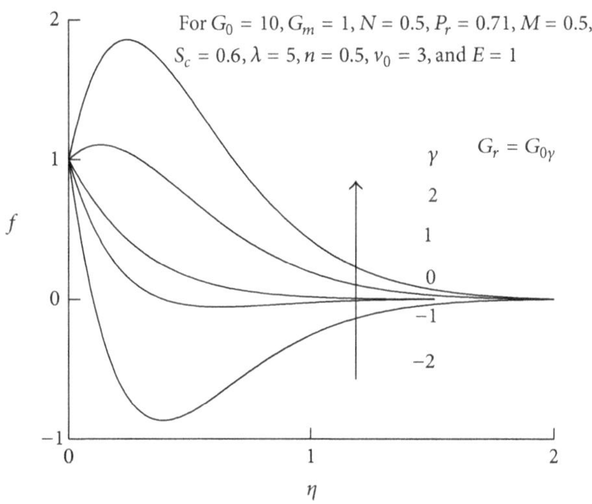

Figure 9: Effects of γ on velocity profiles.

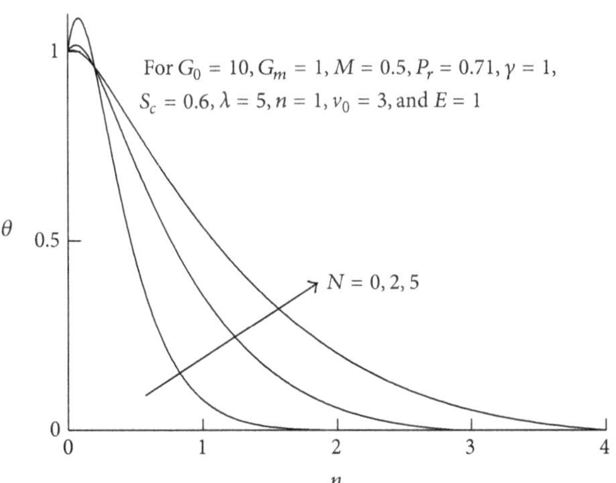

Figure 7: Effects of N on temperature profiles.

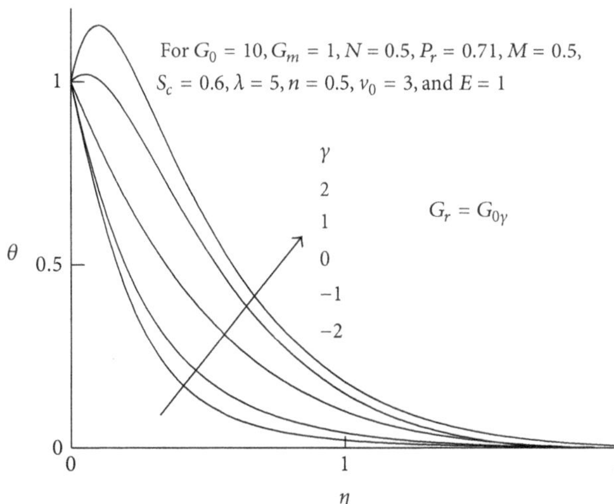

Figure 10: Effects of γ on temperature profiles.

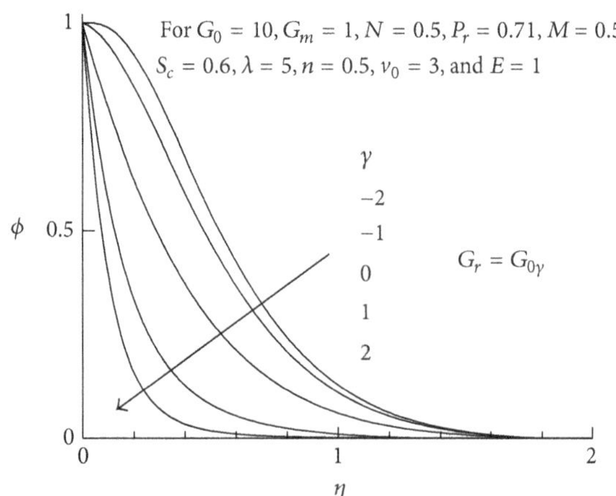

FIGURE 11: Effects of γ on concentration profiles.

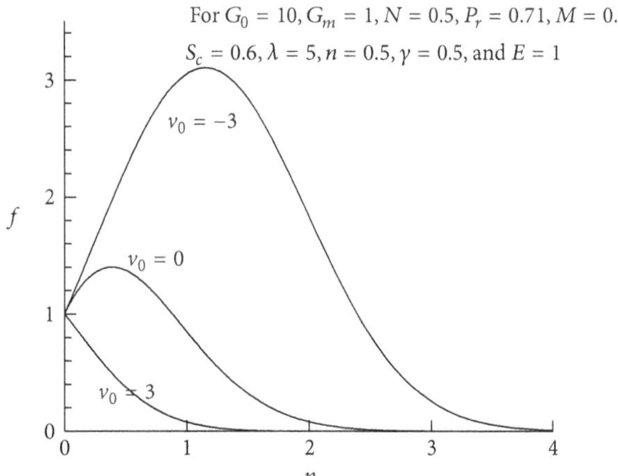

FIGURE 12: Effects of v_0 on velocity profiles.

The heat flux (q_w) and the mass flux (M_w) at the wall are given by

$$q_w = -\kappa\left(\frac{\partial T}{\partial y}\right)_{y=0} = -\kappa\frac{\Delta T}{\delta}\theta'(0),$$
$$M_w = -D_M\left(\frac{\partial C}{\partial y}\right)_{y=0} = -D_M\frac{\Delta C}{\delta}\phi'(0). \tag{30}$$

Hence the Nusselt number (N_u) and the Sherwood number (S_h) are obtained as

$$N_u = \frac{q_w\delta}{\kappa\,\Delta T} = -\theta'(0), \qquad S_h = \frac{M_w\delta}{D_M\Delta C} = -\phi'(0). \tag{31}$$

These previous coefficients are then obtained from the procedure of the numerical computations and are sorted in Table 1.

5. Results and Discussions

The parameters entering into the fluid flow are Grashof number $G_r = G_0\gamma$, Solutal (modified) Grashof number G_m, suction parameter v_0, Prandtl number P_r, the nondimensional chemical reaction rate constant λ^2, Schmidt number

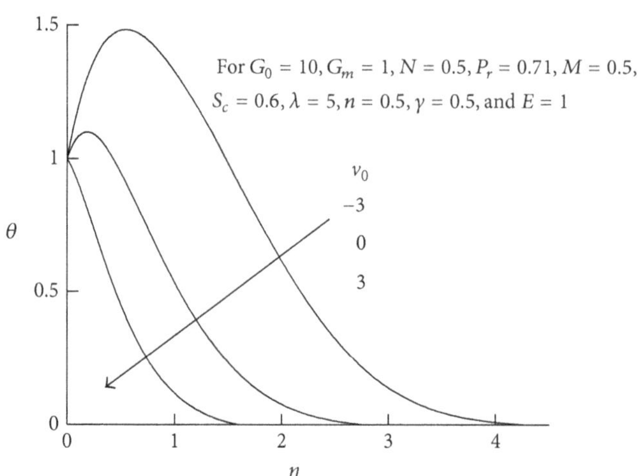

FIGURE 13: Effects of v_0 on temperature profiles.

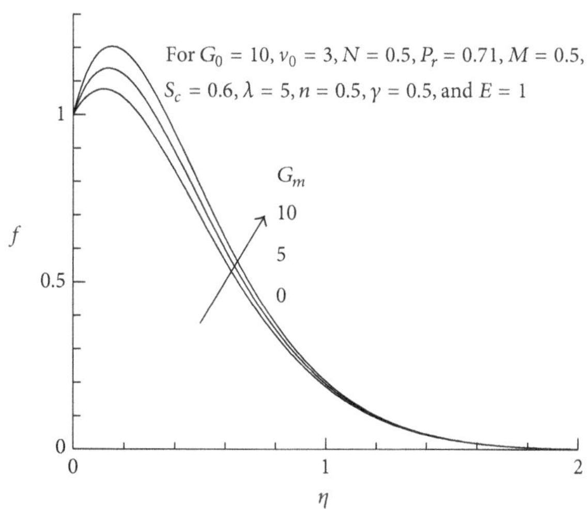

FIGURE 14: Effects of G_m on velocity profiles.

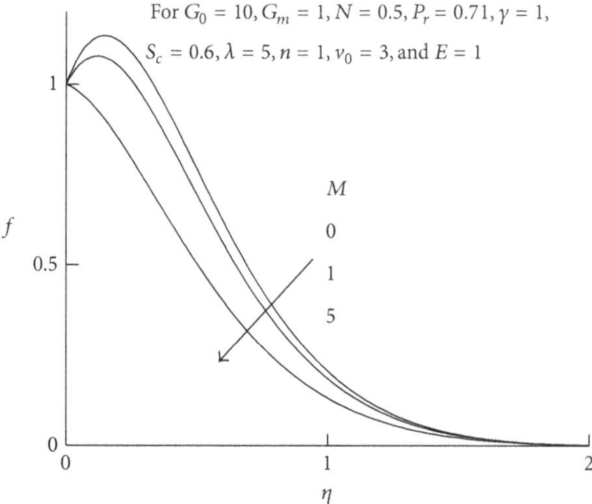

FIGURE 15: Effects of M on velocity profiles.

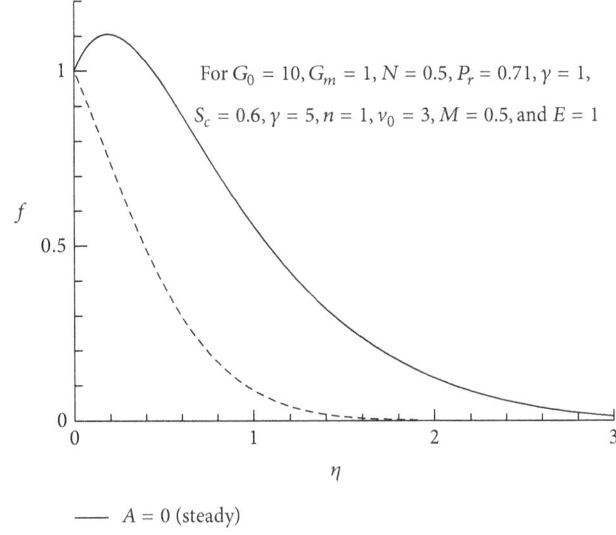

FIGURE 16: Effects of A on velocity profiles.

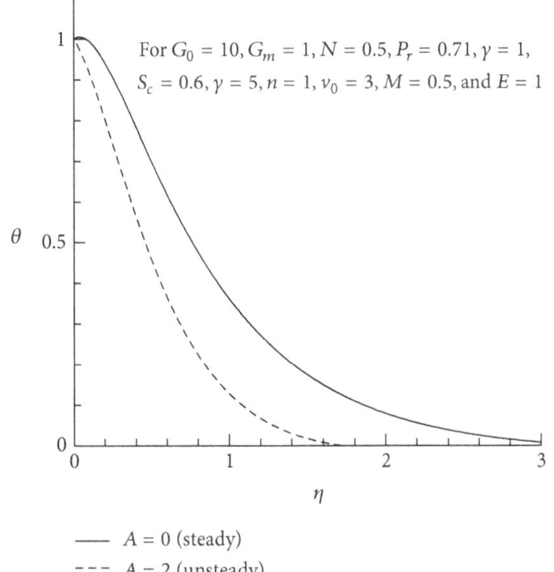

FIGURE 17: Effects of A on temperature profiles.

S_c, the magnetic parameter M, the nondimensional activation energy E, the radiation parameter N, the exothermic/endothermic parameter β, and the temperature relative parameter γ.

It is, therefore, pertinent to inquire the effects of variation of each of them when the others are kept constant. The numerical results are thus presented in the form of velocity profiles, temperature profiles, and concentration profiles in Figures 1–14 for the different values of $G_r, G_m, \lambda, M, E, v_0, \beta$, and γ. The value of γ is taken to be both positive and negative, since these values represent, respectively, cooling and heating of the plate.

The values of G_r is taken to be large ($G_r = 10$, $G_0 = 10$, and $\gamma = 1$), since the value corresponds to

TABLE 1: Effects on skin friction, heat transfer, and mass transfer coefficients. For $G_0 = 10, G_m = 1, N = 0.5, P_r = 0.71, M = 0.5, S_c = 0.6, \lambda = 5, n = 0.5, v_0 = 3, \gamma = 0.5$, and $E = 1$.

	$-f'(0)$	$-\theta'(0)$	$-\phi'(0)$
E for $\beta = 1$			
0.0	−1.65163	−0.91802	5.78634
0.5	−1.60009	−0.88980	5.75523
1.0	−1.54562	−0.85343	5.70174
E for $\beta = -1$			
0.0	1.30241	4.49966	5.55159
0.5	0.98702	4.19573	5.21093
1.0	0.63543	3.87806	4.81246
λ for $\beta = 1$			
1.0	−0.43158	1.73298	2.42627
2.0	−0.63434	1.34699	2.91446
5.0	−1.54562	−0.85343	5.70174
λ for $\beta = -1$			
1.0	−0.27903	1.98971	2.42088
2.0	−0.07554	2.32400	2.84424
5.0	0.63543	3.87806	4.81246

a cooling problem that is generally encountered in nuclear engineering in connection with the cooling of reactors. In air ($P_r = 0.71$) the diluting chemical species of most common interest have Schmidt number in the range from 0.6 to 0.75. Therefore, the Schmidt number $S_c = 0.6$ is considered. In particular, 0.6 corresponds to water vapor that represents a diffusing chemical species of most common interest in air. The values of the suction parameter v_0 are taken to be large. Apart from the figures and tables, the representative velocity, temperature,and concentration profiles and the values of the physically important parameters, that is, the local shear stress, the local rates of heat and mass transfer, are illustrated for uniform wall temperature and species concentration in Figures 1–17 and in Tables 1-2.

5.1. Effects of Activation Energy (E) on the Concentration, the Temperature, and the Velocity Profi es for Exothermic/ Endothermic Chemical Reaction. In chemistry, *activation energy* is defined as the energy that must be overcome in order for a chemical reaction to occur. Activation energy may also be defined as the minimum energy required starting a chemical reaction. The activation energy of a reaction is usually denoted by E_a and given in units of kilojoules per mole.

An exothermic reaction is a chemical reaction that releases energy in the form of light and heat. It is the opposite of an endothermic reaction. It gives out energy to its surroundings. The energy needed for the reaction to occur is less or greater than the total energy released for exothermic or endothermic reaction, respectively. Energy is obtained from chemical bonds. When bonds are broken, energy is required. When bonds are formed, energy is released. Each type of bond has specific bond energy. It can be predicted whether a chemical reaction will release or require heat by using bond energies. When there is more energy used to form

TABLE 2: Effects on skin friction, heat transfer, and mass transfer coefficients. For $G_0 = 10, G_m = 1, N = 0.5, P_r = 0.71, M = 0.5,\ S_c = 0.6, \lambda = 5, n = 0.5, v_0 = 3, \gamma = 0.5$, and $E = 1$.

	$-f'(0)$	$-\theta'(0)$	$-\phi'(0)$
N			
0.0	−0.87260	−1.44125	5.68996
2.0	−3.05961	−0.30027	5.71624
5.0	−4.93466	−0.04756	5.72770
M			
0.0	−1.85010	−0.85309	5.70164
2.0	−0.75246	−0.85443	5.70203
5.0	0.46744	−0.85628	5.70257
γ			
−2.00000	7.32171	3.65294	0.07835
−1.00000	5.66728	3.48636	0.26098
0.00000	3.31180	1.88820	2.28076
1.00000	−1.54562	−0.85343	5.70174
2.00000	−7.85616	−3.67839	9.25776
A			
0.00000	−1.27274	−0.26974	3.97243
2.00000	1.20419	0.56519	3.91549

the bonds than to break the bonds, heat is given off. This is known as an exothermic reaction. When a reaction requires an input or output of energy, it is known as an exothermic or endothermic reaction.

Effects of activation energy (E) and exothermic parameter (β) on the temperature, the concentration and velocity profiles are shown in Figure 1 to Figure 3, respectively. $\beta = 1$ and $\beta = -1$ represent exothermic and endothermic chemical reactions, respectively. Figure 1 shows that a small decreasing effect of temperature profile is found for increasing values of E for exothermic reaction ($\beta = 1$), but mark opposite effects are found for endothermic reaction in Figure 1. That is the temperature profile increases for increasing values of E for endothermic reaction $\beta = -1$. From (1) we observe that chemical reaction rate (K) decreases with the increasing values of activation energy (E_a). We also observe from (20) that increase in activation energy (E) leads to decrease $\lambda^2 \exp(-E/\theta)$ as well as to increase in the concentration profiles shown in Figure 2. Small and reported effects are found for $\beta = 1$ and $\beta = -1$, respectively. The parameter E does not enter directly into the momentum equation, but its influence comes through the mass and energy equations. Figure 3 shows the variation of the velocity profiles for different values of E. From this figure it has been observed that the velocity profile mark increases with the increasing values of E for endothermic chemical reaction. But negligible effects are found for exothermic reaction.

5.2. Effects of λ on the Concentration, the Temperature, and the Velocity Profiles for Exothermic/Endothermic Reaction. Considering chemical reaction rate constant λ^2 is always positive. Figures 4–6 represent the effect of chemical rate constant λ on the velocity, the temperature, and the concentration profiles, respectively.

We observe from Figures 4 and 5 that velocity and temperature profiles increase with the increasing values of λ for exothermic reaction, but opposite effects are found in these figures for endothermic reaction. It is observed from (1) that increasing temperature frequently that causes a marked increase in the rate of reactions is shown in Figure 4. As the temperature of the system increases, the number of molecules that carry enough energy to react when they collide also increases. Therefore, the rate of reaction increases with temperature. As a rule, the rate of a reaction doubles for every 10°C increase in the temperature of the system.

Last part of (20) shows that $\lambda^2 \exp(-E/\theta)$ increases with the increasing values of λ. We also observe from this equation that increase in $\lambda^2 \exp(-E/\theta)$ means that increase in λ leads to the decrease in the concentration profiles. This is in great agreement with Figure 6 for both cases $\beta = \pm 1$. It is also observed from this figure that for exothermic reaction the mass boundary layer is close to the plate other than endothermic reaction.

5.3. Effects of Radiation Parameter N on Temperature and Velocity Profiles. The effects of radiation parameter N on the temperature and the velocity profiles are shown in Figures 7 and 8. From Figure 7, it can be seen that an increase in the values of N leads to a decrease in the values of temperature profiles within the thermal boundary layers $\eta < 0.22$, while outside $\eta > 0.22$, the temperature profile gradually increases with the increase of the radiation parameter N. It is appear from Figure 8 that when $N = 0$, that is, without the radiation, the velocity profile shows its usual trend of gradually decay. As radiation becomes larger the profiles overshoot the uniform velocity close to the boundary.

5.4. Effects of γ on the Velocity, Temperature, and Concentration Profi es. The value of $\gamma = \Delta T/T_\infty$ is taken to be both positive and negative, since these values represent, respectively, cooling and heating of the plate. The effects of temperature relative parameter on velocity, temperature, and concentration profiles are shown in Figures 9–11.

The velocity profiles generated due to impulsive motion of the plate is plotted in Figure 9 for both cooling ($\gamma > 0$) and heating ($\gamma < 0$) of the plate keeping other parameters fixed ($G_0 = 10$, $G_m = 1.0$, $M = E_c = 0.5$, $P_r = 0.71$, $S_c = 0.6$, $\lambda = 5$, $v_0 = 3.0$, $E = 1.0$, and $n = 1$). In Figure 9, velocity profiles are shown for different values of γ. We observe that velocity increases with increasing values of γ for the cooling of the plate. From this figure it is also observed that the negative increase in the temperature relative parameter leads to the decrease in the velocity field. That is, for heating of the plate (Figure 9), the effects of the γ on the velocity field have also opposite effects, as compared to the cooling of the plate. Figure 10 shows the effects of temperature relative parameter γ on the temperature profiles. It has been observed from this figure that for heating plate the temperature profile shows its usual trend of gradually decay, and the thermal boundary layer is close to the plate. But for cooling plate that is temperature relative parameter γ becomes larger, the profiles overshoot the uniform temperature close to the thermal boundary. Opposite effects of temperature profile are found for concentration profiles shown in Figure 11.

5.5. Effects of Suction/Injection (v_0) on the Velocity and the Temperature Profiles. The effects of suction and injection (v_0) for $G_0 = 10$, $G_m = 1.0$, $M = E_c = 0.5$, $P_r = 0.71$, $S_c = 0.6$, $\lambda = 5$, $\gamma = 0.5$, $E = 1.0$, and $n = 1$ on the velocity profiles and temperature profiles are shown, respectively, in Figures 12 and 13. For strong suction ($v_0 > 0$), the velocity and the temperature decay rapidly away from the surface. The fact that suction stabilizes the boundary layer is also apparent from these figures. As for the injection ($v_0 < 0$), from Figures 12 and 13 it is observed that the boundary layer is increasingly blown away from the plate to form an interlayer between the injection and the outer flow regions.

5.6. Effects of G_m and M on the Velocity Profi es. The effects of Solutal Grashof number G_m and magnetic interaction parameter M on velocity profiles are shown in Figures 14 and 15, respectively. Solutal Grashof number $G_m > 0$ corresponds that the chemical species concentration in the free stream region is less than the concentration at the boundary surface. Figure 14 presents the effects of Solutal Grashof number G_m on the velocity profiles. It is observed that the velocity profile increases with the increasing values of Solutal Grashof number G_m.

Imposition of a magnetic field to an electrically conducting fluid creates a drag like force called Lorentz force. The force has the tendency to slow down the flow around the plate at the expense of increasing its temperature. This is depicted by decreases in velocity profiles as magnetic parameter M increases as shown in Figure 15.

5.7. Effects of A on the Velocity and the Temperature Profi es. Here $A = 0$ and $A \neq 0$ represent steady and unsteady flows, respectively. The effects of A on velocity and temperature profiles are shown in Figures 16 and 17, respectively. For unsteady flow ($A \neq 0$) both figures show that the profiles rapidly decay and close to the plate. But for steady flow ($A = 0$) the profiles overshoot the uniform velocity/temperature close to the boundary.

5.8. The Skin-Friction, the Heat Transfer, and the Mass Transfer Coeffici ts. The skin-friction, the heat transfer, and the mass transfer coefficients are tabulated in Tables 1 and 2 for different values of E, β, λ, N, M, γ, and A. We observe from Table 1 that the skin-friction coefficient $(-f'(0))$ and the Nusselt number N_u $(= -\theta'(0))$ increase for increasing values of dimensionless activation parameter E for exothermic chemical reaction, but opposite actions are found for endothermic reaction. That is for endothermic reaction, it is found that both shearing stress and heat transfer coefficients decrease for increasing values of E. For both cases exothermic/endothermic reactions, the mass transfer coefficient $(= -\phi'(0))$ decreases for increasing values of E. We also observe from Table 1 that the skin-friction coefficient $(-f'(0))$ and the Nusselt number N_u $(= -\theta'(0))$ decrease for increasing values of dimensionless reaction rate constant λ for exothermic chemical reaction, but opposite effects are found for endothermic reaction. That is for endothermic reaction, it is found that both shearing stress and heat transfer coefficients increase for increasing values of λ. For both cases exothermic/endothermic reactions, the mass transfer coefficient $(= -\phi'(0))$ remarkably increases for increasing values of λ. From Table 2 it has been observed that the skin-friction coefficient remarkably decreases and Nusselt number remarkable increases for increasing values of radiation parameter N. The skin-friction coefficient increases for increasing values of magnetic parameter M. From Table 2 we also found that skin friction and heat transfer coefficients decrease, but mass transfer coefficient increases for increasing values of temperature relative parameter γ.

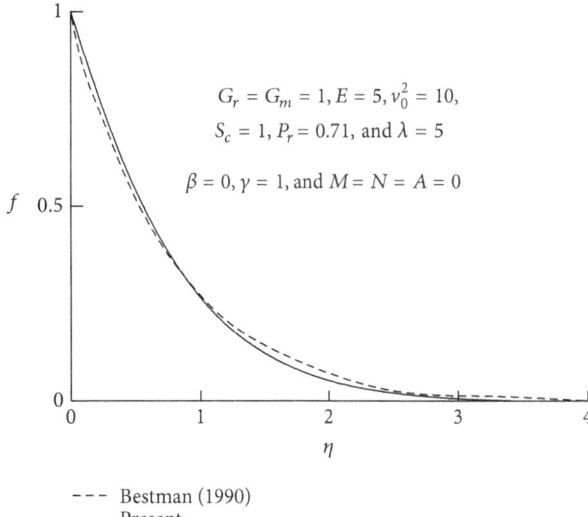

FIGURE 18: Comparison of our calculated velocity profile and the velocity profile of Bestman [7].

5.9. Comparison between Our Numerical Result and the Analytical Result of Bestman [7]. Bestman studied steady natural convective boundary layer flow with large suction. He solved his problem analytically by employing the perturbation technique proposed by Singh and Dikshit [8]. In our present work, take $A = 0$ in (16) for considering steady flow. The values of the suction parameter $v_0^2 = 10$ is taken to see the effects of large suction. $G_r = G_m = 1.0$, $\gamma = 1$, $M = N = 0$, $\lambda = 5$, $E = 5$, and $n = 1$ are also chosen with a view to compare our numerical results with the analytical results of Bestman [7]. The comparison of velocity profiles as seen in Figure 18 highlights the validity of the numerical computations adapted in the present investigation.

6. Conclusions

In this paper, we investigate the effects of chemical reaction rate and Arrhenius activation energy on an unsteady MHD natural convection heat and mass transfer boundary layer flow past a flat porous plate in presence of thermal radiation.

The Nachtsheim and Swigert [18] iteration technique based on sixth-order Runge-Kutta and Shooting method has been employed to complete the integration of the resulting solutions.

The following conclusions can be drawn as a result of the computations.

(1) Velocity increases with increasing values of γ for the cooling of the plate, and the negative increase in the temperature relative parameter γ leads to the decrease in the velocity field. That is, for heating of the plate the effects of the temperature relative parameter γ on the velocity field have also opposite effects, as compared to the cooling of the plate.

(2) Solutal Grashof number $G_m > 0$ corresponds that the chemical species concentration in the free stream region is less than the concentration at the boundary surface. It is observed that the velocity profile increases with the increasing values of Solutal Grashof number G_m.

(3) Increase in λ leads to increase in the velocity, and temperature profiles for exothermic reaction but opposite effects are found for endothermic reaction. For $\beta = \pm 1$ increase in λ leads to the decrease in the concentration profiles, but for exothermic reaction, the mass boundary layer is close to the plate other than endothermic reaction.

(4) The velocity profile mark increases with the increasing values of E for endothermic chemical reaction. But negligible effect is found for exothermic reaction. Temperature decreases for increasing values of E for exothermic reaction, but opposite effects are found for endothermic reaction. Activation energy (E) leads to increase in the concentration profiles, but small and reported effects are found for $\beta = 1$ and $\beta = -1$, respectively.

(5) For strong suction ($v_0 > 0$), the velocity, the temperature, and the concentration profiles decay rapidly away from the surface. As for the injection ($v_0 < 0$), it is observed that the boundary layer is increasingly blown away from the plate to form an interlayer between the injection and the outer flow regions.

(6) Imposition of a magnetic field to an electrically conducting fluid creates a drag like force called Lorentz force. The force has the tendency to slow down the flow around the plate at the expense of increasing its temperature. This is depicted by decreases in velocity profiles as magnetic parameter M increases

(7) An increase in the values of N leads to a decrease in the values of temperature profiles within the thermal boundary layers $\eta < 0.22$, while outside $\eta > 0.22$ the temperature profile gradually increases with the increase of the radiation parameter N.

(8) For unsteady flow ($A \neq 0$) the velocity and the temperature profiles rapidly decay and close to the plate. But for steady flow ($A = 0$) the profiles overshoot the uniform velocity/temperature close to the boundary.

Nomenclature

(u, v): Components of velocity field
c_p: Specific heat at constant pressure
N_u: Nusselt number
P_r: Prandtl number
Q_0: Heat generation/absorption coefficient
S_h: Sherwood number
T: Temperature of the flow field
T_∞: Temperature of the fluid at infinity
T_w: Temperature at the plate
f: Dimensionless similarity functions
E_a: The activation energy
k: The Boltzmann constant
S_c: Schmidt number
E: The nondimensional activation energy
G_r: Grashof number
G_m: Modified (Solutal) Grashof number.

Greek Symbols

θ: Dimensionless temperature
ϕ: Dimensionless concentration
η: Dimensionless similarity variable
δ: Scale factor
ν: Kinematic viscosity
ρ: Density of the fluid
κ: Thermal conductivity
τ: Shear stress,
μ: Fluid viscosity
λ^2: Dimensionless chemical reaction rate constant
β and β^*: The coefficients of volume expansions for temperature and concentration, respectively
γ: The dimensionless heat generation/absorption coefficient.

References

[1] M. Tencer, J. S. Moss, and T. Zapach, "Arrhenius average temperature: the effective temperature for non-fatigue wearout and long term reliability in variable thermal conditions and climates," *IEEE Transactions on Components and Packaging Technologies*, vol. 27, no. 3, pp. 602–607, 2004.

[2] C. Truesdell, "Sulle basi della thermomeccanica," *Rendiconti Lincei*, vol. 22, no. 8, pp. 33–38, 1957.

[3] N. Mills, "Incompressible mixtures of newtonian fluids," *International Journal of Engineering Science*, vol. 4, no. 2, pp. 97–112, 1966.

[4] C. E. Beevers and R. E. Craine, "On the determination of response functions for a binary mixture of incompressible newtonian fluids," *International Journal of Engineering Science*, vol. 20, no. 6, pp. 737–745, 1982.

[5] A. Al-Sharif, K. Chamniprasart, K. R. Rajagopal, and A. Z. Szeri, "Lubrication with binary mixtures: liquid-liquid emulsion," *Journal of Tribology*, vol. 115, no. 1, pp. 46–55, 1993.

[6] S. H. Wang, A. Al-Sharif, K. R. Rajagopal, and A. Z. Szeri, "Lubrication with binary mixtures: liquid-liquid emulsion in an EHL conjunction," *Journal of Tribology*, vol. 115, no. 3, pp. 515–522, 1993.

[7] A. R. Bestman, "Natural convection boundary layer with suction and mass transfer in a porous medium," *International Journal of Energy Research*, vol. 14, no. 4, pp. 389–396, 1990.

[8] A. K. Singh and C. K. Dikshit, "Hydromagnetic flow past a continuously moving semi-infinite plate for large suction," *Astrophysics and Space Science*, vol. 148, no. 2, pp. 249–256, 1988.

[9] A. R. Bestman, "Radiative heat transfer to flow of a combustible mixture in a vertical pipe," *International Journal of Energy Research*, vol. 15, no. 3, pp. 179–184, 1991.

[10] M. A. Alabraba, A. R. Bestman, and A. Ogulu, "Laminar convection in binary mixture of hydromagnetic flow with radiative heat transfer, I," *Astrophysics and Space Science*, vol. 195, no. 2, pp. 431–439, 1992.

[11] R. Kandasamy, K. Periasamy, and K. K. S. Prabhu, "Effects of chemical reaction, heat and mass transfer along a wedge with heat source and concentration in the presence of suction or injection," *International Journal of Heat and Mass Transfer*, vol. 48, no. 7, pp. 1388–1394, 2005.

[12] O. D. Makinde, P. O. Olanrewaju, and W. M. Charles, "Unsteady convection with chemical reaction and radiative heat transfer past a flat porous plate moving through a binary mixture," *Africka Matematika*, vol. 22, no. 1, pp. 65–78, 2011.

[13] O. D. Makinde and P. O. Olanrewaju, "Unsteady mixed convection with Soret and Dufour effects past a porous plate moving through a binary mixture of chemically reacting fluid," *Chemical Engineering Communications*, vol. 198, no. 7, pp. 920–938, 2011.

[14] Kh. Abdul Maleque, "Unsteady natural convection boundary layer flow with mass transfer and a binary chemical reaction," *British Journal of Applied Science and Technology (Sciencedomain International)*, vol. 3, no. 1, pp. 131–149, 2013.

[15] Kh. Abdul Maleque, "Unsteady natural convection boundary layer heat and mass transfer flow with exothermic chemical reactions," *Journal of Pure and Applied Mathematics (Advances and Applications)*, vol. 9, no. 1, pp. 17–41, 2013.

[16] Kh. Abdul Maleque, "Effects of binary chemical reaction and activation energy on MHD boundary layer heat and mass transfer flow with viscous dissipation and heat generation/absorption," *ISRN Thermodynamics*, vol. 2013, Article ID 284637, 9 pages, 2013.

[17] Kh. Abdul Maleque, "Effects of combined temperature- and depth-dependent viscosity and hall current on an unsteady MHD laminar convective flow due to a rotating disk," *Chemical Engineering Communications*, vol. 197, no. 4, pp. 506–521, 2010.

[18] P. R. Nachtsheim and P. Swigert, "Satisfaction of asymptotic boundary conditions in numerical solution of system of nonlinear of boundary layer type," NASA TN-D3004, 1965.

Thermodynamic Modelling of Dolomite Behavior in Aqueous Media

Tadeusz Michałowski[1] and Agustin G. Asuero[2]

[1] Faculty of Engineering and Chemical Technology, Technical University of Cracow, 31-155 Kraków, Poland
[2] Department of Analytical Chemistry, The University of Seville, 41012 Seville, Spain

Correspondence should be addressed to Tadeusz Michałowski, michalot@o2.pl

Academic Editor: Jaime Wisniak

The compact thermodynamic approach to the systems containing calcium, magnesium, and carbonate species is referred to dissolution of dolomite, as an example of nonequilibrium ternary salt when introduced into aqueous media. The study of dolomite is based on all attainable physicochemical knowledge, involved in expressions for equilibrium constants, where the species of the system are interrelated. The species are also involved in charge and concentration balances, considered as constraints put on a closed system, separated from the environment by diathermal walls. The inferences are gained from calculations performed with use of an iterative computer program. The simulated *quasistatic* processes occurred under isothermal conditions, started at a pre-assumed pH_0 value of the solution where dolomite was introduced, and are usually involved with formation of other solid phases. None simplifying assumptions in the calculations were made.

1. Introduction

Dolomite ($CaMg(CO_3)_2$, abbr. pr1) is perceived as an unusual, metastable mineral [1, 2], and its behavior is considered as one of the most exciting topics in geology [3]. Its chemical properties should be put in context with other, most important carbonate minerals: calcite ($CaCO_3$, pr2) and magnesite ($MgCO_3$, pr3) [4]. The trigonal structure of calcite is composed of alternate layers of calcium and carbonate ions. The crystal structure of magnesite is the same as one for calcite, and then magnesite properties are similar to those of calcite. However, a significant difference in Pauling's ionic radii: 99 pm (for Ca^{+2}), 65 pm (for Mg^{+2}), [5] causes an incompatibility of the cations in the same layer of dolomite structure. Crystal lattice of ideal dolomite (M_1 = 184.4 g/mol, ρ_1 = 2.899 g/cm^3) consists of alternating octahedral layers of Ca^{+2} and Mg^{+2} ions, separated by layers of CO_3^{-2} ions [6]. Complete ordering is energetically favourable at lower temperatures. Presumably, it is the principal crystallographic constraint securing nearly ideal (Ca : Mg = 1 : 1) dolomite stoichiometry [6].

Ideal dolomite, $CaMg(CO_3)_2$, is a particular case (x = 0) of the compound $Ca_{1-x}Mg_{1+x}(CO_3)_2$ [7–10], considered as a result of partial disarrangement of Ca^{+2} and Mg^{+2} in structure of real dolomite. This disarrangement is evidenced by weak or diffuse spectral lines in the reflected X-ray diffraction patterns [11]. The composition of sedimentary dolomites is not exactly stoichiometric; it ranges from $Ca_{1.16}Mg_{0.84}(CO_3)_2$ to $Ca_{0.96}Mg_{1.04}(CO_3)_2$ (i.e., $-0.16 < x < 0.04$) [1]. The mineral $Ca_{0.5}Mg_{1.5}(CO_3)_2$ (x = 0.5) is named huntite. Sedimentary dolomites with $x > 0$ are named calcium dolomites, and the ones with $x < 0$ are named magnesian dolomites. The carbonate dominated by calcite (ρ_2 = 2.71 g/cm^3) is called limestone. Other carbonates: ankerite Ca(Fe, Mg, Mn)$(CO_3)_2$ and kutnohorite Ca(Mn,Mg,Fe)$(CO_3)_2$, with bivalent cations: Fe^{+2} and Mn^{+2}, substituting a part of Mg^{+2} ions in the layered structure, enter also the dolomite group.

A rock formed by the replacement of dolomite with calcite [12]

$$CaMg(CO_3)_2 + Ca^{+2} = 2CaCO_3 + Mg^{+2} \quad (1)$$

in dedolomitisation (calcitisation) process is named dedolomite [13]. Calcium ions in this process are provided, for example, by calcium-rich water, and magnesium ions are liberated. Molar volume of dolomite (184.4/2.899 = 63.608 cm^3/mol) is lower than two molar volumes of calcite (2 · 100.09/2.71 = 73.867 cm^3/mol), and, therefore, reaction (1) causes an increase in the solid volume, equal (73.867 − 63.608)/63.608, that is, ca. 16% [4].

Dedolomitisation is a particular case of a diagenesis [14], where an alteration of sediments into sedimentary rocks occurs. This process is accompanied by an increase in porosity, expressed by a percentage pore space. The pores form a void space or are filled with a fluid.

Dolomite is thermodynamically unstable, and dedolomitisation occurs under different conditions. The dedolomitisation in alkaline media is represented by reaction [15]

$$CaMg(CO_3)_2 + Ca(OH)_2 = 2CaCO_3 + Mg(OH)_2. \quad (2)$$

The reverse process (i.e., dolomitisation) occurs, for example, during evaporation of seawater. High temperatures enhance the dolomitisation process [16]. In dolomitisation, magnesium ions from seawater replace calcium ions in calcite, and dolomite is formed [17]. Dolomite growth is favoured by high Mg/Ca ratios and high carbonate contents; this fact is predictable from Le Chatelier-Brown principle [18].

The main objective of this paper is to provide the way for better understanding the dolomite dissolution at different conditions [19] affected by pH, initial concentration of dolomite (C_0) in the system ($[\text{pr1}]_{t=0} = C_0$), and concentration of CO_2. The knowledge of dissolution in acidic media is essential in aspect of improved recovery of oil and gas from sedimentary basins at low temperatures [20], whereas dedolomitisation in alkaline media plays a significant role in deteriorating the concrete structure [15, 21].

As will be stated in what follows, the solubility product value (K_{sp1}) of dolomite is a critical factor in quantitative description of its behavior in the systems where dolomite is put in contact with aqueous solutions, containing dissolved CO_2 and, moreover, a strong acid (e.g., HCl) or a strong base (e.g., KOH). This paper follows the one concerning struvite [22, 23], as another representative of a group of the nonequilibrium precipitates formed by ternary salts.

2. Kinetics of Dolomite Dissolution and Stoichiometry

In aqueous systems, ideal dolomite can be considered as an equimolar mixture of two carbonate components: calcite and magnesite, that is, pr1 = pr2 + pr3. This simplifying assumption deserves some reservation, concerning relative rates of dissolution of the dolomite components; the literature provides ambiguous data in this respect, however. The results obtained according to AAS method by Lund et al. [24] exhibited stoichiometric dissolution of dolomite in relation to Ca and Mg, whereas other experiments [25] showed that pure dolomite dissolves more slowly than pure calcite.

Dolomite exists in a variety of morphological forms [26]. Minerals with greater defect densities dissolve faster since their effective surface areas are greater than more perfect specimens of the same compound. The rate of surface-controlled dolomite dissolution is significantly less than one of calcite [27].

Dissolution rate increases with decreasing grain size [28]. The experiments done for kinetic purposes showed that the mass loss of single dolomite crystals [29] (in response to pH and pC_{CO_2}) or one for finely dispersed dolomite particles [27] (in response to pH) was measured.

The dissolution of ionic crystals is a complex process, involving some surface and transportation phenomena. Ions are transferred from the surface of the solid material to an unsaturated solution [30]. The surface phenomena depend on the morphology (microstructure) of the crystals. The rate of any dissolution process is effected by surface and transport phenomena.

Kinetics of dolomite dissolution has been tested at different pH and temperatures [20]. According to a model by Busenberg and Plummer [29], the dissolution of dolomite is an effect of simultaneous action of H^+, H_2CO_3, and H_2O.

The dissolution studies were usually carried out with suspensions or powdered materials employed, and the resulting concentration changes of Ca and Mg species in the bulk solution were measured [31]. For this purpose, *in situ* (e.g., conductometric, pH-metric, pH-static [32]) or *ex situ* (e.g., titrimetry, AAS [33]) methods of analysis were employed [30]. As an option, a rotating disc (RD) formed of dissolving dolomite attached at the end of rotating disc shaft was applied [34]. A loss in mass of the solid material was also measured [35].

It should be noted that repeated trials to precipitate ideal dolomite under laboratory conditions at room temperature were unsuccessful [36, 37]; dolomite was precipitated at elevated temperatures (150–300°C) [38], for example, by heating calcite with Mg salt in aqueous media, at elevated CO_2 pressures [39]. Dolomite is formed as a result of complex, not well-understood physicochemical phenomena [40, 41], because of the difficulties arising in preparation of stoichiometric dolomites [42]. These difficulties caused, among others, that the solubility product (K_{sp1}) of dolomite in water

$$K_{sp1} = [Ca^{+2}][Mg^{+2}][CO_3^{-2}]^2 \quad (3)$$

measured according to different methods yielded inconsistent and unreliable results. The mechanism of dolomite formation in sedimentary environment (the so-named dolomite problem [43]) is not well understood, as hitherto [44].

3. Principles of Simulation of Dolomite Dissolution

3.1. General Remarks. Simulations are needed to check the models used. In modelling of chemical systems, different computer programs were developed. Among others, the Joint Expert Speciation System (JESS) computer program [45–48] is sometimes applied, for example, in [49]. A new approach, called Generalized Approach to Electrolytic Systems (GATES), was elaborated by Michałowski in 1992 and presented lately in some review papers [50–52] and in

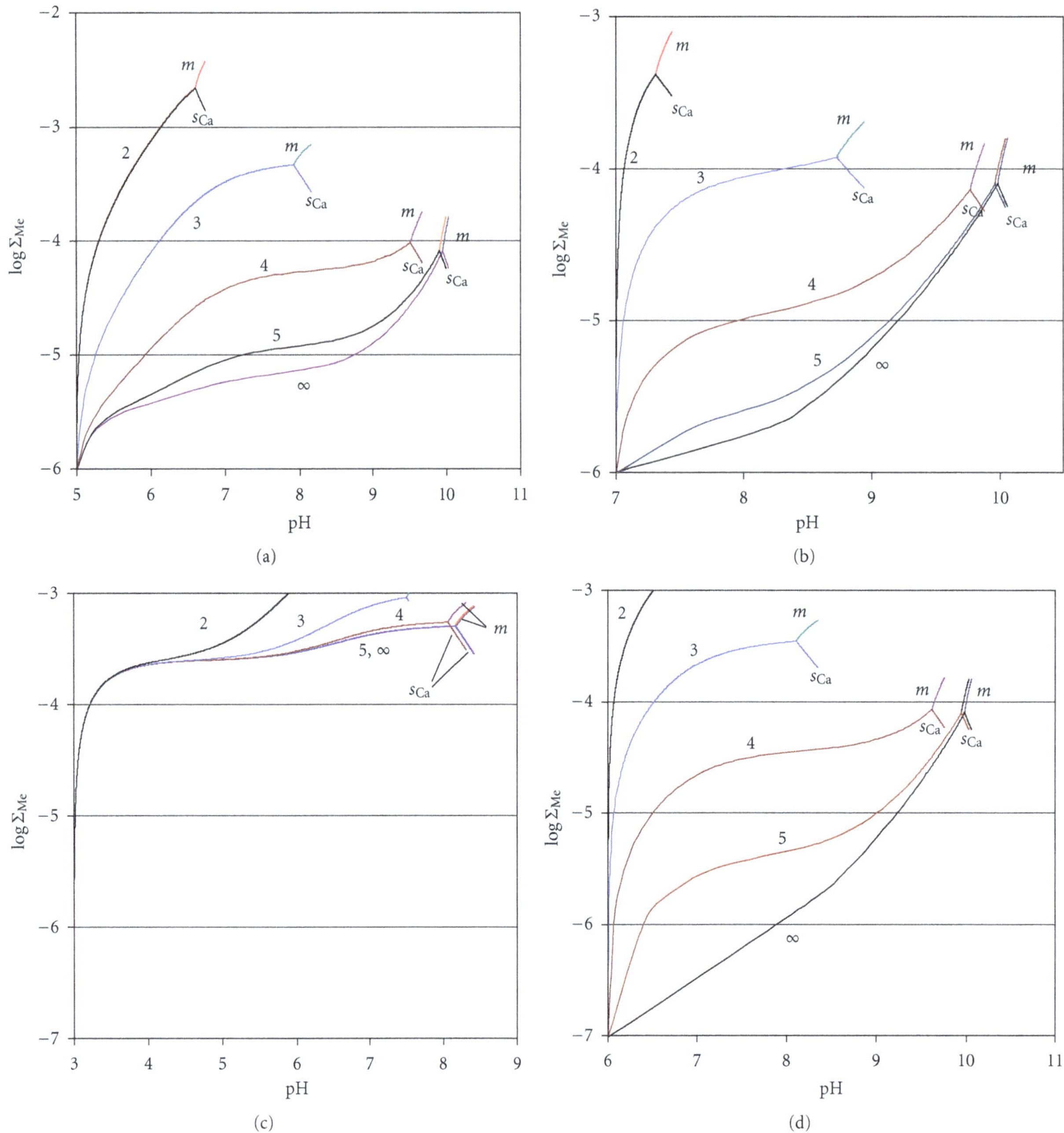

FIGURE 1: The $\log \sum \mathrm{Me}$ versus pH relationships ((9), (10)) plotted for different sets of (pC_0, pC_{CO_2}, pH_0) values: (a) (2, pC_{CO_2}, 5); (b) (2, pC_{CO_2}, 7); (c) (3, pC_{CO_2}, 3); (d) (3, pC_{CO_2}, 6). The pC_{CO_2} values are indicated at the corresponding curves. For further details, see text.

textbooks [53, 54]. For example, the systems with struvite [22, 23] were elaborated according to GATES.

The equilibria in the system with solid carbonates are affected by total concentration (C_{CO_2}) of carbonate species, introduced by CO_2 dissolved in aqueous media, and by presence of NaOH (C_b) or HCl (C_a), used to moderate pH of the solution. At $C_b = C_a$, the effect is practically tantamount with absence of the related species in the solution, provided that C_a and C_b values are small; that is, the effect of ionic strength is negligible.

3.2. Equilibrium Constants. The set of equilibrium constants [1, 55–59] referred to the two-phase system in question is involved with solubility products (K_{spi}) of precipitates, (pri, $i = 1, \ldots, 5$), expressed by (3) and

$$\left[\mathrm{Ca}^{+2}\right]\left[\mathrm{CO}_3^{-2}\right] = K_{sp2}\left(pK_{sp2} = 8.48\right),$$

$$\left[\mathrm{Mg}^{+2}\right]\left[\mathrm{CO}_3^{-2}\right] = K_{sp3}\left(pK_{sp3} = 7.46\right),$$

$$[Mg^{+2}][OH^{-1}]^2 = K_{sp4}(pK_{sp4} = 10.74),$$
$$[Ca^{+2}][OH^{-1}]^2 = K_{sp5}(pK_{sp5} = 5.03), \tag{4}$$

and soluble species (complexes, protonated forms):

$$[MgHCO_3^{+1}] = 10^{1.16} \cdot [Mg^{+2}][HCO_3^{-1}],$$
$$[MgCO_3] = 10^{3.4} \cdot [Mg^{+2}][CO_3^{-2}],$$
$$[CaHCO_3^{+1}] = 10^{1.11} \cdot [Ca^{+2}][HCO_3^{-1}],$$
$$[CaCO_3] = 10^{3.22} \cdot [Ca^{+2}][CO_3^{-2}],$$
$$[HCO_3^{-1}] = 10^{10.33-pH} \cdot [CO_3^{-2}], \tag{5}$$
$$[H_2CO_3] = 10^{6.38-pH} \cdot [HCO_3^{-1}],$$
$$[MgOH^{+1}] = 10^{2.57} \cdot [Mg^{+2}][OH^{-1}],$$
$$[CaOH^{+1}] = 10^{1.3} \cdot [Ca^{+2}][OH^{-1}],$$
$$[H^{+1}][OH^{-1}] = K_w \ (pK_w = 14.0).$$

One should be noted that soluble complex species: $MgCO_3$ and $CaCO_3$, characterized by stability constants, are different from the precipitates: pr2 = $CaCO_3$ and pr3 = $MgCO_3$, characterized by the solubility products (K_{spi}, i = 2, 3) values.

The pK_{sp1} values for dolomite reported in the literature range from ca. 16.5 to 19.35 [37, 55, 56, 60, 61], that is, within ca. 3 units. Some inferences [37] lead to the conclusion that the most probable pK_{sp1} value is 17.2 ± 0.2. In this context, the value 19.35 taken in [55] seems to be excessively high. Such diversity in pK_{sp1} value may reflect the difficulties involved with obtaining stoichiometric dolomite.

The inequality $K_{sp1} < K_{sp2} \cdot K_{sp3}$, that is, $pK_{sp1} > pK_{sp2} + pK_{sp3}$, valid for all K_{sp1} values quoted above, expresses a kind of synergistic effect securing almost perfect arrangement of Mg^{+2} and Ca^{+2} ions in area of the corresponding planes of dolomite crystallographic structure.

The values for stability constants of soluble complexes: *Mg(OH)*$_2$ and *Ca(OH)*$_2$, introduced for calculations in [55], are highly controversial and then were omitted in the related balances formulated below. In this context, the complexes $MgOH^{+1}$ and $CaOH^{+1}$ and their stability constants are well established.

The dissolution of dolomite (pr1) proceeds up to the saturation of the solution against the corresponding precipitate. Provided that the solution is unsaturated against the corresponding solid phase (pri), the expression for the related solubility product (K_{spi}) is not valid, under such conditions. For this purpose, the expressions

$$q_1 = \frac{[Ca^{+2}][Mg^{+2}][CO_3^{-2}]^2}{K_{sp1}},$$
$$q_2 = \frac{[Ca^{+2}][CO_3^{-2}]}{K_{sp2}},$$
$$q_3 = \frac{[Mg^{+2}][CO_3^{-2}]}{K_{sp3}}, \tag{6}$$
$$q_4 = \frac{[Mg^{+2}][OH^{-1}]^2}{K_{sp4}},$$
$$q_5 = \frac{[Ca^{+2}][OH^{-1}]^2}{K_{sp5}}$$

related to all possible precipitates (pri, i = 1,...,5) were considered.

3.3. Formulation of the Dolomite Dissolution Model. We refer to the system obtained after introducing m_1 g of dolomite into V mL of aqueous solution with dissolved CO_2 (C_{CO_2}), NaOH (C_b), and/or HCl (C_a); NaOH and/or HCl are used to moderate pH value of the solution. Assuming that the volume change resulting from addition of pr1 is negligible, and denoting current ($t \geq 0$) concentration of pr1 by [pr1], $[pr1]_{t=0} = C_0 = (10^3 \cdot m_1/M_1)/V$, we get the concentration and charge balances

$$F_1(\mathbf{x}) = [pr1] + \sum Mg - C_0 = 0,$$
$$F_2(\mathbf{x}) = [pr1] + \sum Ca - C_0 = 0,$$
$$\begin{aligned} F_3(\mathbf{x}) = {} & 2[pr1] + [MgHCO_3^{+1}] + [MgCO_3] \\ & + [CaHCO_3^{+1}] + [CaCO_3] + [H_2CO_3] \\ & + [HCO_3^{-1}] + [CO_3^{-2}] - (2C_0 + C_{CO_2}) = 0, \end{aligned} \tag{7}$$
$$\begin{aligned} F_4(\mathbf{x}) = {} & [H^{+1}] - [OH^{-1}] + \Delta + 2[Mg^{+2}] + [MgOH^{+1}] \\ & + 2[Ca^{+2}] + [CaOH^{+1}] + [MgHCO_3^{+1}] \\ & + [CaHCO_3^{+1}] - [HCO_3^{-1}] - 2[CO_3^{-2}] = 0, \end{aligned}$$

where

$$\Delta = C_b - C_a \tag{8}$$

the expressions:

$$\sum Mg = [Mg^{+2}] + [MgOH^{+1}] + [MgHCO_3^{+1}] + [MgCO_3], \tag{9}$$

$$\sum Ca = [Ca^{+2}] + [CaOH^{+1}] + [CaHCO_3^{+1}] + [CaCO_3] \tag{10}$$

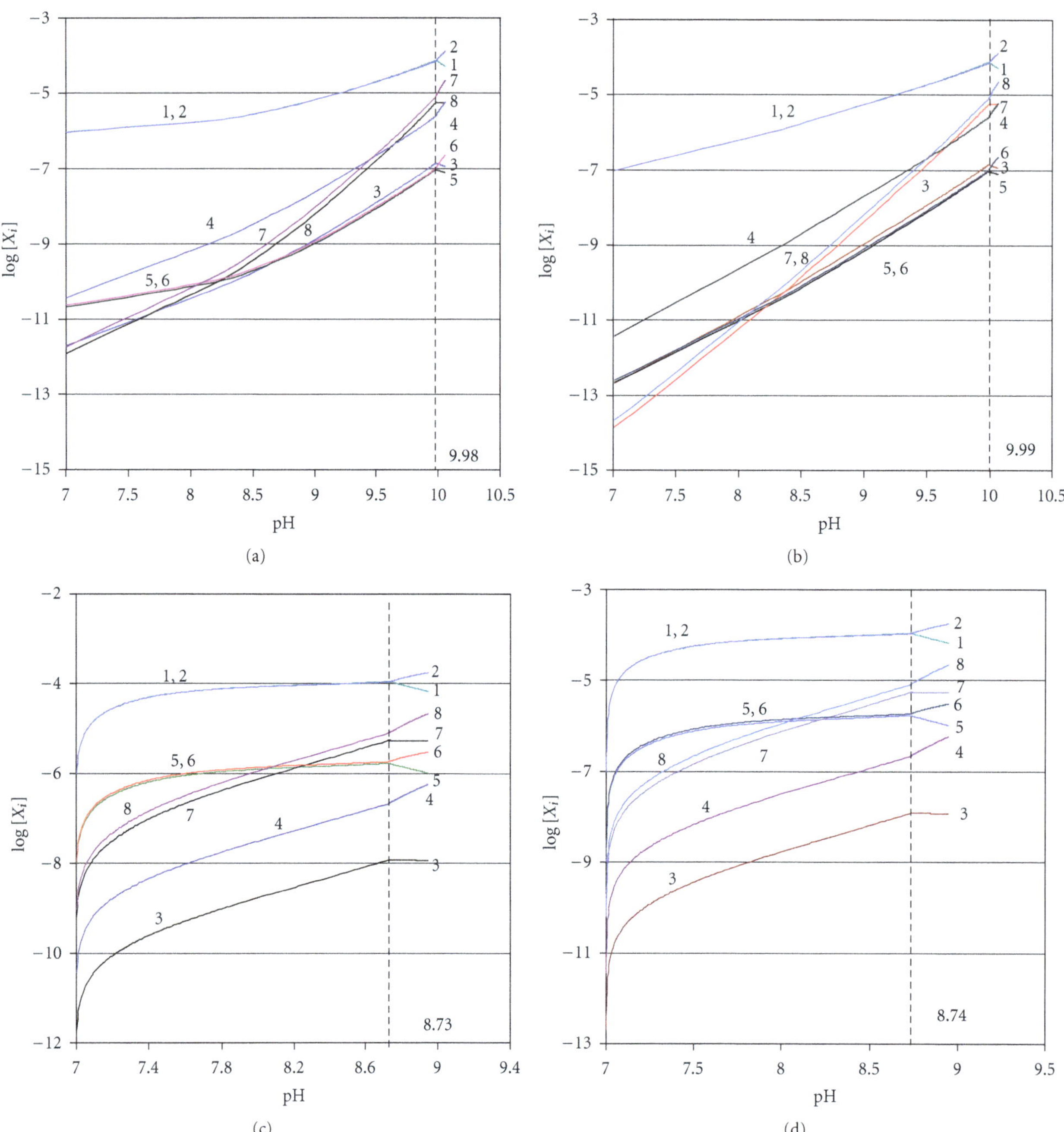

Figure 2: The $\log[X_i]$ versus pH relationships plotted for different sets of (pC_0, pC_{CO_2}, pH_0) values: (a) (2, ∞, 7); (b) (2, 3, 7); (c) (3, ∞, 7); (d) (3, 3, 7) for different soluble species X_i indicated at the corresponding curves: 1—Ca^{+2}, 2—Mg^{+2}, 3—$CaOH^{+1}$, 4—$MgOH^{+1}$, 5—$CaHCO_3{}^{+1}$, 6—$MgHCO_3{}^{+1}$, 7—*$CaCO_3$*, 8—*$MgCO_3$*.

involve all soluble magnesium and calcium species; $\mathbf{x} = [x_1, x_2, x_3, x_4]^{\top}$ is the vector of independent variables specified below. The number of variables forming the vector $\mathbf{x}$ is equal to the number of equations

$$F_i(\mathbf{x}) = 0 \quad (i = 1, \ldots, 4) \tag{11}$$

specified above.

3.4. Calculation Procedure. The relations (11) are valid, if $\mathbf{x}$ is chosen properly. For any other vector $\mathbf{x}'$, $\mathbf{x}' \neq \mathbf{x}$, at least one of the functions (11) differs from zero, and then

$$F(\mathbf{x}') = \sum_{i=1}^{4} [F_i(\mathbf{x}')]^2 > 0. \tag{12}$$

The calculation procedure is based on minimizing principle, quite analogous to one presented in [22, 23]. According to this approach, the sum of squares (12) is taken as the minimized function.

In a dynamic process, as the dissolution of pr1 is, a choice of ppr1 = $-\log[\text{pr1}]$ as the steering variable is advised; the value of [pr1] changes during the pr1 dissolution. The next step is the choice of independent variables, $x_i = x_i(\text{ppr1})$, $i = 1, \ldots, 4$. The x_i values are involved with concentrations of some independent species. In our case, the best choice is $\mathbf{x} = [x_1, x_2, x_3, x_4]^{\top} = [\text{pH}, \text{pHCO}_3, \text{pMg}, \text{pCa}]^{\top}$, where $\text{pH} = -\log[\text{H}^{+1}]$, $\text{pHCO}_3 = -\log[\text{HCO}_3^{-1}]$, $\text{pMg} = -\log[\text{Mg}^{+2}]$, $\text{pCa} = -\log[\text{Ca}^{+2}]$ were considered as independent variables. A choice of pX indices, not concentrations $[X]$, resulted from the fact that $10^{-\text{p}X_i} > 0$ for any real $\text{p}X_i = -\log[X_i]$ value, $\text{p}X_i \in \Re$. The third step is the choice of numerical values for components of the starting vector $\mathbf{x}' = \mathbf{x}^*$; if $\mathbf{x}^* \neq \mathbf{x}$, one can expect that $F(\mathbf{x}^*) > 0$ (12). The x^* value is referred to particular value of the steering variable, $x^* = \mathbf{x}^*(\text{ppr1})$, for example, to $\text{ppr1} = -\log C_0$. The step Δppr1 is needed, and initial steps $\Delta\text{p}X_i$ for $\text{p}X_i$ and lower ($\text{p}X_{i,\text{inf}}$) and upper ($\text{p}X_{i,\text{sup}}$) limits for expected $\text{p}X_i$ values are also required by the iterative computer programs applied in the minimization procedure.

The searching of $\mathbf{x}(\text{ppr1})$ vectors, where $F(\mathbf{x}(\text{ppr1}))$ is close to zero for different ppr1 values, can be made according to MATLAB iterative computer program. The searching procedure satisfies the requirements put on optimal $\mathbf{x}(\text{ppr1})$ values—provided that the $F(\mathbf{x}'(\text{ppr1}))$ value (12), considered as optimal one, is lower than a preassumed, sufficiently small δ-value

$$F(\mathbf{x}'(\text{ppr1})) < \delta. \qquad (13)$$

Then we consider that the equality $\mathbf{x}'(\text{ppr1}) = \mathbf{x}(\text{ppr1})$ is fulfilled. It means that $F_i(\mathbf{x}(\text{ppr1})) \cong 0$ for all $i = 1, \ldots, 4$, that is, (7) are fulfilled simultaneously, within tolerable degree of proximity assumed for all ppr1 values taken from defined ppr1 interval. If pr1 dissolves wholly, the ppr1 covers the interval from $\text{p}C_0$ up to the value resulting from graphical needs, that is, from the scale for ppr1 assumed to plot the related (e.g., speciation) curves. If the solubility product for pr1 is attained at defined point of the dissolution process, then upper value assumed for ppr1 is the value corresponding to this point.

The iterative computer programs are usually designed for the curve-fitting procedures where the degree of fitting of a curve to experimental points is finite. In particular case, the criterion of optimization is based on differences $F(\mathbf{x}(\text{ppr1}(N+1))) - F(\mathbf{x}(\text{ppr1}(N)))$ between two successive (Nth and N+1th) approximations of the F-value, and the optimisation is terminated if the inequality

$$\left| F(\mathbf{x}(\text{ppr1}(N+1))) - F(\mathbf{x}(\text{ppr1}(N))) \right| < \delta \qquad (14)$$

is valid for any preassumed, sufficiently low δ-value, for example, $\delta = 10^{-15}$.

However, one may happen that the condition (14) can be fulfilled at local minimum different from the global minimum. It occurs when the starting values $\mathbf{x}^*(\text{ppr1})$ are too distant from the true value $\mathbf{x}(\text{ppr1})$, where the equality $F(\mathbf{x}(\text{ppr1})) = 0$ is fulfilled. In this case, one should repeat the calculations for a new vector $\mathbf{x}^*(\text{ppr1})$, guessed at a particular ppr1 value chosen at the start for minimisation.

All vectors $\mathbf{x} = \mathbf{x}(\text{ppr1})$ obtained for different ppr1 values are the basis for plotting the speciation curves for all species (X_i) in the system. The curves are usually plotted in the logarithmic diagrams, on 2D plane, with ppr1 on the abscissa and $\log[X_i]$ on the ordinate. Other variables, for example, pH on the abscissa, can also be applied.

4. Simulated Dedolomitisation

4.1. Preliminary Data. We refer first to aqueous solutions obtained before introducing pr1 into it. Any particular solution is characterised by $\text{p}C_{\text{CO}_2} = -\log C_{\text{CO}_2}$ and $\text{pH} = \text{pH}_0$ values, where $\text{p}C_{\text{CO}_2}$ equals 2, 3, 4, 5, or ∞ (the latter value refers to $C_{\text{CO}_2} = 0$) and pH_0 values cover the set of natural numbers within the interval $\langle 3, 12 \rangle$. The pH_0 value corresponds to the presence of a strong acid (HB, C_a mol/L) or strong base (MOH, C_b mol/L), see (8). The solution is then characterised by a defined pair of ($\text{p}C_{\text{CO}_2}$, pH_0) values.

After introducing pr1 into the solution, its initial ($t = 0$) concentration in the two-phase system equals $[\text{pr1}]_{t=0} = C_0$ mol/L. Two values for $\text{p}C_0 = -\log C_0$, equal 2 and 3, were assumed. The set of parameters ($\text{p}C_0$, $\text{p}C_{\text{CO}_2}$, pH_0) assumed involves a multitude of different phenomena occurred in the system considered. The examples presented in the follwing concern only some particular cases.

The volume change of the system, affected by addition of pr1, can be neglected. The volume changes, involved with further dissolution and formation of precipitates, are neglected too. In order to neglect the diffusion effects, the systems were (virtually) mixed (homogenised). The dissolution has been considered as a *quasistatic* process, carried out under isothermal conditions.

4.2. Discussion on Particular Systems. At the first stage of the process that occurred after pr1 addition, the solution is unsaturated against any particular precipitate, that is, $q_i < 1$ ($i = 1, \ldots, 5$) in (6). At defined point, it saturates against another precipitate. As will be stated in the following, two different precipitates: pr2 = CaCO_3 or pr5 = Mg(OH)_2, are formed in the systems considered. Then the solution saturates against pr1 or transforms wholly to the second precipitate; in the latter case, the saturation towards pr1 is not attained. At a relatively high C_{CO_2} value, pr1 dissolves wholly and no other precipitate is formed ($q_i < 1$). The curves expressing the relations (9) and (10) are referred to unsaturated solutions in nonequilibrium systems and termed dissolution curves. The curves expressing the relations (9) and (10), when referred to the solutions saturated against a particular precipitate, are termed the solubility curves ($\sum \text{Me} \leq s_{\text{Me}}$, Me = Ca, Mg). The dissolution curves presented below are then terminated:

(a) by the bifurcation point, where the solubility product ($K_{\text{sp}i}$) for the corresponding precipitate (pr2 or pr5) is attained ($q_2 = 1$ or $q_5 = 1$), or

(b) by the point, where $[\text{pr1}] = 0$ at $q_i < 1$ ($i = 1, \ldots, 5$), that is, pr1 dissolves wholly.

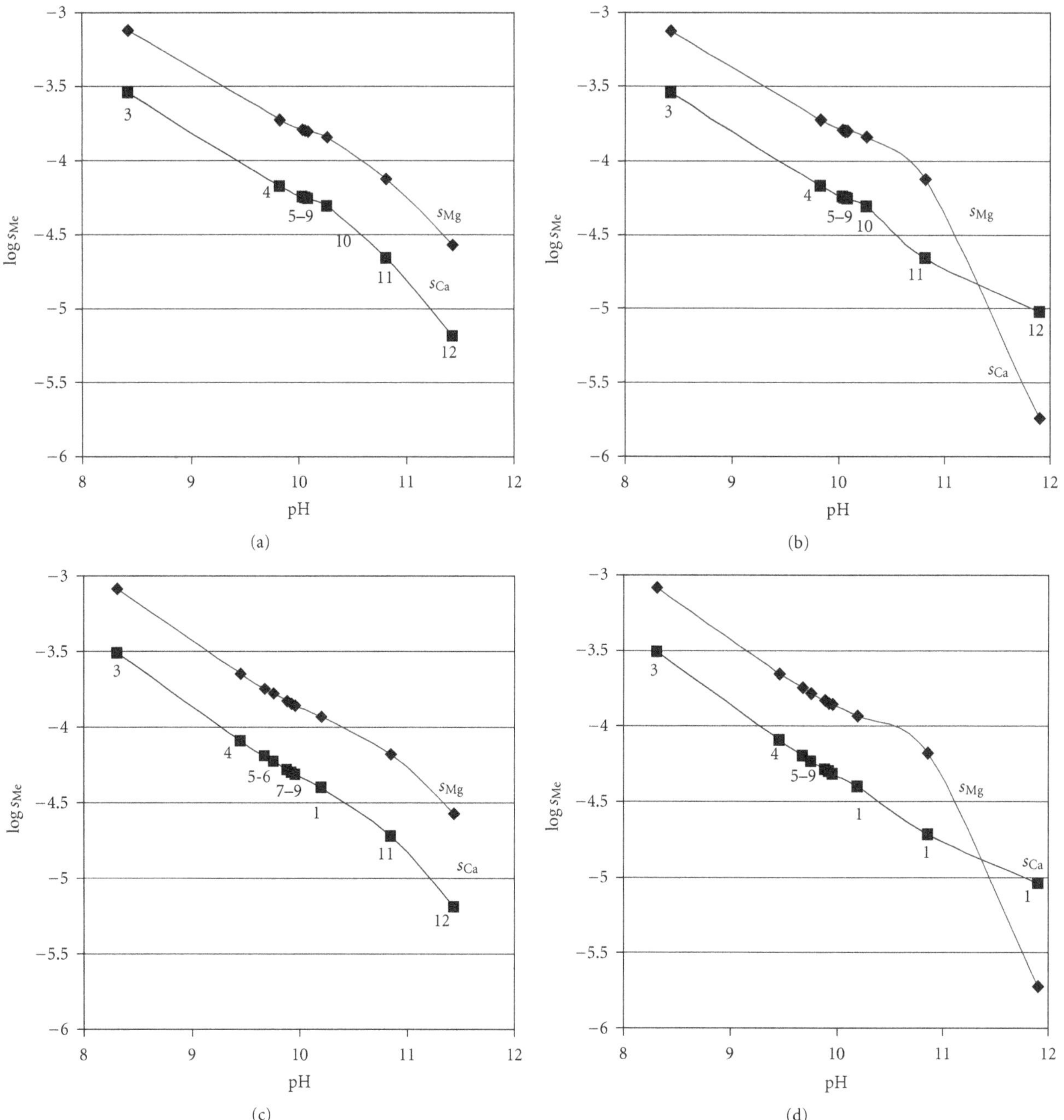

FIGURE 3: The $\log s_{Me}$ versus pH relationships (Me = Ca, Mg) plotted at $pK_{sp1} = 16.54$ for pr1, $\sum Me = s_{Me}$ ((9), (10)) at different (pC_0, pC_{CO_2}) values: (a) (2, ∞); (b) (3, ∞); (c) (2, 4); (d) (3, 4); the numbers at the corresponding points correspond to pH_0 values.

The dissolution/solubility curves, obtained at $pC_0 = 2$ and 3, are plotted in Figures 1(a)–1(d), for different pH_0 and pC_{CO_2} values. The curves plotted in Figures 1(c) and 1(d) refer to different sets of (pC_0, pC_{CO_2}, pH_0) = (3, pC_{CO_2}, pH_0) values. When C_{CO_2} exceeds distinctly the C_0 value ($pC_0 = 3$), pr1 dissolves wholly before the solubility product for pr2 is attained.

The plots for soluble species consisting the expressions for $\sum$ Ca and $\sum$ Mg ((9), (10)) and referred to different sets of (pC_0, pC_{CO_2}, pH_0) values are presented in Figures 2(a) and 2(b) (for $pC_0 = 2$) and Figures 2(c) and 2(d) (for $pC_0 = 3$).

As results from the course of speciation curves plotted in Figure 2, the first (dissolution) step can be represented by reactions:

$$pr1 + 2H_2O = Ca^{+2} + Mg^{+2} + 2HCO_3^{-1} + 2OH^{-1}, \quad (15a)$$

$$pr1 + H_2O = Ca^{+2} + Mg^{+2} + HCO_3^{-1} + CO_3^{-2} + OH^{-1}, \quad (15b)$$

$$pr1 = Ca^{+2} + Mg^{+2} + 2CO_3^{-2}, \quad (15c)$$

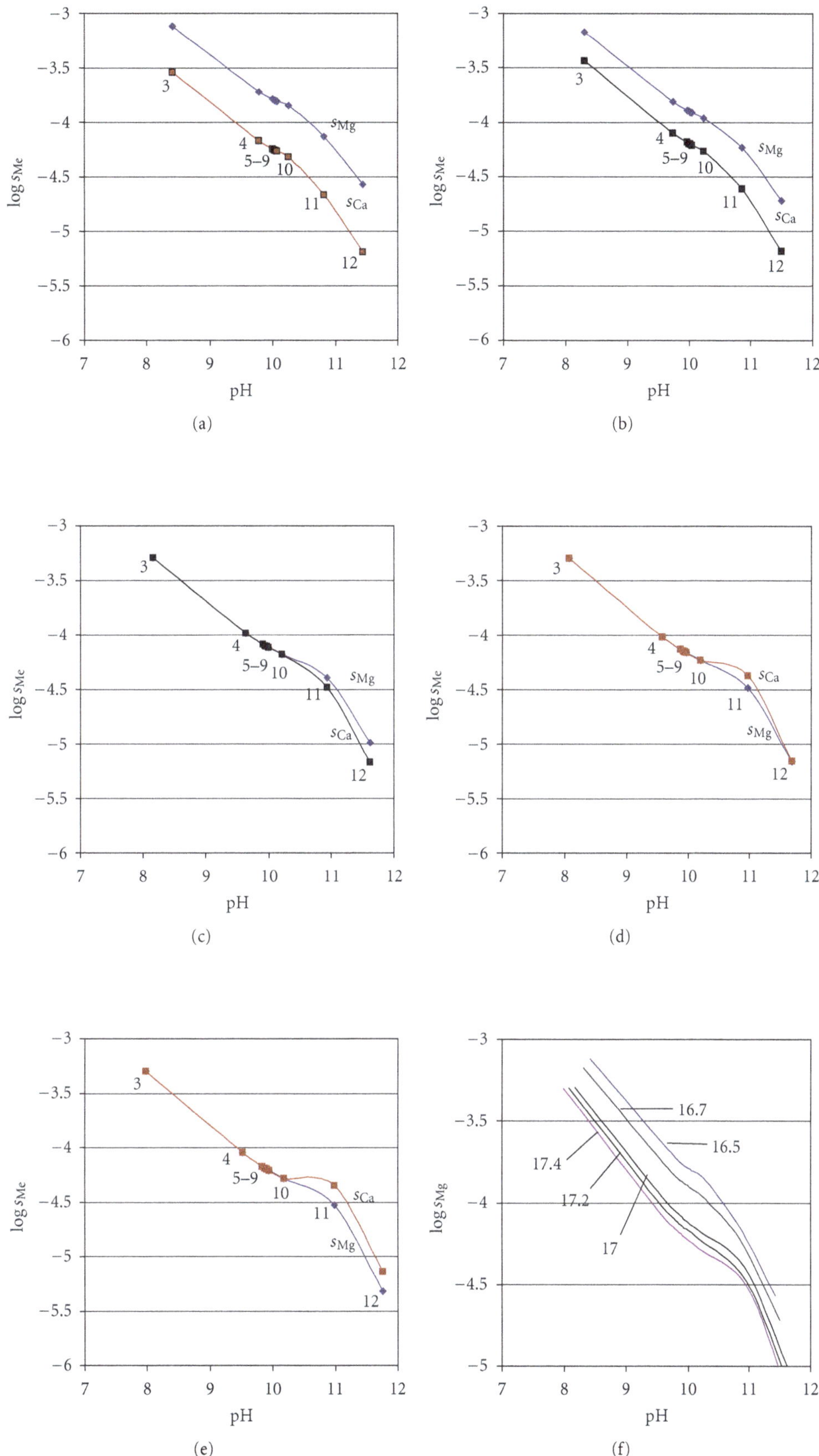

FIGURE 4: The log s_{Me} versus pH relationships ($\sum$ Me = s_{Me} in (9), (10) at (pC_0, pC_{CO_2}) = (2, 5) for different pK_{sp1} values assumed for pr1: (a) 16.54; (b) 16.7; (c) 17.0; (d) 17.2; (e) 17.4; (f) collected effect of K_{sp1} values (indicated at the corresponding curves) on log s_{Mg} value.

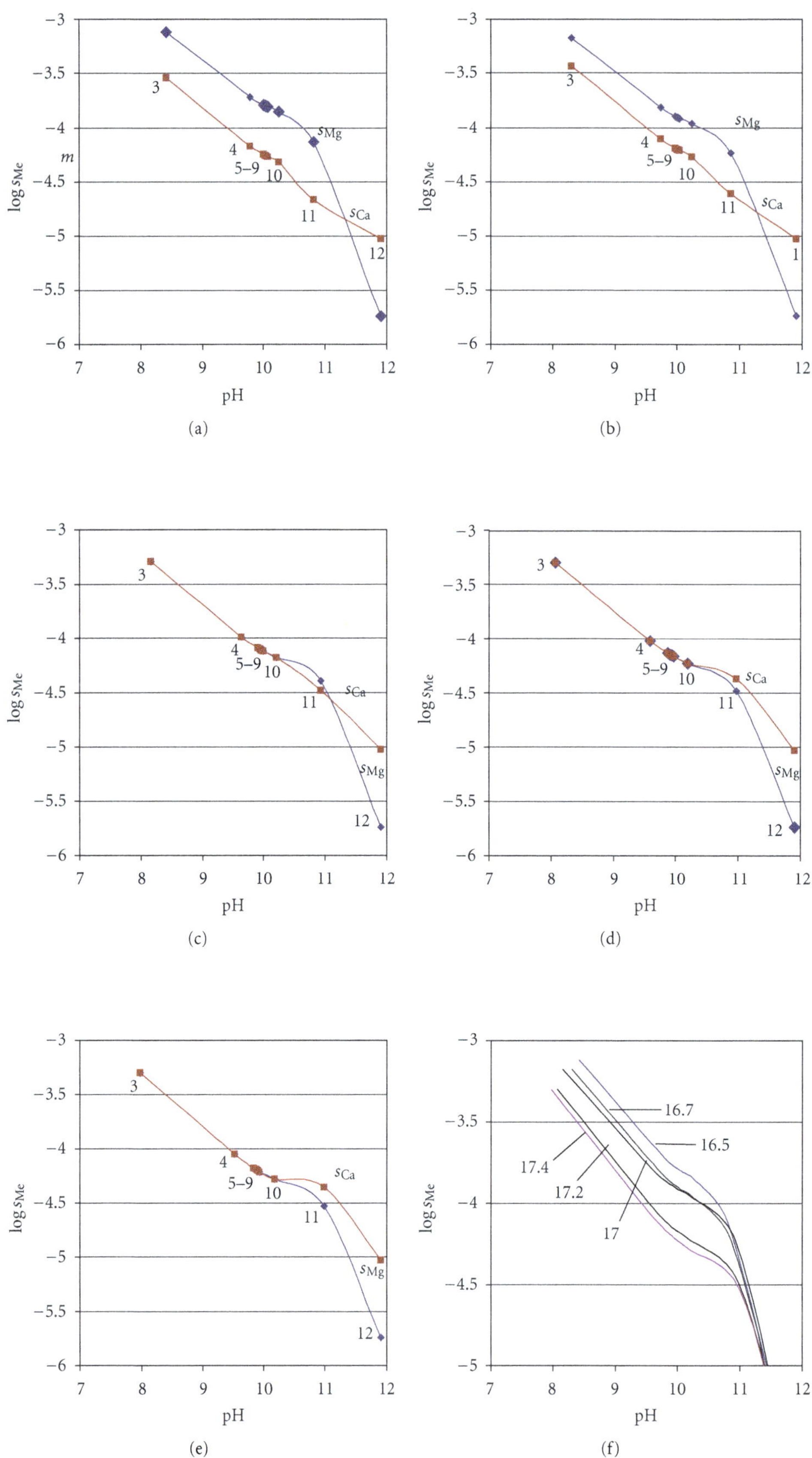

Figure 5: The $\log s_{Me}$ versus pH relationships ($\sum \mathrm{Me} = s_{Me}$ in (9), (10)) at ($\mathrm{p}C_0$, $\mathrm{p}C_{CO_2}$) = (3, 5) for different $\mathrm{p}K_{sp1}$ values assumed for pr1: (a) 16.54; (b) 16.7; (c) 17.0; (d) 17.2; (e) 17.4; (f) collected effect of $\mathrm{p}K_{sp1}$ values (indicated at the corresponding curves) on $\log s_{Mg}$ value.

where soluble species are formed. At the bifurcation points, the solubility product for pr2 is attained ($q_2 = 1$) and then pr2 is precipitated:

$$\text{pr1} + Ca^{+2} = 2\text{pr2} + Mg^{+2}, \quad (16a)$$

$$\text{pr1} + H_2O = \text{pr2} + Mg^{+2} + HCO_3^{-1} + OH^{-1}. \quad (16b)$$

The soluble magnesium species evolved into the solution, as the result of transformation of pr1 into pr2, are represented by m-lines, referred to (9); each of the m-lines terminates at the point where the solubility product (K_{sp1}) for pr1 is attained (i.e., $q_1 = 1$) and further dissolution of pr1 is stopped. As we see, the pH values corresponding to the bifurcation points are lowered with growth of C_{CO_2} value. The growth in C_{CO_2} causes also a significant growth in dissolution/solubility values.

The points where the solubility product ($pK_{sp1} = 16.54$) for pr1 is attained ($q_1 = 1$) at the same pair of (pC_0, pC_{CO_2}) values and different pH_0 values are marked and connected by lines on the corresponding plots in Figure 3. In the middle part of the resulting curves, the points are grouped together, within a small area.

The course of the related plots is affected by pK_{sp1} value, assumed for solubility product of dolomite (pr1), as indicated in Figure 4 (for $pC_0 = 2$) and Figure 5 (for $pC_0 = 3$). The points referred to different pH_0 values were omitted there, for brevity.

4.3. Solubility Curves for Dolomite Plotted at Different Preassumed pK_{sp1} Values. As has been stated above, the values for pK_{spd} attainable in the literature differ in wide range of values: from 16.54 to 19.35. Except the troubles involved with dolomite stoichiometry, these (serious) discrepancies in pK_{sp1} value are affected by (and resulted from) differences in solubility of calcite and magnesite. Namely, calcite constituent of dolomite dissolved more rapidly than magnesite constituent [27]. These effects, together with possible nonstoichiometry of dolomite (i.e., formation of magnesian calcite), make the system with dolomite a highly complicated one.

The speciation curves for dolomite are plotted at $pC_{CO_2} = 5$ and $pC_0 = 2$ and different pK_{sp1} values; it results that at pK_{sp1} equal 16.54 and 16.7, the equilibrium solid phase is calcite. However, when pH of the solution is greater than the boundary (minimal) value, the solid phase contains two equilibrium precipitates: calcite and dolomite. However, for $pK_{sp1} = 17$ at lower pH values, the equilibrium solid phase is dolomite and calcite appears as the second equilibrium solid phase at pH greater than ca. 10.2. At $pK_{sp1} = 19.35$, calcite is not formed. The curves of $\log[X_i]$ versus pH relationships plotted for different species X_i (i.e., precipitates: pr1, pr2 and soluble complexes: Mg^{+2}, Ca^{+2}, $MgHCO_3^{+1}$, $CaHCO_3^{+1}$, $MgCO_3$, $CaCO_3$) are terminated at pH where the solubility product (K_{sp5}) for pr5 is attained. The plots of solubility curves for magnesium and calcium differ significantly at pK_{sp1} equal 16.54 and 16.7. At $pK_{sp1} = 17$, the plots bifurcate at higher pH values. At $pK_{sp1} = 19.35$, both plots overlap.

Some thermodynamic data given above are modified, to some degree, by kinetic effects. Namely, from the data referred to dissolution of dolomite into solutions with different pH_0 [27] or pH_0 and $p(CO_2)$ [29] values, it results that overall effectiveness of dolomite dissolution is largely affected (limited) by dissolution of pr3 component, owing to the fact that pr2 in dolomite dissolves faster than pr3.

5. Final Comments

The paper provides the calculation procedure that enables some changes in speciation in the system with dissolving dolomite to be followed. In each case, the dissolution was considered as a *quasistatic*, isothermal process. The dissolution concept refers to the systems where solubility product of the related precipitate has not been crossed yet. The dissolution has been considered under different conditions, expressed by (a) concentration of the solid phase (C_0 mol/L), (b) concentration (C_{CO_2} mol/L) of CO_2, and (c) concentration of a strong acid (C_a mol/L) or base (C_b mol/L), expressed by the value $\Delta = C_b - C_a$. The procedure applied enables the concentrations of particular species formed at different pH of the solution to be calculated at different moments of the dolomite dissolution. This way, the dissolution ($\sum Me$, mol/L) value was plotted. At the end of the dissolution process, $\sum Me$ assumes its limiting value, equal to the solubility (s_{Me}, mol/L), that is, $\sum Me = s_{Me}$. As results from the examples quoted, in some instances the solubility product of other (different from the dissolving) solid phases is crossed. Higher (relative) concentration of CO_2 in the solution promotes the dissolution of dolomite.

Symbols

C_0: initial concentration (mol/L) of dolomite (pr1) in two-phase system

C_b, C_a: concentrations (mol/L) of KOH or HCl, respectively, in the initial solution
$\Delta = C_b - C_a$

K_{sp1-5}: solubility products for $CaMg(CO_3)_2$ (pr1), $CaCO_3$ (pr2), $MgCO_3$ (pr3), $Ca(OH)_2$ (pr4); $Mg(OH)_2$ (pr5), respectively

M_i: molar mass (g/mol) of pri
$pC_0 = -\log C_0$
$pC_{CO_2} = -\log C_{CO_2}$
$pK_{spi} = -\log K_{spi}$ ($i = 1, \ldots, 5$)

[pri]: molar conc. of pri

s: Solubility (mol/L)

t: time.

References

[1] J. Warren, "Dolomite: occurence, evolution and economically important associations," *Earth Science Reviews*, vol. 52, no. 1–3, pp. 1–81, 2000.

[2] M. Schmidt, S. Xeflide, R. Botz, and S. Mann, "Oxygen isotope fractionation during synthesis of CaMg-carbonate and implications for sedimentary dolomite formation," *Geochimica et Cosmochimica Acta*, vol. 69, no. 19, pp. 4665–4674, 2005.

[3] R. S. Arvidson and F. T. Mackenzie, "Tentative kinetic model for dolomite precipitation rate and its application to dolomite distribution," *Aquatic Geochemistry*, vol. 2, no. 3, pp. 273–298, 1996.
[4] T. Michalowski, M. Borzęcka, M. Toporek, S. Wybraniec, P. Maciukiewicz, and A. Pietrzyk, "Quasistatic processes in non-equilibrium two-phase systems with ternary salts: II. Dolomite + Aqueous media," *Chemia Analityczna*, vol. 54, no. 6, pp. 1203–1217, 2009.
[5] http://www.webelements.com/.
[6] L. Chai, A. Navrotsky, and R. J. Reeder, "Energetics of calcium-rich dolomite," *Geochimica et Cosmochimica Acta*, vol. 59, no. 5, pp. 939–944, 1995.
[7] E. Busenberg and L. Niel Plummer, "Thermodynamics of magnesian calcite solid-solutions at 25°C and 1 atm total pressure," *Geochimica et Cosmochimica Acta*, vol. 53, no. 6, pp. 1189–1208, 1989.
[8] W. D. Bischoff, F. T. Mackenzie, and F. C. Bishop, "Stabilities of synthetic magnesian calcites in aqueous solution: comparison with biogenic materials," *Geochimica et Cosmochimica Acta*, vol. 51, no. 6, pp. 1413–1423, 1987.
[9] M. A. Bertram, F. T. Mackenzie, F. C. Bishop, and W. D. Bischoff, "Influence of temperature on the stability of magnesian calcite," *American Mineralogist*, vol. 76, no. 11-12, pp. 1889–1896, 1991.
[10] J. S. Tribble, R. S. Arvidson, M. Lane, and F. T. Mackenzie, "Crystal chemistry, and thermodynamic and kinetic properties of calcite, dolomite, apatite, and biogenic silica: applications to petrologic problems," *Sedimentary Geology*, vol. 95, no. 1-2, pp. 11–37, 1995.
[11] M. J. Malone, P. A. Baker, and S. J. Burns, "Recrystallization of dolomite: an experimental study from 50–200°C," *Geochimica et Cosmochimica Acta*, vol. 60, no. 12, pp. 2189–2207, 1996.
[12] B. D. Evamy, "Dedolomitization and the development of rhombohedral pores in limestones," *Journal of Sedimentary Petrology*, vol. 37, no. 4, pp. 1204–1215, 1967.
[13] C. Ayora, C. Taberner, M. W. Saaltink, and J. Carrera, "The genesis of dedolomites: a discusion based on reactive transport modeling," *Journal of Hydrology*, vol. 209, no. 1–4, pp. 346–365, 1998.
[14] D. Bernoulli and B. Gunzenhauser, "A dolomitized diatomite in an Oligocene-Miocene deep-sea fan succession, Gonfolite Lombarda group, Northern Italy," *Sedimentary Geology*, vol. 139, no. 1, pp. 71–91, 2001.
[15] E. García, P. Alfonso, E. Tauler, and S. Galí, "Surface alteration of dolomite in dedolomitization reaction in alkaline media," *Cement and Concrete Research*, vol. 33, no. 9, pp. 1449–1456, 2003.
[16] D. A. Budd, "Cenozoic dolomites of carbonate islands: their attributes and origin," *Earth-Science Reviews*, vol. 42, no. 1-2, pp. 1–47, 1997.
[17] C. Reinhold, "Multiple episodes of dolomitization and dolomite recrystallization during shallow burial in Upper Jurassic shelf carbonates: Eastern Swabian Alb, southern Germany," *Sedimentary Geology*, vol. 121, no. 1-2, pp. 71–95, 1998.
[18] P. V. Brady, J. L. Krumhansl, and H. W. Papenguth, "Surface complexation clues to dolomite growth," *Geochimica et Cosmochimica Acta*, vol. 60, no. 4, pp. 727–731, 1996.
[19] I. Stratful, M. D. Scrimshaw, and J. N. Lester, "Conditions influencing the precipitation of magnesium ammonium phosphate," *Water Research*, vol. 35, no. 17, pp. 4191–4199, 2001.
[20] M. Gautelier, E. H. Oelkers, and J. Schott, "An experimental study of dolomite dissolution rates as a function of pH from −0.5 to 5 and temperature from 25 to 80°C," *Chemical Geology*, vol. 157, no. 1-2, pp. 13–26, 1999.
[21] J. E. Gillott, "Alkali-reactivity problems with emphasis on canadian aggregates," *Engineering Geology*, vol. 23, no. 1, pp. 29–43, 1986.
[22] T. Michałowski and A. Pietrzyk, "A thermodynamic study of struvite + water system," *Talanta*, vol. 68, no. 3, pp. 594–601, 2006.
[23] T. Michałowski and A. Pietrzyk, "Quasistatic processes in non-equilibrium two-phase systems with ternary salts: I. Struvite + Aqueous solution (CO2 + KOH)," *Chemia Analityczna*, vol. 53, no. 1, pp. 33–46, 2008.
[24] K. Lund, H. S. Fogler, C. C. McCune, and J. W. Ault, "Acidization-II. The dissolution of calcite in hydrochloric acid," *Chemical Engineering Science*, vol. 30, no. 8, pp. 825–835, 1975.
[25] H. W. Rauch and W. B. White, "Dissolution kinetics of carbonate rocks - 1. Effects of lithology on dissolution rate," *Water Resources Research*, vol. 13, no. 2, pp. 381–394, 1977.
[26] H. A. Wanas, "Petrography, geochemistry and primary origin of spheroidal dolomite from the Upper Cretaceous/Lower Tertiary Maghra El-Bahari Formation at Gabal Ataqa, Northwest gulf of Suez, Egypt," *Sedimentary Geology*, vol. 151, no. 3-4, pp. 211–224, 2002.
[27] L. Chou, R. M. Garrels, and R. Wollast, "Comparative study of the kinetics and mechanisms of dissolution of carbonate minerals," *Chemical Geology*, vol. 78, no. 3-4, pp. 269–282, 1989.
[28] J. S. Herman and W. B. White, "Dissolution kinetics of dolomite: effects of lithology and fluid flow velocity," *Geochimica et Cosmochimica Acta*, vol. 49, no. 10, pp. 2017–2026, 1985.
[29] E. Busenberg and L. N. Plummer, "The kinetics of dissolution of dolomite in CO_2-H_2O systems at 1.5 to 65°C and 0 to 1 atm PCO_2," *American Journal of Science*, vol. 282, no. 1, pp. 45–78, 1982.
[30] J. V. Macpherson and P. R. Unwin, "Recent advances in kinetic probes of the dissolution of ionic crystals," *Progress in Reaction Kinetics*, vol. 20, no. 3, pp. 185–244, 1995.
[31] Z. Zhang and G. H. Nancollas, "Mechanisms of growth and dissolution of sparingly soluble salts," *Reviews in Mineralogy and Geochemistry*, vol. 23, no. 1, pp. 365–396, 1990.
[32] M. B. Tomson and G. H. Nancollas, "Mineralization kinetics: a constant composition approach," *Science*, vol. 200, no. 4345, pp. 1059–1060, 1978.
[33] D. R. Boomer, C. C. McCune, and H. S. Fogler, "Rotating disk apparatus for reaction rate studies in corrosive liquid environments," *Review of Scientific Instruments*, vol. 43, no. 2, pp. 225–229, 1972.
[34] A. F. M. Barton and S. R. McConnel, "Dissolution behaviour of solids: the rotating disc method," *Chemistry in Australia*, vol. 46, pp. 427–433, 1979.
[35] D. Hofmann and F. Moll, "The effect of ultrasonics on the dissolution rate of powdered drugs," *European Journal of Pharmaceutics and Biopharmaceutics*, vol. 40, no. 3, pp. 142–146, 1994.
[36] E. Usdovski, "Synthesis of dolomite and geochemical implications," in *Dolomites: A Volume in Honour of Dolomieu*, B. Purser et al., Ed., pp. 345–360, Blackwell Sci. Publ, Oxford, UK, 1994.
[37] L. A. Sherman and P. Barak, "Solubility and dissolution kinetics of dolomite in Ca-Mg-HCO_3/CO_3solutions at 25°C and 0.1 MPa carbon dioxide," *Soil Science Society of America Journal*, vol. 64, no. 6, pp. 1959–1968, 2000.
[38] D. F. Sibley, S. H. Nordeng, and M. L. Borkowski, "Dolomitization kinetics in hydrothermal bombs and natural settings," *Journal of Sedimentary Research A*, vol. 64, no. 3, pp. 630–637, 1994.

[39] J. A. D. Dickson, "Transformation of echinoid Mg calcite skeletons by heating," *Geochimica et Cosmochimica Acta*, vol. 65, no. 3, pp. 443–454, 2001.

[40] M. El Tabakh, C. Utha-Aroon, J. K. Warren, and B. C. Schreiber, "Origin of dolomites in the Cretaceous Maha Sarakham evaporites of the Khorat Plateau, northeast Thailand," *Sedimentary Geology*, vol. 157, no. 3-4, pp. 235–252, 2003.

[41] C. M. Yoo and Y. I. Lee, "Origin and modification of early dolomites in cyclic shallow platform carbonates, Yeongheung Formation (Middle Ordovician), Korea," *Sedimentary Geology*, vol. 118, no. 1–4, pp. 141–157, 1998.

[42] C. Rodríguez-Navarro, E. Sebastián, and M. Rodríguez-Gallego, "An urban model for dolomite precipitation: authigenic dolomite on weathered building stones," *Sedimentary Geology*, vol. 109, no. 1-2, pp. 1–11, 1997.

[43] M. Schmidt, S. Xeflide, R. Botz, and S. Mann, "Oxygen isotope fractionation during synthesis of CaMg-carbonate and implications for sedimentary dolomite formation," *Geochimica et Cosmochimica Acta*, vol. 69, no. 19, pp. 4665–4674, 2005.

[44] M. El Tabakh, C. Utha-Aroon, J. K. Warren, and B. C. Schreiber, "Origin of dolomites in the Cretaceous Maha Sarakham evaporites of the Khorat Plateau, northeast Thailand," *Sedimentary Geology*, vol. 157, no. 3-4, pp. 235–252, 2003.

[45] P. M. May and K. Murray, "JESS, a joint expert speciation system-I. Raison d'être," *Talanta*, vol. 38, no. 12, pp. 1409–1417, 1991.

[46] A. Murray, M. Halliday, and R. J. Croft, "Popliteal artery entrapment syndrome," *British Journal of Surgery*, vol. 78, no. 12, pp. 1414–1419, 1991.

[47] P. M. May and K. Murray, "JESS, a joint expert speciation system-III. Surrogate functions," *Talanta*, vol. 40, no. 6, pp. 819–825, 1993.

[48] P. M. May and K. Murray, "Database of chemical reactions designed to achieve thermodynamic consistency automatically," *Journal of Chemical and Engineering Data*, vol. 46, no. 5, pp. 1035–1040, 2001.

[49] P. W. Jones, D. M. Taylor, and D. R. Williams, "Analysis and chemical speciation of copper and zinc in wound fluid," *Journal of Inorganic Biochemistry*, vol. 81, no. 1-2, pp. 1–10, 2000.

[50] T. Michałowski, "The generalized approach to electrolytic systems: I. Physicochemical and analytical implications," *Critical Reviews in Analytical Chemistry*, vol. 40, no. 1, pp. 2–16, 2010.

[51] T. Michałowski, A. Pietrzyk, M. Ponikvar-Svet, and M. Rymanowski, "The generalized approach to electrolytic systems: II. The generalized equivalent mass (GEM) concept," *Critical Reviews in Analytical Chemistry*, vol. 40, no. 1, pp. 17–29, 2010.

[52] A. G. Asuero and T. Michalowski, "Comprehensive formulation of titration curves for complex acid-base systems and its analytical implications," *Critical Reviews in Analytical Chemistry*, vol. 41, no. 2, pp. 151–187, 2011.

[53] T. Michałowski, "Calculations in Analytical Chemistry with Elements of Computer Programming, PK, Kraków," 2001, http://www.biblos.pk.edu.pl/bcr&id=1762&ps=-12&dir=MD.MichalowskiT.ObliczeniaChemii.html.

[54] T. Michałowski, "Application of GATES and MATLAB for Resolution of Equilibrium, Metastable and Non-Equilibrium Electrolytic Systems," in *Applications of MATLAB in Science and Engineering*, T. Michałowski, Ed., chapter 1, InTech, 2011.

[55] G. Chen and D. Tao, "Effect of solution chemistry on flotability of magnesite and dolomite," *International Journal of Mineral Processing*, vol. 74, no. 1–4, pp. 343–357, 2004.

[56] W. Stumm and J. J. Morgan, *Aquatic Chemistry. An Introduction Emphasising Chemical Equilibria in Natural Waters*, John Wiley and Sons, New York, NY, USA, 1981.

[57] P. M. May and K. Murray, 2001, http://jess.murdoch.edu.au/jess/jess_home.htm.

[58] J. Y. Gal, J. C. Bollinger, H. Tolosa, and N. Gache, "Calcium carbonate solubility: a reappraisal of scale formation and inhibition," *Talanta*, vol. 43, no. 9, pp. 1497–1509, 1996.

[59] K. M. Udert, T. A. Larsen, and W. Gujer, "Estimating the precipitation potential in urine-collecting systems," *Water Research*, vol. 37, no. 11, pp. 2667–2677, 2003.

[60] L. A. Hardie, "Dolomitization; a critical view of some current views," *Journal of Sedimentary Petrology*, vol. 57, no. 1, pp. 166–183, 1987.

[61] D. K. Nordstrom, L. N. Plummer, D. Langmuir, E. May, H. M. Jones, and D. L. Parkhurst, "Revised chemical equi-librium data for major water mineral reactions and their limitations," in *Chemical Modelling of Aqueous Systems*, D. C. Melchior and R. L. Basset, Eds., ACS Ser. 416, chapter 31, American Chemical Society, Washington, DC, USA, 1990.

17

Thermo-Diffusion and Diffusion-Thermo Effects on MHD Free Convective Heat and Mass Transfer from a Sphere Embedded in a Non-Darcian Porous Medium

B. Vasu,[1] V. R. Prasad,[1] and O. Anwar Bég[2]

[1] *Department of Mathematics, Madanapalle Institute of Technology and Science, Madanapalle 517325, India*
[2] *Department of Engineering and Mathematics, Sheffield Hallam University, Room 4112, Sheaf Building, Sheffield S1 1WB, UK*

Correspondence should be addressed to B. Vasu, bvsmaths1@gmail.com

Academic Editor: Mohammad Al-Nimr

The problem of combined heat and mass transfer by natural convection over a sphere in a homogenous non-Darcian porous medium subjected to uniform magnetic field is numerically studied, taking Soret/Dufour effects into account. The coupled, steady, and laminar partial differential conservation equations of mass, momentum, energy, and species diffusion are normalized with appropriate transformations. The resulting well-posed two-point boundary value problem is solved using the well-tested, extensively validated Keller-Box implicit finite difference method, with physically realistic boundary conditions. A parametric study of the influence of Soret number (Sr), Dufour number (Du), Forchheimer parameter (Λ), Darcy parameter (Da), buoyancy ratio parameter (N), Prandtl number (Pr), Schmidt number (Sc), magnetohydrodynamic body force parameter (M), wall transpiration (f_w) is the blowing/suction parameter, and streamwise variable (ξ) on velocity, temperature, and concentration function evolution in the boundary layer regime is presented. Shear stress, Nusselt number, and Sherwood number distributions are also computed. Applications of the study arise in hydromagnetic flow control of conducting transport in packed beds, magnetic materials processing, geophysical energy systems, and magnetohydrodynamic chromatography technology.

1. Introduction

Magnetohydrodynamic transport phenomena arise in numerous branches of modern chemical engineering. These include crystal magnetic damping control [1], hydromagnetic chromatography [2], conducting flows in trickle-bed reactors [3], and enhanced magnetic filtration control [4]. Numerous studies of both an experimental and numerical nature have been communicated regarding magnetohydrodynamic flow in chemical engineering. In general, in such flows, the Lorentz electromagnetic force arises as a result of the interaction between the magnetic field and the electrical current. This latter is generated hydrodynamically by the bulk flow of the conducting liquid phase. In porous media applications such as packed beds, to sustain a given flow rate of the electrically conducting liquid in the bed, the pressure drop and the liquid holdup will be increased under magnetohydrodynamic conditions compared with the case of nonconducting fluids. In crystal growth applications in porous media, the external magnetic field imposed has been successfully exploited to suppress unsteady flow and also reduce composition nonuniformity. The most extensively studied flows involve static magnetic fields which may be aligned (for which magnetic boundary layers arise) [5, 6], transversal [7–11], radial [12, 13], inclined [14, 15] and so forth. Researchers have also considered the use of rotating magnetic fields to control convection in the solution zone during crystal growth by, for example, the traveling heater method (THM). A wide variety of mathematical models have been employed in simulating magnetohydrodynamic transport in both fluid and also porous media regimes. Rudariah et al. [16] obtained analytical solutions for Hartmann-Couette flow in a porous regime, showing that for mercury flowing over a long permeable channel of width 0.7 cm, the effect of magnetic field (of strength 0.25 Web/m^2, for which Hartmann number = 8.8) in the presence of a porous wall (of permeability = 5×10^{-6} Darcy) is to retard the mass flow and to increase the friction factor relative to the corresponding quantities for nonmagnetic flow. They further identified that transition to turbulence occurs at a higher Reynolds number owing to the presence of a magnetic field. Lioubashevski et al. [17] investigated the effect of a static homogeneous magnetic field orientated parallel to the planar electrode surface using a hydrodynamic boundary layer formation and equivalent mass transfer model for a semiinfinite electrode, showing that there is a decrease of the boundary layer thickness with increasing magnetic field strength or reagent bulk concentration. Boum and Alemany [18] studied mass transfer processes in an electrochemical system under an applied external magnetic field, studying in particular the magnetically induced flow of electrolyte in a short duct. Fahidy [19] employed an approximate method for the estimation of the mean velocity and the root mean square velocity in the natural-convective diffusion boundary layer under weak magnetic field, examining in detail the case of a vertical solid surface where the electric and magnetic field are horizontal and mutually transverse. Al-Nimr and Hader [20] obtained closed form solutions for fully developed hydromagnetic free convection flow in open-ended vertical porous channels using a superposition method, for four different boundary conditions. McWhirter et al. [21] presented an analytical model for magnetohydrodynamic porous medium flow, simulating the regime as a packed bed of uniform spheres. They considered two particular cases—an infinite packed bed and a finite packed bed including wall effects, the latter being modeled with a two-zone porosity model, with a higher porosity wall region inserted between the solid wall and the lower porosity core region. Dahikar and Sonolikar [22] studied experimentally the hydromagnetic flow in scaled magnetically assisted circulating fluidized bed, with a circular cross-section, showing that magnetic field strongly influences fluidization phenomenon and accelerates the transition towards much denser suspension of the fluidized bed. Alchaar et al. [23] studied the stability of free convection in a porous regime under transverse magnetic field using a CFD(computational fluid dynamics) approach. Mansour et al. [24] used a power series expansion method to assess the steady convective heat and mass transfer in micropolar conducting flow on a circular cylinder maintained at uniform heat and mass flux. Postelnicu [25] conducted a theoretical study of magnetohydrodynamic free convection heat and mass transfer from a vertical surface in a porous medium, incorporating Dufour and Soret diffusion effects, showing that increasing magnetic field elevates the local Nusselt and Sherwood numbers. The above studies have not considered the combined thermal convection and mass diffusion in magnetohydrodynamic boundary layer flow from a sphere in a porous medium, despite the importance of such a problem in filtration engineering and materials processing. It is therefore proposed to study this problem in the present paper. Furthermore, the medium is simulated to include non-Darcian and variable porosity effects with blowing/suction also included for the sphere surface. A well-tested implicit finite difference procedure, the Keller-box method is used to solve the transformed boundary layer equations. A detailed discussion of the effects of thermophysical parameters on flow variables is provided.

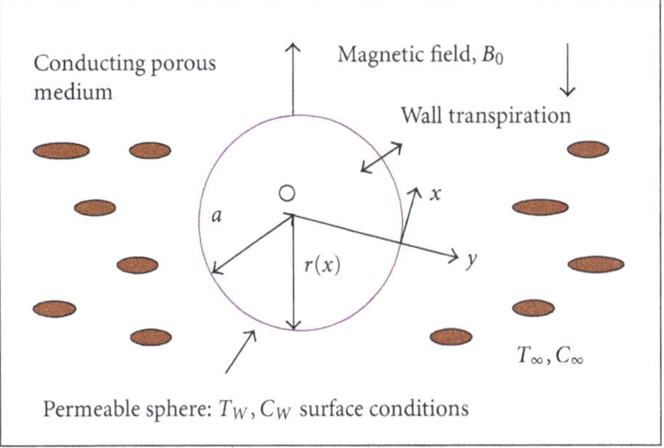

FIGURE 1: Physical model.

2. Mathematical Model

We consider the steady, laminar, two-dimensional, incompressible, electrically conducting, and buoyancy-driven convection heat and mass transfer flow form a permeable isothermal sphere embedded in a non-Darcy porosity medium, under the action of an outwardly directed radial magnetic field and Soret (thermodiffusion) and Dufour (diffusion-thermal) effects. Figure 1 illustrates the physical model and coordinate system. Here x is measured along the surface of the sphere, y is measured normal to the surface, respectively, and r is the radial distance from symmetric axis to the surface. in $r = a\sin(x/a)$, a is the radius of the sphere. Both the sphere and the fluid are maintained initially at the same temperature. Instantaneously they are raised to a temperature $T_w(> T_\infty)$ and concentration $C_w(> C_\infty)$ which remain unchanged.

The fluid properties are assumed to be constant except the density variation in the buoyancy force term. The porous medium is simulated using the well-tested and validated non-Darcian drag force model. This incorporates a linear Darcian drag for low velocity effects (bulk impedance of the porous matrix at low Reynolds numbers) and a quadratic (second order) resistance, the Forchheimer drag, for high velocity flows, as may be encountered in chemical engineering systems operating at higher velocities. The appropriate non-Darcian model, following Bég et al. [26, 27], is therefore

$$\nabla p = -\frac{\mu}{K}V - \frac{\rho b}{K}V^2, \tag{1}$$

where ∇p is the pressure drop across the porous medium, μ is the dynamic viscosity of the fluid, b is the Forchheimer (geometric) inertial drag parameter, K is the permeability of the porous medium (hydraulic conductivity), and V is a general velocity. The impressed electrical field is assumed to be zero and both the induced magnetic and electric fields of the flow are negligible in comparison with the applied magnetic field which corresponds to very small magnetic Reynolds number. The Maxwell electromagnetic field equations, neglecting displacement currents, may be presented, following Sutton and Sherman, thus [28]

$$\nabla \cdot \mathbf{B} = 0$$
$$\nabla \times \mathbf{B} = \mu_0 \mathbf{J}$$
$$\nabla \times \mathbf{E} = -\frac{\partial \mathbf{B}}{\partial t} \quad (2)$$
$$\mathbf{J} = \sigma[\mathbf{E} + \mathbf{V} \times \mathbf{B}],$$

where $\mathbf{B}$ is the uniform magnetic field vector, $\mathbf{J}$ is the current density vector, μ_0 is magnetic permeability, σ is the electrical conductivity of the fluid, $\mathbf{E}$ is the electrical field vector, $\mathbf{V}$ is the linear (translational) fluid velocity vector, and t is time. As is customary in engineering magnetohydrodynamics [28], we consider the case for which magnetic lines of force are fixed relative to the fluid. Further, we neglect Hall currents, ionslip currents, and assume steady-state incompressible flow. Since magnetic Reynolds number $\mathrm{Re}_m = \mathrm{Re} = VL/\eta^*$ (based on a general velocity component, V, where η^* is the magnetic diffusivity) is much lower than unity in the present problem, for such scenarios, $\mathrm{Re}_m << 1$, advection is relatively unimportant, and therefore the magnetic field will tend to relax towards a purely diffusive state, determined by the boundary conditions rather than the flow. Magnetic diffusion will be much greater than viscous diffusion and magnetic induction will therefore not arise. The flow regime is also not influenced by the magnetization of the fluid in the applied magnetic field; electrical current is therefore present and the fluid is electrically conducting [28, 29]. In two-dimensional hydromagnetic boundary layer flows, the applied uniform magnetic field vector, $\mathbf{B}$, defined in the Maxwell relations, (2), is assumed to lie in the x-y plane. For steady boundary layer flows, electrical field $\mathbf{E}$ will vanish, as emphasized by Hughes and Young [30]. The Lorentzian hydromagnetic drag will therefore have two components, F_x and F_y, and there will be two components of the applied magnetic field, B_x and B_y. Giving

$$F_x = -\sigma\left(uB_y^2 - vB_xB_y\right)$$
$$F_y = \sigma\left(uB_xB_y - vB_x^2\right). \quad (3)$$

It is customary [28, 30] to assume that B_y varies only with the transverse coordinate, y. For a two-dimensional, steady, uniform applied magnetic field, the Maxwell equation in (1) will reduce to the following, from $\nabla \cdot \mathbf{B} = 0$:

$$\frac{\partial B_x}{\partial x} + \frac{\partial B_y}{\partial y} = 0. \quad (4)$$

Further, assuming

$$B_x = o\left(B_y\right)$$
$$\frac{\partial}{\partial x} = o(1) \quad (5)$$
$$\frac{\partial}{\partial y} = o(\delta^{-1}),$$

it follows that the change in B_y across the boundary layer (of thickness δ) will be of the order of δ and may be neglected, so that effectively the hydromagnetic Lorentzian drag force in the x-momentum equation becomes

$$F_x \sim \sigma\, B_y^2(u). \quad (6)$$

In light of the above approximations, and under the usual Boussinesq and boundary layer approximations, the equations for mass continuity, momentum, energy, and concentration can be written as follows:

$$\frac{\partial(ru)}{\partial x} + \frac{\partial(rv)}{\partial y} = 0 \quad (7)$$

$$\left(u\frac{\partial u}{\partial x} + v\frac{\partial u}{\partial y}\right) = \nu\frac{\partial^2 u}{\partial y^2} + g\beta(T - T_\infty)\sin\left(\frac{x}{a}\right) + g\beta^*(C - C_\infty)\sin\left(\frac{x}{a}\right) - \frac{\sigma B_0^2}{\rho}u - \frac{\nu}{K}u - \Gamma u^2 \quad (8)$$

$$u\frac{\partial T}{\partial x} + v\frac{\partial T}{\partial y} = \alpha\frac{\partial^2 T}{\partial y^2} + \frac{D_m K_T}{c_s c_p}\frac{\partial^2 C}{\partial y^2} \quad (9)$$

$$u\frac{\partial C}{\partial x} + v\frac{\partial C}{\partial y} = D_m\frac{\partial^2 C}{\partial y^2} + \frac{D_m K_T}{T_m}\frac{\partial^2 T}{\partial y^2}. \quad (10)$$

The boundary conditions are defined as follows:

$$y = 0: \quad u = 0, \quad v = V_w, \quad T = T_w, \quad C = C_w,$$
$$y \longrightarrow \infty: \quad u = 0, \quad T = T_\infty, \quad C = C_\infty, \quad (11)$$

where u and v denote the velocity components in the x- and y-directions, respectively, K and B are, respectively, the regime permeability and the Forchheimer inertial drag coefficient of the porous medium, ν is the kinematic viscosity, β and β^* are the coefficients of thermal expansion and concentration expansion, respectively, T and C are the temperature and concentration, respectively, σ is the electrical conductivity, B_0 is the externally imposed radial magnetic field (i.e., applied in the y-direction), ρ is the density, D_m is the mass diffusivity, α is the thermal diffusivity, c_p is the specific heat capacity, c_s is the concentration susceptibility, α is the thermal diffusivity, T_m is the mean fluid temperature, K_T is the thermal diffusion ratio, T_∞ is the free stream temperature, C_∞ is the free stream concentration, and V_w is the uniform blowing/suction velocity at the sphere surface. In the momentum equation (8), the first term on the right hand side simulates the viscous shear, the second is the thermal buoyancy term, the third is the species buoyancy, the fourth is the Lorentzian linear hydromagnetic drag, the fifth term on the right hand side is the porous medium Darcian drag force representing pressure loss due to the presence of the porous medium, and the final term on the right hand side designates the second-order inertial porous medium drag force (also referred to as the Forchheimer impedance) which accounts for additional pressure drop resulting from interpore mixing appearing at high velocities. The stream function ψ is defined by $ru = \partial(r\psi)/\partial y$ and $rv = -\partial(r\psi)/\partial x$; therefore, the continuity equation is automatically satisfied. Proceeding with the analysis we introduce the following dimensionless variables:

$$\xi = \frac{x}{a}, \qquad \eta = \frac{y}{a}\sqrt[4]{\mathrm{Gr}},$$
$$f(\xi,\eta) = \frac{\psi}{\nu\xi\sqrt[4]{\mathrm{Gr}}}, \qquad \theta(\xi,\eta) = \frac{T - T_\infty}{T_w - T_\infty}, \tag{12}$$
$$\phi(\xi,\eta) = \frac{C - C_\infty}{C_w - C_\infty}, \qquad \mathrm{Gr} = \frac{g\beta(T_w - T_\infty)a^3}{\nu^2}.$$

Substituting (12) into (7)–(11), we obtain

$$f''' + (1+\xi\cot\xi)ff'' - (1+\xi\Lambda)f'^2 + \frac{\sin\xi}{\xi}(\theta + N\phi) - \left(M + \frac{1}{\mathrm{Da}}\right)f' = \xi\left(f'\frac{\partial f'}{\partial \xi} - f''\frac{\partial f}{\partial \xi}\right), \tag{13}$$

$$\frac{\theta''}{\mathrm{Pr}} + (1+\xi\cot\xi)f\,\theta' + \mathrm{Du}\,\phi'' = \xi\left(f'\frac{\partial \theta}{\partial \xi} - \theta'\frac{\partial f}{\partial \xi}\right), \tag{14}$$

$$\frac{\phi''}{\mathrm{Sc}} + (1+\xi\cot\xi)f\,\phi' + \mathrm{Sr}\,\theta'' = \xi\left(f'\frac{\partial \phi}{\partial \xi} - \varphi'\frac{\partial f}{\partial \xi}\right). \tag{15}$$

The transformed dimensionless boundary conditions are

$$\eta = 0: \quad f' = 0, \quad f = f_w, \quad \theta = 1, \quad \phi = 1,$$
$$\eta \longrightarrow \infty: \quad f' = 0, \quad \theta = 0, \quad \phi = 0. \tag{16}$$

In the above equations, the primes denote the differentiation with respect to η, the dimensionless radial coordinate, f is dimensionless stream function, θ is dimensionless temperature function, ϕ is dimensionless concentration function, $\Lambda = \Gamma a$ is the local inertia coefficient (Forchheimer parameter), $\mathrm{Da} = (K\sqrt{\mathrm{Gr}})/a^2$ is a Darcy parameter, $N = (\beta * (C_w - C_\infty))/(\beta(T_w - T_\infty))$ is the concentration to thermal buoyancy ratio parameter, $\mathrm{Pr} = (\rho\nu c_p)/k$ is the Prandtl number, $\mathrm{Sc} = \nu/D_m$ is the Schmidt number, $M = \sigma B_0^2 a^2/\rho\nu\sqrt{\mathrm{Gr}}$ is the magnetohydrodynamic body force parameter, $\mathrm{Du} = (D_m K_T (C_w - C_\infty))/(c_s c_p \nu(T_w - T_\infty))$ is the Dufour number, $\mathrm{Sr} = (D_m K_T(T_w - T_\infty))/(\nu T_m(C_w - C_\infty))$ is the Soret number, $f_w = -V_w a/\nu\sqrt[4]{\mathrm{Gr}}$ is the transpiration (blowing/suction) parameter, and Gr is the Grashof (free convection) parameter. $f_w < 0$ for $V_w > 0$ (the case of injection), and $f_w > 0$ for $V_w < 0$ (the case of suction). The important case of a solid sphere surface is retrieved for $f_w = 0$. The chemical engineering design quantities of physical interest include the skin-friction coefficient, Nusselt number, and Sherwood number, which are given by

$$\frac{C_f}{\rho(\nu/a)^2(\mathrm{Gr})^{3/4}} = \xi f''(\xi,0) \tag{17}$$

$$\frac{\mathrm{Nu}}{\sqrt[4]{\mathrm{Gr}}} = -\theta'(\xi,0) \tag{18}$$

$$\frac{\mathrm{Sh}}{\sqrt[4]{\mathrm{Gr}}} = -\phi'(\xi,0). \tag{19}$$

3. Numerical Method

In this study, the efficient Keller-Box implicit difference method has been employed to solve the general flow model defined by (13) to (15) with boundary conditions (16). This method, originally developed for low speed aerodynamic boundary layers by Keller [31], has been employed in a diverse range of nonlinear magnetohydrodynamics and coupled heat transfer problems. This method is chosen since it seems to be most flexible of the common methods, being easily adaptable for solving equations of any order by Cebeci and Bradshaw [32]. For the sake of brevity, the numerical method is not described. Computations were carried out with $\Delta\xi = 0.1$; the first step size $\Delta\eta = 0.02$. The requirement that the variation of the velocity, temperature, and concentration distribution is less than 10^{-5} between any two successive iterations is employed as the criterion convergence. These include Takhar et al. [33] who studied geothermal hydromagnetic gas convection flow and omitted here for conservation of space. Effectively the complete linearized system is formulated as a block matrix system where each element in the coefficient matrix is a matrix itself. The numerical results are affected by the number of mesh points in both directions. The edge of the boundary-layer y_∞ was adjusted for different ranges of parameters.

4. Results and Discussion

Comprehensive solutions have been obtained and are presented in Figures 2 to 12. The numerical problem comprises **2** independent variables (ξ,η), **3** dependent fluid dynamic variables (f, θ, ϕ) and 10 thermophysical and body force "control" parameters Pr, M, N, Λ, Da, Sc, f_w, ξ, Sr, and Du. The present code has been extensively validated in numerous problems and benchmarked against shooting quadrature and finite element methods [34–38] and also more recently with Crank-Nicolson difference methods [39–41]. It is extremely reliable and the solutions presented have been rigorously checked. Comparisons are therefore excluded here for conservation of space. In the present computations, the following default parameters are prescribed (unless otherwise stated): $\mathrm{Pr} = 0.71$, $M = 1.0$, $N = 1.0$, $\Lambda = 0.1$, $\xi = 1.0$, $\mathrm{Da} = 0.1$, $\mathrm{Sc} = 0.25$, $f_w = 0.5$ $\mathrm{Du} = 0.2$, and $\mathrm{Sr} = 0.25$. These correspond to weak hydromagnetic convection in electrolytic fluid ($\mathrm{Pr} = 0.71$) flowing through a high permeability ($\mathrm{Da} = 0.1$), high regime with weak blowing ($f_w = 0.5$) at the sphere surface, weak inertial drag ($\Lambda = 0.1$), air diffusing in the electrolytic solution ($\mathrm{Sc} = 0.25$), and

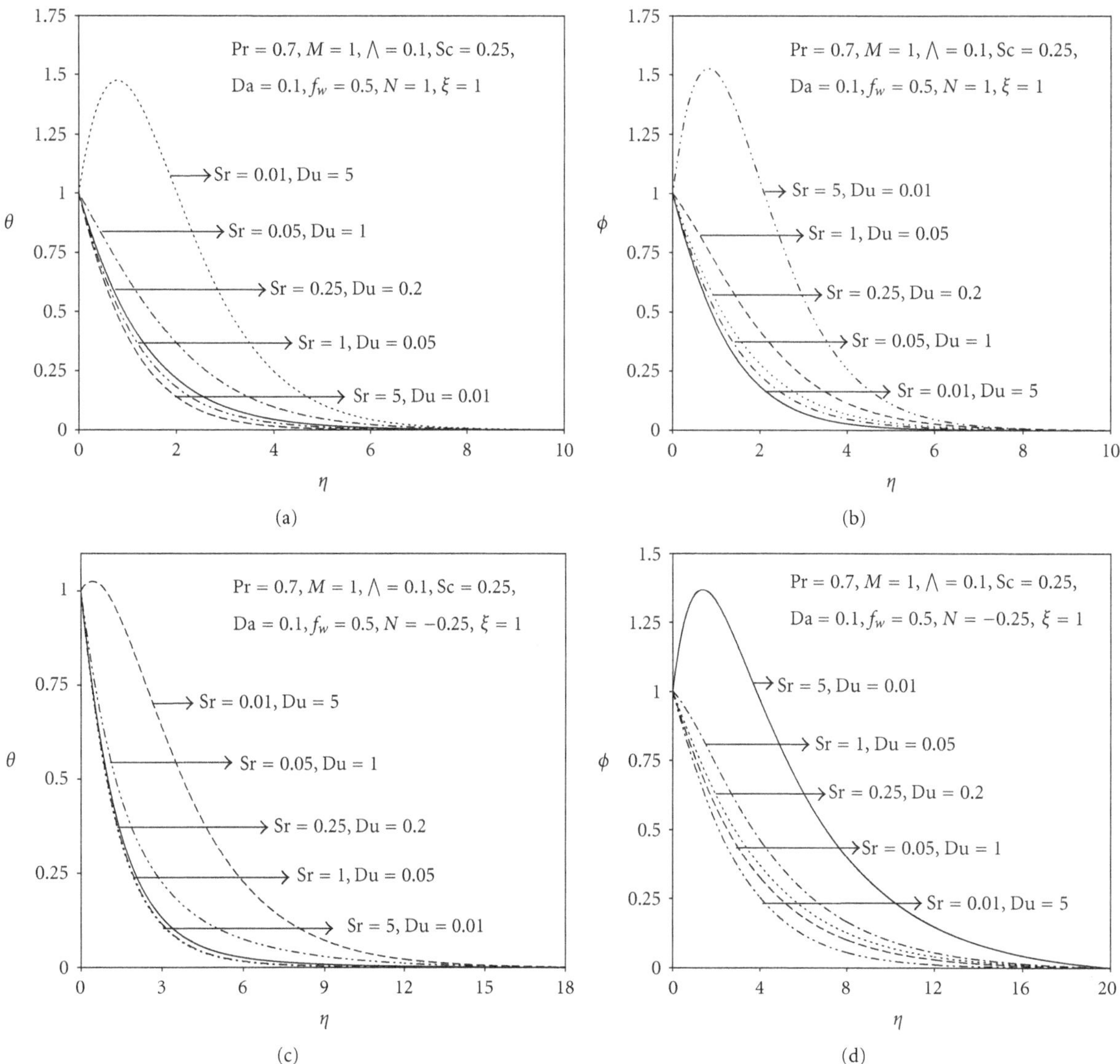

FIGURE 2: (a) Effect of the Sr and Du on the temperature profiles. (b) Effect of the Sr and Du on the concentration profiles. (c) Effect of the Sr and Du on the temperature profiles with $N = -0.25$. (d) Effect of the Sr and Du on the concentration profiles with $N = -0.25$.

a balance of thermal and species buoyancy forces ($N = 1$), at a general location along the sphere curvature ($\xi = 1$). In addition, we also consider the effect of the streamwise coordinate location on flow dynamics. The value of the parameter ξ is extremely important since in two extreme cases it corresponds to stagnation-point flows. For $\xi \sim 0$, the location is in the vicinity of the lower stagnation point on the sphere. The governing dimensionless equations (13) to (15) in this case reduce to the following ordinary differential equations:

$$f''' + ff'' - f'^2 + (\theta + N\phi) - \left(M + \frac{1}{\mathrm{Da}}\right)f' = 0 \quad (20)$$

$$\mathrm{Pr}^{-1}\theta'' + f\theta' + \mathrm{Du}\phi'' = 0 \quad (21)$$

$$\mathrm{Sc}^{-1}\phi'' + f\phi' + \mathrm{Sr}\theta'' = 0. \quad (22)$$

We note, following Bég et al. [10], the buoyancy term is retained since $(\sin\xi)/\xi \rightarrow 0/0$, that is, 1, so that $(\sin\xi/\xi)(\theta + N\phi) \rightarrow (\theta + N\phi)$; however, the Forchheimer drag force term vanishes at $\xi = 0$, whereas the Darcian drag force and Lorentzian hydromagnetic drag force are retained. The equations however remain strongly coupled. The other extreme case is $\xi \sim \pi$, which physically corresponds to the upper stagnation point on the sphere surface (diametrically opposite to the lower stagnation point).

In Figures 2(a) and 2(b), the influence of Sr and Du on temperature and concentration distributions is illustrated, near the leading edge ($\xi = 0.1$) for the buoyancy-assisted case ($N = 1$). Soret number (Sr) defines the effect of temperature gradients inducing significant mass diffusion effects. Dufour number (Du) measures the contribution of concentration gradients to thermal energy flux in the flow domain. Du and Sr are varied together so that their product

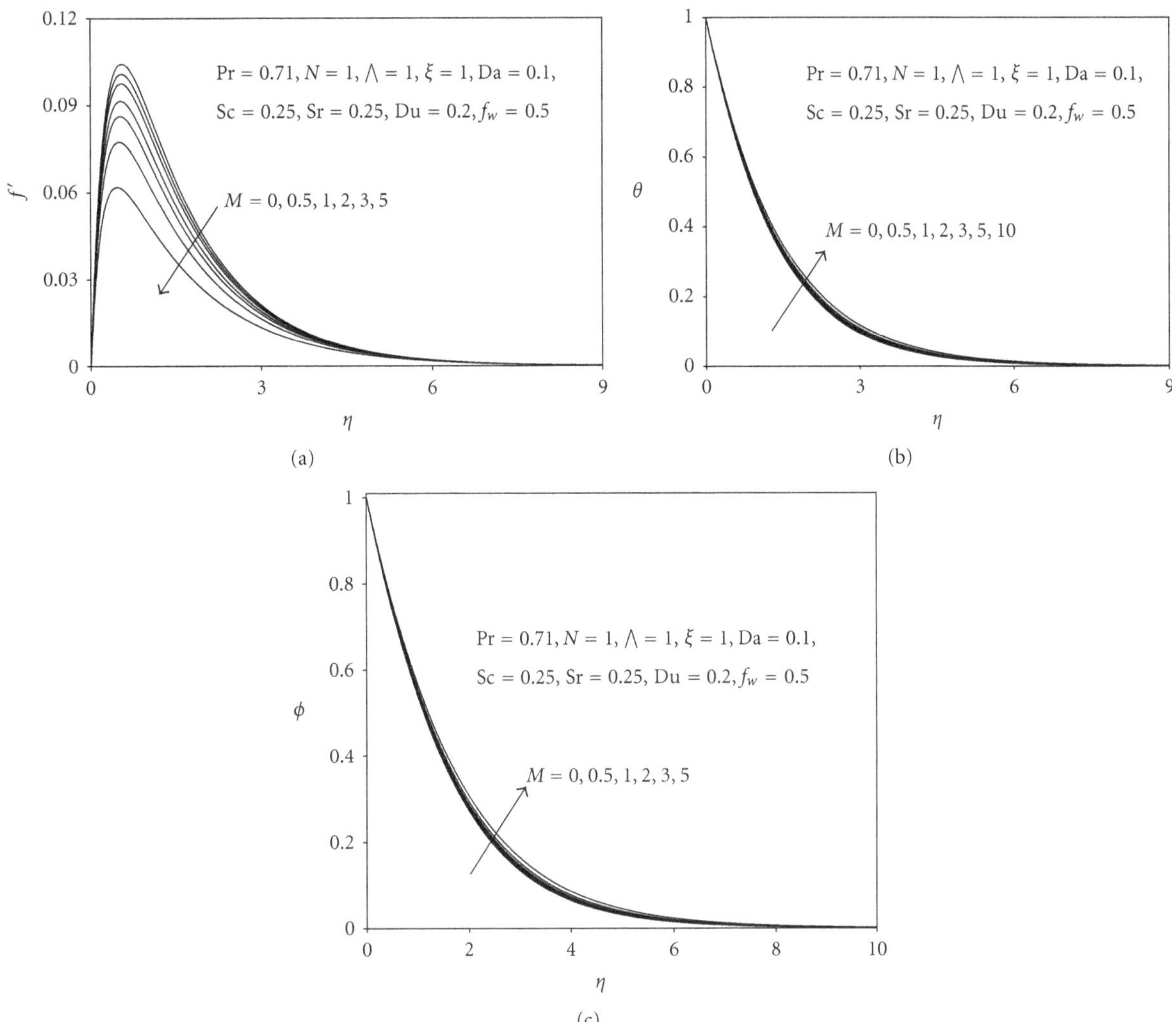

FIGURE 3: (a) Effect of the M on the velocity profiles. (b) Effect of the M on the temperature profiles. (c) Effect of the M on the concentration profiles.

remains constant: for example, Du = 5.0, Sr = 0.01; Du = 0.01, and Sr = 5.0. An increase in Du (and concurrent decrease in Sr) considerably increases temperature (θ) in the boundary layer, as observed in Figure 2(a). For Du $\leq$ 1, profiles decay smoothly from the wall to zero in the free stream. However, for Du $>$ 1, that is, Du = 5.0, a distinct velocity overshoot exists near the sphere, and thereafter the profile falls to zero at the edge of the boundary layer. Increasing concentration gradients, that is, increasing Du value therefore strongly heats the flow. Increasing Soret number has the exact reverse affect, that is, cooling the boundary layer regime. Figure 2(b) shows that an increase in Soret number due to the contribution of temperature gradients to species diffusion, increases concentration (ϕ) values for all η increasing Du values have the opposite effect. No concentration overshoot is however observed for any values of Soret number which is varied from 0.01 through 0.05, 0.25, and 1.0 to 5.0. The influence of the Soret and Dufour terms will be relatively weak on the velocity fields, and these are therefore not plotted.

For $N > 0$, thermal and concentration buoyancy forces act in unison. For $N = 0$, buoyancy forces are absent, that is, forced convection arises in the regime. For $N < 0$, both buoyancy forces oppose each other. When $N < 0$, we have the case of opposing buoyancy. Particularly, for $N = -0.25$, Figures 2(c) and 2(d), show the temperature and concentration distributions with collective variation in Soret number (Sr) and Dufour number (Du) for the case of buoyancy opposition (opposing buoyancy force, $N <$ 0). Therefore, from the sphere surface, negative N, that is, opposing buoyancy is beneficial to the flow regime, whereas closer to the sphere surface it has a retarding effect. A much more consistent response to a change in the N parameter is observed that temperature throughout the boundary layer is strongly reduced as increase in Sr (and concurrent decrease in Du). We have considered a few cases of larger Sr, Du—this was done to indicate that a concentration overshoot will arise in, for example, Figure 2(d); similarly, with a large Sr = 5.0, we observe that in Figure 2(c) that the temperature distribution deviates noticeably from the profiles for smaller

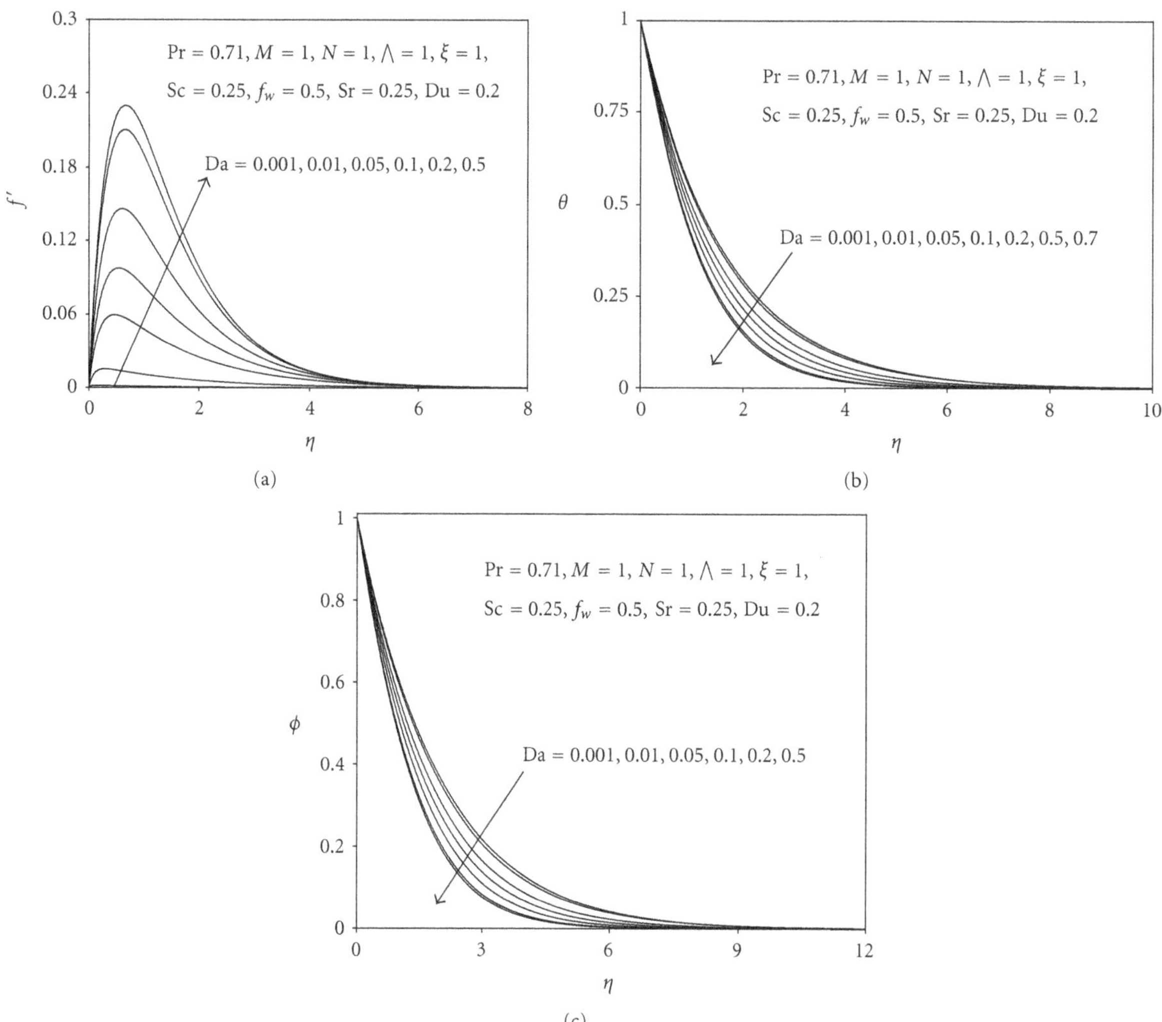

FIGURE 4: (a) Effect of the Da on the velocity profiles. (b) Effect of the Da on the temperature profiles. (c) Effect of the Da on the concentration profiles.

Sr values. A number of authors have considered the effect of Sr and Du, mainly to address the cases of very strong thermal diffusion encountered in chromatographic applications.

In Figures 3(a) to 3(b) present the response of velocity (f'), temperature (θ), and concentration (ϕ) to magnetohydrodynamic body force parameter (M). $M = \sigma B_0^2 a^2/\rho\nu\sqrt{\mathrm{Gr}}$, and signifies the ratio of Lorentz hydromagnetic body force to viscous hydrodynamic force. Increasing M from 0 (non-conducting case) to 1.0 (magnetic body force and viscous force equal) through to 10.0 (very strong magnetic body force) induces a distinct reduction in velocities as shown in Figure 3(a). With higher M values since the magnetic body force, $-Mf'$ in the momentum equation (13) is amplified; this serves to increasingly retard the flow. The imposition of a radial magnetic field is therefore a powerful mechanism for inhibiting flow in the regime. The maximum velocities as before arise close to the sphere surface, a short distance from it (at the sphere surface, $\eta = 0$, and velocity vanishes in consistency with the no-slip condition, that is, $f' = 0$); with further distance into the boundary layer, the profiles converge, that is, the magnetic body force has a weaker effect in the far field regime than in the near-field regime. Conversely with increasing M, temperature, (Figure 3(b)) is observed to be markedly increased. This is physically explained by the fact that the extra work expended in dragging the fluid against the magnetic field is dissipated as thermal energy in the boundary layer, as elucidated by Sutton and Sherman [28], Pai [29], and Hughes and Young [30]. This results in heating of the boundary layer and an ascent in temperatures, an effect which is maximized some distance away from the sphere surface. The magnetic field influence on temperatures while noticeable is considerably less dramatic than that on the velocity field, since the Lorentz body force only arises in the momentum equation (13) and influences the temperature (θ) and concentration (ϕ) fields, only via the thermal and buoyancy terms, $(\sin\xi/\xi)(\theta + N\phi)$. Magnetic effects do not feature in either the temperature (14) or species diffusion equations (15). The deceleration in flow serves to enhance species diffusion in the regime and this causes a rise also in the concentration profiles

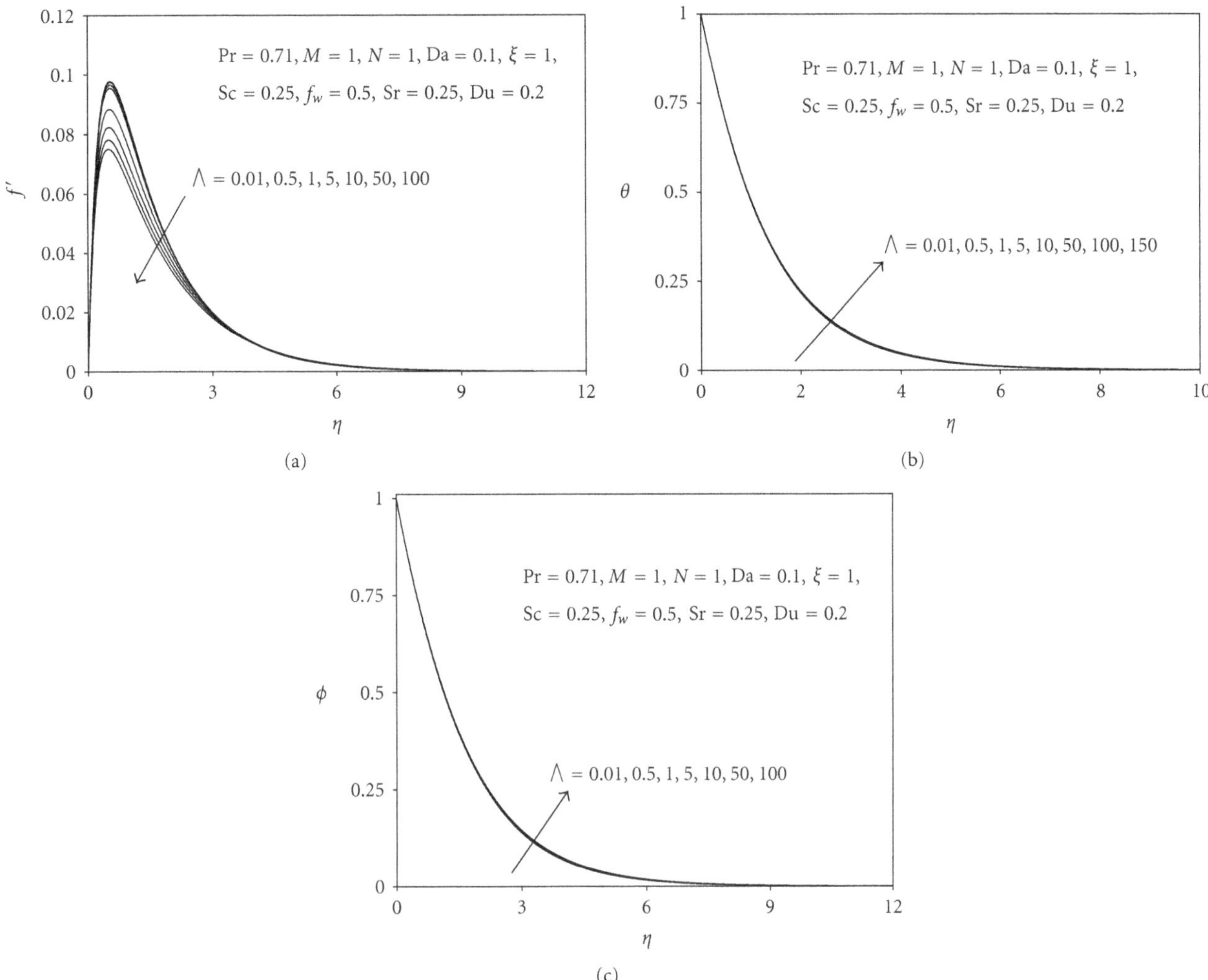

FIGURE 5: (a) Effect of the Λ on the velocity profiles. (b) Effect of the Λ on the temperature profiles. (c) Effect of the Λ on the concentration profiles.

(Figure 3(c)) with increasing magnetic parameter. Applied radial magnetic field, therefore, while counteracting the momentum development in the regime, serves to enhance the heat and species diffusion, and this is of immense benefit in chemical engineering operations, where designers may wish to elevate transport in a regime without accelerating the flow.

In Figures 4(a) to 4(c), the variation of velocity (f'), temperature (θ), and species concentration (ϕ) with transverse coordinate (η) over a wide range of Darcy parameters is illustrated. The Darcian body force, $(1/\mathrm{Da})f'$ features only in the momentum equation (13). This body force although linear is inversely proportional to the Darcy parameter, $Da = (K\sqrt{\mathrm{Gr}})/a^2$ which itself is a measure of the permeability of the regime, that is, hydraulic transmissivity of the porous medium to fluid percolation. Increasing Da from 0.001 (extremely low permeability) through 0.01, 0.05, 0.1, 0.15 to the maximum value of 0.2, clearly substantially enhances the flow velocity in the boundary layer. With higher Da values, there will be a corresponding reduction in the Darcian drag force, and this will serve to effectively accelerate the flow in the medium adjacent to the sphere. In all the velocity profiles, the peak velocity is located close to the sphere surface; with an increase in Da, this peak is displaced progressively away from the sphere surface. Temperature (θ), however, as shown in Figure 4(b), is observed to progressively decrease with an increase in Darcy number (Da). With increasing Da, there is a decrease in the density of solid matrix fibers in the regime (permeability is increased). As such, thermal conduction is depressed and this reduces the temperatures in the boundary layer. The greatest reduction in temperatures arises at some distance form the sphere surface. In all cases, the profiles decay smoothly to zero in the free stream. The quality of the Keller-box mesh is emphasized by the smooth asymptotic tendency of the profiles far from the wall. The specification of an adequate distance for infinity boundary conditions is therefore confirmed. Concentration profiles (ϕ) are depressed even more strongly than temperature profiles, with an increase in Darcy number, as shown in Figure 4(c). With N set to unity, both thermal and species buoyancy

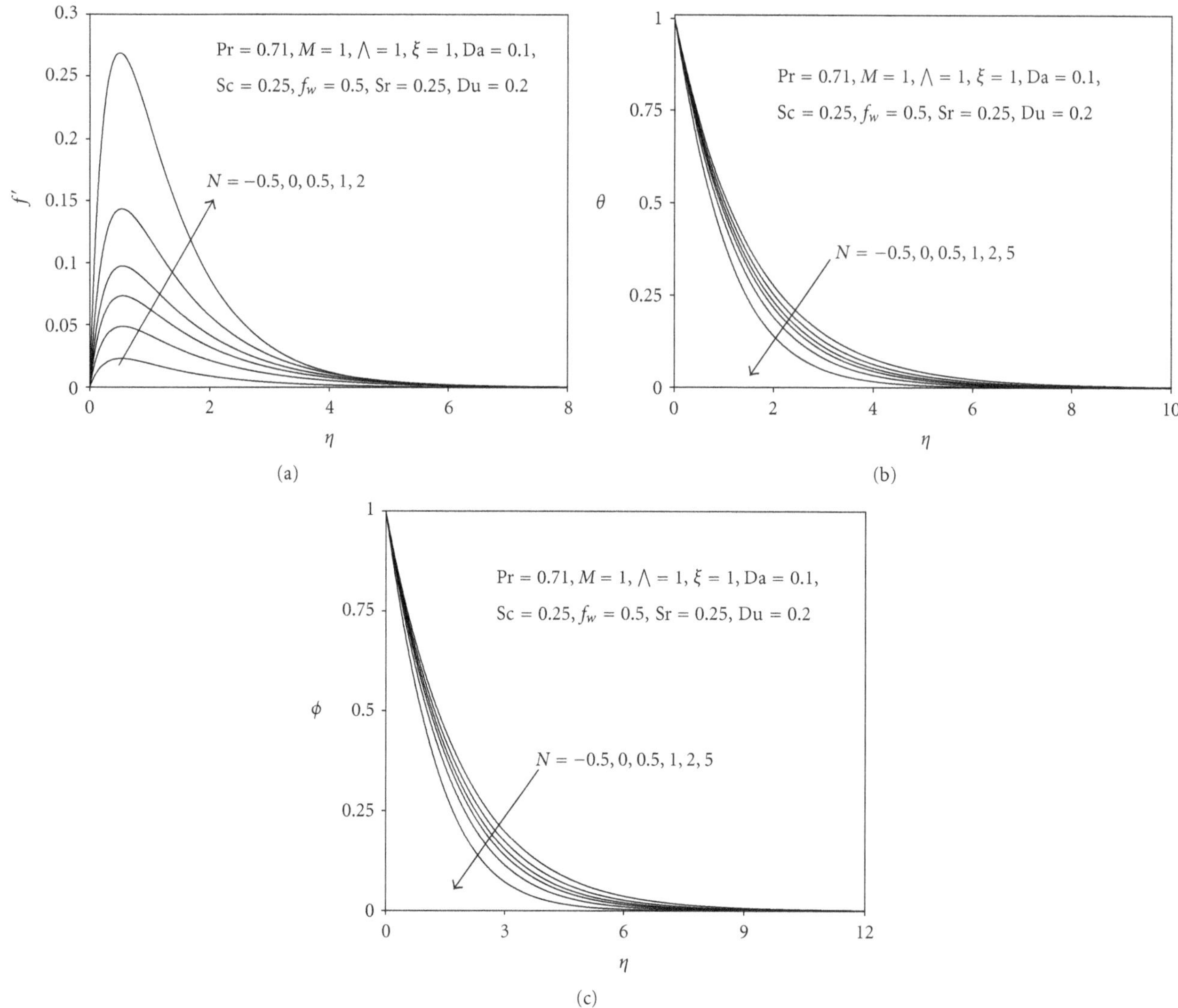

Figure 6: (a) Effect of the N on the velocity profiles. (b) Effect of the N on the temperature profiles. (c) Effect of the N on the concentration profiles.

forces in the momentum equation have the same magnitude. With lesser fibers in the regime, the diffusion of species would be expected to be enhanced. However, it has been documented by numerous experimental studies that lower permeability regimes in fact enhance species diffusion, as opposed to counteracting it. Our results are in excellent agreement with the findings of Vafai and Tien [42].

Figures 5(a) to 5(c) show the influence of the Forchheimer inertial parameter (Λ) on the flow variables. This parameter is associated with the second-order Forchheimer resistance term, $\xi\Lambda(f')^2$, in (13). Forchheimer drag is directly proportional to the parameter, Λ. An increase in Λ evidently strongly retards the flow, as illustrated in Figure 5(a), for some considerable distance into the in boundary layer, transverse to the sphere surface. Beyond a certain point however negligible effects are observed. Bear [43] has highlighted that Forchheimer effects are associated with higher velocities in porous media transport. Forchheimer drag however is quadratic and the increase in this "form" drag swamps the momentum development, effectively decelerating the flow. We note that the deviation from the linear, the nonlinear behavior in porous media is gradual unlike the sharp change from laminar to turbulence flow in the case of fluid flow in conduits. As such, there are no sudden changes in velocity profile with increasing Forchheimer parameter, that is, no fluctuations in the velocity field. This confirms the modern perspective of many researchers [44, 45] that the main cause of deviation from the Darcian behavior is related to other factors different than those which contribute to turbulence. The drag force model employed in this study is physically logical since it implies that non-Darcy flow is occurring at high spatial velocity in porous media as a consequence of geometrical factors rather than attributing this behavior to "turbulence" phenomena. In this context, the term "non-Darcian" does not allude to a different regime of flow, but to the amplified effects of Forchheimer drag at

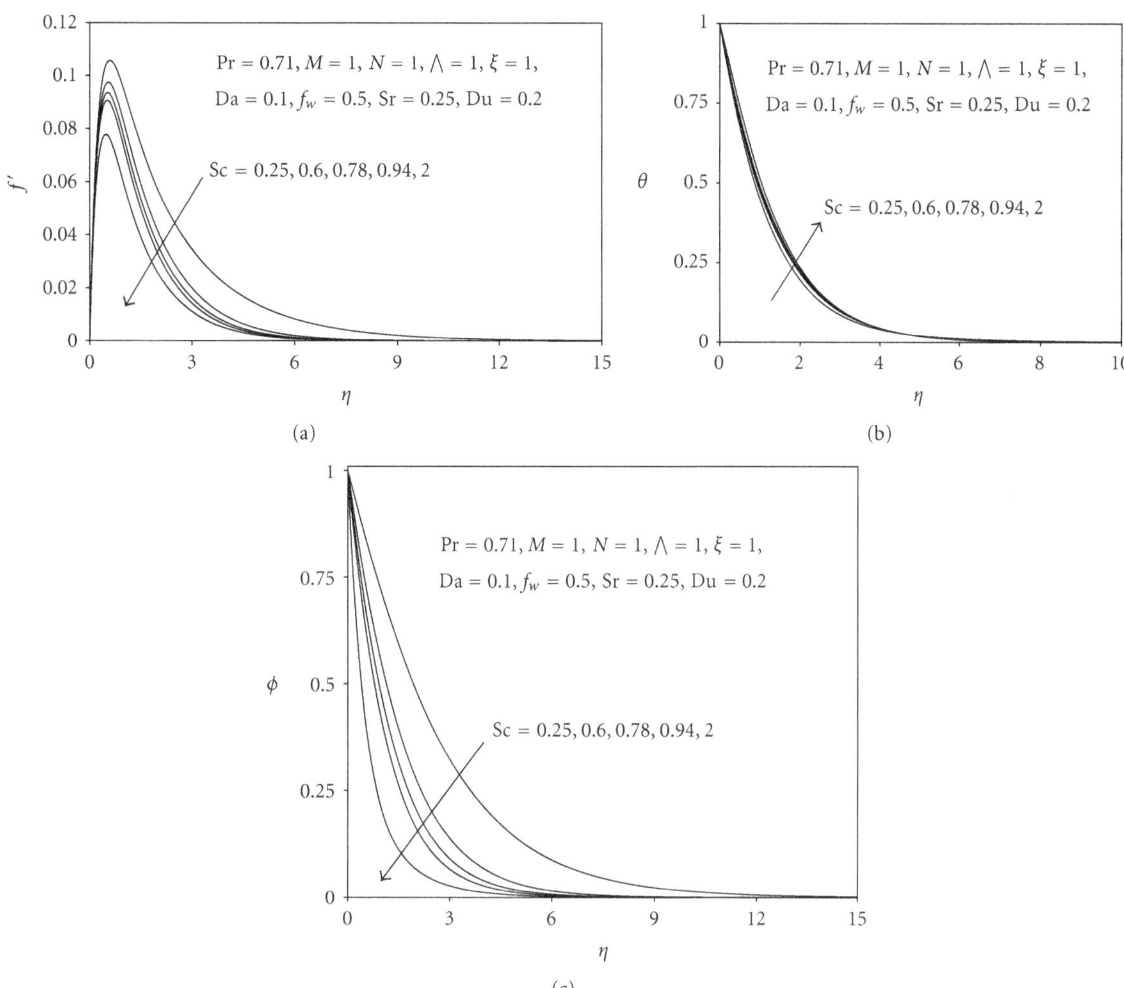

FIGURE 7: (a) Effect of the Sc on the velocity profiles. (b) Effect of the Sc on the temperature profiles. (c) Effect of the Sc on the concentration profiles.

higher velocities, as elaborated by a number of other studies [46]. With a dramatic increase in Λ, there is a very slight elevation in temperatures (Figure 5(b)) in the regime. The deceleration in the flow results in thinner velocity boundary layers which serve to enhance energy diffusion. The influence on the concentration (species diffusion) field (Figure 5(c)) is similar to that of the temperature field but more pronounced. Species concentration (ϕ) is slightly increased, especially at some distance from the sphere surface, with an increase in Forchheimer parameter, Λ.

Figures 6(a) to 6(c) depict the effect of the buoyancy ratio parameter (N) on the velocity, temperature, and species concentration variables. $N = (\beta * (C_w - C_\infty))/(\beta(T_w - T_\infty))$ is a key parameter controlling the transport phenomena in the regime. For $N > 0$ thermal and concentration buoyancy forces act in unison. For $N = 0$, buoyancy forces are absent, that is, forced convection arises in the regime. For $N < 0$, both buoyancy forces oppose each other. Inspection of Figure 6(a) indicates that for $N > 0$, that is, aiding buoyancy forces, the flow velocities are greatly enhance. Positive N therefore as expected accelerates the flow. With an increase in N, the velocity peaks progressively migrate further from the sphere surface. Figures 6(b) and 6(c) reveal that temperature, θ, and concentration, ϕ, exhibit the converse response to a positive increase in buoyancy ratio, N; both are considerably reduced. Although the buoyancy parameter does not arise in either the thermal boundary layer (heat conservation) (14) or the concentration boundary layer (species diffusion conservation) (15), via coupling with the term, $(\sin\xi/\xi)(\theta + N\phi)$ in the momentum equation (13), N exerts a strong influence on both energy and species diffusion in the boundary layer. For all values of N there is a smooth decay in both θ and ϕ profiles from a maximum at the sphere surface to the free stream. The buoyancy effect can clearly be exploited to control effectively both temperature and concentration distributions in such a regime, and this again is of considerable utility to chemical engineering designers involved in packed bed transport phenomena systems. We further note that since N is directly coupled with the species field in the buoyancy term, it will exert a more pronounced effect on the concentration distributions than on the temperature profiles. This explains the greater

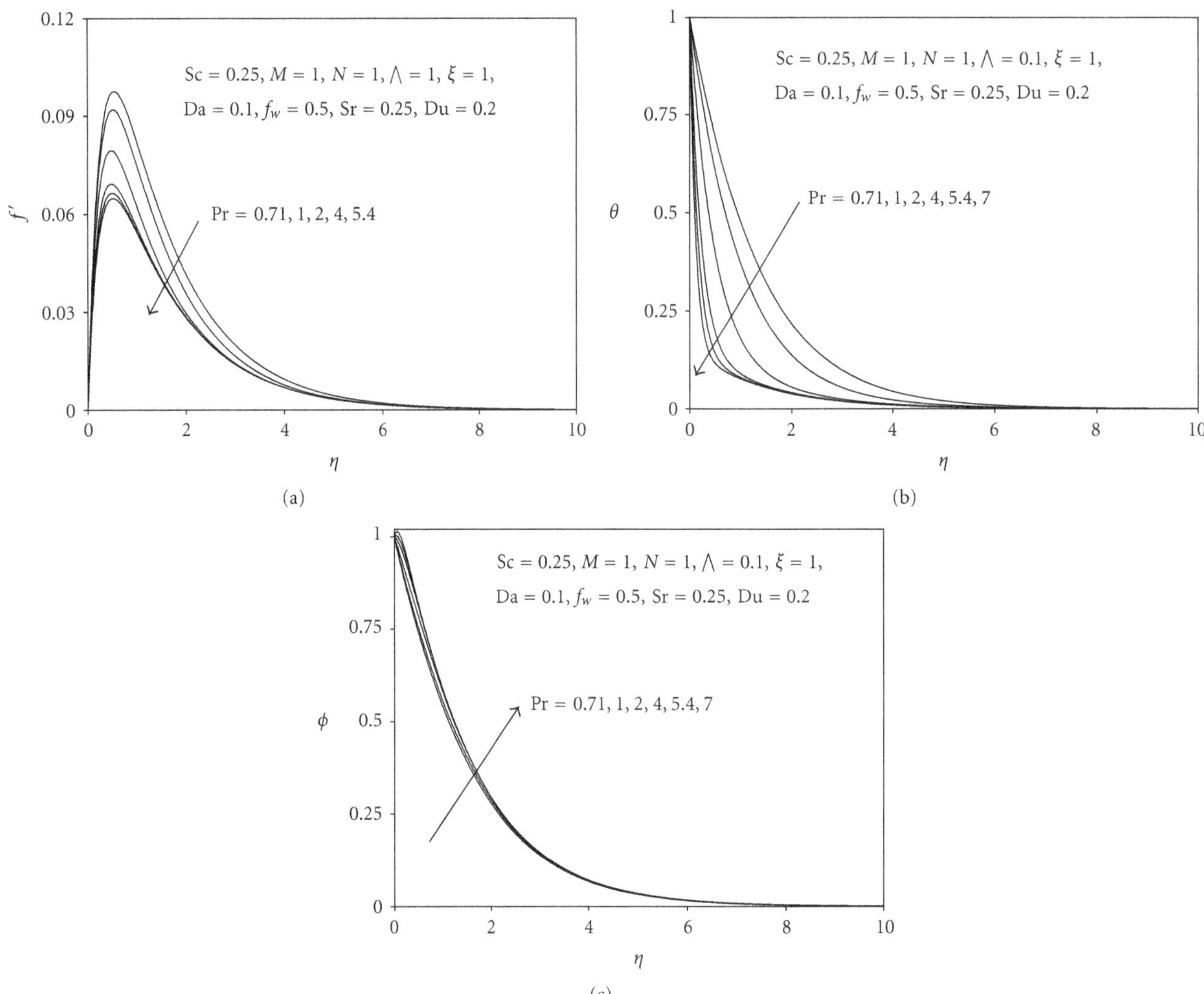

FIGURE 8: (a) Effect of the Pr on the velocity profiles. (b) Effect of the Pr on the temperature profiles. (c) Effect of the Pr on the concentration profiles.

decreases in concentration values (Figure 6(c)) over the same range of N increase (from 0 through 0.1, 0.5, 1.0 to 2.0), compared with the reduction in the temperature profiles (Figure 6(b)).

Figures 7(a) to 7(c) illustrate the influence of Schmidt number (Sc) on velocity (f'), temperature (θ), and species concentration (ϕ). Velocity (Figure 7(a)) is strongly reduced with an increase in Sc from 0.25 through 0.6, 0.78, and 0.94 to 2.0. Sc represents the ratio of the mass (species) and viscous diffusion time sales. It is also the ratio of momentum diffusivity to species diffusivity. For Sc $<$ 1, the momentum diffusivity is lower than the species (mass) diffusivity and the species diffusion rate exceeds the momentum diffusion rate. For Sc $>$ 1, this scenario is reversed. Higher values of Sc correspond to higher density species diffusing in air for example, Sc = 1.0 corresponds to methanol diffusing in electrically conducting air, Sc = 2.0 implies ethylbenzene diffusing in air. Increasing Sc lowers the chemical molecular diffusivity of the species. As Sc is increased, the concentration boundary layer will become relatively thinner than the viscous (momentum) boundary layer. Velocity will therefore be reduced. For the special case of Sc = 1.0, the velocity and concentration boundary layers will be of the same thickness and both momentum and species will be diffused at the same rate. Inspection of Figure 7(b) shows that temperatures (θ) are enhanced with an increase in Sc; however, the alteration in profiles is not dramatic. In Figure 7(c), the concentration profiles are, as expected, found to be much more markedly affected by a rise in Sc values. ϕ values are strongly reduced with increasing Sc values. With thinner concentration boundary layers, the concentration gradients will be enhanced causing a decrease in concentration of species in the boundary layer. For Sc $<$ 1, species diffusivity exceeds momentum diffusivity and this accounts for the greater concentration values for Sc = 0.25, 0.6, 0.78, 0.94 compared with the minimized concentration profile for Sc = 2.0, since in this latter case Sc $>$ 1, that is, momentum diffusivity exceeds mass (species) diffusivity. The implication

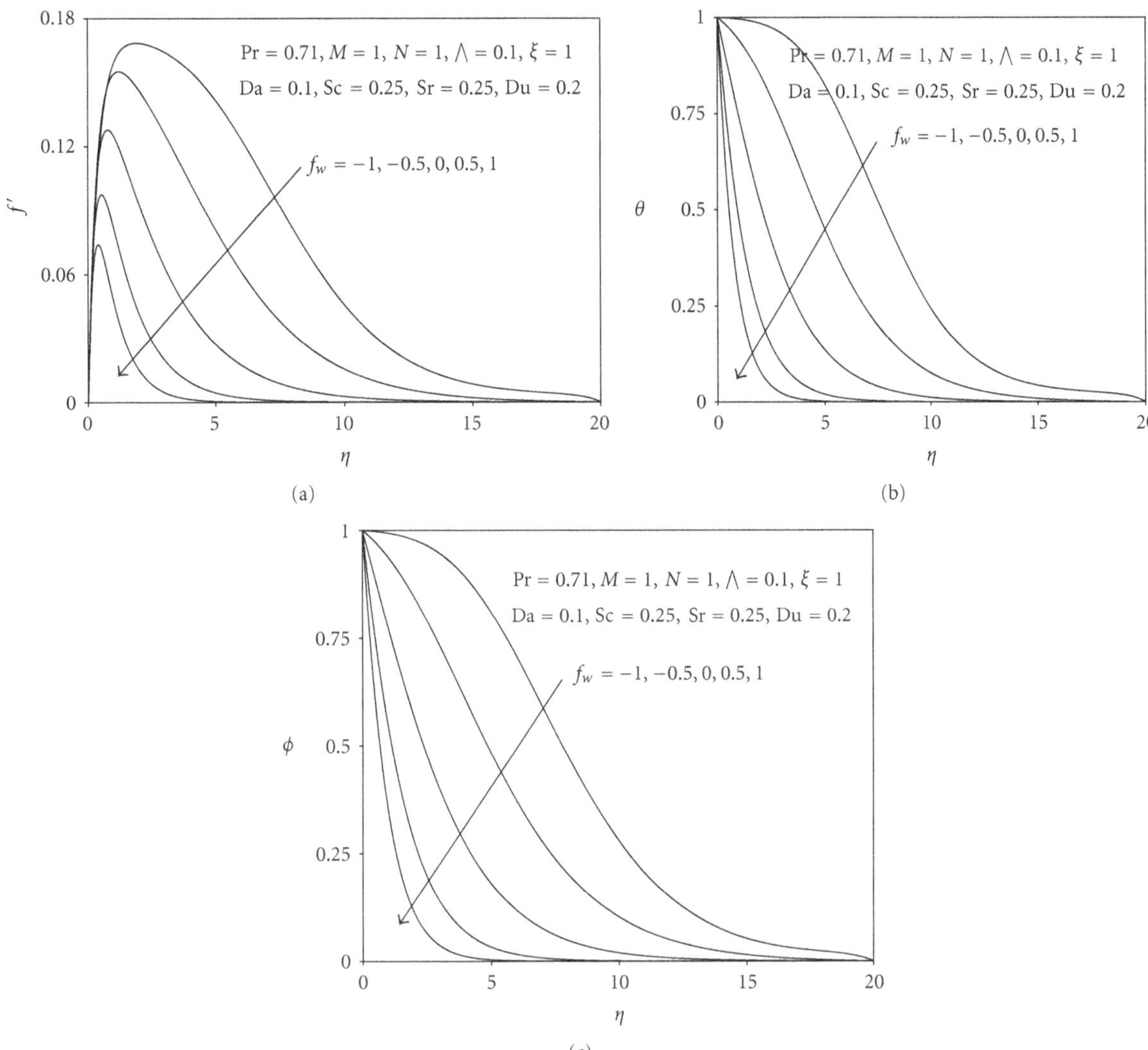

FIGURE 9: (a) Effect of the f_w on the velocity profiles. (b) Effect of the f_w on the temperature profiles. (c) Effect of the f_w on the concentration profiles.

for chemical engineering designers is that in such a regime, a lower Schmidt number diffusing species must be employed to enhance concentration distributions in the medium.

Figures 8(a) to 8(c) present the effect of Prandtl number (Pr) on the primitive flow variables of velocity, temperature, and concentration. Prandtl number signifies the ratio of viscous diffusion to thermal diffusion in the boundary layer regime. With greater Pr values, viscous diffusion rate exceeds thermal diffusion rate. An increase in Pr from 0.71 (air) through 1.0, 2.0, 4.0, 5.4 to 7.0 (conducting water e.g., saline solution) strongly depresses velocities (Figure 8(a)) in the regime. For Pr $<$ 1, thermal diffusivity exceeds momentum diffusivity, that is, heat will diffuse faster than momentum. For Pr = 1.0, both the viscous and energy diffusion rates will be the same as will the thermal and velocity boundary layer thicknesses. For Pr $>$ 1, momentum diffusivity will exceed thermal diffusivity. As such for lower Pr fluids, velocities we expect to be maximized and this is indeed testified to by Figure 8(a). Higher Pr fluids will correspond to lower velocities. Similarly temperatures (θ), as shown in Figure 8(b), are also considerably lowered with an increase in Pr. Prandtl number also represents the product of dynamic viscosity and specific heat capacity divided by thermal conductivity of the primary fluid (Pr $= (\rho \nu c_p)/k \equiv (\mu c_p)/k$). Higher Pr fluids (e.g., Pr = 7.0 for electrically conducting water) will therefore possess a much lower thermal conductivity and this will result in a significant decrease in temperatures in the boundary layer. Conversely lower Pr fluids will possess a much greater thermal conductivity (e.g., Pr = 0.7 for conducting air) and will generate much higher temperatures. All temperatures are observed to decay smoothly from a maximum at the wall to zero in the free stream of the boundary layer. In Figure 8(c), the concentration profiles are seen to be enhanced in magnitude with increasing Pr values. Prandtl number is a thermophysical property at a given temperature and pressure. As such, it is associated with actual liquids. It will exert a similar influence on the temperature distribution whether in the pure free convection or pure forced convection scenarios. This is disadvantageous to the temperature field but beneficial to the species diffusion

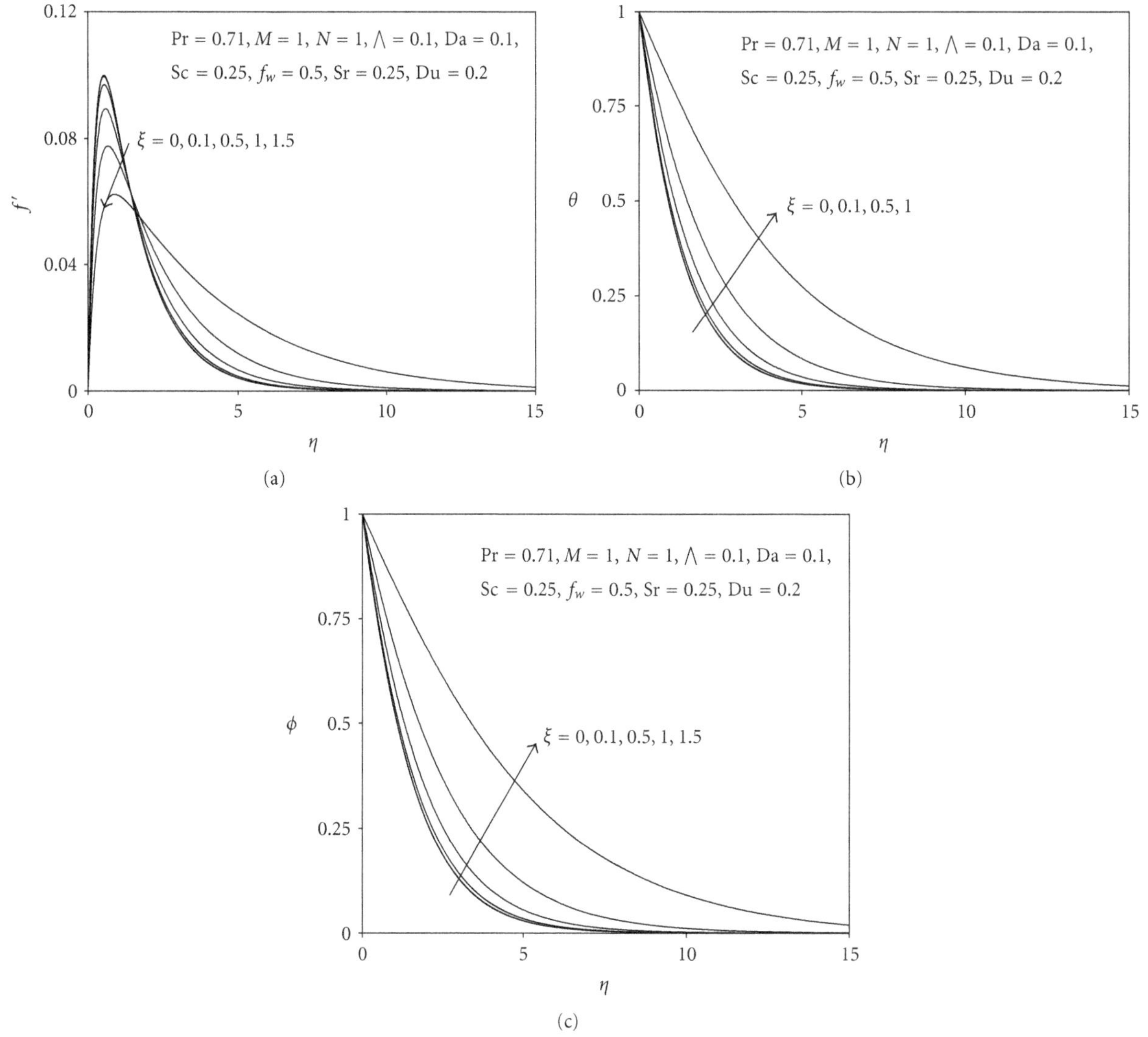

(a) (b)

(c)

FIGURE 10: (a) The velocity profiles with various values of ξ. (b) The temperature profiles with various values of ξ. (c) The concentration profiles with various values of ξ.

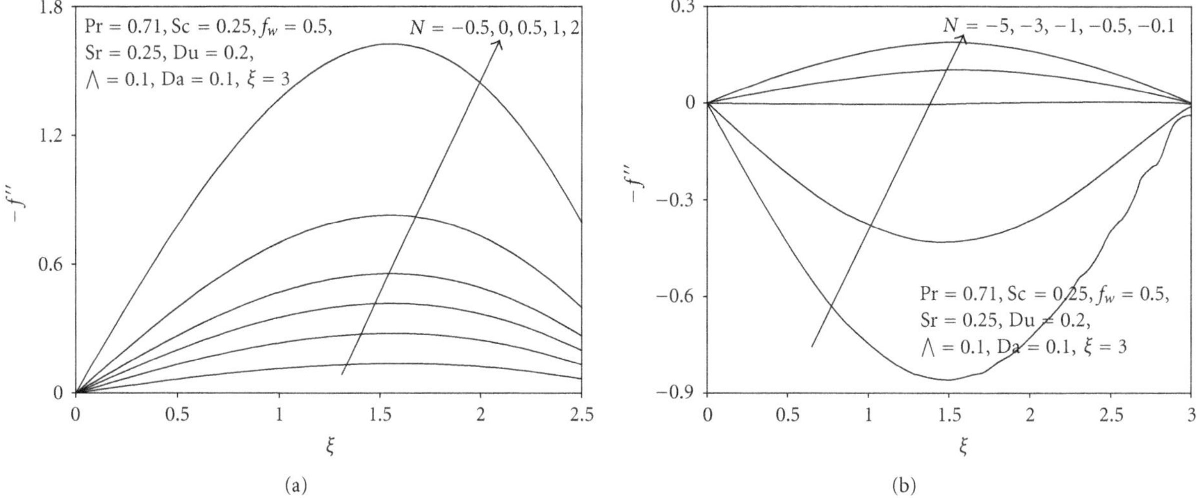

FIGURE 11: (a) Skin friction coefficient results for various values of N. (b) Skin friction coefficient results for negative values of N.

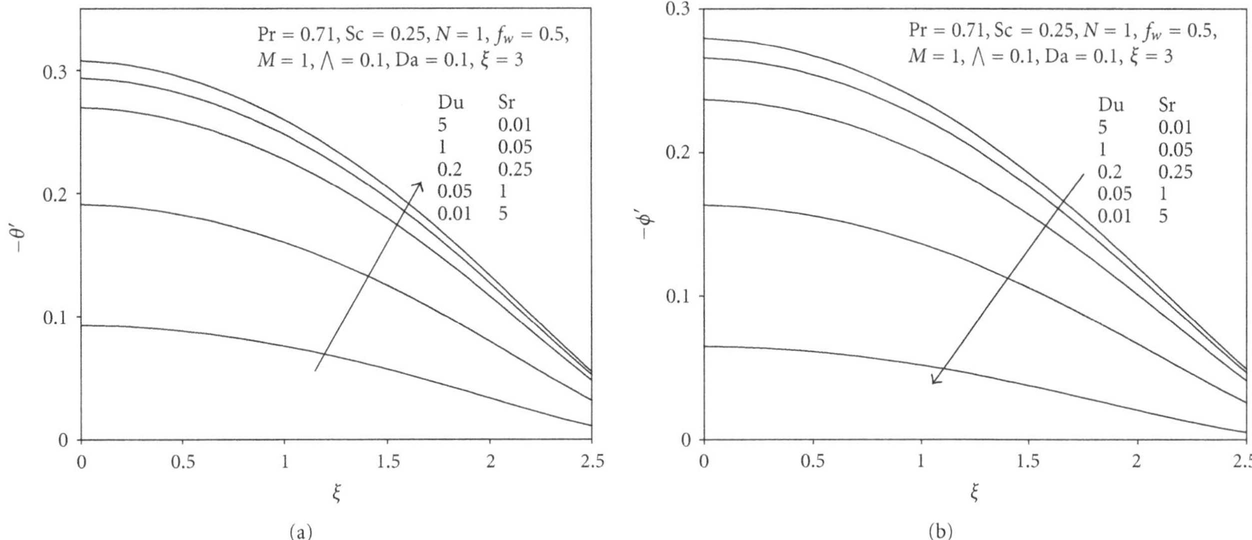

FIGURE 12: (a) Local Nusselt number results for various values of Sr and Du. (b) Local Sherwood number results for various values of Sr and Du.

field since greater momentum diffusion aids the advection of mass in the flow. In chemical engineering design applications (e.g., chromatographical transport phenomena), therefore to achieve a better distribution of species across the boundary layer (transverse to the sphere surface), higher Prandtl number liquids will be more logical than lower Prandtl numbers. The interplay of momentum, thermal, and species diffusion will imply inevitably that all three variables cannot be simultaneously maximized, irrespective of the magnitude of the buoyancy forces involved. A strategic approach is therefore required in selecting primary fluids which posses high thermal diffusivities (low Prandtl numbers) for temperature enhancement or low thermal diffusivities (high Prandtl numbers) for species diffusion enhancement. Such aspects would clearly require more rigorous experimental observations to which the present theoretical and numerical study is complimentary.

Figures 9(a) to 9(c) depict the distributions of velocity, temperature and concentration with wall transpiration parameter (f_w). We have restricted attention here only to the case of suction (lateral mass withdrawal through the sphere surface out of boundary layer regime). With an increase in suction ($f_w > 0$), the velocity is clearly decreased, that is, flow decelerated. Velocity is therefore maximized for the solid sphere case ($f_w = 0$). Increasing suction causes the boundary layer to adhere closer to the flow and destroys momentum transfer; it is therefore an excellent control mechanism and has been exploited in numerous technologies including aerodynamics where flow control is imperative [47]. Temperature, θ, and concentration, ϕ, as depicted in Figures 9(b) and 9(c), respectively, are also markedly stifled with increased suction at the sphere wall and depressed with increased suction. The temperature and concentration profiles, once again assume a continuous decay from the wall to the free stream, whereas the velocity field (Figure 9(a)) initially ascends, peaks, and then decays into the free stream. The strong influence of wall transpiration on all the flow variables is clearly identified and again such a mechanism (as with magnetic field, porous media drag forces and buoyancy forces) discussed earlier is greatly beneficial in allowing flow control and regulation of heat and mass transfer characteristics in such regimes.

Figures 10(a) to 10(b) depict the velocity, temperature, and species concentration distributions transverse to the sphere wall for various streamwise coordinate values, ξ. Velocity is clearly decelerated with increasing migration from the leading edge, that is, larger ξ values (Figure 10(a)) for some distance into the boundary layer, transverse to the wall ($\eta \sim 20$). However, closer to the free stream, this effect is reversed and the flow is accelerated with increasing distance along the sphere surface. Conversely a very strong increase in temperature (θ) and concentration (ϕ), as shown in Figures 10(b) and 10(c), occurs with increasing ξ values. Also unlike the velocity response which ascends from the surface of the spheres, and peaks and then decreases further into the boundary layer (Figure 10(a)), the temperature and concentration fields both decrease continuously across the boundary layer transverse to the wall. Temperature and concentration are both minimized at the leading edge and maximized with the greatest distances along the sphere surface from the leading edge.

The influence of buoyancy ratio, N, on skin friction, that is, local shear stress at the sphere surface is presented in Figures 11(a) and 11(b) for negative and positive, respectively. Positive N values clearly accentuate $-f''$ values, whereas negative N values cause a depression in skin friction. As elucidated earlier, aiding buoyancy forces ($N > 0$) serve to accelerate the flow and this will increase skin friction at the sphere wall; opposing thermal and species buoyancy forces will exert the opposite effect and cause a flow deceleration leading to a decay in skin friction magnitudes.

Figure 12(a) shows the variation of local Nusselt number, $-\theta'$ $(\xi, 0)$ with the combined effects of Soret and Dufour number. Increasing Soret number (Sr) and simultaneously reducing Dufour (Du) number greatly boosts the local heat transfer rate at the sphere surface. With increasing distance from the leading edge ($\xi = 0$), however, the profiles all decrease.

Figure 12(b) presents the local Sherwood number distribution with streamwise coordinate, ξ, for various Soret and Dufour numbers. Increasing Dufour number and decreasing Soret number strongly enhance the mass transfer rate at the wall, that is, boosts $-\phi'$ $(\xi, 0)$ values. Generally with greater distance along the sphere surface, that is, with increasing ξ values, the local Sherwood number decreases. However, for very low Dufour numbers (Du = 0.05, 0.01) and very high Soret numbers (Sr =1.0 and 5.0, resp.) there is a slight upturn in $-\phi'$ $(\xi, 0)$ at large distances from the leading edge. The species cross-diffusion term, Du ϕ'', in the energy equation (14) and the temperature cross-diffusion term, Sr θ'', in the species equation (15) clearly exert a significant influence on both heat transfer and mass transfer rates at the sphere surface in the porous media regime and should not be ignored in advanced studies of importance in materials processing.

5. Conclusions

A detailed mathematical study of the steady, laminar, incompressible hydromagnetic buoyancy-driven convective boundary layer heat and mass transfer from a spherical body immersed in a saturated non-Darcy porous medium with Soret/Dufour effects has been conducted. Numerical solutions have been obtained for the normalized conservations equations. It has been shown that increasing magnetic field generally decelerates the flow but increases temperatures and concentration values in the regime. Increasing porosity serves to accelerate the flow but reduce temperatures and concentration values. Increasing Forchheimer (second order) porous form drag tends to strongly retard the flow but enhance temperatures and concentration values. The present numerical code based on the robust, implicit Keller-box finite difference method has been shown to produce excellent results. Very good correlation between the present computations and the trends of other previous studies has been identified. Further investigations will consider transient effects and also employ the Keller-box method to consider more complex chemical engineering phenomena including electrophoretic deposition, nanofluids, and thermophoresis.

Acknowledgment

The authors are grateful to both reviewers for their constructive comments which have helped to improve the present paper.

References

[1] S. Yesilyurt, L. Vujisic, S. Motakef, F. R. Szofran, and M. P. Volz, "Numerical investigation of the effect of thermoelectromagnetic convection (TEMC) on the Bridgman growth of Ge1-xSix," *Journal of Crystal Growth*, vol. 207, no. 4, pp. 278–291, 1999.

[2] J. C. T. Eijkel, C. Dalton, C. J. Hayden, J. P. H. Burt, and A. Manz, "A circular ac magnetohydrodynamic micropump for chromatographic applications," *Sensors and Actuators, B*, vol. 92, no. 1-2, pp. 215–221, 2003.

[3] I. Iliuta and F. Larachi, "Magnetohydrodynamics of trickle bed reactors: mechanistic model, experimental validation and simulations," *Chemical Engineering Science*, vol. 58, no. 2, pp. 297–307, 2003.

[4] T. L. Sanders, *Magnetohydraulic flow through a packed bed of electrically conducting spheres [Ph.D. thesis]*, University of Texas at Austin, Austin, Tex, USA, 1985.

[5] T.-B. Chang, O. Anwar Bég, and E. Kahya, "Numerical study of laminar incompressible velocity and magnetic boundary layers along a flat plate with wall effects," *International Journal of Applied Mathematics and Mechanics (IJAMM)*, vol. 6, pp. 99–118, 2010.

[6] S. Rawat, R. Bhargava, R. Bhargava, and O. Anwar Bég, "Transient magneto-micropolar free convection heat and mass transfer through a non-Darcy porous medium channel with variable thermal conductivity and heat source effects," *Proceedings of the Institution of Mechanical Engineers, Part C*, vol. 223, no. 10, pp. 2341–2355, 2009.

[7] O. Anwar Bég, A. Y. Bakier, and V. R. Prasad, "Numerical study of free convection magnetohydrodynamic heat and mass transfer from a stretching surface to a saturated porous medium with Soret and Dufour effects," *Computational Materials Science*, vol. 46, no. 1, pp. 57–65, 2009.

[8] O. D. Makinde and O. Anwar Bég, "On inherent irreversibility in a reactive hydromagnetic channel flow," *Journal of Thermal Science*, vol. 19, no. 1, pp. 72–79, 2010.

[9] S. K. Ghosh, O. Anwar Bég, and M. Narahari, "Hall effects on MHD flow in a rotating system with heat transfer characteristics," *Meccanica*, vol. 44, no. 6, pp. 741–765, 2009.

[10] O. Anwar Bég, J. Zueco, R. Bhargava, and H. S. Takhar, "Magnetohydrodynamic convection flow from a sphere to a non-Darcian porous medium with heat generation or absorption effects: network simulation," *International Journal of Thermal Sciences*, vol. 48, no. 5, pp. 913–921, 2009.

[11] R. Bhargava, R. Sharma, and O. Anwar Bég, "A numerical solution for the effect of radiation on micropolar flow and heat transfer past a horizontal stretching sheet through porous medium," in *Proceedings of the 5th IASME/WSEAS International Conference on Continuum Mechanics (CM '10)*, University of Cambridge, Cambridge, UK, February 2010.

[12] O. Anwar Bég, A. Y. Bakier, V. R. Prasad, J. Zueco, and S. K. Ghosh, "Nonsimilar, laminar, steady, electrically-conducting forced convection liquid metal boundary layer flow with induced magnetic field effects," *International Journal of Thermal Sciences*, vol. 48, no. 8, pp. 1596–1606, 2009.

[13] O. D. Makinde, O. Anwar Bég, and H. S. Takhar, "Magnetohydrodynamic viscous flow in a rotating porous medium cylindrical annulus with an applied radial magnetic field," *Journal of Applied Mathematics and Mechanics*, vol. 5, no. 6, pp. 68–81, 2009.

[14] S. K. Ghosh, O. Anwar Bég, J. Zueco, and V. R. Prasad, "Transient hydromagnetic flow in a rotating channel permeated by an inclined magnetic field with magnetic induction

and Maxwell displacement current effects," *Zeitschrift fur Angewandte Mathematik und Physik*, vol. 61, no. 1, pp. 147–169, 2010.

[15] O. Anwar Bég, S. K. Ghosh, and M. Narahari, "Mathematical modeling of oscillatory MHD couette flow in a rotating highly permeable medium permeated by an oblique magnetic field," *Chemical Engineering Communications*, vol. 198, no. 2, pp. 235–254, 2011.

[16] N. Rudraiah, B. K. Ramaiah, and B. M. Rajasekhar, "Hartmann flow over a permeable bed," *International Journal of Engineering Science*, vol. 13, no. 1, pp. 1–24, 1975.

[17] O. Lioubashevski, E. Katz, and I. Willner, "Magnetic field effects on electrochemical processes: a theoretical hydrodynamic model," *Journal of Physical Chemistry B*, vol. 108, no. 18, pp. 5778–5784, 2004.

[18] N. G. B. Boum and A. Alemany, "Numerical simulations of electrochemical mass transfer in electromagnetically forced channel flows," *Electrochimica Acta*, vol. 44, no. 11, pp. 1749–1760, 1999.

[19] T. Z. Fahidy, "On the magnetohydrodynamics of natural convective diffusion boundary layers in coupled horizontal electric and magnetic fields," *Chemical Engineering Journal*, vol. 72, no. 1, pp. 79–82, 1999.

[20] M. A. Al-Nimr and M. A. Hader, "MHD free convection flow in open-ended vertical porous channels," *Chemical Engineering Science*, vol. 54, no. 12, pp. 1883–1889, 1999.

[21] J. D. McWhirter, M. E. Crawford, D. E. Klein, and T. L. Sanders, "Model for inertialess magnetohydrodynamic flow in packed beds," *Fusion Technology*, vol. 33, no. 1, pp. 22–30, 1998.

[22] S. K. Dahikar and R. L. Sonolikar, "Influence of magnetic field on the fluidization characteristics of circulating fluidized bed," *Chemical Engineering Journal*, vol. 117, no. 3, pp. 223–229, 2006.

[23] S. Alchaar, P. Vasseur, and E. Bilgen, "Effect of an electromagnetic field on natural convection in a porous medium," in *Proceedings of the 7th International Symposium on Transport Phenomena in Manufacturing Processes,*, pp. 275–280, Acapulco, Mexico, 1994.

[24] M. A. Mansour, M. A. El-Hakiem, and S. M. El Kabeir, "Heat and mass transfer in magnetohydrodynamic flow of micropolar fluid on a circular cylinder with uniform heat and mass flux," *Journal of Magnetism and Magnetic Materials*, vol. 220, no. 2, pp. 259–270, 2000.

[25] A. Postelnicu, "Influence of a magnetic field on heat and mass transfer by natural convection from vertical surfaces in porous media considering Soret and Dufour effects," *International Journal of Heat and Mass Transfer*, vol. 47, no. 6-7, pp. 1467–1472, 2004.

[26] O. Anwar Bég, R. Bhargava, S. Rawat, K. Halim, and H. S. Takhar, "Computational modeling of biomagnetic micropolar blood flow and heat transfer in a two-dimensional non-Darcian porous medium," *Meccanica*, vol. 43, no. 4, pp. 391–410, 2008.

[27] O. Anwar Bég, J. Zueco, and H. S. Takhar, "Laminar free convection from a continuously-moving vertical surface in thermally-stratified non-Darcian high-porosity medium: network numerical study," *International Communications in Heat and Mass Transfer*, vol. 35, no. 7, pp. 810–816, 2008.

[28] G. W. Sutton and A. S. Sherman, *Engineering Magnetohydrodynamics*, MacGraw-Hill, New York, NY, USA, 1965.

[29] S. I. Pai, *Magnetogasdynamics and Plasma Dynamics*, Springer, Berlin, Germany, 1962.

[30] W. F. Hughes and F. J. Young, *The Electromagnetodynamics of Fluids*, John Wiley & Sons, New York, NY, USA, 1966.

[31] H. B. Keller, "A new difference method for parabolic problems," in *Numerical Methods for Partial Differential Equations*, J. Bramble, Ed., 1970.

[32] T. Cebeci and P. Bradshaw, *Physical and Computational Aspects of Convective Heat Transfer*, Springer, New York, NY, USA, 1984.

[33] H. S. Takhar, O. Anwar Bég, and M. Kumari, "Computational analysis of coupled radiation-convection dissipative non-gray gas flow in a non-darcy porous medium using the keller-box implicit difference scheme," *International Journal of Energy Research*, vol. 22, no. 2, pp. 141–159, 1998.

[34] H. S. Takhar and O. Anwar Bég, "Effects of transverse magnetic field, prandtl number and reynolds number on non-darcy mixed convective flow of an incompressible viscous fluid past a porous vertical flat plate in a saturated porous medium," *International Journal of Energy Research*, vol. 21, no. 1, pp. 87–100, 1997.

[35] O. Anwar Bég, V. Prasad, H. S. Takhar, and V. M. Soundalgekar, "Thermoconvective flow in a saturated, isotropic, homogeneous porous medium using Brinkman's model: numerical study," *International Journal of Numerical Methods for Heat and Fluid Flow*, vol. 8, no. 5-6, pp. 559–589, 1998.

[36] H. S. Takhar, O. Anwar Bég, and M. Kumari, "Computational analysis of coupled radiation-convection dissipative non-gray gas flow in a non-darcy porous medium using the keller-box implicit difference scheme," *International Journal of Energy Research*, vol. 22, no. 2, pp. 141–159, 1998.

[37] A. J. Chamkha, H. S. Takhar, and O. Anwar Bég, "Radiative free convective non-newtonian fluid flow past a wedge embedded in a porous medium," *International Journal of Fluid Mechanics Research*, vol. 31, no. 2, pp. 101–115, 2004.

[38] O. Anwar Bég, H. S. Takhar, G. Nath, and M. Kumari, "Computational fluid dynamics modeling of buoyancy-induced viscoelastic flow in a porous medium under transverse magnetic field," *International Journal of Applied Mechanics and Engineering*, vol. 6, no. 1, pp. 187–210, 2001.

[39] V. Ramachandra Prasad, B. Vasu, and O. Anwar Bég, "Numerical modeling of transient dissipative radiation free convection heat and mass transfer from a non-isothermal cone with variable surface conditions," *Elixir-Applied Mathematics*, vol. 41, pp. 5592–5603, 2011.

[40] V. R. Prasad, B. Vasu, O. Anwar Bég, and R. D. Prashad, "Unsteady free convection heat and mass transfer in a Walters-B viscoelastic flow past a semi-infinite vertical plate: a numerical study," *Thermal Science*, vol. 15, no. 2, supplement, pp. S291–S305, 2011.

[41] V. R. Prasad, B. Vasu, O. Anwar Bég, and D. R. Parshad, "Thermal radiation effects on magnetohydrodynamic free convection heat and mass transfer from a sphere in a variable porosity regime," *Communications in Nonlinear Science and Numerical Simulation*, vol. 17, no. 2, pp. 654–671, 2012.

[42] K. Vafai and C. L. Tien, "Boundary and inertia effects on convective mass transfer in porous media," *International Journal of Heat and Mass Transfer*, vol. 25, no. 8, pp. 1183–1190, 1982.

[43] J. Bear, *Dynamics of Fluids in Porous Media*, Dover, New York, NY, USA, 1988.

[44] H. Belhaj, J. Biazar, S. Butt, and R. Islam, "Adomian solution of Forchheimer model to describe porous media flow," in *Proceedings of the SPE/DOE Symposium on Improved Oil Recovery*, Tulsa, Okla, USA, April 2004.

[45] S. Whitaker, "The Forchheimer equation: a theoretical development," *Transport in Porous Media*, vol. 25, no. 1, pp. 27–61, 1996.
[46] Y. Qin and J. Chadam, "Nonlinear convective stability in a porous medium with temperature-dependent viscosity and inertial drag," *Studies in Applied Mathematics*, vol. 96, no. 3, pp. 273–288, 1996.
[47] H. Schlichting, *Boundary-Layer Theory*, MacGraw-Hill, New York, NY, USA, 8th edition, 2000.

18

Stagnation Point Flow of a Nanofluid toward an Exponentially Stretching Sheet with Nonuniform Heat Generation/Absorption

A. Malvandi,[1] F. Hedayati,[2] and G. Domairry[3]

[1] *Department of Mechanical Engineering, Amirkabir University of Technology (Tehran Polytechnic), 424 Hafez Avenue, P.O. Box 15875-4413, Tehran, Iran*
[2] *Department of Mechanical Engineering, Islamic Azad University, Sari Branch, Sari, Iran*
[3] *Department of Mechanical Engineering, Babol Noshirvani University of Technology, Babol 47148-71167 Iran*

Correspondence should be addressed to A. Malvandi; amirmalvandi@aut.ac.ir

Academic Editor: Mohammad Al-Nimr

This paper deals with the steady two-dimensional stagnation point flow of nanofluid toward an exponentially stretching sheet with nonuniform heat generation/absorption. The employed model for nanofluid includes two-component four-equation nonhomogeneous equilibrium model that incorporates the effects of Brownian diffusion and thermophoresis simultaneously. The basic partial boundary layer equations have been reduced to a two-point boundary value problem via similarity variables and solved analytically via HAM. Effects of governing parameters such as heat generation/absorption λ, stretching parameter ε, thermophoresis N_t, Lewis number Le, Brownian motion N_b, and Prandtl number Pr on heat transfer and concentration rates are investigated. The obtained results indicate that in contrast with heat transfer rate, concentration rate is very sensitive to the abovementioned parameters. Also, in the case of heat generation $\lambda > 0$, despite concentration rate, heat transfer rate decreases. Moreover, increasing in stretching parameter leads to a gentle rise in both heat transfer and concentration rates.

1. Introduction

For years, many researchers have paid much attention to viscous fluid motion near the stagnation region of a solid body, where "body" corresponds to either fixed or moving surfaces in a fluid. This multidisciplinary flow has frequent applications in high speed flows, thrust bearings, and thermal oil recovery. Hiemenz [1] developed the first investigation in this field. He applied similarity transformation to collapse two-dimensional Navier-Stokes equations to a nonlinear ordinary differential one and then presented its exact solution. Extension of this study was carried out with a similarity solution by Homann [2] to the case of axisymmetric three-dimensional stagnation point flow. After these original studies, many researchers have put their attention on this subject [3–9]. Besides stagnation point flow, stretching surfaces have a wide range of applications in engineering and several technical purposes particularly in metallurgy and polymer industry, for instance, gradual cooling of continuous stretched metal or plastic strips which have multiple applications in mass production. Crane [10] was the first to present a similarity solution in the closed analytical form for steady two-dimensional incompressible boundary layer flow caused by the stretching plate whose velocity varies linearly with the distance from a fixed point on the sheet. The combination of stretching surface and stagnation point flow was analyzed by Yao et al. [11]. Different types fluids such as viscoelastic [12] or micropolar ones [13] past a stretching sheet have been studied later. The popularity of stretching surfaces can be gauged from the researches done by scientists for its frequent applications and can be found in the literature, for example, [14–19]. Recently, Nadeem and Lee [20] have investigated the boundary layer over exponentially stretching surfaces analytically.

Improving the technology, limit in enhancing the performance of conventional heat transfer is a main issue [21]

owing to low thermal conductivity of the most common fluids such as water, oil, and ethylene-glycol mixture. Since the thermal conductivity of solids is often higher than that of liquids, the idea of adding particles to a conventional fluid to enhance its heat transfer characteristics emerged. Among all dimensions of particles such as macro, micro, and nano, due to some obstacles in the pressure drop through the system or keeping the mixture homogenous, nanoscaled particles have attracted more attention. These tiny particles are fairly close in size to the molecules of the base fluid and thus can realize extremely stable suspensions with slight gravitational settling over long periods of time. Coined term "nanofluid" was proposed by Choi [22] to point out engineered colloids composed of nanoparticles dispersed in a base fluid. Following the seminal study of this concept by Masuda et al. [23], a considerable amount of research in this field has risen exponentially. Meanwhile, theoretical studies emerged to model the nanofluids behaviors for which the proposed models are twofold: homogeneous flow models and dispersion models. Buongiorno [24] indicated that the homogeneous models tend to underpredict the nanofluid heat transfer coefficient, and due to nanoparticle size, the dispersion effect is completely negligible. Hence, Buongiorno developed an alternative model to explain the abnormal convective heat transfer enhancement in nanofluids and eliminate the shortcomings of the homogenous and dispersion models. He considers seven slip mechanisms, including inertia, Brownian diffusion, thermophoresis, diffusiophoresis, Magnus, fluid drainage, and gravity, and claimed that, of these seven, only Brownian diffusion and thermophoresis are important slip mechanisms in nanofluids. Moreover, Buongiorno concluded that turbulence is not affected by nanoparticles. Based on this finding, he proposed a two-component four-equation nonhomogeneous equilibrium model for convective transport in nanofluids. The aforementioned model has recently been used by Kuznetsov and Nield [25] to study the influence of nanoparticles on natural convection boundary layer flow past a vertical plate. Then, a comprehensive survey of convective transport of nanofluids in the boundary layer flow was conducted by Khan and Pop [26], Bachok et al. [27], Alsaedi et al. [28], and Rana and Bhargava [29]. Some very recent review papers are written by Daungthongsuk and Wongwises [30], Wang and Mujumdar [31], and Kakaç and Pramuanjaroenkij [32].

In order to study the aforementioned issues, there is a certain need to model them mathematically and solve them with an appropriate technique. For years, numerical approaches have been developed, but due to some restrictions [33], analytical solutions have been considered as alternative ways by scientists. Perturbation technique is one of the most common methods in this field which is widely applied in science and engineering [34]. A drawback of perturbation techniques is that they strongly depend upon small/large physical parameters, so they cannot be applied to strongly nonlinear problems. Hence, nonperturbation techniques such as Adomian decomposition method [35] and variational iteration method [36–38] appear in order to omit the dependency on small/large parameters. It must be noted that these methods cannot ensure the convergence of series solution. On the other hand, the homotopy analysis method (HAM) proposed by Liao [39–41] is a general analytical approach to get series solutions of strongly nonlinear equations [42–44] with great freedom options to ensure the convergence of solutions series. Moreover, in contrast with numerical methods, it can be implemented with far field boundary conditions. Needless to say that for boundary layer problems, the physical domain is unbounded, whereas the computational domain has to be finite, so similarity variable at infinity must be evaluated with the aid of previous studies [45]. Newly, Liao [46] has presented the optimal HAM to guarantee the convergence of solution. He defined a new kind of averaged residual error to find the optimal convergence-control parameters which can accelerate the convergence of series.

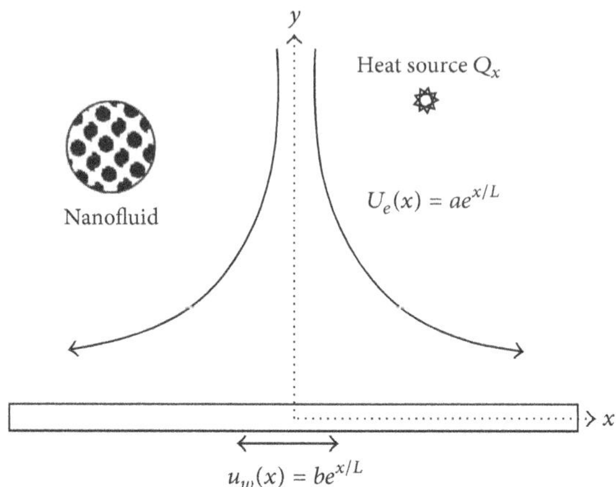

FIGURE 1: Geometry of physical model and coordinates system.

This paper deals with the analytical study of boundary layer stagnation point flow of nanofluid toward an exponentially stretching surface with nonuniform heat generation/absorption which is the extension of Hassani and coworkers' study [47] and the mentioned Nadeem and Lees' one [20]. The studied model incorporates the effects of suction injection parameter α, Lewis number Le, the Brownian motion parameter N_b, and thermophoresis parameter N_t. It is hoped that the obtained results will not only present useful information for applications, but also serve as a complement to the previous studies.

2. Governing Equations

Consider the steady laminar two-dimensional flow of nanofluids near the stagnation point at a stretching sheet in the presence of heat generation/absorption as shown in Figure 1. The coordinates x and y are taken with the origin O at the stagnation point. Two opposite forces are applied along the x-axis similarly so that the wall is stretched whilst keeping the position of the origin fixed. The free stream fluid's velocity and the stretching one are assumed to vary nonlinearly, which corresponded to $U_e(x) = ae^{x/L}$ and $u_w(x) = be^{x/L}$, respectively. It is to be said that a and b are constants and always positive; that is, $a, b > 0$. It is also assumed that the temperature and concentration at the surface have constant values of T_w and C_w, respectively, while the ambient temperature and concentration beyond boundary layer have constant values T_∞ and C_∞, respectively. The continuity, momentum, and energy equations in the Cartesian coordinates for this flow can be expressed as [48]

$$\frac{\partial u}{\partial x} + \frac{\partial v}{\partial y} = 0, \quad (1)$$

$$u\frac{\partial u}{\partial x} + v\frac{\partial u}{\partial y} = U_e\frac{dU_e}{dx} + \nu\frac{\partial^2 u}{\partial y^2}, \quad (2)$$

$$u\frac{\partial T}{\partial x} + v\frac{\partial T}{\partial y} = \alpha\frac{\partial^2 T}{\partial y^2} + \tau\left[D_B\frac{\partial C}{\partial y}\frac{\partial T}{\partial y} + \frac{D_T}{T_\infty}\left(\frac{\partial T}{\partial y}\right)^2\right] + \frac{Q_x}{\rho c_p}\left(T - T_\infty\right), \quad (3)$$

$$u\frac{\partial C}{\partial x} + v\frac{\partial C}{\partial y} = D_B\frac{\partial^2 C}{\partial y^2} + \frac{D_T}{T_\infty}\frac{\partial^2 T}{\partial y^2}, \quad (4)$$

subject to the boundary conditions

$$\begin{aligned} &v = 0, \quad u = be^{x/L}, \quad T = T_w, \quad C = C_\infty \quad \text{at } y = 0, \\ &u = u_e(x) = ae^{x/L}, \quad T = T_\infty, \quad C = C_\infty \quad \text{as } y \longrightarrow \infty. \end{aligned} \quad (5)$$

Here, u and v are the velocity components along the x- and y-directions, respectively, and T is the temperature. D_B is the Brownian diffusion coefficient, D_T is the thermophoretic diffusion coefficient, τ is the ratio between the effective heat capacity of the nanoparticle material and heat capacity of the fluid, μ_{nf} is the viscosity of nanofluid, $Q_x = \lambda(b/2L)e^{x/L}$ is the nonuniform heat generation/absorption where $\lambda > 0$ and $\lambda < 0$ stand for heat generation and absorption, respectively, ρ_{nf} is the density of nanofluid, and α_{nf} is the thermal diffusivity of nanofluid. In order to find a similarity solution of (1)–(4), we employed the following dimensionless parameters:

$$\begin{aligned} &\eta = \sqrt{\frac{b}{2\nu L}}e^{x/2L}y, \qquad \psi = \sqrt{2b\nu L}e^{x/2L}f(\eta), \\ &\theta(\eta) = \frac{T - T_\infty}{T_w - T_\infty}, \qquad C(\eta) = \frac{C - C_\infty}{C_w - C_\infty}. \end{aligned} \quad (6)$$

Here, η is the similarity variable and ψ is the usual stream function; that is, $u = \partial\psi/\partial y$ and $v = -\partial\psi/\partial x$. The governing equations (1)–(3) and the boundary conditions of (5) collapse into

$$f''' + ff'' + 2\left(\varepsilon^2 - f'^2\right) = 0, \quad (7)$$

$$\theta'' + \Pr\theta'\left(N_b\phi' + N_t\theta' + f\right) + \Pr\lambda\theta = 0, \quad (8)$$

$$\phi'' + \text{Le}\,f\theta' + \frac{N_t}{N_b}\theta'' = 0. \quad (9)$$

The appropriate dimensionless forms of boundary condition of (5) are

$$\begin{aligned} &\text{At } \eta = 0: \quad f = 0, \quad f' = 1, \quad \theta = 1, \quad \phi = 1 \\ &\text{At } \eta = \infty: \quad f'(\eta) = \varepsilon, \quad \theta(\eta) = 0, \quad \phi(\eta) = 0, \end{aligned} \quad (10)$$

where prime denotes differentiation with respect to η and Pr, Le, N_b, N_t, and ε denote Prandtl number, Lewis number, Brownian motion, thermophoresis, and stretching parameter, respectively. The physical quantities of interest in this study are

$$\begin{aligned} \text{Cf}_x &= \frac{\tau_w|_{y=0}}{\rho U_e^{\,2}}, \\ \text{Nu}_x &= -\frac{x}{(T_w - T_\infty)}\left.\frac{\partial T}{\partial y}\right|_{y=0}, \\ \text{Sh}_x &= -\frac{x}{(C_w - C_\infty)}\left.\frac{\partial C}{\partial y}\right|_{y=0} \end{aligned} \quad (11)$$

and can be expressed as

$$\begin{aligned} \sqrt{2\text{Re}_x}\text{Cf}_x &= f''(0), \\ \frac{\text{Nu}_x}{\sqrt{2\text{Re}_x}} &= -\sqrt{\frac{x}{2L}}\theta'(0), \\ \frac{\text{Sh}_x}{\sqrt{2\text{Re}_x}} &= -\sqrt{\frac{x}{2L}}\phi'(0), \end{aligned} \quad (12)$$

where $\text{Re}_x = u_w x/\nu$ is the local Reynolds number.

3. Analytical Solution by Homotopy Analysis Method

In order to solve the equations by means of HAM, we have to choose the initial guesses and auxiliary linear operators which are assumed as follows:

$$\begin{aligned} f_0(\eta) &= \varepsilon\eta + (1 - \varepsilon)\left(1 - e^{-\eta}\right), \\ \theta_0(\eta) &= e^{-\eta}, \\ \phi_0(\eta) &= e^{-\eta}, \end{aligned} \quad (13)$$

$$\begin{aligned} L(f) &= f''' - f', \\ L(\theta) &= \theta'' - \theta, \\ L(\phi) &= \phi'' - \phi. \end{aligned} \quad (14)$$

Then, we have proceeded the solution of HAM for which, for the sake of brevity, the details are skipped here and can be found in [49–51]. As pointed out by Liao [40], the convergence rate of approximation for the HAM solution strongly depends on the value of auxiliary parameter, c_i ($i = 1, 2, 3$). In order to seek the permissible values of c_i, the considered functions have to be plotted for a specific physical point at an appropriate order of approximations. Obtaining the suitable

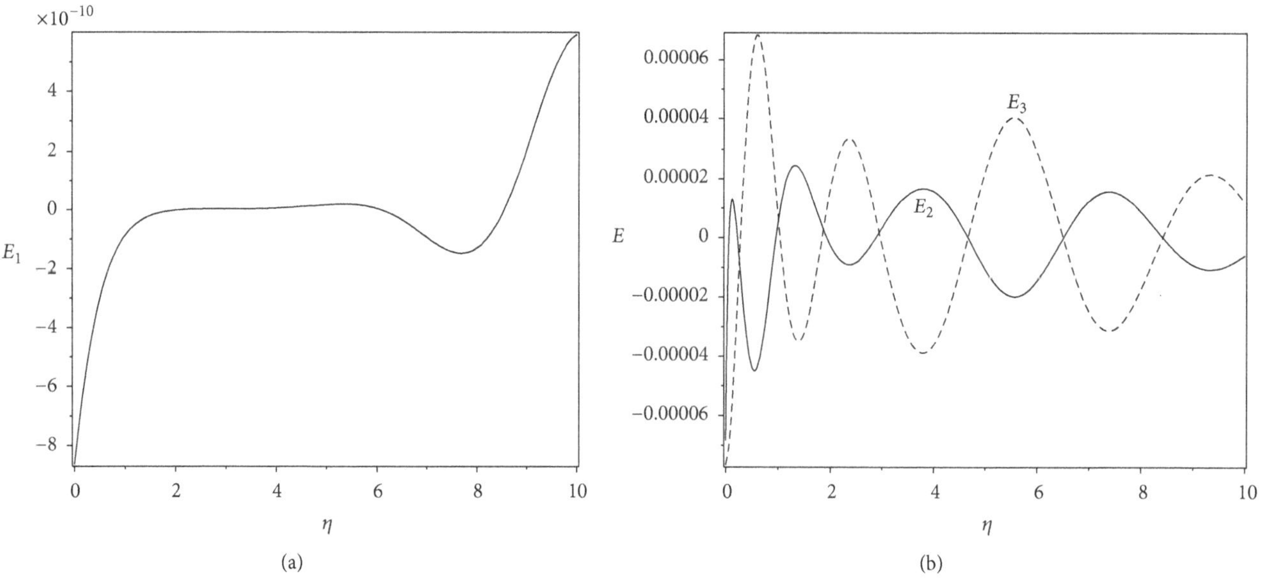

FIGURE 2: Average residual errors which are defined by (15).

c_i, we followed Liao and defined the averaged residual errors to find the optimal convergence-control parameters

$$E_1 = \int_0^{\infty} \left(N\left[\sum_{i=0}^{m} F_i(\xi) \right] \right)^2 d\xi,$$
$$E_2 = \int_0^{\infty} \left(N\left[\sum_{i=0}^{m} \theta_i(\xi) \right] \right)^2 d\xi, \quad (15)$$
$$E_3 = \int_0^{\infty} \left(N\left[\sum_{i=0}^{m} \phi_i(\xi) \right] \right)^2 d\xi.$$

The more quickly E_i $(i = 1, 2, 3)$ decreases to zero, the faster the corresponding homotopy-series solution converges. So, at a proper order of approximation m, the corresponding optimal values of the convergence-control parameter will be obtained by the minimum values of E_i which can be estimated as

$$\frac{\partial E_1}{\partial c_1} = 0,$$
$$\frac{\partial E_2}{\partial c_2} = 0, \quad (16)$$
$$\frac{\partial E_3}{\partial c_3} = 0.$$

It is worth mentioning that since (8) and (9) are coupled, for minimizing the E_2 and E_3, we should apply either a least square technique or the method introduced by Yabushita et al. [52], where, following [53], we have applied the second one.

4. Result and Discussion

The system of (7)–(9) with boundary conditions of (10) has been solved analytically via homotopy analysis method (HAM). The best accuracy of our results has been obtained and shown in Figure 2 where residual values of (16) are plotted in the solution domain. In addition, for selective parameters, we have compared our analytical outcomes with numerical ones which are cited in Table 1. In comparison with regular fluids, the present study involves three more parameters: Le, N_b, and N_t; hence, contour plots have been presented instead of regular diagrams. The advantage of contour lines is that they illustrate more physical interpretation rather than common plots and are appropriate for problems with many effective parameters. Here, in all contours, the negative values have been shown with dashed lines. Needless to say that higher density of the contour lines demonstrates wide range of variations of the understudied parameter, and for sparse sets of contours, it is vice versa.

N_t-N_b contour lines of the physical interests, that is, the heat transfer and concentration rates for different values of unsteadiness parameter (A), stretching parameter (ε), Lewis number (Le), heat generation/absorption parameter (λ), and Prandtl number (Pr), are shown in Figures 3 and 6. As the eye sees, the behavior of heat transfer rate $-\theta'(0)$ reveals its straightforward dependency on almost all parameters except for higher values of Pr for which variations of N_t and N_b have stronger effects and should be considered in depth. Clearly, increase in the values of N_t and N_b leads to the decrease in the values of heat transfer rate, that is, $-\theta'(0)$. In contrast to heat transfer, we can observe that the concentration rate $-\phi'(0)$ is very sensitive to all parameters and its variations are not predictable at a glance. A rise in N_t at lower values of Le leads to a drop in concentration rate, while for higher Lewis numbers, concentration rate takes an increasing trend except for some values of low thermophoresis and Brownian

TABLE 1: Comparison of our analytical results with numerical results via shooting method.

N_b	Le	$-\theta'(0)$ HAM	$-\theta'(0)$ Numerical	Error %	$-\phi'(0)$ HAM	$-\phi'(0)$ Numerical	Error %
0.1	1	0.644835	0.645023	−0.029148	−1.52771	−1.52922583	−0.099355
	2	0.589582	0.5883122	0.2153212	−0.65237	−0.64663798	0.886434
0.5	1	0.458667	0.4586645	0.0005066	0.312874	0.31278339	0.0289599
	2	0.373846	0.3735257	0.0857389	0.752468	0.75300072	−0.070857
1	1	0.284093	0.2838612	0.0815335	0.514955	0.51509636	−0.027359
	2	0.197857	0.1985839	−0.367558	0.886216	0.88559894	0.0696118

motion. Furthermore, as N_b increases, the $-\phi'(0)$'s variation is dwindled down which is more suppressed as Lewis number increases.

Considering Figure 3, we can observe that the increase in stretching parameter ε causes a gentle rise in both heat transfer and concentration rates. Additionally, in the case of heat absorption, that is, $\lambda < 0$, the values of the heat transfer rate increase while concentration rate decreases; a reversed trend can be observed for the case of heat generation which is supported by Figure 4. According to Figure 5, it can be observed that despite the concentration rate, a rise in Le number decreases the heat transfer rate which is more obvious for lower values of Le. An interesting feature that can be observed in Figure 5 is that with lower Lewis number and Brownian motion, $-\phi'(0)$ is negative; that is, reverse concentration rate occurs. Effects of Pr have been considered in Figure 6 which shows that the increase in Pr leads to a rise in the values of heat transfer rate at lower values of Brownian motion; however, for higher values of Brownian motion, increasing in Prandtl number decreases the heat transfer rate. In contrast, a rise Prandtl number increases the concentration rate especially with higher value of N_t.

Finally, sample profiles of boundary layer including velocity, temperature, and concentration for different values of ε have been presented in Figures 7, 8, and 9. It is noteworthy that these profiles have the same form with regular fluids. Increasing the stretching parameter, due to less difference between the sheet and free stream velocities, the momentum boundary layer gets thinner. This causes an increase of the momentum at the surface. So, it is not surprising that heat transfer and concentration rates climb up both (Figure 3).

5. Conclusion

This paper deals with an analytical study of boundary layer stagnation point flow of nanofluid toward an exponentially stretching surface with nonuniform heat generation/absorption. The governing PDE equations including continuity, momentum, and energy have been transformed into ODE ones with similarity solution and are solved with HAM. The main outcomes of the paper can be summarized as follows.

(i) Heat transfer rate has simple dependency on almost all parameters except for higher values of Pr. Increasing the values of thermophoresis (N_t), Brownian motion (N_b), and Lewis number (Le) results in a reduction in heat transfer rate.

(ii) In contrast to heat transfer rate, concentration rate is sensitive to the parameters of thermophoresis (N_t), Brownian motion (N_b), Lewis (Le), and Prandtl numbers (Pr). A rise in N_t at lower values of Le leads to a drop in concentration rate, while for higher Lewis numbers, concentration rate takes an increasing trend except for some values of low thermophoresis and Brownian motion parameters.

(iii) Heat generation $\lambda > 0$ reduces the heat transfer rate while increasing the amounts of concentration rate. A reversed behavior can be observed for heat absorption $\lambda < 0$. Also, we can see that increasing the stretching parameter ε causes a gentle rise in both heat transfer and concentration rates.

(iv) Increasing Pr leads to a rise in the values of heat transfer rate at lower values of Brownian motion; however, for higher values of Brownian motion, increasing Prandtl number decreases the heat transfer rate.

Nomenclature

a, b: Positive constants
C: Nanoparticle volume fraction
Cf_x: Local skin friction coefficient
D_B: Brownian diffusion coefficient
D_T: Thermophoresis diffusion coefficient
f: Dimensionless stream function
L: Characteristic length
Le: Lewis number
N_b: Brownian motion parameter
N_t: Thermophoresis parameter
Nu_x: Local Nusselt number
p: Pressure
Pr: Prandtl number
Q_x: Nonuniform heat generation/absorption
Re_x: Local Reynolds number
Sh_x: Local Sherwood number
T: Temperature
u, v: Velocity components along the x- and y-directions, respectively
x, y: Cartesian coordinates system.

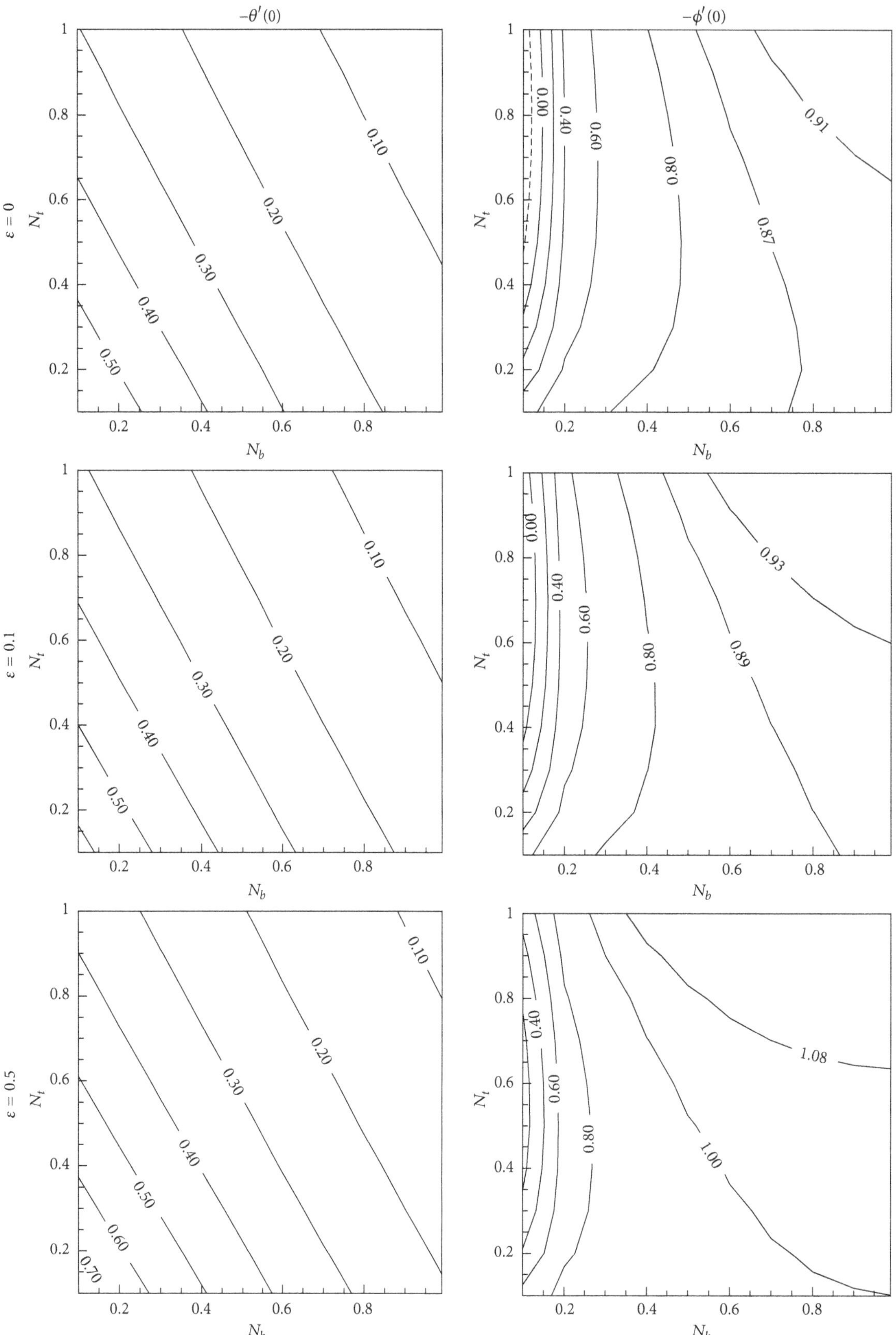

FIGURE 3: Contour lines of the heat transfer and concentration rates for different values of e when Le = Pr = 2, $\lambda = 0.1$.

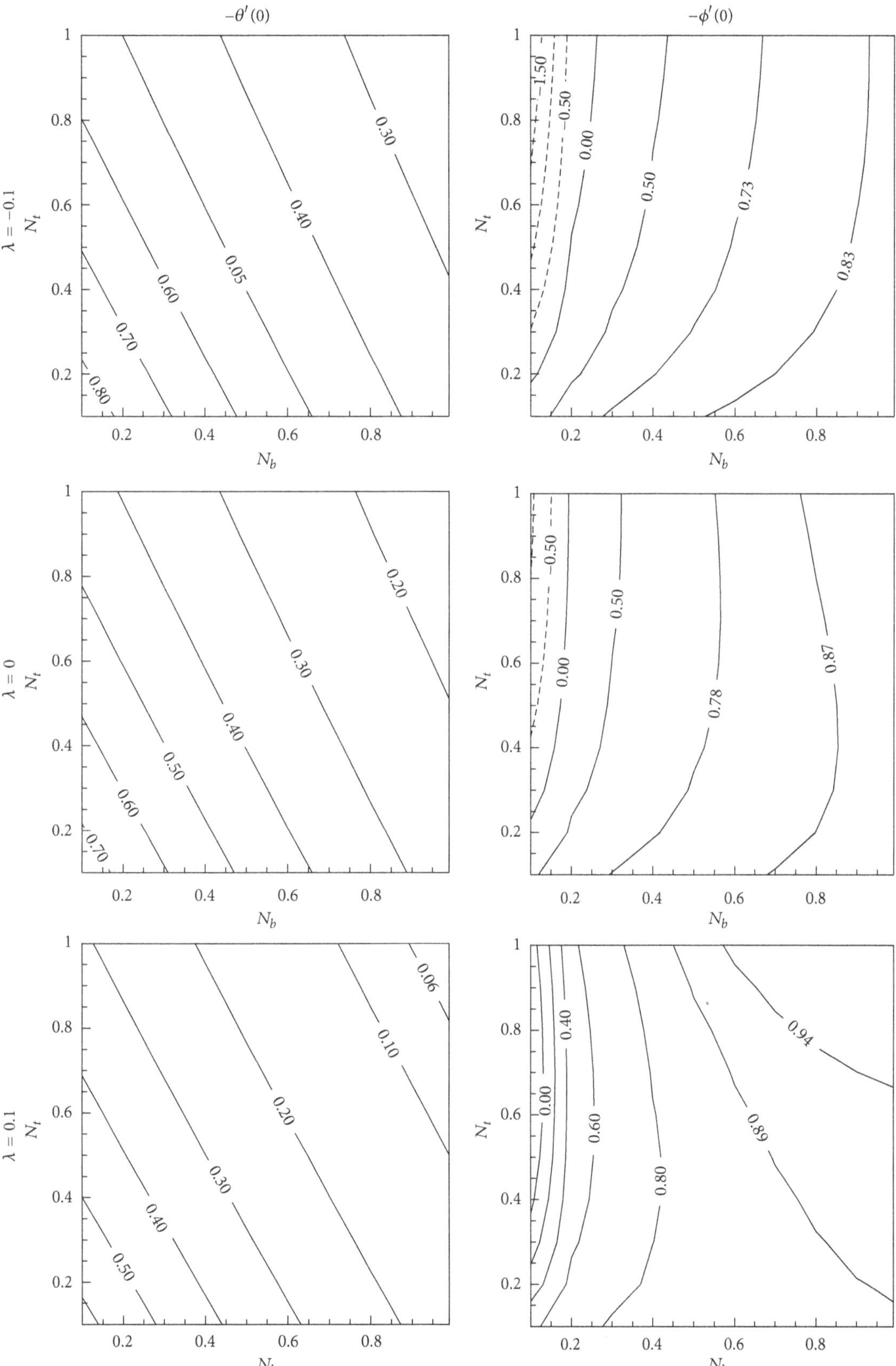

FIGURE 4: Contour lines of heat transfer and concentration rates for different values of λ when Le = Pr = 2, $\varepsilon = 0.1$.

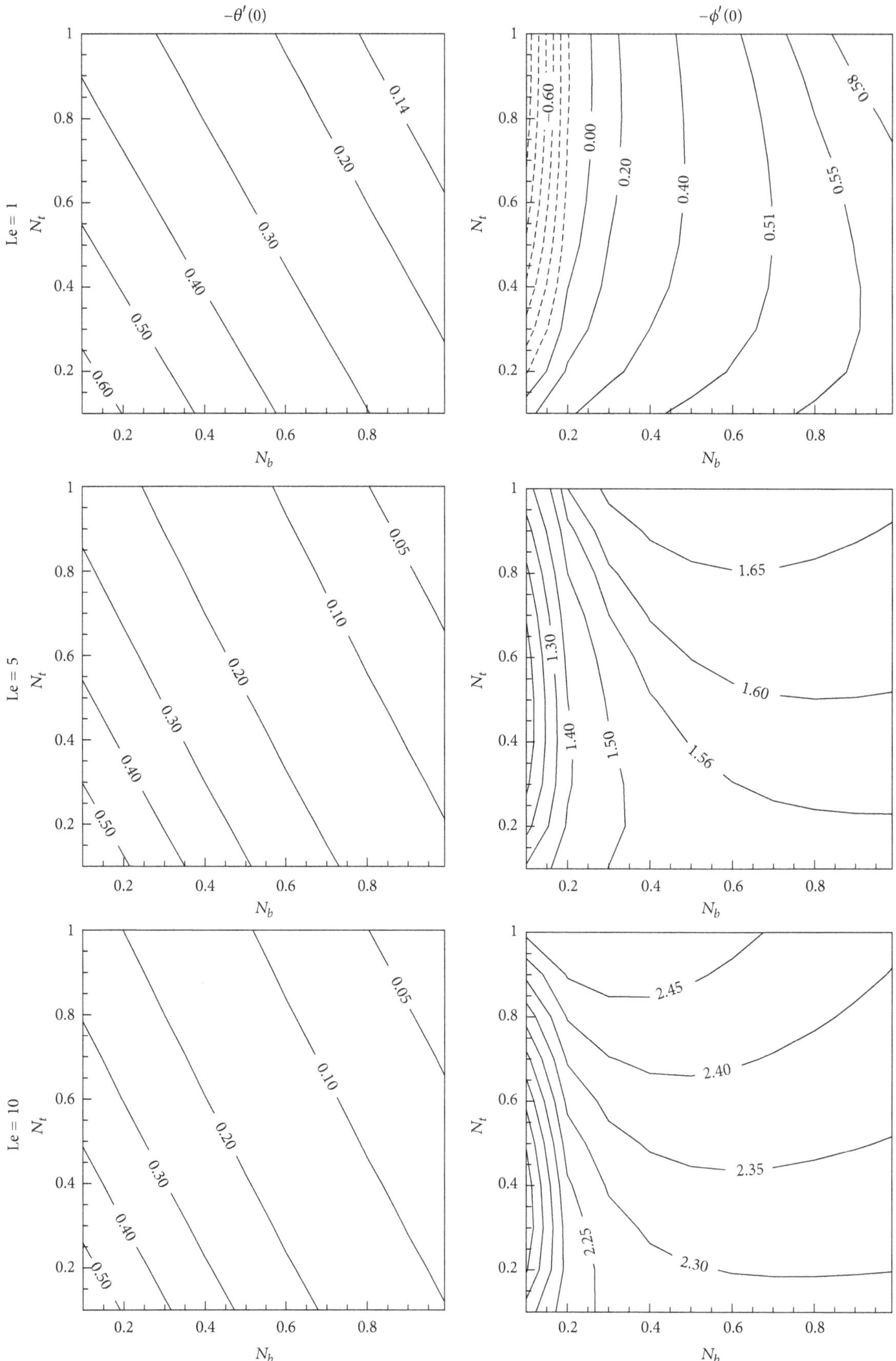

Figure 5: Contour lines of heat transfer and concentration rates for different values of Le when Pr = 2, $\lambda = \varepsilon = 0.1$.

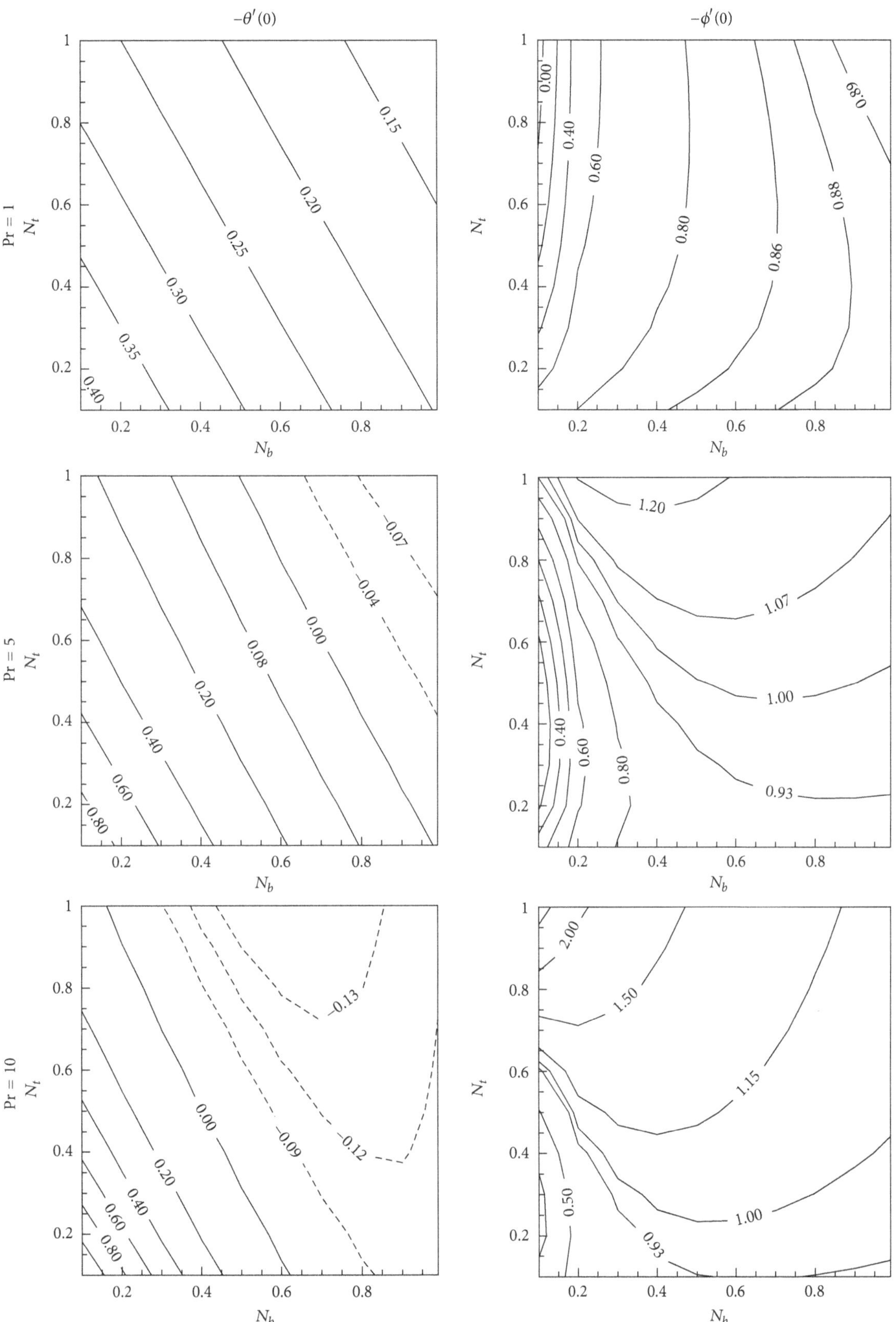

FIGURE 6: Contour lines of heat transfer and concentration rates for different values of Pr when Le = 2, $\lambda = \varepsilon = 0.1$.

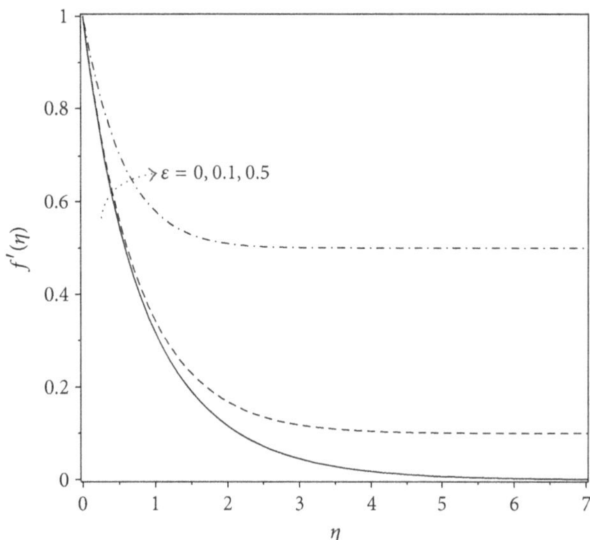

FIGURE 7: Hydrodynamic boundary layer for different values of stretching parameter.

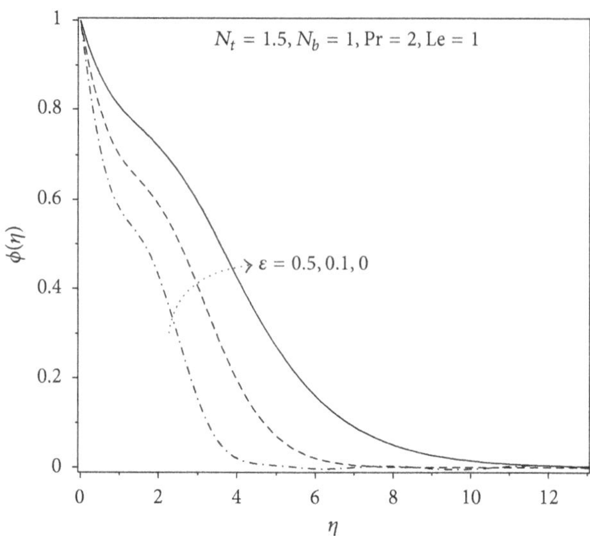

FIGURE 9: Concentration profiles for different values of stretching parameter.

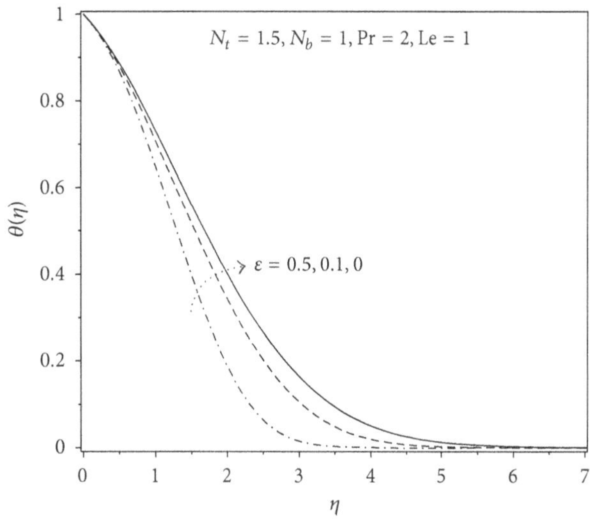

FIGURE 8: Temperature profiles for different values of stretching parameter.

Greek Symbols

α: Thermal diffusivity
ε: Stretching sheet parameter
ϕ: Rescaled nanoparticle volume fraction
η: Similarity variable
μ: Dynamic viscosity
ν: Kinematic viscosity
ρ: Density
$(\rho c_p)_f$: Heat capacity of the fluid
$(\rho c_p)_p$: Effective heat capacity of the nanoparticle
λ: Dimensionless heat generation/absorption
τ: Parameter defined by $(\rho c_p)_p/(\rho c_p)_f$
ψ: Stream function.

Subscripts

nf: Nanofluid
f: Fluid
w: Condition on the sheet
∞: Ambient conditions.

References

[1] K. Hiemenz, "Die Grenzschicht an einem in den gleichformingen Flussigkeitsstrom eingetauchten graden Kreiszylinder," *Dingler's Polytechnic Journal*, vol. 326, pp. 321–324, 1911.

[2] F. Homann, "Der Einfluß großer Zähigkeit bei der Strömung um den Zylinder und um die Kugel," *Zeitschrift für Angewandte Mathematik und Mechanik*, vol. 16, no. 3, pp. 153–164, 1936.

[3] H. A. Attia, "Homann magnetic flow and heat transfer with uniform suction or injection," *Canadian Journal of Physics*, vol. 81, no. 10, pp. 1223–1230, 2003.

[4] K. Bhattacharyya and K. Vajravelu, "Stagnation-point flow and heat transfer over an exponentially shrinking sheet," *Communications in Nonlinear Science and Numerical Simulation*, vol. 17, no. 7, pp. 2728–2734, 2012.

[5] A. Malvandi, "The unsteady flow of a nanofluid in the stagnation point region of a time-dependent rotating sphere," *THERMAL SCIENCE*, 2013.

[6] V. Kumaran, R. Tamizharasi, and K. Vajravelu, "Approximate analytic solutions of stagnation point flow in a porous medium," *Communications in Nonlinear Science and Numerical Simulation*, vol. 14, no. 6, pp. 2677–2688, 2009.

[7] M. A. A. Hamad and M. Ferdows, "Similarity solution of boundary layer stagnation-point flow towards a heated porous stretching sheet saturated with a nanofluid with heat absorption/generation and suction/blowing: a Lie group analysis," *Communications in Nonlinear Science and Numerical Simulation*, vol. 17, no. 1, pp. 132–140, 2012.

[8] Z. Ziabakhsh, G. Domairry, and H. Bararnia, "Analytical solution of non-Newtonian micropolar fluid flow with uniform

suction/blowing and heat generation," *Journal of the Taiwan Institute of Chemical Engineers*, vol. 40, no. 4, pp. 443–451, 2009.

[9] Z. Ziabakhsh, G. Domairry, and H. R. Ghazizadeh, "Analytical solution of the stagnation-point flow in a porous medium by using the homotopy analysis method," *Journal of the Taiwan Institute of Chemical Engineers*, vol. 40, no. 1, pp. 91–97, 2009.

[10] L. J. Crane, "Flow past a stretching plate," *Zeitschrift für Angewandte Mathematik und Physik*, vol. 21, no. 4, pp. 645–647, 1970.

[11] S. Yao, T. Fang, and Y. Zhong, "Heat transfer of a generalized stretching/shrinking wall problem with convective boundary conditions," *Communications in Nonlinear Science and Numerical Simulation*, vol. 16, no. 2, pp. 752–760, 2011.

[12] K. R. Rajagopal, T. Y. Na, and A. S. Gupta, "Flow of a viscoelastic fluid over a stretching sheet," *Rheologica Acta*, vol. 23, no. 2, pp. 213–215, 1984.

[13] R. Nazar, N. Amin, D. Filip, and I. Pop, "Stagnation point flow of a micropolar fluid towards a stretching sheet," *International Journal of Non-Linear Mechanics*, vol. 39, no. 7, pp. 1227–1235, 2004.

[14] M. A. A. Hamad, "Analytical solution of natural convection flow of a nanofluid over a linearly stretching sheet in the presence of magnetic field," *International Communications in Heat and Mass Transfer*, vol. 38, no. 4, pp. 487–492, 2011.

[15] Z. Ziabakhsh, G. Domairry, M. Mozaffari, and M. Mahbobifar, "Analytical solution of heat transfer over an unsteady stretching permeable surface with prescribed wall temperature," *Journal of the Taiwan Institute of Chemical Engineers*, vol. 41, no. 2, pp. 169–177, 2010.

[16] S. Nadeem, A. Hussain, and M. Khan, "HAM solutions for boundary layer flow in the region of the stagnation point towards a stretching sheet," *Communications in Nonlinear Science and Numerical Simulation*, vol. 15, no. 3, pp. 475–481, 2010.

[17] M. M. Nandeppanavar, K. Vajravelu, M. Subhas Abel, S. Ravi, and H. Jyoti, "Heat transfer in a liquid film over an unsteady stretching sheet," *International Journal of Heat and Mass Transfer*, vol. 55, no. 4, pp. 1316–1324, 2012.

[18] A. Malvandi, F. Hedayati, and D. D. Ganji, "Thermodynamic optimization of fluid flow over an isothermal moving plate," *Alexandria Engineering Journal*, 2013.

[19] A. K. Singh, "Heat source and radiation effects on magneto-convection flow of a viscoelastic fluid past a stretching sheet: analysis with Kummer's functions," *International Communications in Heat and Mass Transfer*, vol. 35, no. 5, pp. 637–642, 2008.

[20] S. Nadeem and C. Lee, "Boundary layer flow of nanofluid over an exponentially stretching surface," *Nanoscale Research Letters*, vol. 7, article 94, pp. 1–15, 2012.

[21] M. R. H. Nobari and A. Malvandi, "Torsion and curvature effects on fluid flow in a helical annulus," *International Journal of Non-Linear Mechanics*, vol. 57, pp. 90–101, 2013.

[22] S. U. S. Choi, "Enhancing thermal conductivity of fluids with nanoparticles," in *Developments and Applications of Non-Newtonian Flows*, D. A. Siginer and H. P. Wang, Eds., pp. 99–105, ASME, 1995.

[23] H. Masuda, A. Ebata, K. Teramae, and N. Hishinuma, "Alteration of thermalconductivity and viscosity of liquid by dispersing ultra-fine particles. Dispersion of Al_2O_3, SiO_2 and TiO_2 ultra-fine particles," *Netsu Bussei*, vol. 7, no. 4, pp. 227–233, 1993.

[24] J. Buongiorno, "Convective transport in nanofluids," *Journal of Heat Transfer*, vol. 128, no. 3, pp. 240–250, 2006.

[25] A. V. Kuznetsov and D. A. Nield, "Natural convective boundary-layer flow of a nanofluid past a vertical plate," *International Journal of Thermal Sciences*, vol. 49, no. 2, pp. 243–247, 2010.

[26] W. A. Khan and I. Pop, "Boundary-layer flow of a nanofluid past a stretching sheet," *International Journal of Heat and Mass Transfer*, vol. 53, no. 11-12, pp. 2477–2483, 2010.

[27] N. Bachok, A. Ishak, and I. Pop, "Boundary-layer flow of nanofluids over a moving surface in a flowing fluid," *International Journal of Thermal Sciences*, vol. 49, no. 9, pp. 1663–1668, 2010.

[28] A. Alsaedi, M. Awais, and T. Hayat, "Effects of heat generation/absorption on stagnation point flow of nanofluid over a surface with convective boundary conditions," *Communications in Nonlinear Science and Numerical Simulation*, vol. 17, no. 11, pp. 4210–4223, 2012.

[29] P. Rana and R. Bhargava, "Flow and heat transfer of a nanofluid over a nonlinearly stretching sheet: a numerical study," *Communications in Nonlinear Science and Numerical Simulation*, vol. 17, no. 1, pp. 212–226, 2012.

[30] W. Daungthongsuk and S. Wongwises, "A critical review of convective heat transfer of nanofluids," *Renewable and Sustainable Energy Reviews*, vol. 11, no. 5, pp. 797–817, 2007.

[31] X.-Q. Wang and A. S. Mujumdar, "Heat transfer characteristics of nanofluids: a review," *International Journal of Thermal Sciences*, vol. 46, no. 1, pp. 1–19, 2007.

[32] S. Kakaç and A. Pramuanjaroenkij, "Review of convective heat transfer enhancement with nanofluids," *International Journal of Heat and Mass Transfer*, vol. 52, no. 13-14, pp. 3187–3196, 2009.

[33] S. Liao, "On the homotopy analysis method for nonlinear problems," *Applied Mathematics and Computation*, vol. 147, no. 2, pp. 499–513, 2004.

[34] B. Wu and H. Zhong, "Summation of perturbation solutions to nonlinear oscillations," *Acta Mechanica*, vol. 154, no. 1–4, pp. 121–127, 2002.

[35] G. Adomian, "A review of the decomposition method in applied mathematics," *Journal of Mathematical Analysis and Applications*, vol. 135, no. 2, pp. 501–544, 1988.

[36] J.-H. He, "Variational iteration method—a kind of non-linear analytical technique: some examples," *International Journal of Non-Linear Mechanics*, vol. 34, no. 4, pp. 699–708, 1999.

[37] F. Hedayati, D. Ganji, S. Hamidi, and A. Malvandi, "An analytical study on a model describing heat conduction in rectangular radial fin with temperature-dependent thermal conductivity," *International Journal of Thermophysics*, vol. 33, no. 6, pp. 1042–1054, 2012.

[38] A. Malvandi, D. D. Ganji, F. Hedayati, M. H. Kaffash, and M. Jamshidi, "Series solution of entropy generation toward an isothermal flat plate," *Th rmal Science*, vol. 16, no. 5, pp. 1289–1295, 2012.

[39] S. Liao, "On the relationship between the homotopy analysis method and Euler transform," *Communications in Nonlinear Science and Numerical Simulation*, vol. 15, no. 6, pp. 1421–1431, 2010.

[40] S. Liao, "Homotopy analysis method: a new analytical technique for nonlinear problems," *Communications in Nonlinear Science and Numerical Simulation*, vol. 2, no. 2, pp. 95–100, 1997.

[41] S.-I. Liao, "A short review on the homotopy analysis method in fluid mechanics," *Journal of Hydrodynamics*, vol. 22, no. 5, pp. 839–841, 2010.

[42] D. G. Domairry, A. Mohsenzadeh, and M. Famouri, "The application of homotopy analysis method to solve nonlinear

differential equation governing Jeffery-Hamel flow," *Communications in Nonlinear Science and Numerical Simulation*, vol. 14, no. 1, pp. 85–95, 2009.

[43] G. Domairry and M. Fazeli, "Homotopy analysis method to determine the fin efficiency of convective straight fins with temperature-dependent thermal conductivity," *Communications in Nonlinear Science and Numerical Simulation*, vol. 14, no. 2, pp. 489–499, 2009.

[44] G. Domairry and N. Nadim, "Assessment of homotopy analysis method and homotopy perturbation method in non-linear heat transfer equation," *International Communications in Heat and Mass Transfer*, vol. 35, no. 1, pp. 93–102, 2008.

[45] S.-J. Liao, "An explicit, totally analytic approximate solution for Blasius' viscous flow problems," *International Journal of Non-Linear Mechanics*, vol. 34, no. 4, pp. 759–778, 1999.

[46] S. Liao, "An optimal homotopy-analysis approach for strongly nonlinear differential equations," *Communications in Nonlinear Science and Numerical Simulation*, vol. 15, no. 8, pp. 2003–2016, 2010.

[47] M. Hassani, M. Mohammad Tabar, H. Nemati, G. Domairry, and F. Noori, "An analytical solution for boundary layer flow of a nanofluid past a stretching sheet," *International Journal of Th rmal Sciences*, vol. 50, no. 11, pp. 2256–2263, 2011.

[48] N. Bachok, A. Ishak, and I. Pop, "On the stagnation-point flow towards a stretching sheet with homogeneous-heterogeneous reactions effects," *Communications in Nonlinear Science and Numerical Simulation*, vol. 16, no. 11, pp. 4296–4302, 2011.

[49] G. Domairry and Z. Ziabakhsh, "Solution of boundary layer flow and heat transfer of an electrically conducting micropolar fluid in a non-Darcian porous medium," *Meccanica*, vol. 47, no. 1, pp. 195–202, 2012.

[50] A. A. Joneidi, G. Domairry, and M. Babaelahi, "Analytical treatment of MHD free convective flow and mass transfer over a stretching sheet with chemical reaction," *Journal of the Taiwan Institute of Chemical Engineers*, vol. 41, no. 1, pp. 35–43, 2010.

[51] Z. Ziabakhsh, G. Domairry, H. Bararnia, and H. Babazadeh, "Analytical solution of flow and diffusion of chemically reactive species over a nonlinearly stretching sheet immersed in a porous medium," *Journal of the Taiwan Institute of Chemical Engineers*, vol. 41, no. 1, pp. 22–28, 2010.

[52] K. Yabushita, M. Yamashita, and K. Tsuboi, "An analytic solution of projectile motion with the quadratic resistance law using the homotopy analysis method," *Journal of Physics A*, vol. 40, no. 29, pp. 8403–8416, 2007.

[53] Z. Niu and C. Wang, "A one-step optimal homotopy analysis method for nonlinear differential equations," *Communications in Nonlinear Science and Numerical Simulation*, vol. 15, no. 8, pp. 2026–2036, 2010.

19

Relativistic Accretion into a Reissner-Nordström Black Hole Revisited

J. A. de Freitas Pacheco

Laboratoire Cassiopée-UMR 6202, Observatoire de la Côte d'Azur, University of Nice-Sophia Antipolis, BP 4222, 06304 Nice, Cedex 4, France

Correspondence should be addressed to J. A. de Freitas Pacheco, pacheco@oca.eu

Academic Editor: Y. Peles

The accretion of relativistic and nonrelativistic fluids into a Reissner-Nordström black hole is revisited. The position of the critical point, the flow velocity at this point, and the accretion rate are only slightly affected with respect to the Schwarzschild case when the fluid is nonrelativistic. On the contrary, relativistic fluids cross the critical point always subsonically. In this case, the sonic point is located near the event horizon, which is crossed by the fluid with a velocity less than the light speed. The accretion rate of relativistic fluids by a Reissner-Nordström black hole is reduced with respect to those estimated for uncharged black holes, being about 60% less for the extreme case (charge-to-mass ratio equal to one).

1. Introduction

The steady relativistic spherical flow of a perfect gas into a black hole has been intensely investigated in the past thirty years and a comprehensive review on the subject can be found, for instance, in [1]. For a Schwarzschild black hole, the basic relativistic equations of the inflow were discussed by Michel [2], who has derived the relations involving the sound and flow velocities at the critical point. These early investigations have shown that, for an adiabatic flow of a perfect gas with $\gamma < 5/3$ into a Schwarzschild black hole, the only critical point of the flow lying outside the horizon is that corresponding to the Bondi solution. A similar situation occurs in the case of a weakly interacting gas supposed to model dark matter. Assuming that during the inflow the phase space density is conserved, the authors in [3] derived a solution for the critical point, which is located at a distance of about 30–150 times the horizon radius for conditions expected to be present in typical dark matter halos.

Under adiabatic conditions (the cooling time is longer than the free-fall timescale), the gas is compressed as it approaches the horizon, its temperature increases, and X-rays are emitted from the inner accreting envelope. This emission preheats the infalling gas, reducing considerably the accretion rate and the radiative efficiency [4]. Higher efficiencies can be obtained if dissipative turbulent motions and magnetic field line reconnection effects are included in the treatment of the inflow [5]. A consistent relativistic analysis of radiation effects on the flow was investigated in the pioneering work by Thorne and collaborators [6]. For an optically thick inflow, the distance of the critical point to the black hole horizon is reduced as well as the accretion efficiency [7]. In fact, when radiation effects are included, two branches appear in the diagram "luminosity versus accretion rate" [8, 9] but even in the most favourable cases, corresponding to the high luminosity branch, the radiation efficiency is at maximum 2×10^{-4}, three orders of magnitude smaller than the "canonical" value (of ten percent, derived from an accreting disk) usually adopted in cosmological simulations intended to explain the presence of supermassive black holes in the early universe by the growth of primordial seeds [10].

How the inflow properties are affected if the black hole has an electrical charge? Although the existence of charged black holes in the universe may be contested, some authors have hypothesized that such objects could play an important role in some astrophysical processes. For instance, the creation of positron-electron pairs in the "dyadosphere" of a Reissner-Nordström (RN) black hole was investigated in [11] (and references therein) as a possible mechanism to drive gamma-ray bursts. If the observed acceleration of the expansion of the universe is due to a phantom field, the interaction with a supermassive Schwarzschild black hole was analyzed in [12], while the interaction with an RN black hole was considered by [13, 14]. As a consequence of this process, the mass of the black hole decreases but its charge remains constant and after a finite timescale, the black hole reaches the extreme state and may eventually produce a naked singularity [13].

As mentioned above, the formation of a charged black hole during the gravitational collapse and, in particular, the formation of a "dyadosphere" present several difficulties whose discussion is beyond the scope of the present paper (see, however, criticisms by [15]). Nevertheless, the accretion process by a charged black hole offers the possibility of studying different aspects of gravity and of accretion flows in extreme conditions. In the present paper, the accretion of fluids with different equations of state into an RN black hole is revisited. We will show that for a nonrelativistic baryonic fluid, the physical properties of the flow do not differ considerably from the Schwarzschild case, since the critical point is only slightly modified by the presence of a charge even in the extreme case. However, the situation is considerably different if the fluid is relativistic. In this case, the critical point is situated near the horizon and the flow velocity at this position is subsonic. Moreover, also the horizon crossing occurs at a subsonic velocity, contrary to what happens when the flow is nonrelativistic, case in which the horizon is crossed with a flow velocity equal to the light velocity. No solutions for the flow were found beyond the critical point if a naked singularity is present, confirming the conclusion by [13]. For both cases, relativistic and nonrelativistic flows, corrections to the accretion rate due to the presence of a charge are given. This paper is organized as follows: in Section 2, the equations of the flow are presented; in Section 3, the accretion of a nonrelativistic fluid is discussed, while the accretion of relativistic fluids is analyzed in Section 4. Finally, in Section 5, the main conclusions are given.

2. Equations of the Accretion Flow

The metric describing an RN black hole is given by

$$ds^2 = -B(r)dt^2 + B^{-1}(r)dr^2 + r^2 d\varpi^2, \tag{1}$$

where $d\varpi^2 = d\theta^2 + \sin^2\theta d\phi^2$ and the lapse function is defined by

$$B(r) = 1 - \frac{r_g}{r} + \frac{Q^2}{r^2}, \tag{2}$$

where $r_g = 2M$ is the gravitational radius, M and Q are, respectively, the mass and the charge of the black hole. The zeros of the lapse function $B(r)$ define two horizons given by

$$r_\pm = \frac{r_g}{2}\left(1 \pm \sqrt{1-\beta^2}\right), \tag{3}$$

and we have defined $\beta = Q/M$. The sign "+" corresponds to the outer or to the event horizon while the sign "−" corresponds to the inner horizon. When the charge satisfies the condition $Q = M$ (or, equivalently, $\beta = 1$), both horizons coincide and this case corresponds to an extreme RN black hole.

We assume that a steady spherical inflow is set up inside the influence radius of the black hole, defined by the equality between the gravitational potential of the black hole and the mean kinetic energy of particles constituting the fluid far away from the horizon. Denoting by "∇" the covariant derivative, the conservation equations are the mass flux conservation:

$$\nabla_k J^k = 0, \tag{4}$$

where $J^k = mnu^k$ is the mass-current density. The second equation is the energy-momentum flux conservation:

$$\nabla_k T_i^k = 0, \tag{5}$$

where the energy-momentum tensor is that of an ideal fluid, that is, $T_i^k = (P+\varepsilon)u^k u_i - P\delta_i^k$, with P, ε, and n being, respectively, the proper pressure, the proper energy density, and the proper particle number density. The only nonnull components of the 4-vector velocity are $u = u^1 = dr/ds$ and $u^0 = dt/ds$. In the one hand, under spherical symmetry and steady state conditions, from the time-space component of (5) and the expression for the stress-energy tensor, one obtains

$$(P+\varepsilon)u_0 u^1 \sqrt{-g} = C_1, \tag{6}$$

where $-g$ is the metric determinant of (1) and C_1 is an arbitrary integration constant. Using the normalization condition $u_i u^i = -1$, one obtains trivially that

$$u_0 = \left(1 - \frac{r_g}{r} + \frac{Q^2}{r^2} + u^2\right)^{1/2}. \tag{7}$$

On the other hand, the integration of the space component of (4) gives

$$nu^1\sqrt{-g} = C_2, \tag{8}$$

where C_2 is another arbitrary integration constant. Substituting (7) into (6) and performing the ratio with (8) one obtains

$$\frac{(P+\varepsilon)}{n}\left[1 - \frac{r_g}{r} + \frac{Q^2}{r^2} + u^2\right]^{1/2} = \Delta, \tag{9}$$

where $\Delta = C_1/C_2$ is a new constant. Notice that if $Q = 0$, one obtains the same result as in [2, equation (9)] but notice that here the particle number density n appears in the denominator instead of the fluid energy density. Deriving (9) with respect to the radial coordinate r and using the mass conservation equation (8), one obtains, after some algebra, the following equation:

$$\frac{d\lg u}{d\lg r}\left[u^2 - V^2 F(r,u)\right] = \left[2V^2 F(r,u) - \left(\frac{r_g}{2r} - \frac{Q^2}{r^2}\right)\right]. \tag{10}$$

In the above equation, we have introduced, respectively,

$$F(r,u) = \left(1 - \frac{r_g}{r} + \frac{Q^2}{r^2} + u^2\right), \tag{11}$$

$$V^2 = \frac{d\lg(P+\varepsilon)}{d\lg n} - 1. \tag{12}$$

The critical point of the flow occurs when both bracketed factors in (10) vanish simultaneously. It should be emphasized that the critical point, contrary to usual assertions found in the literature (see, e.g, [14]), does not necessarily coincide with the sonic point, in which a transition from subsonic to supersonic flow occurs. In fact, this is the situation occurring in outflows present in atmospheres of massive stars, driven by radiative forces [16, 17]. For an accreting relativistic fluid, as we will see below, the crossing of the critical point always occurs while the flow is still subsonic.

The conditions at the critical point (coordinate r_c and velocity u_c) are derived from the relations below, which are similar to those obtained in [14]:

$$4u_c^2 = \frac{r_g}{r_c} - \frac{2Q^2}{r_c^2}, \tag{13}$$

$$u_c^2 = V_c^2 F(r_c, u_c). \tag{14}$$

From these two equations, one obtains for the critical radius:

$$r_c = \frac{r_g(1+3V_c^2)}{8V_c^2}\left\{1 \pm \left[1 - \frac{8V_c^2\beta^2(1+V_c^2)}{(1+3V_c^2)^2}\right]^{1/2}\right\}. \tag{15}$$

As pointed out by [13], two solutions for the critical radius are possible. The first corresponds to the sign "+", which locates the critical radius outside the event horizon and represents a true physical solution. The other possibility corresponds to the sign "−" locates the critical radius between the inner and the outer horizon. This last solution will be discarded in the present analysis.

3. Accretion of a Nonrelativistic Fluid

For a baryonic nonrelativistic fluid with an equation of state $P = Kn^\gamma$, the energy density is given by

$$\varepsilon = mc^2 n + \frac{P}{(\gamma - 1)}. \tag{16}$$

Notice that the first term on the right side of (16) corresponds to the rest energy (usually neglected) while the second represents the interaction energy among the fluid particles. Using these relations and (12), one obtains after some algebra:

$$V^2 = \frac{a^2}{[1 + a^2/(\gamma - 1)]}, \tag{17}$$

where $a^2 = \gamma P/mnc^2$ is the square of the adiabatic sound velocity measured in units of the velocity of light. Notice that solutions with $\gamma = 1$ are excluded since in this case we have a null velocity for the flow. The constant Δ in (9) can be calculated by imposing a zero flow velocity ($u = 0$) when $r \to \infty$ or, in other words, at distances far away from the influence sphere of the black hole. Under these conditions, one obtains trivially

$$\Delta = mc^2\left[1 + \frac{a_\infty^2}{(\gamma - 1)}\right], \tag{18}$$

where a_∞ is the adiabatic sound velocity well beyond the influence radius of the black hole. Using this result and evaluating now (9) at the critical point with the help of (13), one obtains a relation between the adiabatic sound velocity at the critical point and that at "infinity," that is,

$$\frac{a_c}{a_\infty} \approx \sqrt{\frac{2}{(5 - 3\gamma)}}. \tag{19}$$

The flow velocity at the critical point can be evaluated from (11), (14), and (17) under the approximation $a << 1$ (the sound velocity is much less than the velocity of light). Performing a series expansion up to third order, one obtains

$$u_c \approx a_c - \frac{3}{4}\frac{(5-3\gamma)}{(\gamma - 1)}a_c^3. \tag{20}$$

Inspection of this equation reveals two distinct and important aspects: firstly, as we have mentioned before, the flow velocity at the critical point is not exactly coincident with the sound velocity, since they differ at least by a term of third order in the sound velocity. Secondly, at the considered order, the flow velocity at the critical point is not affected by the black hole charge. However, this is not the case for the critical radius. Substituting the relations above into (15), one obtains for the critical radius up to second-order terms in β

$$\frac{r_c}{r_g} \approx \frac{(5-3\gamma)}{8a_\infty^2}\left[1 + \frac{(6 - 4\beta^2)a_\infty^2}{(5 - 3\gamma)}\right]. \tag{21}$$

Notice that, as in the Schwarzschild case, the existence of a physical solution requires $\gamma < 5/3$ and that the electrical charge reduces slightly the distance of the critical point with respect to the black hole horizon.

The accretion rate can now be computed from the conditions at the critical radius, that is,

$$\frac{dM_{\rm bh}}{dt} = 4\pi r_c^2 T_t^r = \pi\,\Gamma(\gamma)(GM_{\rm bh})^2 f(\beta)\rho_\infty a_\infty^{-3}. \tag{22}$$

In the equation above, $\rho_\infty = mn_\infty$ is the mass density of the baryonic fluid at "infinity" (beyond the influence radius of the black hole), $M_{\rm bh}$ is the black hole mass, and the functions $\Gamma(\gamma)$ and $f(\beta)$ are defined, respectively, as

$$\Gamma(\gamma) = \left[\frac{2}{(5-3\gamma)}\right]^{(5-3\gamma)/2(\gamma-1)}, \tag{23}$$

$$f(\beta) = \left[1 + \frac{(12 - 8\beta^2)a_\infty^2}{(5 - 3\gamma)}\right]. \tag{24}$$

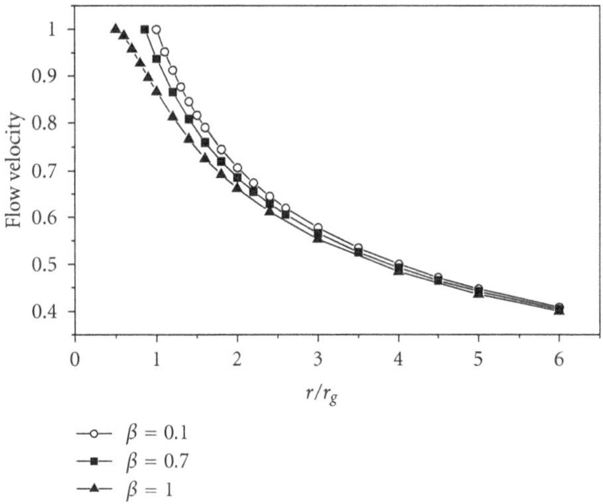

FIGURE 1: Radial velocity profiles for a nonrelativistic fluid being accreted by a Reissner-Nordström black hole for different charge-to-mass ratios. The radial coordinate is given in terms of the gravitational radius.

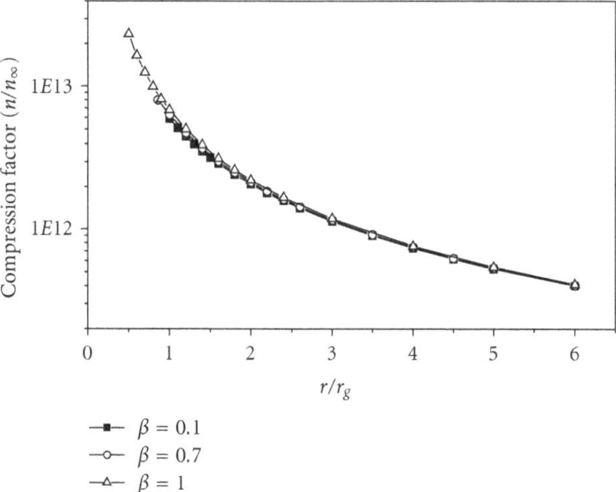

FIGURE 2: Compression factor profiles for an inflow of a nonrelativistic gas as a function of the radial coordinate in units of the gravitational radius for different charge-to-mass ratios.

3.1. Some Numerical Results. In order to derive the radial velocity and the density profiles of the flow, let us define the dimensionless variables $x = r/r_g$ and $y = n/n_\infty$, which measure, respectively, the radial distance in terms of the gravitational radius r_g and the particle number density in terms of its value at "infinity." Recall that the flow velocity u is given in units of the velocity of light. Under these conditions, (9) can be written as

$$\left(1+\lambda_\infty^2 y^{\gamma-1}\right)\left(1-\frac{1}{x}+\frac{1}{4}\frac{\beta}{x^2}+u\right)^{1/2} = \left(1+\lambda_\infty^2\right), \quad (25)$$

where $\lambda_\infty^2 = a_\infty^2/(\gamma-1)$ and, as mentioned above, the adiabatic sound velocity is also given in terms of the velocity of light. Using the same notation, the particle density conservation (8) can be written as

$$y = (\gamma-1)^{1/2}\lambda_\infty\left(\frac{x_c}{x}\right)^2\left(\frac{a_c}{a_\infty}\right)^{(\gamma+1)/(\gamma-1)}. \quad (26)$$

The constant in (8) was calculated by applying the particle conservation equation at the critical point and using the previous result on the ratio between the adiabatic sound velocity at "infinity" and at the critical point. Equations (25) and (26) constitute an algebraic system of nonlinear equations, which can be solved numerically for the fluid velocity u and the ratio n/n_∞ for a given value of $x = r/r_g$. The parameters characterizing the flow are the fluid temperature (or the sound velocity) at "infinity" and the adiabatic coefficient, while the parameter $\beta = Q/M$ affects the position of the critical point. The radial velocity profile of the flow (in units of the velocity of light) derived numerically from the equations above is shown in Figure 1. The velocity profile was computed by assuming a fluid temperature of $10^4\,K$ at "infinity" and an adiabatic coefficient $\gamma = 1.2$, adequate for an ionized gas. Solutions were obtained for different values of the charge, that is, $\beta = 0.1$, 0.7, and 1.0 (extreme Reissner-Nordström case). For these models, the event horizon is located, respectively, at 0.9975, 0.8571, and 0.5000 times the gravitational radius of a Schwarzschild black hole. Notice that for all these cases, when the charge-to-mass ratio is in the range $0 < \beta \leq 1$, the flow crosses the event horizon always at the velocity of light. In these examples, the critical radius is very far from the event horizon ($r_c \approx 1.575 \times 10^8 r_g$) and the flow velocity at the critical point is about 12 kms^{-1}.

Figure 2 shows the particle density profile in terms of the density at "infinity" (compression ratio) as a function of the radial coordinate, measured in units of the gravitational radius and for different values of the charge-to-mass ratio. The compression ratio at the event horizon increases as β increases, reaching a factor of about 2.352×10^{13} for an extreme RN black hole. Since the temperature varies as $T \propto n^{\gamma-1}$, it may attain values of the order of 10^6–10^7 K near the horizon. It should be emphasized that these values correspond to an adiabatic flow and that radiative transfer effects may change appreciably these results.

When $V_c << 1$ (nonrelativistic flow), (21) admits real solutions only if the charge-to-mass ratio satisfies the condition:

$$\beta^2 \leq \frac{\left(1+3V_c^2\right)^2}{8V_c^2\left(1+V_c^2\right)} \approx \frac{1}{8V_c^2}. \quad (27)$$

The extreme case $\beta^2 \approx 1/(8V_c^2) >> 1$ corresponds to a situation of a naked singularity. The critical radius is located at $r_c \approx r_g/(8V_c^2)$, close to the sonic point within the considered approximation. However, no solution for the flow beyond the critical point was found under these circumstances. In [13], the authors suggested that when $\beta > 1$, an ideal fluid does not accrete at all onto a naked singularity, instead a static "atmosphere" will be formed. In this case, the fluid should obey the hydrostatic equilibrium equation:

$$\frac{\partial P}{\partial r} + (P + \varepsilon)\frac{\partial \lg\sqrt{B(r)}}{\partial r} = 0. \quad (28)$$

The solution of this equation indicates that the matter density increases as the singularity is approached, attains a maximum, and then decreases toward the limit $\rho \to 0$ as $r \to 0$. This configuration, including a density inversion, is clearly unstable (Rayleigh-Taylor instability). The observed behaviour can be explained by the fact that the effective acceleration $g_{ef} = \partial \lg\sqrt{B(r)}/\partial r$ is always negative outside the horizon but becomes positive for $r < 0.5\beta^2 r_g$. The repulsive force due to the modification of the space curvature near the singularity is a consequence of the presence of the electric charge. Past investigations [18] on the gravitational collapse of a uniform sphere constituted by charged dust led to a similar conclusion. Once the horizon is crossed by the surface of the sphere, the gravitational attraction is replaced by a repulsive force and, consequently, the surface of the sphere never reaches the singularity. Therefore, in a certain sense, the singularity is "protected" by the repulsive force induced by the electric charge.

4. Accretion of a Relativistic Fluid

In the case of a relativistic fluid, the equation of state is simply given by $P = \varepsilon/3$ and the pressure is related to the particle number density by $P \propto n^{4/3}$. Under these conditions, from (12), one obtains trivially that $V^2 = 1/3$. Replacing this result into (15), one obtains for the critical radius:

$$r_c = \frac{3}{4} r_g \left[1 \pm \left(1 - \frac{8}{9}\beta^2\right)^{1/2}\right]. \quad (29)$$

This relation indicates that a critical point exists only if $\beta \leq 3/\sqrt{8}$. This leaves open the possibility for the existence of relativistic flows in the presence of a naked singularity. However, even in this case, no solutions were found when $\beta > 1$ for the same reasons already mentioned.

When $\beta = 0$ (the case of a Schwarzschild black hole), contrary to what happens for a nonrelativistic fluid, the critical radius is located near the horizon, that is, $r_c/r_g = 3/2$ and the radial velocity of the flow at this point is $u_c = 1/\sqrt{6}$, implying that the crossing occurs subsonically. The flow becomes supersonic only at $r_s \approx 1.07754 r_g$, quite close to the horizon, which is crossed with a velocity of $\sim$0.62 c. Using (9) and the conditions at the critical point, the ratio between the density or the temperature ($T \propto n^{1/3}$) at this point and the value at "infinity" can be easily computed, for example,

$$\left(\frac{n_c}{n_\infty}\right)^{1/3} = \left(\frac{T_c}{T_\infty}\right) = \sqrt{2}. \quad (30)$$

When $\beta << 1$, the critical radius is given approximately by

$$r_c \approx \frac{3}{2} r_g \left(1 - \frac{2}{9}\beta^2\right). \quad (31)$$

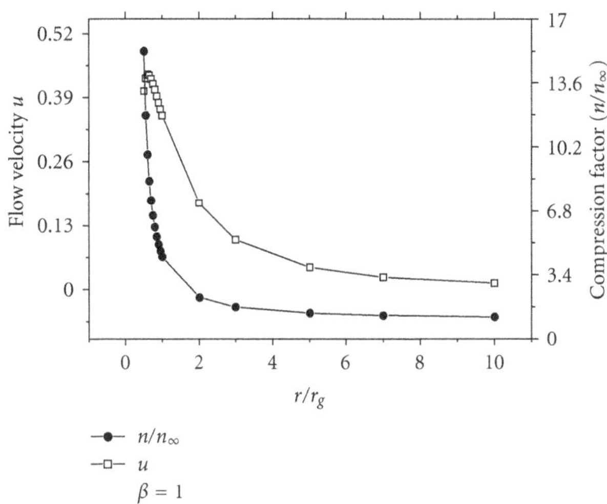

FIGURE 3: The radial velocity profile (left ordinate) for an accreting relativistic fluid as a function of the radial coordinate, in units of the gravitational radius, for an extreme RN black hole. The profile of the compression ratio (right ordinate) is also shown.

Using this result and (13), the flow velocity at the critical point is

$$u_c^2 \approx \frac{1}{6}\left(1 - \frac{1}{9}\beta^2\right). \quad (32)$$

Notice that the flow velocity at the critical point is reduced by the effect of the charge and the critical radius approaches the horizon.

For the extreme case ($\beta = 1$), two solutions for the critical radius exist. The first locates the critical radius at a distance twice the horizon and the flow velocity at this point is $u_c = 1/\sqrt{8}$. The other solution is not physically acceptable, since the critical radius coincides with the horizon, resulting in a null flow velocity at this position. The velocity and the density profiles of the flow can be derived as explained in Section 3.1, taking into account that for a relativistic fluid $(P + \epsilon)/n \propto n^{1/3}$. In Figure 3, the radial velocity profile is shown as a function of the radial coordinate given in units of the gravitational radius for an extreme RN black hole. After crossing the critical point, the flow reaches a maximum velocity of about $u_{\max} \approx 0.4367$ at $r_{\max} \approx 0.60 r_g$. Then the velocity decreases and the flow crosses the horizon with a subsonic velocity of about 0.4029c. Compression ratios attained by a relativistic fluid are always considerably smaller than those derived for the nonrelativistic case. For an extreme RN black hole, the compression factor near the horizon is about 15.3, about twelve orders of magnitude less than that obtained for a nonrelativistic fluid. Since in the relativistic case, the temperature varies as $T \propto n^{1/3}$, near the horizon, due to the adiabatic compression, the temperature is only about 2.48 times the value at "infinity."

The accretion rate can be computed from the flow conditions at the critical point by following the same steps as before. In this case, one obtains

$$\frac{dM_{\rm bh}}{dt} = 32\sqrt{3}\pi\frac{(GM_{\rm bh})^2}{c^5}\varepsilon_\infty\left(1 + O(\beta^4)\right), \quad (33)$$

where ε_∞ is the energy density of the fluid at "infinity." This relation differs from the Schwarzschild case only in terms of the order of β^4. It is worth mentioning that the accretion rate of relativistic particles by a black hole is usually computed by using the capture cross-section $\sigma_{\text{cap}} = 27\pi r_g^2/4$, leading to a numerical factor in (33) equal to 27 instead of $32\sqrt{3}$ here obtained. The simple use of the capture cross-section neglects hydrodynamical and relativistic effects that increase the accretion efficiency by about a factor of two. In the case of an extreme RN black hole, the accretion rate is given by

$$\frac{dM_{\text{bh}}}{dt} = \frac{512\sqrt{3}\pi}{27}\frac{(GM_{\text{bh}})^2}{c^5}\varepsilon_\infty. \quad (34)$$

It can be easily verified that in this case the accretion efficiency decreases by almost 60% with respect to an uncharged black hole.

5. Conclusions

The steady and spherically symmetric accretion of nonrelativistic and relativistic fluids into a Reissner-Nordström black hole was revisited. A charged black hole modifies slightly the properties of the inflow of a nonrelativistic fluid since the critical radius, the radial velocity at the critical point, and the accretion rate are affected only by second-order terms in the charge-to-mass ratio.

The situation is rather different for the inflow of relativistic fluids. Firstly, the critical point occurs closer to the horizon, what is not the case for the inflow of nonrelativistic fluids. Secondly, the crossing of the critical point occurs always in a subsonic regime, even if the black hole is uncharged, contrary to what is usually stated in the literature. The sonic point is reached only very near the horizon and the fluid crosses the horizon with a velocity less than the light speed. In nonrelativistic flows, the ratio between the particle density near the horizon and at "infinity" may attain values of the order of 10^{13}, if the flow is adiabatic. This is not the case when relativistic flows are considered, since in this situation, the compression ratio varies from about 2.83 for a weakly charged black hole up to 4.35 for the extreme case ($\beta = 1$). Finally, the accretion rate of a relativistic fluid by an extreme RN black hole is about 60% smaller than that expected for a Schwarzschild black hole.

When $\beta > 1$, the singularity is "naked" and, depending on the value of the charge-to-mass ratio, mathematical solutions for the critical point exist either for nonrelativistic or for relativistic fluids. However, no solutions for the flow beyond the critical point were found. Some authors [13] suggested that beyond that point, a "static" solution is possible. These solutions present an inversion in the mass density profile, consequence of a repulsive force due to the black hole charge, which is manifested for distances less than $\beta^2 GM_{\text{bh}}/c^2$. Such an inversion observed in the mass density profile is probably unstable against the Rayleigh-Taylor instability, suggesting that such static solutions cannot exist.

References

[1] S. K. Chakrabarti, "Accretion processes on a black hole," *Physics Report*, vol. 266, no. 5-6, pp. 229–390, 1996.

[2] F. C. Michel, "Accretion of matter by condensed objects," *Astrophysics and Space Science*, vol. 15, no. 1, pp. 153–160, 1972.

[3] S. Peirani and J. A. de Freitas Pacheco, "Dark matter accretion into supermassive black holes," *Physical Review D*, vol. 77, no. 6, Article ID 064023, 2008.

[4] J. P. Ostriker, R. McCray, R. Weaver, and A. Yahil, "A new luminosity limit for spherical accretion onto compact X-ray sources," *The Astrophysical Journal*, vol. 208, pp. L61–L65, 1976.

[5] P. Meszaros, "Radiation from spherical accretion onto black holes," *Astronomy & Astrophysics*, vol. 44, pp. 59–68, 1975.

[6] K. S. Thorne, R. A. Flammang, and A. N. Zytkow, "Stationary spherical accretion into black holes—I: equations of structure," *Monthly Notices of the Royal Astronomical Society*, vol. 194, pp. 475–484, 1981.

[7] R. A. Flammang, "Stationary spherical accretion into black holes—II: theory of optically thick accretion," *Monthly Notices of the Royal Astronomical Society*, vol. 199, pp. 833–867, 1982.

[8] M. G. Park, "Self-consistent models of spherical accretion onto black holes. II. Two-temperature solutions with pairs," *The Astrophysical Journal*, vol. 354, no. 1, pp. 83–97, 1990.

[9] L. Nobili, R. Turolla, and L. Zampieri, "Spherical accretion onto black holes: a complete analysis of stationary solutions," *The Astrophysical Journal*, vol. 383, no. 1, pp. 250–262, 1991.

[10] T. Di Matteo, N. Khandai, C. DeGraf et al., "Cold gas flows and the first quasars in cosmological simulations," http://arxiv.org/abs/1107.1253v1.

[11] C. Cherubini, A. Geralico, J. A. Rueda, and R. Ruffini, "e^-e^+ pair creation by vacuum polarization around electromagnetic black holes," *Physical Review D*, vol. 79, no. 12, Article ID 124002, 15 pages, 2009.

[12] J. A. de Freitas Pacheco and J. E. Horvath, "Generalized second law and phantom cosmology," *Classical and Quantum Gravity*, vol. 24, no. 22, pp. 5427–5433, 2007.

[13] E. Babichev, V. I. Dokuchaev, and Yu. N. Eroshenko, "Perfect fluid and scalar field in the Reissner-Nordstrom metric," *Journal of Experimental and Theoretical Physics*, vol. 112, pp. 784–793, 2011.

[14] M. Jamil, A. Qadir, and M. A. Rashid, "Charged black holes in phantom cosmology," *European Physical Journal C*, vol. 58, no. 2, pp. 325–329, 2008.

[15] D. N. Page, "Evidence against macroscopic astrophysical dyadospheres," *The Astrophysical Journal*, vol. 653, no. 2, pp. 1400–1409, 2006.

[16] J. I. Castor, D. C. Abbott, and R. I. Klein, "Radiation-driven winds in of stars," *The Astrophysical Journal*, vol. 195, pp. 157–174, 1975.

[17] F. X. Araujo and J. A. de Freitas Pacheco, "Radiatively driven winds with azimuthal symmetry—application to Be stars," *Monthly Notices of the Royal Astronomical Society*, vol. 241, pp. 543–557, 1989.

[18] I. Novikov, "The replacement of relativistic gravitational contraction by expansion, and the physical singularities during contraction," *Soviet Astronomy. AJ*, vol. 10, no. 5, p. 731, 1967.

20

Thermodynamic Properties of Real Porous Combustion Reactor under Diesel Engine-Like Conditions

M. Weclas,[1] J. Cypris,[1] and T. M. A. Maksoud[2]

[1] *Institute of Vehicle Technology (IFZN), Faculty of Mechanical Engineering, Georg Simon Ohm University of Applied Sciences, Nuremberg, Kesslerplatz 12, 90489 Nuremberg, Germany*
[2] *Faculty of Advanced Technology, University of Glamorgan, Pontypridd CF37 1DL, UK*

Correspondence should be addressed to M. Weclas, miroslaw.weclas@ohm-hochschule.de

Academic Editor: L. De Goey

Thermodynamic conditions of the heat release process under Diesel engine-like conditions in a real porous combustion reactor simulated in a special combustion chamber were analyzed. The same analyses were performed for a free volume combustion chamber, that is, no porous reactor is applied. A common rail Diesel injection system was used for simulation of real engine fuel injection process and mixture formation conditions. The results show that thermodynamic of the heat release process depends on reactor heat capacity, pore density, specific surface area, and pore structure, that is, on heat accumulation in solid phase of porous reactor. In real reactor, the gas temperature and porous reactor temperature are not equal influenced by initial pressure and temperature and by reactor parameters. It was found that the temperature of gas trapped in porous reactor volume during the heat release process is less dependent on air-to-fuel-ratio than that observed for free volume combustion chamber, while the maximum combustion temperature in porous reactor is significantly low. As found this temperature depends on reactor heat capacity, mixture formation conditions and on initial pressure. Qualitative behavior of heat release process in porous reactors and in free volume combustion chamber is similar, also the time scale of the process.

1. Introduction

One of novel combustion technologies is combustion in porous reactors (PM). It is possible to perform a homogeneous and flameless heat release process inside a stationary operating porous reactor resulting in a nearly zero-emission level [1–6]. The combustion process is characterized by a high power density in large dynamic range under stable combustion conditions. Such features of combustion process would be much advantaged in application to internal combustion engine. However, the process conditions in engine are much more complex: the process is nonstationary and is performed under high pressure with requirement of mixture formation directly inside the combustion reactor. Durst and Weclas [7–9] proposed engine concept entails mixture formation and combustion in porous reactor characterized by a nearly zero-emission level and indicating potential for increasing cycle efficiency. Furthermore, the effect of reactor heat capacity resulting in lowering of combustion temperature as well as internal heat recuperation during the engine cycle changes the thermodynamic conditions of the process as compared to conventional system, for example, Otto and Diesel cycles [8–13]. In a real engine, the fuel must directly be injected into combustion reactor and must be distributed throughout the reactor volume together with vaporization and mixing with combustion air. The hot reactor is used as a three-dimensional "hot-spot" igniter, and the heat release process should be performed in the reactor volume only. This requires that the fuel should be injected in a short period of time close to the TDC of compression stroke. Such complex process conditions make on the one hand description of the process itself and on the other hand a practical realization of engine with combustion in porous reactor very difficult. The reactor temperature and its distribution in porous medium volume (temperature of solid phase of the reactor and temperature of the gas trapped inside reactor volume) together with mixture formation inside the reactor define the conditions for thermal ignition

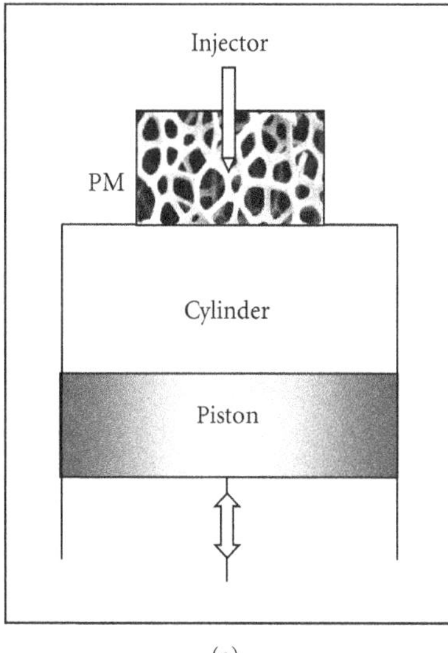

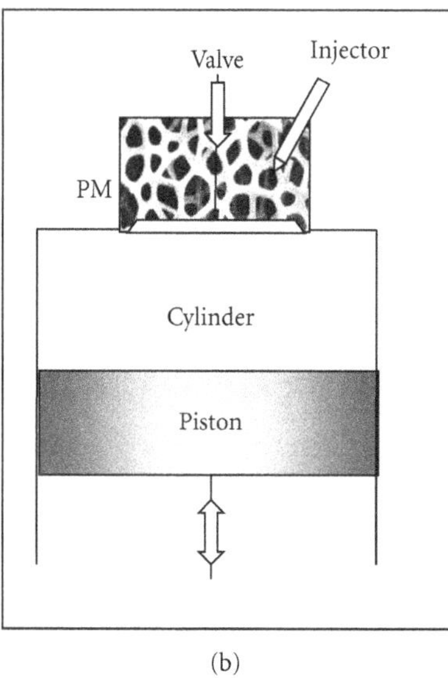

FIGURE 1: Principle of internal combustion engine with mixture formation and combustion in porous reactor (as proposed by Durst and Weclas [8]): (a) reactor with permanent contact with cylinder content; (b) reactor with periodic contact with cylinder content.

and following heat release process. This is because the reactor heat capacity and large specific surface area for interphase heat transfer inside the reactor volume change the thermodynamic conditions of the process [14, 15]. Numerical simulations of heat release process in porous reactors under engine-like conditions usually use simplified conditions and cannot properly describe conditions found in a real combustion reactor [6, 10, 12, 13]. The present work describes thermodynamic of real porous reactor under Diesel engine-like combustion conditions. The fuel is directly injected into a porous reactor resulting in real engine mixture formation conditions. The effect of reactor heat capacity, pore density, and pore structure is also considered in this investigation. The authors concentrate on describing the thermodynamic conditions of the porous reactor with a heat release process under Diesel engine-like conditions. This is compared to a free volume Diesel combustion. The focus of this work is to analyse reactor thermodynamics, but the detailed analysis of the heat release process itself is outside the scope of this work (more information is available in [14–16]).

An engine concept with mixture formation and combustion in a porous reactor is described in Section 2.1. Definition and description of considered thermodynamic systems are given in Section 2.2, where a free volume system is compared to the ideal and real porous reactors. An engine simulator (special combustion chamber) used in present investigation for simulation of real reactor operation under Diesel engine-like conditions is also presented in this section. The next, Section 2.3, presents description of thermodynamic conditions of all considered systems. In Section 2.4, the authors present simplified analysis of heat recuperated (accumulated) in the reactor solid phase as compared to free volume system. Results of experimental investigation for a real combustion reactor in comparison to free volume Diesel engine conditions are presented and discussed in Section 3. Section 4 summarizes presented results.

2. Engine Concept with Heat Release in Porous Reactor and Corresponding Thermodynamic Systems Describing Engine Conditions

2.1. Short Description of Engine Concept with Combustion in Porous Reactor. Durst and Weclas [7–9] proposed engine concept with mixture formation and combustion processes in porous reactor. Application of a combustion porous reactor to engine allows realization of homogeneous and flameless combustion process characterized by a near-zero-emission level. Heat recuperation in porous reactor may allow increase of engine-cycle efficiency resulting in reduction of CO_2 emissions. Heat accumulation in porous reactor results in significantly lowered combustion temperature permitting nearly zero NO_x level. This is the most important difference between the combustion processes in conventional Diesel engine and in engine with combustion in porous reactor. The latter kind of engine may be realized in two different ways [8]: engine in which the PM reactor has permanent contact with the cylinder content (open chamber Figure 1(a)) and engine in which the PM reactor has periodic contact with the cylinder content (closed chamber Figure 1(b)). In the case of a closed chamber, it is assumed that the valve in the PM chamber opens near to the TDC of compression and remains open during the expansion stroke in the cylinder. In the open-chamber system, as considered in the present work, the porous reactor has permanent contact with the cylinder content (gas) and the most important are processes occurring close to the TDC of compression in a constant volume of porous reactor. These processes are fuel injection into reactor, fuel distribution throughout reactor volume, fuel vaporization and mixing with air, as well as thermal ignition and heat release. First of all, the fuel must directly be injected into porous reactor and distributed throughout reactor volume. The authors performed extensive investigations on

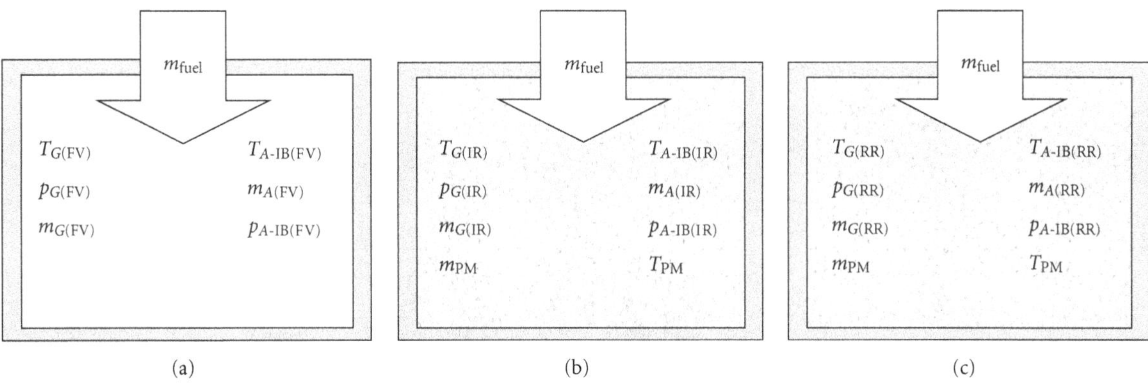

FIGURE 2: Model of considered systems without heat losses through the system walls (description in text): (a) free volume system, (b) ideal porous reactor, and (c) real porous reactor.

TABLE 1

Parameter	System/reactor		
	FV	IR	RR
Mass of air supplied to the system before fuel injection starts	$m_{A(FV)}$	$m_{A(IR)}$	$m_{A(RR)}$
Mass of gas trapped in the system after fuel injection	$m_{G(FV)}$	$m_{G(IR)}$	$m_{G(RR)}$
Mass of porous reactor	—	m_{PM}	m_{PM}
Temperature of porous reactor at the moment of fuel injection	—	$T_{PM\text{-IB}}$	$T_{PM\text{-IB}}$
Temperature of porous reactor after fuel injection	—	T_{PM}	T_{PM}
Air temperature at the moment of fuel injection	$T_{A\text{-IB}(FV)}$	$T_{A\text{-IB}(IR)}$	$T_{A\text{-IB}(RR)}$
Air pressure at the moment of fuel injection	$p_{A\text{-IB}(FV)}$	$p_{A\text{-IB}(IR)}$	$p_{A\text{-IB}(RR)}$
Gas temperature after fuel injection	$T_{G(FV)}$	$T_{G(IR)}$	$T_{G(RR)}$
Gas pressure after fuel injection	$p_{G(FV)}$	$p_{G(IR)}$	$p_{G(RR)}$

the Diesel jet/spray interaction with porous structure and described a multijet splitting as a basic process responsible for three-dimensional spray spreading [17–22]. Owing to the complexity of this process and to the role of reactor structure, pore size and pore geometry in interaction of Diesel spray with porous-structure, extended experimental investigations are still required for better understanding of the process. This extension concerns direct measurements of spray distribution inside porous reactor volume under high reactor temperatures (reactors are optically not transparent).

2.2. Definition of Considered Thermodynamic Systems. There are three system considered in this analysis, Figure 2 and Table 1.

(i) Free volume system (Figure 2(a)): at the moment of fuel injection, the system consists of mass of air trapped in the combustion chamber volume.

(ii) Ideal porous reactor (Figure 2(b)): at the moment of fuel injection, the system consists of mass of air trapped in the porous reactor volume having mass m_{PM}; at any instant of time, the gas temperature is equal to porous reactor temperature $T_{G(IR)} = T_{PM}$.

(iii) Real porous reactor (Figure 2(c)): at the moment of fuel injection, the system consists of mass of air trapped in the porous reactor volume having mass m_{PM}; at any instant of time after fuel injection starts, the gas temperature is not equal to porous reactor temperature $T_{G(RR)} \neq T_{PM}$.

Both porous reactors are made of a highly porous structure having porosity higher than 80% (in real case more than 90%). In all considered cases, the fuel is directly supplied into the combustion chamber/reactor using a common-rail Diesel injection system with electronically controlled injector. The mass of injected fuel results from the combination of injection pressure and injection duration. For both models with PM, it is assumed that the mass of reactor is much higher than the mass of gas trapped in the reactor volume $m_{PM} \gg m_{G(IR),(RR)}$. A liquid fuel directly injected in free volume combustion system corresponds to the mixture formation conditions typical for Diesel engine conditions (no air motion is considered). Direct fuel injection into porous reactor changes fuel distribution in space [20–22] resulting in different mixture formation conditions. This process, however, is significantly dependent on pore density, pore size, pore structure, as well as injection conditions (especially injection pressure).

In the present investigation, a real engine with mixture formation and combustion in porous reactor, as proposed by Durst and Weclas [8], is considered and simulated in a special combustion chamber as shown in Figure 3. A principle of internal combustion engine with mixture formation and

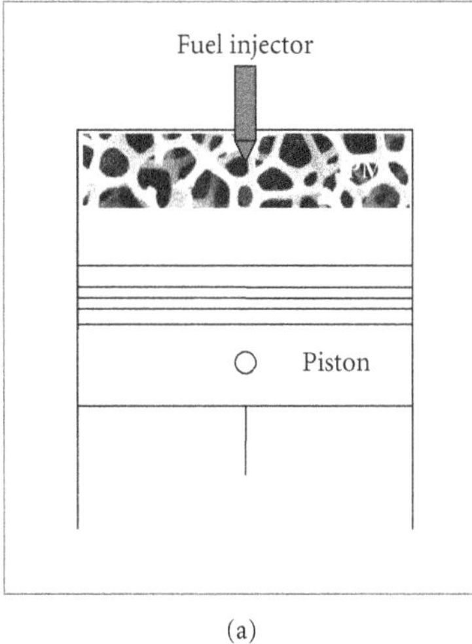

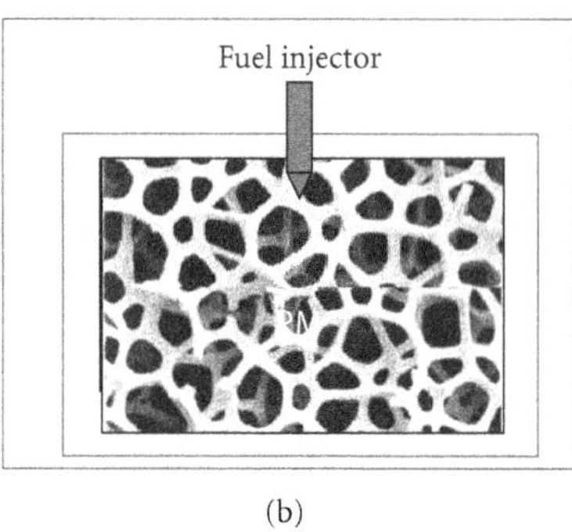

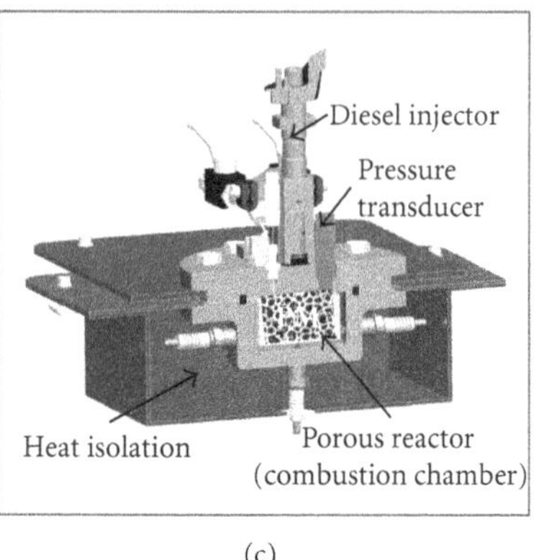

(a) (b) (c)

FIGURE 3: Modeling of real engine realizing combustion in porous reactor (a) with a constant volume reactor (b) and engine simulator used in preset investigation (c).

combustion in porous reactor assumes that, at the TDC of compression, the only available volume in the cylinder is the volume of porous reactor (Figure 3(a)). Fuel is directly injected into porous reactor and is completely trapped in its volume. For investigating the mixture formation and heat release process in engine, it is possible to reduce the system to the reactor volume corresponding to the engine conditions at TDC of compression (Figure 3(b)). For experimental investigation, these conditions have been realized in a special engine simulator built in the form of an isolated constant volume high-temperature and high-pressure combustion chamber equipped with common rail Diesel injection system and electronically controlled fuel injector. The fuel is then directly injected into porous reactor volume [16]. The high-pressure combustion chamber with porous reactor is mounted inside a low-pressure chamber to act as heat isolation (Figure 3(c)). Reactor can be heated electrically, and the current is supplied using electrodes of spark plugs mounted in the chamber. The combustion (dry) air is supplied to the chamber under selected pressure, and the flow rate control valve permits supply of required mass of air (for control of air excess ratio λ). Pressure transducer mounted in the combustion chamber allows very precise and highly time-resolved measurements of pressure changes after fuel injection starts. This pressure history allows the reconstruction of heat release process (combustion) and corresponding temperatures. It is assumed (and experimentally verified) that, during the heat release process, the heat losses through the combustion chamber walls are negligible.

2.3. Thermodynamics of Free Volume System and Porous Reactor. The mass of fuel m_{fuel} injected into system without heat losses through the system walls (see Figure 2) represents amount of energy supplied to the system E_{in} at initial thermodynamic conditions at the time instance of injection starts T_{IB} and p_{IB}:

$$E_{\text{in}} = m_{\text{fuel}} \cdot H_u, \tag{1}$$

where H_u is the heating value of injected fuel. During combustion process, this energy is converted into heat Q_{in}:

$$Q_{\text{in}} = \eta \cdot m_{\text{fuel}} \cdot H_u, \tag{2}$$

where η is the energy conversion efficiency. For the reason of the following analysis, it is assumed that, for the same amount of injected fuel, the conversion efficiency in free volume system η_{FV} and real porous reactor η_{RR} is the same.

In the case of free volume system with no heat losses through the system walls (Figure 2(a)) heat supplied to the system results in increasing internal energy $\Delta U_{\text{G(FV)}}$ of the gas trapped in this volume:

$$Q_{\text{in}} = \Delta U_{\text{G(FV)}}. \tag{3}$$

Increase of internal energy corresponds to the change in the gas temperature $\Delta T_{\text{G(FV)}}$ and gas pressure $\Delta p_{\text{G(FV)}}$ after injection starts (for initial air conditions $T_{\text{A-IB(FV)}}$ and $p_{\text{A-IB(FV)}}$):

$$\begin{aligned} \Delta T_{\text{G(FV)}} &= T_{\text{G(FV)}} - T_{\text{A-IB(FV)}}, \\ \Delta p_{\text{G(FV)}} &= p_{\text{G(FV)}} - p_{\text{A-IB(FV)}}. \end{aligned} \tag{4}$$

This pressure which is directly measured in an engine simulator (see Figure 3) is dependent on direct heat release rate.

In the case of system with porous reactor, the increase of internal energy as a result of heat supplied to the system Q_{in} is distributed in two components:

$$Q_{\text{in}} = \Delta U_{\text{G(PM)}} + \Delta U_{\text{PM}}, \tag{5}$$

where $\Delta U_{\text{G(PM)}}$ is the change in internal energy of the gas trapped in the porous reactor after fuel injection starts and ΔU_{PM} is the internal energy change of porous reactor. Internal energy change of the gas results in change of the gas temperature $\Delta T_{\text{G(PM)}}$ trapped in the reactor volume, and correspondingly the change in internal energy of the porous reactor results in ΔT_{PM}. In the case of ideal reactor (Figure 2(b)), it is assumed that the gas and reactor temperatures are the same at any instant of time after energy supply. From this assumption follows

$$\Delta T_{\mathrm{G(IR)}} \backsimeq \Delta T_{\mathrm{PM}}. \tag{6}$$

Under consideration of a high heat capacity reactor, which is many times higher than heat capacity of the gas trapped in the reactor volume (by a factor of 10^3 and more) the reactor temperature can only slightly change during the heat release process. This means that the gas temperature in an ideal reactor changes also very slightly. For almost constant gas temperature, the pressure changes in the ideal reactor volume resulting from the heat release process are very small $\Delta p_{\mathrm{G(IR)}} \sim 0$.

For real reactor conditions (see Figure 2(c)), the temperature of gas trapped in the reactor volume is not immediately equal the reactor temperature ($\Delta T_{\mathrm{G(RR)}} \neq \Delta T_{\mathrm{PM}}$). Thus, the change in internal energy of the system can be expressed as:

$$\begin{gathered} Q_{\mathrm{in}} = m_{\mathrm{G(RR)}} \cdot c_{\mathrm{VG}} \cdot \Delta T_{\mathrm{G(RR)}} + m_{\mathrm{PM}} \cdot c_{\mathrm{VPM}} \cdot \Delta T_{\mathrm{PM}}, \\ \Delta T_{\mathrm{G(RR)}} = \frac{Q_{\mathrm{in}} - (m_{\mathrm{PM}} \cdot c_{\mathrm{VPM}} \cdot \Delta T_{\mathrm{PM}})}{m_{\mathrm{G(RR)}} \cdot c_{VG}} \longrightarrow \Delta p_{\mathrm{G(RR)}}. \end{gathered} \tag{7}$$

Please note that the heat capacities c_{VG} and c_{VPM} are considered as mean values of those at initial and final temperatures.

According to the assumptions made in Figure 2 and to above analysis, it is possible to express the following relations:

$$\begin{gathered} \Delta T_{\mathrm{G(FV)}} > \Delta T_{\mathrm{G(RR)}} \gg \Delta T_{\mathrm{G(IR)}}, \\ \Delta p_{\mathrm{G(FV)}} > \Delta p_{\mathrm{G(RR)}} \gg \Delta p_{\mathrm{G(IR)}}. \end{gathered} \tag{8}$$

For a real reactor, it is possible also to express that

$$\Delta T_{\mathrm{G(RR)}} = \Delta T_{\mathrm{G(FV)}} - (K \cdot \Delta T_{\mathrm{PM}}), \tag{9}$$

where K is the factor characterizing heat capacity of the reactor and heat transfer between gas and reactor solid phase.

In all considered models (FV, IR, and RR), it was assumed that the fuel is homogeneously distributed throughout the combustion chamber volume. This is the reason why the assumption of equal energy conversion efficiency in FV and RR systems was made. In a real case, however, fuel injection conditions in a free volume and in porous reactor are different, especially regarding fuel distribution in space, vaporization, and mixing with air [16, 20–22]. Complete and quantitative description of real engine conditions is at the present time too complex; however, these differences are considered in analyses presented below.

2.4. Heat Accumulation in Porous Reactor. The following analysis gives insight into amount of heat transferred to the gas and to porous reactor based on the maximum temperature after energy supply (fuel injection starts). Real reactor conditions are considered with a simplification assumption that the energy conversion rate into heat is constant and is the same for free volume system and for porous combustion reactor. For a free volume system, the increase of internal energy based on the maximum combustion temperature is expressed as:

$$\Delta U_{\mathrm{G(FV)}} = U_{\mathrm{G(T_{max})}} - U_{\mathrm{G(T_{IB})}}, \tag{10}$$

where $U_{\mathrm{G(T_{max})}} = m_{\mathrm{G(FV)}} \cdot c_{\mathrm{VG}} \cdot T_{\mathrm{G(FV)}}$ and $U_{\mathrm{G(T_{IB})}} = m_{\mathrm{A(FV)}} \cdot c_{\mathrm{VA}} \cdot T_{\mathrm{A\text{-}IB(FV)}}$.

The authors have defined the efficiency η of the total supplied energy conversion into heat based on the maximum temperature of the heat release process, that is, until the time instance of achieving this maximum temperature:

$$\begin{gathered} \eta_{\mathrm{FV}} = \frac{\Delta U_{\mathrm{G(FV)}}}{E_{\mathrm{in}}}, \\ \eta_{\mathrm{FV}} = \frac{m_{\mathrm{G(FV)}} \cdot c_{\mathrm{VG}} \cdot T_{\mathrm{G(FV)}} - m_{\mathrm{A(FV)}} \cdot c_{\mathrm{VA}} \cdot T_{\mathrm{A\text{-}IB(FV)}}}{m_{\mathrm{fuel}} \cdot H_u} \cdot 100\%, \end{gathered} \tag{11}$$

where $m_G = m_{\mathrm{fuel}}(1 + 14.5 \cdot \lambda)$.

It must be noted that, in both free volume and in porous reactor systems, the fuel is directly injected into the system and the resulting mixture is not homogeneous in space. The time taken to achieve maximum temperature is not necessarily the same time where all the energy supplied is turned into heat. In porous reactor (depending on the injection parameters), this homogeneity can be improved. In the case of porous reactor, the system consists of gas trapped in porous reactor having much higher heat capacity as the gas has. In this case, the amount of heat transferred to the reactor solid phase based on the comparison to total amount of heat supplied under free volume conditions can be expressed as (based on $T_{\max}$)

$$\eta_{\mathrm{RR(PM)}} = \frac{\Delta U_{\mathrm{RR(PM)}}}{Q_{\mathrm{in}}} \cdot 100\% \tag{12}$$

under simplified assumption that $Q_{\mathrm{in(PM)}} = Q_{\mathrm{in(FV)}}$; thus,

$$\eta_{\mathrm{RR(PM)}} = \frac{\Delta U_{\mathrm{G(FV)}} - \Delta U_{\mathrm{G(PM)}}}{\Delta U_{\mathrm{G(FV)}}} \cdot 100\%. \tag{13}$$

3. Heat Release in a Real Porous Reactor under Diesel Engine-Like Conditions

Examples of pressure history and corresponding temperature for free volume system (Diesel engine-like conditions) are shown in Figure 4. Data are plotted for initial gas temperature $T_{\mathrm{A\text{-}IB(FV)}} = 500°\mathrm{C}$ at different initial pressures $p_{\mathrm{A\text{-}IB(FV)}}$. The mass of injected fuel is 23.8 mg. The pressure history shows a strong dependence of the heat release process on the initial pressure. Generally, the higher the initial chamber pressure is the faster is the heat release process and the shorter ignition delay period [16, 23]. Also the corresponding pressure peak level increases with initial pressure. This is not the case for combustion temperature which does not up with increasing pressure. This is due to the fact that with increasing initial pressure, air excess ratio also increases, which controls the combustion temperature. Similar analysis

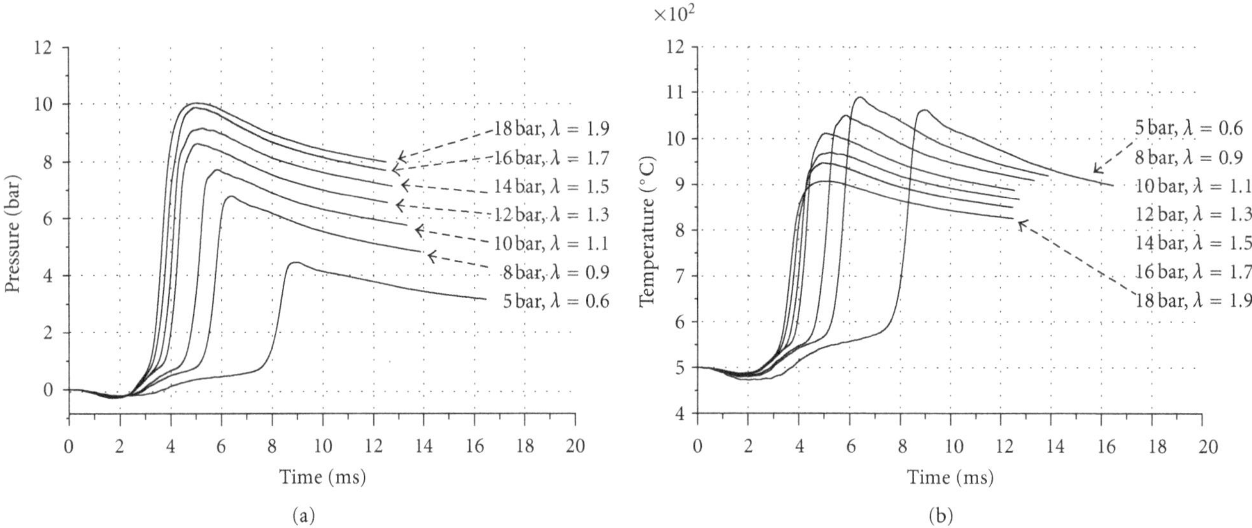

FIGURE 4: Pressure distribution (a) and temperature history (b) of the process in a free volume in time after Diesel injection starts at initial temperature 500°C and different initial pressures (mass of injected fuel is constant 23.8 mg).

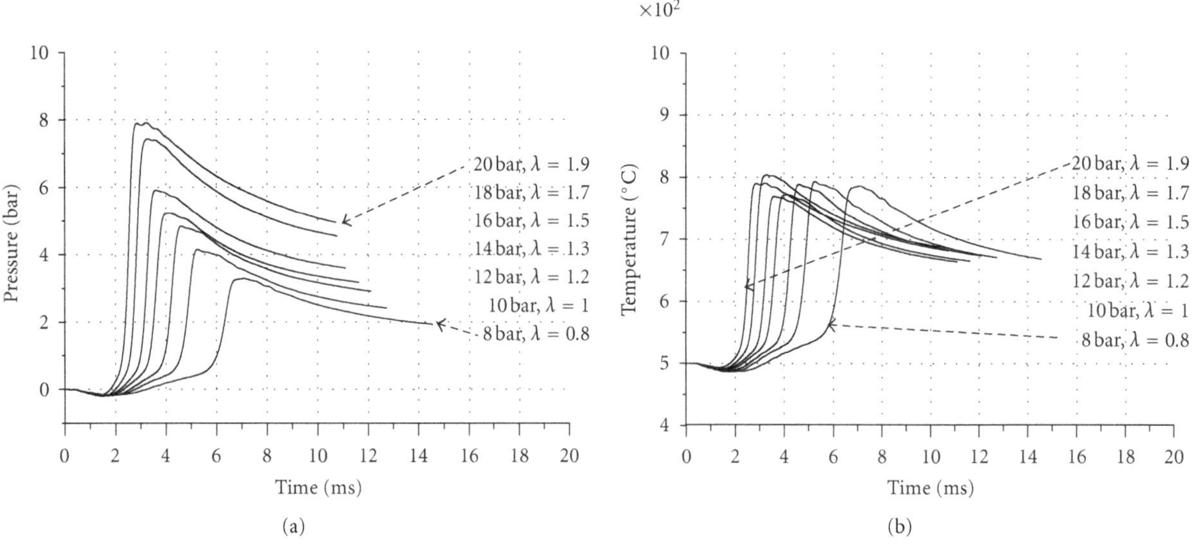

FIGURE 5: Pressure distribution (a) and temperature history (b) of the process in a SiC porous reactor having pore density 8 ppi in time after Diesel injection starts at initial reactor temperature 500°C and different initial pressures (mass of injected fuel is constant 23.8 mg)

is performed for porous reactor (see Figure 5). The data are plotted at $T_{\text{PM-IB}} = 500°\text{C}$ for different initial pressures $p_{\text{A-IB(RR)}}$. Pressure history shows reduced pressure peaks as compared to free volume system, and heat release process is qualitatively similar to free volume heat combustion. Reduced pressure peaks correspond to the heat accumulated in the porous reactor. Also, the corresponding combustion temperatures are reduced. The maximum temperature peaks in the porous reactor are much less dependent on air excess ratio than in a free volume system. This is due to the effect of heat accumulation in the reactor according to reactor heat capacity and heat transfer to the reactor solid phase. According to discussion made in Section 2, higher pore density and larger specific surface area of the porous reactor should result in more effective heat transfer to the reactor and increased heat accumulation in the solid phase of porous reactor. This effect can be observed in Figure 6 showing distribution of maximum combustion temperature after fuel injection in the case of free volume system and maximum gas temperature in porous reactors having different pore densities (8 ppi and 30 ppi). For reactor of higher pore density, the maximum temperature is significantly reduced at all investigated initial pressures. Almost constant temperature recorded in real porous reactors independently of initial pressure indicates the role of heat capacity and heat transfer conditions defining amount of heat accumulated in the porous reactors. This effect is observed for all investigated initial temperatures.

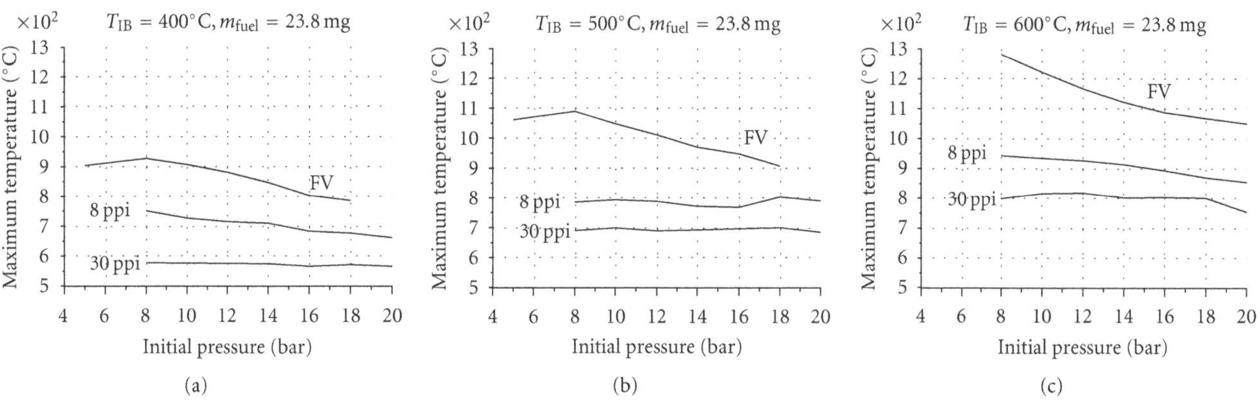

FIGURE 6: Distribution of maximum combustion temperature versus initial pressure in free volume system (FV) and in two porous reactors (8 ppi and 30 ppi) at three initial temperatures 400°C, 500°C, and 600°C for a constant mass of injected fuel.

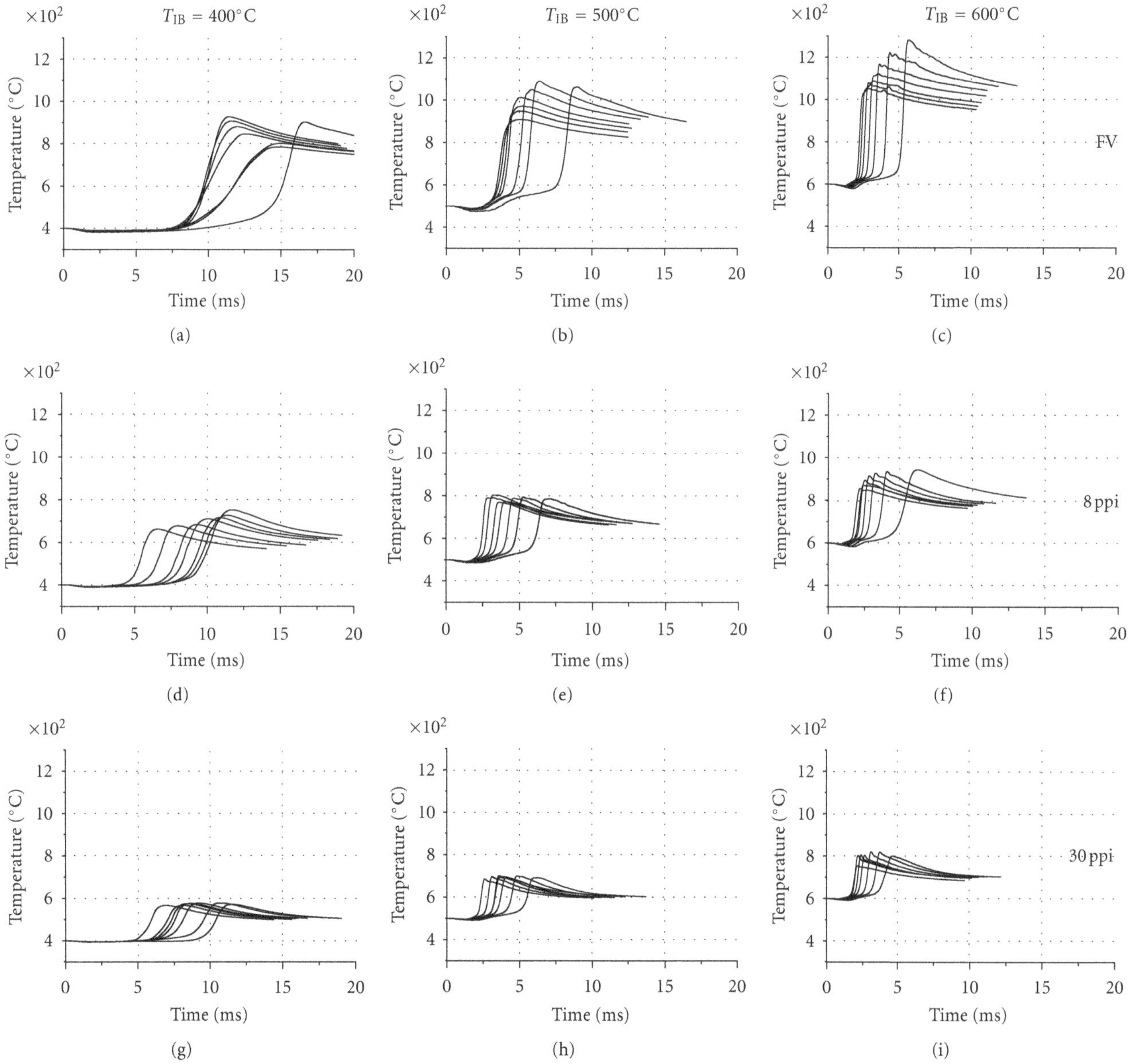

FIGURE 7: Temperature history after Diesel injection starts in free volume system (top), in low-pore-density SiC reactor 8 ppi (middle), and in high-pore-density SiC reactor 30 ppi at three initial temperatures 400°C, 500°C, and 600°C for different initial pressures from 8 to 20 bar (mass of injected fuel is constant 23.8 mg).

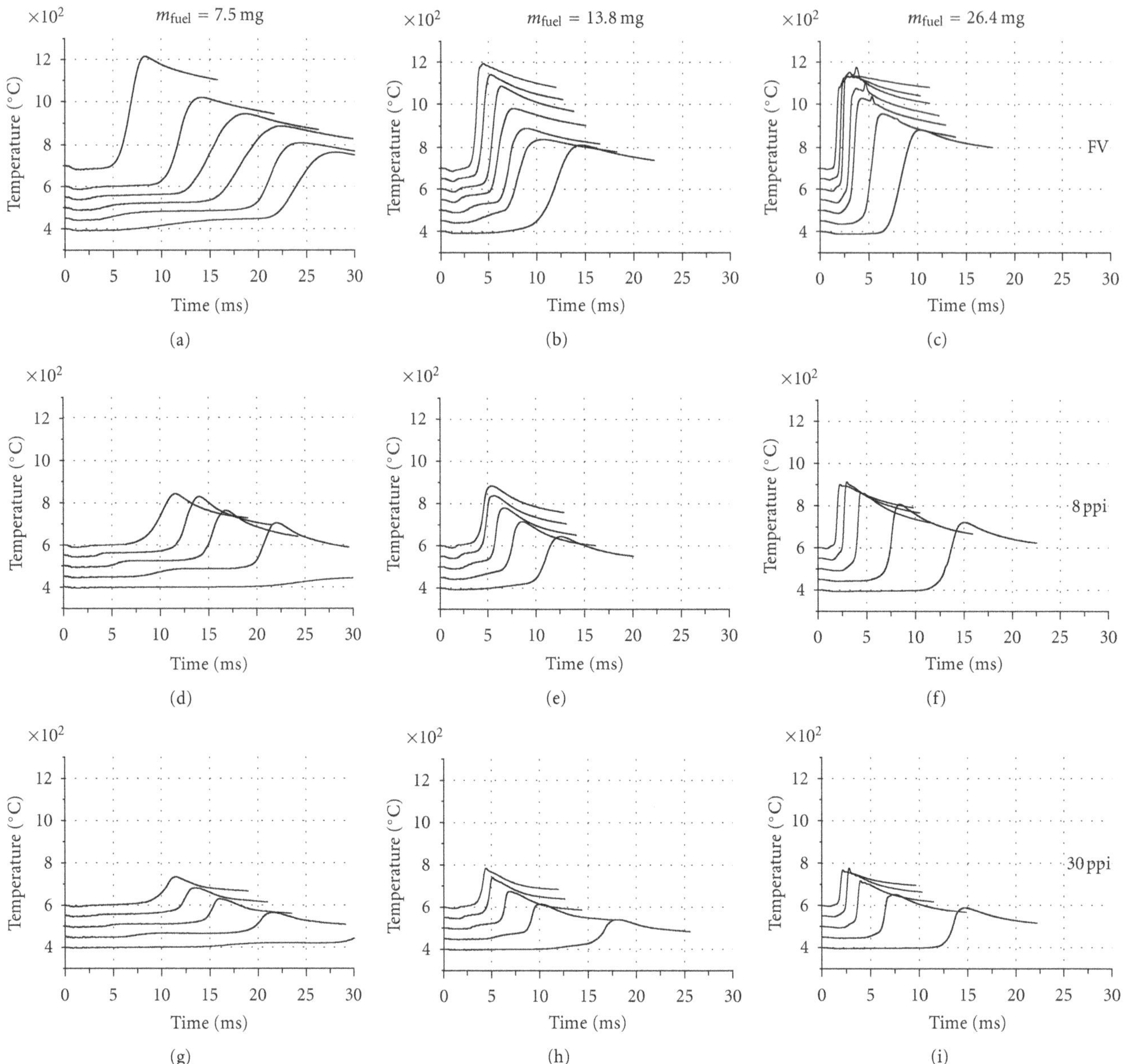

FIGURE 8: Temperature history after Diesel injection starts in free volume system (top), in low-pore-density SiC reactor 8 ppi (middle), and in high-pore-density SiC reactor 30 ppi for three mass of injected fuel (7.5 mg, 13.8 mg, and 26.4 mg) for a constant air excess ratio (initial temperatures and initial pressures varies).

The effect of initial temperature on the combustion temperature histories in free volume as well as in two porous reactors is investigated in Figure 7. Individual curves correspond to particular initial chamber pressure and to a given air excess ratio. As already indicated for free volume combustion, the process is significantly dependent upon initial pressure and air excess ratio. An "irregular" dependence of the combustion temperature on the initial chamber pressure is observed at lowest presented temperature (400°C). This "irregular" dependence means that the heat release process not necessarily gradually becomes faster with increasing pressure but is characterized by a range of pressures where the process is the fastest. Weclas et al. [16] have defined a positive pressure coefficient (PPC) corresponding to slightly lower temperatures and described nonregular dependence of reaction rate and delay time on the initial chamber pressure. Heat release process in porous reactors is faster and especially at lower initial temperature significantly accelerated as compared to free volume combustion (delay time). The combustion temperature (gas temperature according to the discussion in Section 2.2) in the real porous reactor is much less dependent on air excess ratio as a result of energy transport from the gas phase to the solid phase of the reactor. Significantly reduced temperature level is due to large heat capacity of the reactor and effective heat transfer (accumulation) in porous reactor. This gas temperature reduces with heat capacity and heat transfer (pore density and specific surface area) of the reactor. A nearly constant

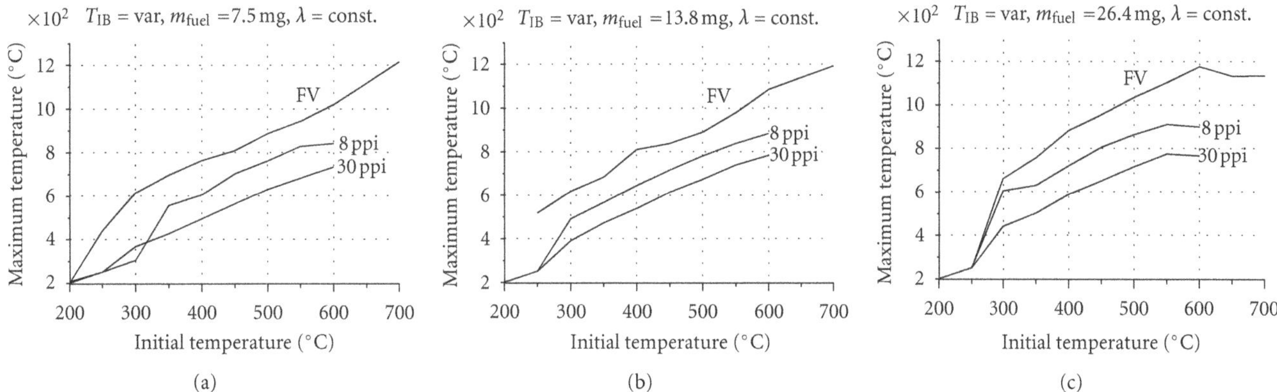

FIGURE 9: Distribution of maximum combustion temperature versus initial temperature in free volume system (FV) and in two porous reactors (8 ppi and 30 ppi) for three mass of injected fuel 7.5 mg, 13.8 mg, and 26.4 mg and a constant air excess ratio.

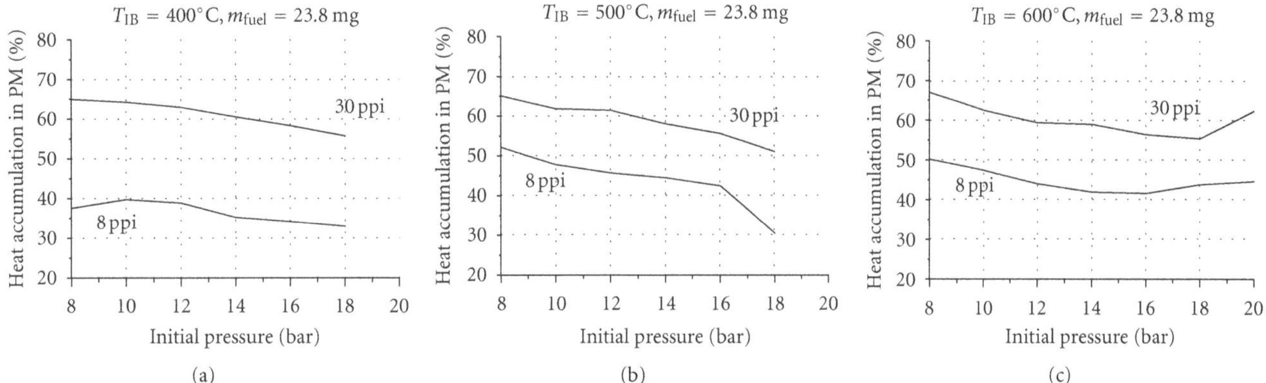

FIGURE 10: Distribution of amount of heat accumulated in porous reactors (8 ppi and 30 ppi) versus initial pressure at three initial reactor temperatures 400°C, 500°C, and 600°C for a constant mass of injected fuel.

gas temperature in porous reactor at 600°C indicates the effect of limited amount of heat to be transferred to real reactor (contrary to assumption made for ideal reactor). A critical influence on heat accumulation in porous reactor and for resulting combustion temperature is the amount of supplied heat. This amount of supplied heat is directly controlled by the mass of fuel injected into the combustion system. Figure 8 shows comparison of the gas temperature for three different mass of fuel injected to the free volume combustion chamber or to porous reactor. Data are plotted for different initial temperatures at different initial pressures. Air excess ratio for a free volume system is 1.5 and for porous reactors is 1.35. The process is analyzed in a limited period of time corresponding to initial 30 ms after fuel injection starts. The effect of amount energy supplied to the system together with heat accumulation in the reactor is clearly indicated. Combination of low amount of heat (low mass of injected fuel) with high pore density of the porous reactor results in only small change of the gas temperature as comparing to a free volume system. In the former case, most of energy is accumulated in a solid phase of the porous reactor significantly reducing the gas temperature change. The maximum temperature distribution is presented in Figure 9. Generally, the maximum combustion temperature gradually increases with increasing initial chamber temperature and is significantly reduced in porous reactors as compared to free volume combustion. This reduced temperature is a result of heat accumulation in a porous reactor, and the maximum temperature reduces with increasing reactor heat capacity, pore density, and specific surface area.

According to simplified analysis presented in Section 2.4, the corresponding amount of heat accumulated in porous reactor is shown in Figure 10. With increasing initial pressure, the heat release process is faster, and, under real reactor conditions, less heat can be transferred to the reactor. Generally, reactor having higher pore density can accumulate more heat under the same conditions. This analysis must, however, consider differences in fuel injection and mixture formation conditions inside the porous reactors having different pore densities. Such differences in process conditions may result in different efficiency of energy conversion into heat.

4. Concluding Remarks

A real porous reactor was considered under Diesel engine-like conditions simulated in a special combustion chamber. Thermodynamic conditions of the heat release process in combustion reactor have been analyzed. For a comparison, the same analysis was performed for a free volume (Diesel) combustion chamber. In all presented investigations, a real Diesel engine fuel injection and mixture formation conditions have been applied. Such a complex nonhomogeneous and time-dependent conditions are expected in a real engine with combustion in porous reactor. First of all, the effect of reactor heat capacity, pore density, specific surface area, and pore structure on the thermodynamic of the heat release process has been indicated. It was observed that heat accumulated in a solid phase of porous reactor significantly influences the thermodynamic conditions of the process. In an ideal reactor, it was assumed that the gas trapped inside reactor volume has the same temperature as the reactor temperature at any instant of time after fuel injection starts. Results obtained confirmed that it is important to consider real reactor conditions under which the gas and porous reactor temperatures are not the same. This effect is influenced by initial pressure and temperature (modeling conditions at TDC of compression) and by reactor parameters (heat capacity, pore density, specific surface area, pore structure). The temperature of the gas trapped in porous reactor volume during the heat release process is less dependent on air excess ratio as it is the case for free volume system. According to the reactor heat capacity and mixture formation conditions (fuel distribution, vaporization, and mixing in reactors having different pore densities), the maximum combustion temperature is significantly lowered as compared to free volume combustion chamber. This temperature, in all considered systems, is dependent on initial pressure. As expected from the analytical model of the porous reactor, the maximum pressure change during the heat release process in porous reactor is much less than in a free volume system. However, it must be underlined that the qualitative behavior of the heat release process in porous reactors and in free volume combustion chamber is similar. Also, the time scale of the process is similar. These similarities may support applicability of this kind of combustion to real engine conditions utilizing differences in thermodynamic conditions of the process.

Acknowledgments

M. Weclas thanks the German Ministry of Education and Research (BMBF) and Research Council (AiF) for financial support of the presented investigation (Project no. 17N2207). The authors thank Mr. Björn Leykauf and Mr. Iker Igartua García for their contribution to this investigation.

References

[1] V. S. Babkin, A. A. Korzhavin, and V. A. Bunev, "Propagation of premixed gaseous explosion flames in porous media," *Combustion and Flame*, vol. 87, no. 2, pp. 182–190, 1991.

[2] V. V. Martynenko, R. Echigo, and H. Yoshida, "Mathematical model of self-sustaining combustion in inert porous medium with phase change under complex heat transfer," *International Journal of Heat and Mass Transfer*, vol. 41, no. 1, pp. 117–126, 1998.

[3] A. A. M. Oliveira and M. Kaviany, "Nonequilibrium in the transport of heat and reactants in combustion in porous media," *Progress in Energy and Combustion Science*, vol. 27, no. 5, pp. 523–545, 2001.

[4] D. Trimis and F. Durst, "Combustion in a porous medium-advances and applications," *Combustion Science and Technology*, vol. 121, no. 1–6, pp. 153–168, 1996.

[5] F. Durst, M. Keppler, and M. Weclas, "Air-assisted nozzle applied to very compact, ultra-low emission porous medium oil-burner," in *the 3rd International Workshop on SPRAY*, Lampoldshausen, Germany, 1997.

[6] M. A. Mujeebu, M. Z. Abdullah, A. A. Mohamad, and M. Z. A. Bakar, "Trends in modeling of porous media combustion," *Progress in Energy and Combustion Science*, vol. 36, no. 6, pp. 627–650, 2010.

[7] F. Durst and M. Weclas, "Porous Medium (PM) combustion technology and its application to internal combustion engines: a new concept for a near-zero emission engine," in *Applied Optical Measurements*, M. Lehner and D. Mewes, Eds., Springer, 1999.

[8] F. Durst and M. Weclas, "A new type of internal combustion engine based on the porous-medium combustion technique," *Proceedings of the Institution of Mechanical Engineers, Part D*, vol. 215, no. 1, pp. 63–81, 2001.

[9] F. Durst and M. Weclas, "A new concept of I.C. engine with homogeneous combustion in Porous Medium (PM)," in *the 5th International Symposium on Diagnostics and Modeling of Combustion in Internal Combustion Engines (COMODIA '01)*, Nagoya, Japan, 2001.

[10] M. Kaviany, "In cylinder-thermal regeneration: Porous-Foam engine regenerator," in *Principles of Heat Transfer in Porous Media*, Springer, New York, NY, USA, 1999.

[11] C. W. Park and M. Kaviany, "Evaporation-combustion affected by in-cylinder, reciprocating porous regenerator," *Journal of Heat Transfer*, vol. 124, no. 1, pp. 184–194, 2002.

[12] M. Polasek and J. Macek, "Homogenization of Combustion in Cylinder of CI Engine Using Porous Medium," SAE Technical Paper 2003-01-1 085, 2003.

[13] H. Liu, M. Xie, and D. Wu, "Simulation of a porous medium (PM) engine using a two-zone combustion model," *Applied Thermal Engineering*, vol. 29, no. 14-15, pp. 3189–3197, 2009.

[14] L. Schlier, W. Zhang, N. Travitzky, P. Greil, J. Cypris, and M. Weclas, "Macro-cellular silicon carbide reactors for non-stationary combustion under piston engine-like conditions," *International Journal of Applied Ceramic Technology*, vol. 8, no. 5, pp. 1237–1245, 2011.

[15] M. Weclas, "Potential of porous media combustion technology as applied to internal combustion engines," *Journal of Thermodynamics*, vol. 2010, Article ID 789262, 39 pages, 2010.

[16] M. Weclas, J. Cypris, and T. M. A. Maksoud, "Combustion of Diesel sprays under real-engine like conditions: analysis of low- and high-temperature oxidation processes, ILASS – Europe 2010," in *the 23rd Annual Conference on Liquid Atomization and Spray Systems*, Brno, Czech Republic, September 2010, Paper no.ID ILASS10-40.

[17] M. Weclas, "Porous media in internal combustion engines," in *Cellular Ceramics-Structure, Manufacturing, Properties and Applications*, M. Scheffler and P. Colombo, Eds., Wiley, 2005.

[18] M. Weclas, "High velocity CR Diesel jet impingement on to porous structure and its utilization for mixture homogenization in I.C. engines," in *the Drop/wall interaction: Industrial applications, Experiments and Modeling Workshop (DITICE '06)*, Bergamo, Italy, May 2006.

[19] M. Weclas and R. Faltermeier, "Diesel jet impingement on small cylindrical obstacles for mixture homogenization by late injection strategy," *International Journal of Engine Research*, vol. 8, no. 5, pp. 399–413, 2007.

[20] M. Weclas, "Some fundamental observations on the diesel jet destruction and spatial distribution in highly porous structures," *Journal of Porous Media*, vol. 11, no. 2, pp. 125–144, 2008.

[21] M. Weclas, "Homogenization of liquid distribution in space by Diesel jet interaction with porous structures and small obstacles," in *the 22nd European Conference on Liquid Atomization and Spray Systems*, Como, Italy, September 2008, Paper no. ID ILASS08-A003.

[22] M. Weclas and J. Cypris, ""Distribution-nozzle" concept: a method for Diesel spray distribution in space for charge homogenization by late injection strategy, ILASS–Europe 2010," in *the 23rd Annual Conference on Liquid Atomization and Spray Systems*, Brno, Czech Republic, September 2010, Paper no. ID ILASS10-39.

[23] M. Weclas and J. Cypris, "Characterization of low- and high-temperature oxidation processes under non-premixed Diesel-engine like conditions," submitted to *International Journal of Engine Research*.

Modified Lennard-Jones Potentials with a Reduced Temperature-Correction Parameter for Calculating Thermodynamic and Transport Properties: Noble Gases and Their Mixtures (He, Ne, Ar, Kr, and Xe)

Seung-Kyo Oh

Pharmaceutics and Biotechnology Department, Konyang University, 121 University Road, Nonsan, Chungnam 320-711,Republic of Korea

Correspondence should be addressed to Seung-Kyo Oh; sunkist@konyang.ac.kr

Academic Editor: Bill Acree

The three-parameter Lennard-Jones (12-6) potential function is proposed to calculate thermodynamic property (second virial coefficient) and transport properties (viscosity, thermal conductivity, and diffusion coefficient) of noble gases (He, Ne, Ar, Kr, and Xe) and their mixtures at low density. Empirical modification is made by introducing a reduced temperature-correction parameter τ to the Lennard-Jones potential function for this purpose. Potential parameters (σ, ε, and τ) are determined individually for each species when the second virial coefficient and viscosity data are fitted together within the experimental uncertainties. Calculated thermodynamic and transport properties are compared with experimental data by using a single set of parameters. The present study yields parameter sets that have more physical significance than those of second virial coefficient methods and is more discriminative than the existing transport property methods in most cases of pure gases and of gas mixtures. In particular, the proposed model is proved with better results than those of the two-parameter Lennard-Jones (12-6) potential, Kihara Potential with group contribution concepts, and other existing methods.

1. Introduction

Accurate representation of thermodynamic and transport properties is essential to process engineers to design and optimize equipment and chemical processes. Second virial coefficient is an important quantity which is useful in calculating vessel size from volumetric data, heating requirements from calorimetric data, and stage requirements from phase equilibrium data. Transport properties such as viscosity, thermal conductivity, and diffusion coefficient are critically important parameters in many engineering applications: for the determination of pipeline, heatexchanger and separation equipment size, mass transfer efficiency of reservoir of oils, and the power required to pump fluid [1].

The intermolecular forces are of great importance to scientists in a wide field of disciplines as information of these interactions provides the progress of collisions between molecules and determines the bulk properties of substances. Approximation of thermodynamic and transport properties from statistical mechanics requires a realistic intermolecular potential [2]. The theoretical basis in statistical mechanics for the virial equation is one of its attractions. The viral equation truncated after the second term is a popular tool to calculate accurate thermodynamic properties at low or moderate densities. A number of investigators have emphasized the determination of second virial coefficient through experiments and correlations. When Chapman-Enskog gas kinetic theory [3] allows the prediction of transport properties, the potential energy of molecular interactions is known as a function of intermolecular separation and orientation. A description of the spherically symmetric potential as a function of intermolecular separation, averaged over all molecular orientations, suffices to calculate dilute gas viscosities, thermal conductivities, and diffusion coefficients of monoatomic gases.

A realistic intermolecular potential allows the calculation of thermodynamic and transport properties. A lot of studies have focused on individual properties like second virial coefficient or viscosity for the determination of intermolecular potential parameters [2]. Potential parameters of any given model that give the best fit for thermodynamic and transport properties (e.g., second virial coefficient, viscosity, thermal conductivity, and diffusion coefficient) are generally different. Therefore, for a simple model such as Lennard-Jones potential, there is one specific set of potential parameters suitable for each property, producing significantly different results [4].

Several investigators [2, 5–7] have used statistical mechanics and kinetic theory of gases to represent thermodynamic and transport properties with a single set of molecular parameters, namely, those appearing in an intermolecular potential function. The Lennard-Jones (12-6) potential has been widely used for the representation of thermodynamic and transport properties of normal fluids. In one particularly interesting study of Tee et al. [6], a single set of molecular parameters was evaluated from the Lennard-Jones (12-6) potential for each species; in this procedure, viscosity data for each substance was fitted first by least-squares analysis, second virial coefficient was fitted next, and the data on second virial coefficient and viscosity were statistically analyzed simultaneously to develop corresponding states correlations with a single set of potential parameters for each substance. They concluded that when second virial coefficient and viscosity data were fitted together, their sets of molecular parameters give the best overall fit to the data for each species and tend to be least affected by experimental errors; beside that their results are quite comparable to those determined individually from viscosity. Potential parameters obtained in this manner were proved to be successful in predicting second virial coefficients and dilute gas viscosities for molecules ranging in shape from spherical to chains as long as n-heptane with good result. Hence, the Lennard-Jones potential is still attractive for its simplicity and capability of predicting noble gas properties if its weak point is compensated for and its accuracy is improved.

The objective of this study is to represent thermodynamic property (second virial coefficient) and transport properties (viscosity, thermal conductivity, and diffusion coefficient) of noble gases (He, Ne, Ar, Kr, and Xe) and their binary mixtures at low density using a single set of modified Lennard-Jones (12-6) potential parameters. For this purpose, a temperature-correction parameter was introduced to the reduced temperature T^* in the Lennard-Jones (12-6) potential function. A set of potential parameters was determined when the second virial coefficient and viscosity data are fitted simultaneously within their experimental errors, separately for each noble gas; parameters obtained in this manner were used in all subsequent calculations of properties such as thermal conductivity and diffusion coefficient, in which data were not supplied to parameter estimations, and in mixture property computations. Validity of the modified Lennard-Jones (12-6) potential with a reduced temperature-correction parameter was tested with good results in comparison with other existing methods.

2. Theory

In this paper, special focus was placed on the Lennard-Jones (12-6) intermolecular potential for the computations of noble gas properties, even though more accurate potentials exist. A form of this potential was first established by Lennard-Jones [8] and is a mathematically simple model that approximates the interaction between a pair of neutral atoms or molecules. The most common expression of the Lennard-Jones (12-6) potential has the form

$$U(r) = 4\varepsilon\left[\left(\frac{\sigma}{r}\right)^{12} - \left(\frac{\sigma}{r}\right)^{6}\right], \tag{1}$$

where $U(r)$ is the intermolecular potential energy as a function of the separation distance between a pair of molecules, ε is the depth of the potential well in Joule, and σ is the finite distance in angstrom Å at which the interparticle potential becomes zero.

These potential parameters can be fitted by least-squares analysis and lead to provide accurate calculations of dilute gas thermodynamic property (second virial coefficient) and transport properties (viscosity, thermal conductivity, and diffusion coefficient) of noble gases, as summarized below.

2.1. Second Virial Coefficient. From statistical mechanics, the relations between second virial coefficient and intermolecular potential functions were theoretically derived; in particular, explicit expression of the second virial coefficient for the Lennard-Jones (12-6) potential was rigorously derived by Hirschfelder et al. [4] for a computational use:

$$B(T) = -\frac{2}{3}\pi N_A \sigma^3 \sum_{n=0}^{\infty} \frac{2^{n+1/2}}{4n!}\Gamma\left(\frac{2n-1}{4}\right) T^{*-(2n+1)/4} \tag{2}$$

in which N_A is the Avogadro's constant $6.022{\cdot}10^{23}$ mol^{-1}, T^* is the reduced (dimensionless) temperature $k_B T/\varepsilon$, and κ_B is the Boltzmann constant $1.3806488{\cdot}10^{-23}$ JK^{-1}. However, Kojima [9] observed that it is quite effective for calculating virial coefficients from the Stockmayer intermolecular potential model with an aid of introducing a reduced temperature-correction parameter of τ as $(T - \tau)$ instead of using temperature T. This temperature-correction parameter was also proved in developing a new virial equation of state by Ichikura et al. [10].

In the present study, our observations on the accurate approximation not only of thermodynamic property (second virial coefficient), but also of transport properties (viscosities, thermal conductivities, and diffusion coefficients) in the dilute gaseous phase of noble gases were made by introducing temperature-correction parameter τ to the reduced temperature T^* shown in the Lennard-Jones (12-6) potential function:

$$T^* = \frac{\kappa_B (T - \tau)}{\varepsilon}. \tag{3}$$

Then the second virial coefficient can be calculated from (2) and (3) when three potential parameters (σ, ε, and τ) of the modified Lennard-Jones (12-6) potential are fitted together to second virial coefficient and viscosity data separately for each substance.

For interaction of different molecular species, the combining rules are used for the molecular distance, energy, and reduced temperature-correction parameter for mixture computations:

$$\sigma_{12} = \frac{\sigma_1 + \sigma_2}{2}, \quad (4)$$

$$\varepsilon_{12} = \left(\varepsilon_1 \varepsilon_2\right)^{1/2}, \quad (5)$$

$$\tau_{12} = \frac{\tau_1 + \tau_2}{2}. \quad (6)$$

For spherical molecules, these equations are of high accuracy for the prediction of second cross-virial coefficient B_{12} required in (2) and other mixture properties discussed later.

2.2. Viscosity. Transport properties (viscosity, thermal conductivity, and diffusion coefficient) at low density can be calculated by using Chapman-Enskog kinetic theory [3], which has been applied rigorously to monatomic gases in a number of studies [11]. The dilute gas viscosity can be well presented by the Chapman-Enskog approximation derived from the kinetic theory:

$$\eta = 26.693 \frac{\sqrt{MT}}{\sigma^2 \Omega^{(2,2)^*}(T^*)}, \quad (7)$$

where M is the molecular weight (gram mol^{-1}), T is the absolute temperature in Kelvin, and the viscosity η is in micropoise μP. For mixture viscosity calculations, the formula proposed by Hirschfelder et al. [4] was used in this study, in which the interaction quantity η_{12} must be determined to employ this method in advance:

$$\eta_{12} = 26.693 \frac{\sqrt{2M_1 M_2 T/(M_1 + M_2)}}{\sigma^2{}_{12} \Omega^{(2,2)^*}(T^*_{12})} \quad (8)$$

in which M_1 and M_2 are the molecular weights of the components 1 and 2, respectively, and $T^*{}_{12}$ is the modified reduced temperature for mixture calculations, $k_B(T - \tau_{12})/\varepsilon_{12}$. The composition dependence of viscosity on the binary gas mixture is defined as follows:

$$\frac{1}{\eta_{\text{mix}}} = \frac{X_\eta + Y_\eta}{1 + Z_\eta}, \quad (9)$$

$$X_\eta = \frac{y_1^2}{\eta_1} + \frac{2y_1 y_2}{\eta_{12}} + \frac{y_2^2}{\eta_2}, \quad (10)$$

$$Y_\eta = \frac{3}{5} A^*{}_{12} \left[\frac{y_1^2}{\eta_1}\left(\frac{M_1}{M_2}\right) + \frac{2y_1 y_2}{\eta_{12}} \left[\frac{(M_1 + M_2)^2}{4M_1 M_2}\right] \times \left(\frac{\eta_{12}^2}{\eta_1 \eta_2}\right) + \frac{y_2^2}{\eta_2}\left(\frac{M_2}{M_1}\right)\right], \quad (11)$$

$$Z_\eta = \frac{3}{5} A^*{}_{12} \left[y_1^2 \left(\frac{M_1}{M_2}\right) + 2y_1 y_2 \times \left[\frac{(M_1 + M_2)^2}{4M_1 M_2} \left(\frac{\eta_{12}}{\eta_1} + \frac{\eta_{12}}{\eta_2} - 1\right)\right] + y_2^2 \left(\frac{M_2}{M_1}\right)\right], \quad (12)$$

$$A^*{}_{12} = \left(\frac{\Omega^{(2,2)^*}}{\Omega^{(1,1)^*}}\right), \quad (13)$$

where η_{mix} is the mixture viscosity, and y_i, M_i, and η_i are the mole fractions, the molecular weights, and the viscosities at the mixture temperature of the pure components i (i = 1, 2). And the quantities $\Omega^{(2,2)^*}$ and $\Omega^{(1,1)^*}$ are the collision integrals for viscosity and diffusion coefficient, respectively, and are defined as a function of the reduced temperature T^* which depends on the intermolecular potential selected. Neufeld et al. [12] proposed analytical approximations to transport collision integrals for the Lennard-Jones (12-6) potential in the range $0.3 < T^* < 100$, being convenient for easy computer application:

$$\Omega^{(2,2)^*}(T^*) = \frac{1.16145}{T^{*0.14874}} + \frac{0.52487}{\exp(0.77320T^*)} + \frac{2.16178}{\exp(2.43787T^*)}, \quad (14)$$

$$\Omega^{(1,1)^*}(T^*) = \frac{1.06036}{T^{*0.15610}} + \frac{0.1930}{\exp(0.47635T^*)} + \frac{1.03587}{\exp(1.52996T^*)} + \frac{1.76474}{\exp(3.89411T^*)}. \quad (15)$$

2.3. Thermal Conductivity. Since the Chapman-Enskog gas kinetic theory uses a common basis for the evaluation of viscosity and thermal conductivity, the statistical expression for the thermal conductivity involves the same collision integral as does the viscosity. For a pure monoatomic gas at low density, which has no rotational or vibrational degrees of freedom, thermal conductivity was calculated through a rigorous analysis by Brokaw [13]:

$$\lambda = 0.0026693 \frac{\sqrt{T/M}}{\sigma^2 \Omega^{(2,2)^*}(T^*)} = \frac{15}{4}\left(\frac{R}{M}\right)\eta, \quad (16)$$

where the thermal conductivity λ is microwatts per meter per degree Kelvin in mW m^{-1} K^{-1} and R is the universal gas constant 83.14 cm^3 bar mol^{-1} k^{-1}. Thus, the thermal conductivity of the noble gas can be calculated from the estimated potential parameters or from experimental viscosity data which are generally available for the molecule. In the present work, pure viscosity value obtained from (7) was used for the calculation of thermal conductivity of pure substance. It thus implies that a check on the thermal conductivity serves as a cross-check between the viscosity and thermal conductivity

TABLE 1: Potential parameters of Lennard-Jones (12-6) Potential with a reduced temperature-correction parameter.

Group	σ/Å	(ε/κ_B)/K	τ/K
Helium	2.628	5.465	−0.836
Neon	2.775	36.831	−2.468
Argon	3.401	116.81	5.642
Krypton	3.601	164.56	11.41
Xenon	4.055	218.18	13.09

data and not necessarily as a check on the potential function chosen [14].

Various prediction methods for estimating mixture thermal conductivity have appeared, one of which is essentially empirical and it is reduced to some form of the Wassiljewa equation [15]:

$$\lambda_{\text{mix}} = \sum_{i=1}^{n} \frac{y_i \lambda_i}{\sum_{j=1}^{n} y_j A_{ij}}, \tag{17}$$

where λ_{mix} is the mixture thermal conductivity and A_{ij} is a combinational factor. This factor is empirically expressed by Mason and Saxena [16] as

$$A_{ij} = \frac{1.065}{2\sqrt{2}} \frac{\left[1 + \left(\lambda_i/\lambda_j\right)^{1/2} \left(M_i/M_j\right)^{1/4}\right]^2}{\left(1 + M_i/M_j\right)^{1/2}}, \tag{18}$$

where λ_i are the thermal conductivities of pure component determined from (16), which are sufficient to predict mixture thermal conductivity when its measurements are not available in the literature.

2.4. Diffusion Coeffici t. The Chapman-Enskog expression for binary diffusion coefficient of dilute gas is presented by Hirschfelder et al. [4]:

$$D_{12} = 0.0026693 T^{3/2} \frac{\sqrt{(M_1 + M_2)/2M_1M_2}}{P\sigma^2{}_{12}\Omega^{(1,1)^*}(T^*{}_{12})}, \tag{19}$$

where D_{12} is in cm^2 sec^{-1} and P is in bar. And $\Omega^{(1,1)^*}$ is the collision integral for diffusion coefficient of nonpolar Lennard-Jones (12-6) potential given by (15). When molecules 1 and 2 are identical, (19) becomes expression for the self-diffusion coefficient:

$$D = 0.0026693 T^{3/2} \frac{\sqrt{1/M}}{P\sigma^2 \Omega^{(1,1)^*}(T^*)}. \tag{20}$$

3. Results and Discussion

3.1. Pure Noble Gases. As a part of systematic program of our researches, modified Lennard-Jones (12-6) potential function with a reduced temperature-correction parameter was applied to noble gases (He, Ne, Ar, Kr, and Xe) for the computation of thermodynamic (second virial coefficient) and transport properties (viscosity, thermal conductivity, and self-diffusion coefficient) at low density.

Using (2), (3), and (7), three potential parameters (σ, ε, and τ) were evaluated from the simultaneous regression of second virial coefficient and viscosity data separately for each species. For instance, three parameter values ($\sigma_{\text{He}}, \varepsilon_{\text{He}}$, and τ_{He}) for helium gas were determined from second virial coefficient and viscosity data of pure helium gas. Potential parameters of other noble gases were evaluated in an analogous manner.

A critical review of the literature on second virial coefficient was achieved by Dymond et al. [17]. They provided the recommended values of virial coefficients for each compound fitted to a smoothing function of temperature by the least-squares criterion. This smoothing function for second virial coefficient is the polynomial of reciprocal temperature with usually three terms. In particular, two different smoothing correlations of helium were given in the temperature ranges between 1.59 K and 35.1 K and between 35.1 K and 1473.15 K. We have used their comprehensive compilation of the second virial coefficient data as our data source. And pure viscosity data required for the potential parameter determinations were all taken from Stephan and Lucas [18], even though other recommended or various sets of data exist. Nonlinear least-squares parameter estimation subroutine based on the Levenberg-Marquardt algorithm supplied by IMSL STAT/library [19] was used in this data regression, in which each data point was weighted by its estimated experimental uncertainty taken from the corresponding Refs. A set of potential parameters individually for each substance can then be estimated when the following objective function is minimized:

$$\Phi = \sum_{i=1}^{n_B} \left(\frac{B_{i,\text{obsd}} - B_{i,\text{calc}}}{\delta_{i,B}} \right)^2 + \sum_{i=1}^{n_\eta} \left(\frac{\eta_{i,\text{obsd}} - \eta_{i,\text{calc}}}{\eta_{i,\text{obsd}} \delta_{i,\eta}} \right)^2 \tag{21}$$

in which δ_B is the observed uncertainty of second virial coefficient in cm^3 mol^{-1} and δ_η is of dimensionless fractional uncertainty of viscosity in %. It is noted that the first term is designated for second virial coefficient and the second for viscosity.

Table 1 summarizes determined parameter values of modified Lennard-Jones (12-6) potential (σ, ε, and τ) of noble gases (He, Ne, Ar, Kr, and Xe). In Table 2, resulting deviations between observed and regressed second virial coefficient data are given on an RMSD (root-mean-square deviation) basis in cm^3 mol^{-1}, which is defined by

$$\text{RMSD} = \sqrt{\frac{1}{n_B} \sum_{i=1}^{n_B} \left(B_{i,\text{obsd}} - B_{i,\text{calc}}\right)^2}. \tag{22}$$

Comparisons of the proposed method with other existing methods are shown in Table 2 along with their data sources and observed temperature ranges. The average RMSD between a total of observed and calculated 735 second virial coefficient data of five noble gases by the present method was 3.33 cm^3 mol^{-1}, indicating that the proposed method compares very well with the Dymond's correlations

TABLE 2: Deviations between experimental and calculated second virial coefficients of pure noble gases.

Compound	Number of data points	Average RMSD in B (cm^3 mol^{-1}) Present study	Dymond et al.	L-J	Kihara	Tsonopoulos	ΔT (K)	Reference
			Regression results					
Helium	304	5.44	5.07[a]	10.9	10.8	39.1	2–1473	[2]
Neon	55	0.28	0.34	0.45	1.38	1.56	44–873	[2]
Argon	105	0.40	0.53	2.43	1.73	1.88	75–773	[2]
Krypton	157	2.87	3.18	5.75	3.67	6.90	107–873	[2]
Xenon	114	2.50	2.67	4.32	2.53	3.51	160–973	[2]
Average	735	3.33	3.29	6.79	5.99	18.6		
			Prediction results					
Helium	8	4.13	1.08	10.8	4.29	14.9	343–374	[27]
Xenon	9	0.50	1.37	2.57	2.11	2.57	206–273	[28]
Average	17	2.21	1.23	6.44	3.14	8.37		

[a]Two different smoothing functions were used in the temperature ranges between 1.59 K and 35.1 K, and between 35.1 K and 1473.15 K.

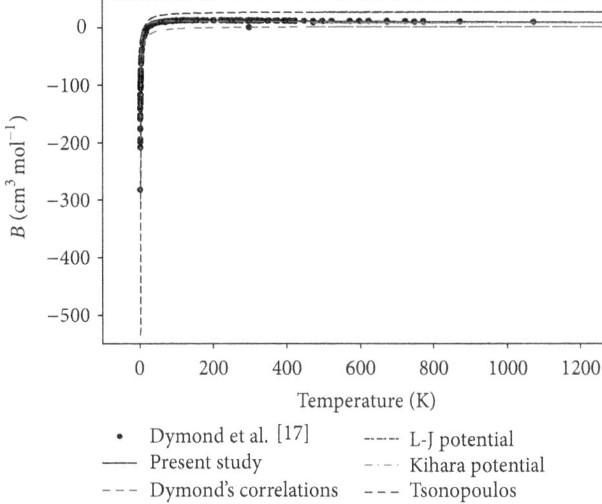

FIGURE 1: Comparison of measured and calculated second virial coefficients for He.

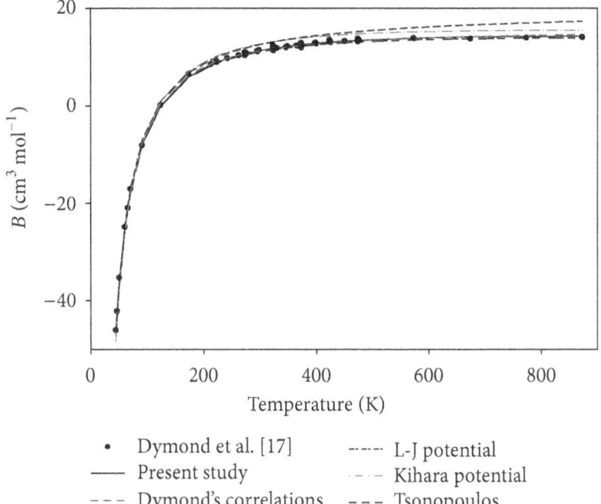

FIGURE 2: Comparison of measured and calculated second virial coefficients for Ne.

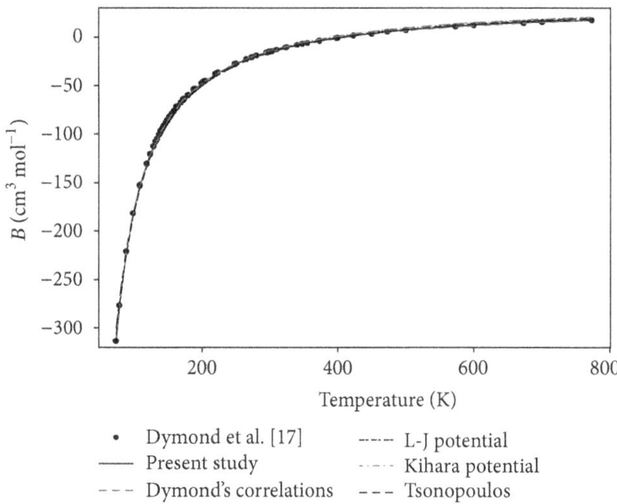

FIGURE 3: Comparison of measured and calculated second virial coefficients for Ar.

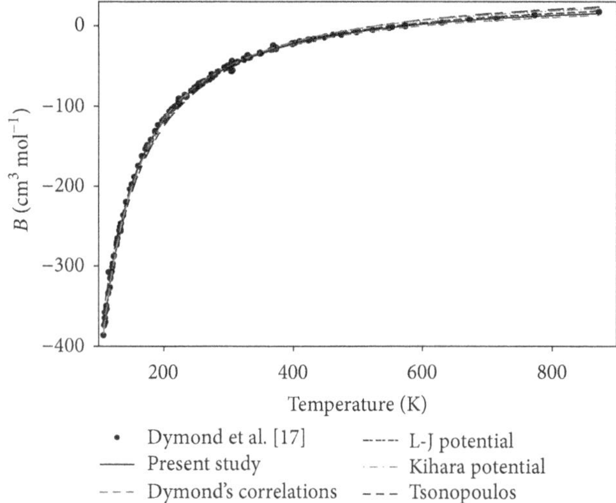

FIGURE 4: Comparison of measured and calculated second virial coefficients for Kr.

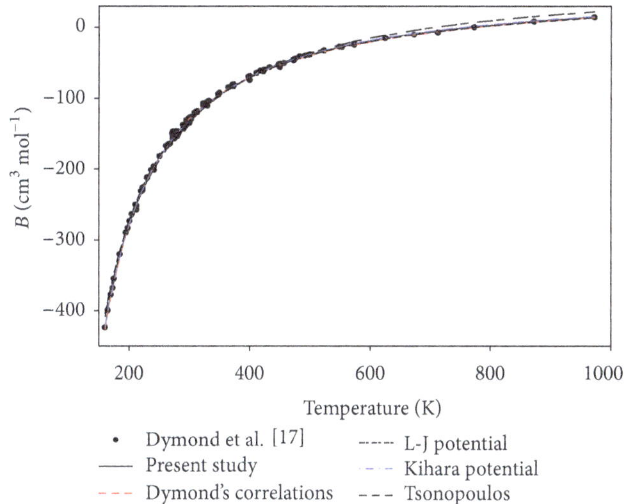

Figure 5: Comparison of measured and calculated second virial coefficients for Xe.

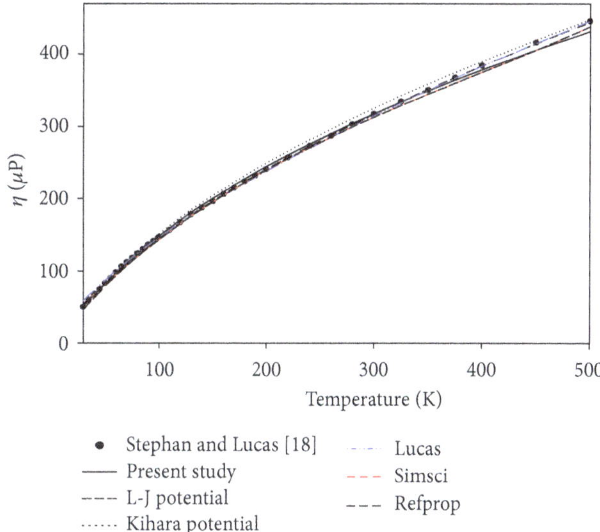

Figure 7: Comparison of measured and calculated viscosities for Ne.

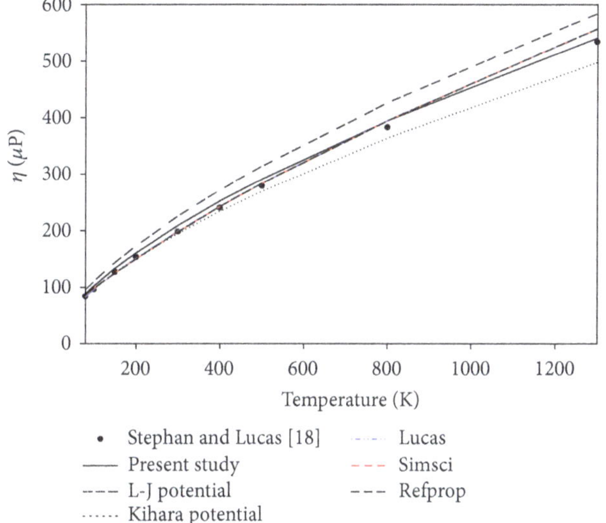

Figure 6: Comparison of measured and calculated viscosities for He.

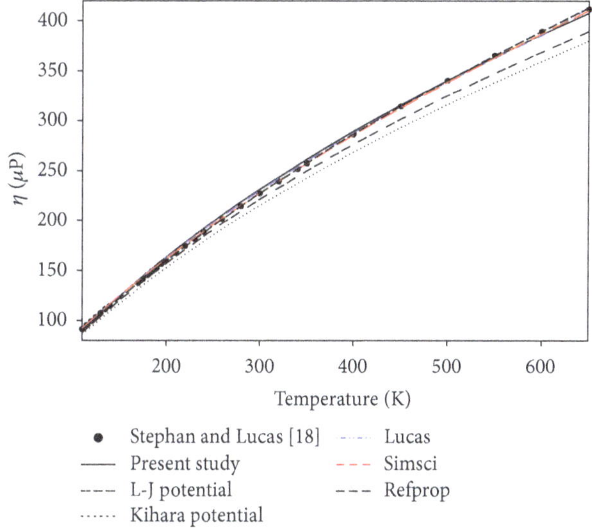

Figure 8: Comparison of measured and calculated viscosities for Ar.

[17] 3.29 cm^3 mol^{-1} and yields more accurate results than other existing methods: original two-parameter Lennard-Jones (12-6) potential [4] 6.79 cm^3 mol^{-1}, Kihara potential with group contribution concept [20] 5.99 cm^3 mol^{-1}, and the corresponding states method of Tsonopoulos [21] 18.6 cm^3 mol^{-1}. A comparison of the measured and calculated second virial coefficients from Dymond et al. [17] for pure noble gases (He, Ne, Ar, Kr, and Xe) is presented in Figures 1, 2, 3, 4, and 5, in order. Also included in Table 2 are prediction results from other second virial coefficient data of helium and xenon available in the literature, showing similar results.

In Table 3, deviations of regression results between observed and calculated viscosity data are presented on an RMSD$_r$ (root-mean-square deviation, relative) basis in %, which is defined by

$$\mathrm{RMSD_r} = \sqrt{\frac{1}{n_\eta}\sum_{i=1}^{n_\eta}\left(\frac{(\eta_{i,\mathrm{obsd}} - \eta_{i,\mathrm{calc}})}{\eta_{i,\mathrm{obsd}}}\right)^2}. \qquad (23)$$

The average RMSD$_r$ value between a total of observed and regressed 117 viscosity data for five noble gases was found to be 1.90%, indicating that the present work is quite comparable to the Refprop Database correlations [22] 1.78% and is in better agreement with experimental data than any other existing methods: original Lennard-Jones (12-6) potential 6.60%, Kihara potential with group contribution concept 5.74%, the corresponding states method of Lucas [23] 2.25%, and the Simsci Database correlations [24] 2.08%.

TABLE 3: Deviations between experimental and calculated viscosities of pure noble gases.

Compound	Number of data points	Average $RMSD_r$ in η (%)						ΔT (K)	Reference
		Present study	L-J	Kihara	Lucas	Simsci	Refprop		
Regression results									
Helium	9	5.27	18.3	13.2	3.23	2.05	2.83	80–1300	[24]
Neon	36	1.50	1.86	3.76	2.80	4.29	2.48	30–500	[24]
Argon	30	1.84	2.72	5.33	1.73	1.32	0.29	10–650	[24]
Krypton	19	2.87	12.2	8.56	1.23	0.91	1.85	150–600	[24]
Xenon	23	0.49	9.88	4.10	2.50	0.57	2.13	300–430	[24]
Average	117**	1.90	6.60	5.74	2.25	2.08	1.78		
Prediction results									
Helium	22	4.11	8.33	2.99	4.95	0.98	3.94	50–973	[26]
	31	4.62	9.03	2.96	3.92	0.60	10.0	20–1000	[29]
	1	4.44	8.82	0.76	2.87	0.64	1.32	300	[30]
	8	2.34	6.52	4.03	6.74	0.91	0.84	298–1010	[31]
	1	4.83	9.93	0.14	2.26	0.06	0.75	303	[32]
	13	3.15	7.32	4.24	6.36	1.31	1.43	198–473	[33]
	8	2.90	6.86	4.17	6.42	0.40	0.53	298–973	[34]
	8	2.71	6.65	4.27	6.54	0.23	0.42	298–973	[35]
	6	2.98	6.87	4.37	6.55	0.36	0.57	298–973	[36]
	9	3.26	7.90	2.47	4.72	0.33	0.49	298–678	[37]
	9	2.51	6.68	3.83	6.17	0.26	0.36	298–778	[38]
	5	4.05	8.85	1.32	3.48	0.09	0.41	298–473	[39]
	1	2.91	7.66	2.66	5.3	3.68	3.46	767	[40–42]
	6	4.24	8.91	1.37	3.54	0.35	0.32	298–512	[43]
	1	5.28	10.2	0.08	2.03	0.09	0.62	292	[44]
	1	5.72	10.5	0.77	1.34	0.76	0.03	291	[45, 46]
	2	6.29	11.1	1.40	0.98	1.39	0.76	291	[47]
	2	5.06	10.1	12.7	2.12	0.13	0.60	293–303	[48]
	4	5.48	10.4	0.95	1.82	0.51	0.81	237–308	[49]
	6	5.91	10.7	1.51	0.36	1.50	2.50	72–291	[50]
	10	5.01	9.61	4.31	1.77	3.08	30.8	14–293	[51]
	5	5.30	10.2	0.46	2.23	1.84	1.83	297–523	[52]
	2	5.07	10.1	0.13	2.11	0.12	0.59	293–303	[53]
	1	4.85	9.99	0.02	2.16	0.14	0.54	303	[54]
	10	4.94	9.75	1.01	2.86	1.40	1.38	293–523	[55]
	39	6.84	10.4	2.73	6.18	5.94	22.0	14–973	[56]
	74	3.40	7.42	3.73	5.97	0.87	0.58	100–1000	[57]
	72	5.77	10.3	6.75	4.09	15.7	62.2	200–1000	[58]
	8	2.69	6.63	4.28	6.55	0.24	0.44	298–973	[59]
	11	4.70	9.22	1.89	3.51	0.46	0.49	100–600	[60]
	19	5.32	9.72	3.05	4.30	0.35	1.68	80–1000	[61]
	7	1.32	4.25	5.50	7.98	3.10	3.45	250–1000	[62]
	13	3.83	8.54	1.89	4.34	2.64	2.53	373–973	[63]
Average	415*	4.49	8.76	3.75	4.82	3.97	15.1		
Neon	27	2.04	1.54	8.00	3.14	3.04	15.2	50–973	[26]
	6	1.58	1.22	7.79	2.45	1.47	2.20	100–600	[29]
	8	1.58	0.72	8.36	3.65	0.57	27.4	298–973	[64]
	13	1.83	1.04	8.60	3.54	1.04	21.2	298–973	[33]
	8	2.07	1.24	8.91	4.21	0.92	27.1	298–973	[35]
	6	2.22	1.37	8.97	4.30	1.09	29.8	298–973	[36]

TABLE 3: Continued.

Compound	Number of data points	Average RMSD$_r$ in η (%)						ΔT (K)	Reference
		Present study	L-J	Kihara	Lucas	Simsci	Refprop		
	8	2.27	1.39	9.05	4.36	1.08	26.9	298–973	[65]
	5	1.13	0.46	7.91	1.82	1.07	2.72	298–473	[39]
	7	1.80	1.00	8.66	3.64	0.91	11.3	298–778	[38]
	9	1.67	0.90	8.47	2.90	1.15	3.52	298–673	[37]
	3	1.07	2.10	5.89	0.82	0.66	0.47	298–348	[66]
	3	4.46	5.53	2.74	4.21	3.99	2.91	298–348	[67, 68]
	3	0.49	1.50	6.45	0.29	0.13	1.07	298–348	[69]
	6	1.03	2.01	6.02	0.81	0.70	0.71	298–348	[40–42]
	3	0.27	1.17	6.78	0.22	0.58	1.39	298–348	[70]
	3	0.19	1.17	6.76	0.22	0.33	1.39	298–348	[71]
	3	0.75	0.75	7.36	0.97	1.00	2.11	298–348	[72]
	3	0.56	0.52	7.38	0.81	0.97	2.04	298–348	[73]
	3	0.49	0.56	7.32	0.74	0.92	1.98	298–348	[74]
	3	1.09	0.89	7.56	1.30	1.28	2.38	298–348	[75]
	3	0.58	0.50	7.39	0.83	0.99	2.06	298–348	[76]
	4	0.65	0.305	7.45	0.64	1.61	1.71	237–308	[49]
	2	0.65	0.328	7.50	0.78	1.35	1.94	293–303	[53]
	5	2.20	1.986	8.31	2.24	2.10	3.38	72–291	[77]
	5	1.08	2.994	8.15	1.92	9.59	7.60	20–293	[51]
	6	2.08	2.176	7.54	2.44	2.05	2.91	273–523	[52]
	2	0.65	0.328	7.50	0.78	1.35	1.94	293–303	[53]
	2	1.12	2.129	5.76	1.14	0.47	0.17	291	[47]
	2	0.52	0.947	6.98	0.51	0.92	1.39	291	[45, 46]
	52	0.98	1.568	7.83	1.68	6.64	13.4	20–1000	[56]
	32	1.24	0.991	7.71	3.11	0.68	22.1	100–950	[57]
	92	4.80	4.116	5.86	4.23	12.4	7.89	27–320	[58]
	8	2.08	1.242	8.92	4.21	0.93	27.1	298–973	[59]
	11	1.77	2.693	5.38	1.53	1.83	1.92	100–600	[60]
	19	1.35	1.439	7.15	2.68	1.10	20.4	80–1000	[61]
	1	1.28	2.29	5.60	1.30	0.60	0.14	291	[48]
	4	1.04	2.09	5.92	0.57	1.51	0.71	293–523	[78]
Average	380*	2.25	2.07	7.24	2.87	4.83	12.0		
Argon	22	2.37	2.77	7.56	2.17	2.23	2.14	50–973	[26]
	63	2.78	2.25	5.85	2.38	2.98	1.94	90–500	[79]
	28	1.63	1.99	6.96	1.73	1.48	1.06	87–1000	[29]
	19	2.45	2.04	6.46	2.02	1.55	8.98	173–1597	[80]
	1	1.71	0.48	6.00	0.18	0.17	0.02	298	[32]
	8	1.14	2.59	8.39	2.39	0.58	1.47	298–973	[64]
	11	0.88	2.64	8.05	1.96	0.87	0.90	298–473	[33]
	6	1.20	2.92	8.25	2.23	0.97	0.61	298–767	[81]
	9	1.48	3.43	8.85	2.86	1.20	3.30	298–1124	[31]
	8	1.17	3.03	8.45	2.43	0.63	1.54	298–973	[35]
	7	1.21	2.97	8.24	2.36	0.73	1.41	298–973	[65]
	5	0.92	1.99	7.36	1.33	0.95	0.62	298–473	[82]
	5	0.85	1.95	7.36	1.29	0.92	0.60	300–473	[83]
	2	8.16	6.93	11.4	5.32	6.43	7.39	55–90	[84]
	1	2.01	0.12	5.79	0.34	0.31	0.07	273	[85]
	1	3.90	1.56	4.81	1.25	6.76	26.7	1373	[86]
	1	1.98	0.21	5.77	0.41	0.40	0.19	293	[87]
	1	2.99	0.62	5.40	0.89	5.27	12.3	1100	[40–42]

TABLE 3: Continued.

Compound	Number of data points	Average RMSD$_r$ in η (%)						ΔT (K)	Reference
		Present study	L-J	Kihara	Lucas	Simsci	Refprop		
	1	5.90	3.55	3.33	2.25	9.28	80.4	1873	[88]
	9	0.94	2.79	8.22	2.08	0.80	0.49	298–773	[38]
	6	1.18	2.84	8.17	2.16	0.85	0.45	298–770	[89]
	8	0.93	2.68	8.07	1.95	0.93	0.46	298–673	[37]
	21	2.21	0.83	5.73	0.83	0.89	0.39	202–394	[90]
	5	12.8	13.9	17.4	13.6	12.2	11.4	775–1053	[40–42]
	7	12.0	12.7	15.9	12.5	11.7	29.9	775–1838	[76]
	5	3.75	1.70	5.30	1.20	5.79	36.5	569–1838	[91]
	6	5.64	4.66	8.52	3.63	3.25	3.96	72–291	[50]
	2	1.79	0.44	11.2	0.27	0.26	0.11	293–303	[48]
	1	2.19	0.01	5.59	0.60	0.58	0.37	291	[45, 46]
	5	5.22	4.39	8.44	3.36	3.00	3.72	72–291	[77]
	4	2.00	0.48	5.92	0.55	1.08	1.46	293–523	[52]
	2	1.72	0.50	6.00	0.23	0.22	0.09	293–303	[53]
	2	1.90	0.31	5.83	0.37	0.36	0.20	293–303	[47]
	8	2.09	0.49	5.83	0.64	1.10	1.47	293–523	[55]
	46	13.3	13.4	15.0	13.3	13.3	13.3	90–1000	[56]
	124	1.55	2.99	8.44	2.59	0.90	6.64	100–1600	[57]
	68	2.21	1.35	5.62	1.55	2.17	1.04	87–700	[58]
	135	1.53	2.67	7.81	2.21	1.46	1.34	80–1000	[92]
	63	1.60	1.43	6.50	1.13	1.50	0.57	90–500	[93]
	1	1.52	0.66	6.17	0.01	0.00	0.14	300	[94]
	8	1.17	3.04	8.81	2.44	0.64	1.53	298–973	[59]
	11	1.65	1.53	6.50	1.24	1.43	0.56	100–600	[60]
	20	4.41	3.96	8.23	3.19	3.18	3.52	60–1000	[61]
	7	0.85	1.43	7.06	0.82	1.29	2.98	250–1000	[62]
	5	2.19	2.35	7.11	1.97	1.7	1.72	323–523	[95]
	5	0.81	1.48	6.98	0.76	0.58	0.47	298–423	[96]
	4	1.59	1.10	6.34	0.59	0.50	0.38	298–373	[97]
Average	787*	2.75	3.14	7.85	2.84	2.44	3.76		
	29	3.41	6.47	8.26	3.40	4.00	59.7	50–3273	[26]
	6	2.96	4.24	6.53	1.85	1.84	3.40	100–600	[29]
	1	3.90	1.95	4.82	0.83	0.50	0.91	293	[73]
	2	3.40	2.59	5.36	0.58	0.51	1.53	293–300	[30]
	8	1.70	6.11	7.76	2.28	1.88	3.05	298–973	[64]
	6	1.92	5.77	7.54	2.15	1.83	3.18	298–767	[81]
	8	1.89	6.25	7.86	2.51	2.07	3.20	298–973	[65]
	8	1.88	6.07	7.67	2.25	1.83	2.97	298–973	[35]
	9	1.60	6.61	8.20	2.80	2.40	3.27	298–1151	[31]
Krypton	9	1.87	5.82	7.53	2.05	1.71	3.10	298–778	[38]
	6	1.98	5.68	7.45	2.06	1.75	3.09	298–770	[89]
	1	5.59	0.36	3.27	2.47	2.13	0.70	293	[67, 68]
	1	5.08	0.84	3.74	1.97	1.64	0.21	293	[66]
	1	5.42	0.52	3.42	2.30	1.95	0.54	293	[76]
	12	2.01	5.43	7.27	1.83	1.57	2.93	298–773	[98]
	4	2.59	3.19	6.23	1.14	1.28	2.56	237–308	[49]
	2	4.93	0.82	3.82	1.96	1.63	0.43	273–293	[52]
	2	3.80	2.13	4.93	0.70	0.37	1.02	273–293	[54]
	2	3.72	2.10	4.40	0.69	0.68	1.10	273–293	[45, 46]

Table 3: Continued.

Compound	Number of data points	Average RMSD$_r$ in η (%)						ΔT (K)	Reference
		Present study	L-J	Kihara	Lucas	Simsci	Refprop		
	1	4.34	1.49	3.82	1.27	0.90	0.48	293	[47]
	15	2.54	5.66	7.18	2.0	1.87	5.45	293–1600	[56]
	78	2.12	5.77	7.43	2.06	1.69	2.60	300–1300	[57]
	71	4.64	4.98	7.24	3.83	3.77	4.28	119–950	[58]
	235	2.02	6.98	8.64	3.61	3.57	12.5	80–2000	[92]
	51	8.49	7.44	8.66	7.50	7.45	7.40	125–500	[93]
	8	1.86	6.08	7.69	2.26	1.84	2.99	298–973	[59]
	10	4.62	2.00	4.19	2.0	2.09	0.26	150–600	[60]
	26	1.89	6.29	7.95	3.19	3.20	13.4	140–2000	[61]
Average	612*	3.03	6.18	7.94	3.42	3.32	10.2		
	22	5.51	7.31	5.84	2.10	9.51	12.3	50–973	[26]
	6	5.06	6.39	5.33	1.86	6.59	12.0	100–600	[29]
	17	2.09	4.22	4.46	2.18	1.34	6.71	173–797	[80]
	8	3.13	7.60	5.87	1.42	2.11	7.94	298–973	[35]
	9	3.04	7.43	5.86	1.34	2.24	4.06	298–778	[38]
	7	1.24	2.65	3.75	2.51	0.46	5.94	202–298	[90]
	4	3.75	5.22	6.18	2.20	3.08	2.30	237–308	[49]
	3	0.53	2.62	2.99	2.32	0.74	2.90	273–373	[99]
Xenon	2	0.26	2.42	3.24	2.83	0.66	3.45	291	[45, 46]
	9	2.45	6.88	5.27	1.07	1.42	8.06	293–1000	[56]
	41	2.88	7.43	5.61	1.22	1.72	16.1	300–1300	[57]
	67	2.32	5.41	4.68	1.97	1.32	12.0	165–1250	[58]
	130	3.77	6.66	9.69	1.43	3.44	8.68	105–1000	[92]
	47	3.54	5.07	5.27	3.74	3.34	6.56	170–500	[93]
	8	3.11	7.52	5.77	1.60	2.08	7.96	298–973	[59]
	9	1.99	3.64	3.42	3.04	1.80	3.59	200–600	[60]
	14	2.45	5.87	4.93	1.85	1.42	8.25	180–1000	[61]
Average	403*	3.21	6.10	6.57	1.93	2.87	9.44		
Overall	2597**	3.09	5.06	6.92	3.15	3.31	9.19		

*Number of data points for each noble gas. **Total number of data points for all noble gases.

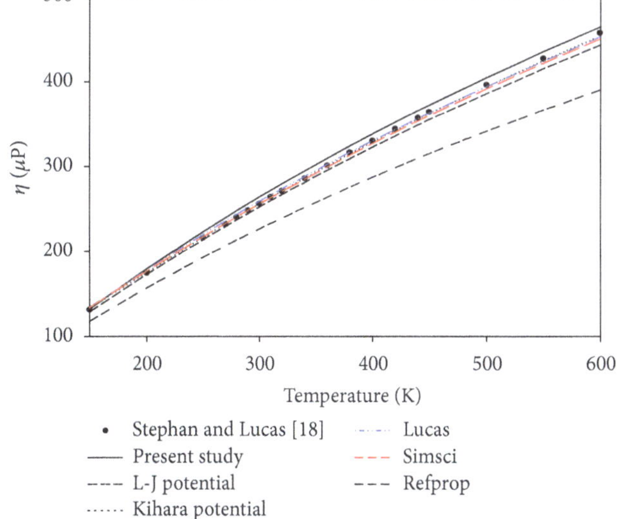

Figure 9: Comparison of measured and calculated viscosities for Kr.

Comparisons of the measured and calculated viscosities for pure noble gases (He, Ne, Ar, Kr, and Xe) are presented in Figures 6, 7, 8, 9, and 10, in order. Using the same set of potential parameters, other 2597 viscosity data available in the literature other than Stephan and Lucas [18] were reproduced with an overall average RMSD$_r$ 3.09% for all noble gases, noting that these results are more accurate than other existing investigations; original Lennard-Jones (12-6) potential 5.06%, Kihara potential with group contribution method 6.92%, Lucas method 3.15%, Simsci correlations 3.31%, and Refprop correlations 9.19%.

It is mentioned here that for each dilute noble gas at 0.1 MPa pressure, the Refprop Database provides selected viscosity data valid in specific temperature ranges, for example, 4–2219 K for helium, 27–1037 K for neon, 87–2992 K for argon, 119–1124 K for krypton, and 164–1100 K for xenon. In this work, their recommended viscosity data were fitted separately for each substance by least-squares analysis to obtain smoothing functions, usually reciprocal temperature expressions of third order, which were used to represent

Table 4: Deviations between experimental and calculated thermal conductivities of pure noble gases.

Compound	Number of points	Average $RMSD_r$ in λ (%)				ΔT (K)	Reference
		Present study	L-J	Simsci	Refprop		
Helium	64	7.77	10.7	2.94	3.52	70–1000	[56]
	8	2.61	5.16	0.78	1.03	298–1010	[31]
	8	2.33	4.40	1.68	1.32	298–973	[34]
	8	2.41	4.42	1.75	1.27	298–973	[36]
	27	5.04	7.95	1.09	1.81	275–725	[57]
	93	6.10	9.10	7.20	20.1	20–1000	[58]
	8	2.47	4.67	1.30	0.95	298–973	[59]
	4	10.2	13.5	4.98	4.16	303–363	[100]
	2	7.72	11.2	2.35	1.25	303	[101]
	25	3.66	6.21	1.41	4.52	50–973	[26]
	6	4.01	6.76	1.19	1.91	100–600	[29]
	11	4.95	7.65	1.44	1.95	100–600	[60]
	29	6.90	9.60	2.50	2.80	80–1000	[102]
	102	6.80	9.50	4.70	12.5	20–1000	[103]
	14	0.41	0.64	0.24	0.31	350–1000	[104]
	6	7.94	11.0	3.37	3.18	303–589	[105]
	52	5.90	8.60	5.50	17.4	20–1000	[106]
	5	11.0	14.0	6.88	7.20	250–600	[62]
	1	5.40	8.76	0.09	1.04	596	[107]
	1	7.24	10.3	1.72	0.53	291	[108]
	4	10.4	13.6	5.11	4.19	308–363	[109]
	2	4.68	7.32	1.32	0.07	302–793	[110]
	1	6.33	9.56	0.92	0.09	311	[111]
	1	6.57	10.0	1.20	0.08	302	[112]
	13	2.16	4.39	1.06	0.29	373–973	[63]
Average	495*	5.90	8.59	3.94	9.47		
Neon	8	1.52	0.62	1.07	2.91	298–973	[64]
	8	2.50	1.16	0.66	3.87	298–973	[65]
	1	1.04	0.54	1.10	2.53	291	[47]
	8	2.74	1.41	0.66	4.11	298–973	[36]
	29	3.50	2.53	0.94	4.11	90–1300	[56]
	44	1.36	0.86	1.65	2.40	275–1275	[57]
	111	2.83	2.65	5.95	12.7	27–1000	[58]
	8	2.55	1.17	0.76	3.92	298–973	[59]
	4	1.67	3.25	2.10	0.84	303–363	[100]
	3	1.25	0.76	1.02	2.14	313–363	[113]
	22	2.40	1.70	2.40	4.36	50–973	[26]
	6	1.81	1.34	1.09	3.25	100–600	[29]
	11	1.48	3.03	2.50	2.04	100–600	[60]
	15	2.92	1.63	0.53	4.01	273–1100	[102]
	111	2.17	2.35	1.59	4.48	27–1500	[114]
	24	2.92	1.41	1.01	5.94	350–1500	[104]
	4	1.39	2.96	1.50	0.37	303–318	[105]
	1	1.36	2.90	1.53	0.14	311	[115]
	1	0.20	1.39	0.32	1.73	296	[107]
	1	1.04	0.53	1.11	2.54	291	[108]
	2	5.86	4.27	4.28	7.50	302–793	[110]
Average	422*	2.41	2.08	2.63	6.20		

TABLE 4: Continued.

Compound	Number of points	Average $RMSD_r$ in λ (%)				ΔT (K)	Reference
		Present study	L-J	Simsci	Refprop		
Argon	21	2.09	2.72	1.07	1.00	202–394	[90]
	3	2.15	3.97	0.48	5.08	90–273	[108]
	1	1.75	2.95	4.90	4.23	1373	[116]
	1	0.14	4.35	1.09	2.98	291	[47]
	9	1.06	5.27	0.53	1.24	298–1124	[31]
	8	1.21	5.11	0.49	1.17	298–973	[64]
	8	1.53	5.71	0.66	0.99	298–973	[65]
	8	1.48	5.66	0.77	1.09	298–973	[35]
	66	1.76	4.50	1.25	1.60	90–2000	[56]
	14	2.25	6.41	2.36	2.57	331–645	[117]
	97	1.74	4.98	0.80	0.99	280–1390	[57]
	122	2.55	6.17	1.26	5.23	87–3300	[58]
	235	2.10	5.97	1.02	2.20	80–2000	[92]
	63	2.74	3.09	3.08	5.05	90–500	[79]
	63	3.82	4.80	3.40	3.53	90–500	[93]
	8	1.36	5.50	0.54	1.10	298–973	[59]
	3	0.77	4.70	1.37	2.70	313–363	[113]
	29	2.63	5.64	2.48	7.10	50–3273	[26]
	6	1.92	4.11	2.46	5.20	100–600	[29]
	11	1.65	3.37	1.39	3.07	100–600	[60]
	54	2.02	5.78	1.07	2.55	80–2000	[102]
	118	1.67	4.82	1.21	2.14	88–2500	[103]
	24	1.70	6.20	0.50	0.90	350–1500	[104]
	4	2.73	3.57	1.65	0.77	273–593	[105]
	4	0.49	4.15	0.68	2.20	308–363	[118]
	3	1.08	3.52	0.21	1.81	300–340	[119]
	8	0.79	4.42	1.20	2.51	295–420	[120]
	60	2.03	4.95	1.16	2.04	90–2000	[106]
	7	1.41	3.86	1.09	1.43	250–1000	[62]
	1	1.40	3.19	0.25	1.52	311	[115]
	1	1.14	3.41	0.06	1.93	296	[107]
	4	0.50	4.15	0.67	2.20	308–363	[109]
	2	3.10	7.10	2.40	2.70	302–793	[110]
	1	1.40	3.20	0.25	1.52	311	[111]
	22	1.46	4.21	0.84	3.31	87–1000	[29]
Average	1089*	2.11	5.22	1.31	2.82		
Krypton	9	1.90	14.8	0.82	2.93	298–1149	[31]
	8	1.80	14.8	0.88	3.10	298–973	[64]
	8	1.65	15.1	1.24	3.44	298–973	[65]
	18	3.20	19.5	4.18	6.57	125–1300	[56]
	5	4.88	10.6	1.94	0.94	291–318	[57]
	101	2.33	14.0	1.71	3.19	119–1900	[58]
	6	3.12	13.4	2.25	3.45	100–600	[26]
	235	2.60	15.5	2.50	4.50	80–2000	[92]
	51	2.50	12.5	0.96	2.50	125–500	[93]
	8	1.75	14.9	1.10	3.28	298–973	[59]
	3	0.81	15.5	3.68	5.50	313–363	[113]
	27	3.50	14.6	2.70	7.74	50–2273	[26]
	10	4.62	10.8	2.46	0.64	150–600	[60]
	109	5.80	14.4	5.33	6.25	120–2000	[114]

Table 4: Continued.

Compound	Number of points	Average $RMSD_r$ in λ (%)				ΔT (K)	Reference
		Present study	L-J	Simsci	Refprop		
	24	3.00	15.9	2.50	3.50	350–1500	[104]
	5	5.76	9.80	2.67	0.78	303–318	[105]
	4	4.24	11.3	1.35	1.06	308–363	[118]
	1	4.38	11.0	1.30	0.65	311	[115]
	1	4.16	10.9	1.17	0.81	291	[108]
	4	4.21	11.3	1.34	1.11	308–363	[109]
	2	4.06	17.4	5.03	6.98	302–793	[110]
Average	639*	3.22	14.7	2.70	4.35		
	7	1.29	14.2	9.68	2.32	202–298	[90]
	8	3.28	14.2	2.65	1.43	298–973	[35]
	24	7.30	17.7	1.02	4.55	175–1300	[56]
	1	3.53	12.7	0.60	1.95	291	[57]
	1	3.36	12.6	0.77	1.78	291	[108]
	134	5.36	15.7	5.80	10.1	165–5000	[58]
	230	4.87	15.3	4.83	1.96	105–2000	[92]
	47	3.33	12.0	2.10	0.71	170–500	[93]
	8	3.29	14.2	2.64	1.43	298–973	[59]
	4	3.39	13.1	0.57	2.48	303–363	[100]
Xenon	3	8.35	17.6	4.75	7.55	313–363	[113]
	29	5.84	14.9	8.18	2.90	50–3273	[26]
	6	5.50	12.8	5.96	2.82	100–600	[29]
	9	1.46	11.2	3.60	0.92	200–600	[60]
	95	6.40	16.7	1.74	3.93	165–1500	[114]
	24	8.00	19.0	1.20	5.10	350–1500	[104]
	6	4.20	7.60	7.87	5.20	303–318	[105]
	67	5.64	16.1	5.97	10.2	165–5000	[106]
	2	9.00	19.0	3.90	7.01	302–793	[110]
	1	0.80	9.10	4.95	2.06	311	[111]
	1	7.30	16.3	3.43	6.00	302	[112]
Average	707*	5.26	15.4	4.39	4.76		
Overall	3352**	3.59	9.28	2.78	4.93		

*Number of data points for each noble gas. **Total number of data points for all noble gases.

viscosity data at the same temperature as those of experimental data for the reasonable comparisons. Prediction results from the Refprop correlations are observed to be not in reliable agreement with measured viscosity data, especially near upper and lower limits of temperature ranges specified previously, as shown in Table 3. The Simsci Database [24] provides the smoothing viscosity function with four coefficients c_i ($i = 1$ to 4)

$$\eta = \frac{c_1 T^{c_2}}{(1 + c_3/T + c_4/T)} \tag{24}$$

valid in specific temperature ranges: 20–2000 K for helium, 30–3272 K for neon, 83–3273 K for argon, 100–1500 K for krypton, and 100–1600 K for xenon.

The next stage of this work is to calculate other properties such as thermal conductivity and self-diffusion coefficient, not used for parameter determinations, using the same set of potential parameters determined earlier. As shown in Table 4, the overall average $RMSD_r$ value of 3.59% between a total of 3352 experimental and calculated thermal conductivities obtained by the proposed model is somewhat less reliable to the Simsci correlations 2.78%, but compares very well with the original Lennard-Jones (12-6) potential 9.28% and the Refprop correlations 4.93%.

Like the case of viscosities, the Refprop Database [22] provides dilute gas thermal conductivity data at 0.1 MPa suitable in specific temperature ranges: 4–1100 K for helium, 27–1039 K for neon, 87–2968 K for argon, 119–1100 K for krypton, and 164–1101 K for xenon. The procedure to produce thermal conductivity data is the same as that of viscosity. The Simsci Database also provides the same type of soothing thermal conductivity function, (24), in specific temperature ranges: 30–2000 K for helium, 30–3272 K for neon, 90–3273 K for argon, 120–2000 K for krypton, and 165–1500 K for xenon. Comparisons of the measured and calculated thermal conductivities for pure noble gases (He, Ne, Ar,

TABLE 5: Deviations between experimental and calculated self-diffusion coefficients of pure noble gases.

Compound	Number of points	Average RMSD$_r$ in D (%) Present study	L-J	Fuller	ΔT (K)	Reference
		Prediction results				
Helium	17	4.72	2.04	10.6	50–623	[26]
	7	11.0	8.44	13.1	14–296	[56]
Average	*24**	*6.55*	*3.90*	*11.3*		
Neon	29	3.36	2.25	8.07	50–3273	[26]
	17	6.85	4.93	13.2	77–6000	[56]
	20	4.72	2.78	6.50	295–1219	[121]
	1	4.55	7.77	5.96	293	[122]
Average	*67**	*4.67*	*3.17*	*8.88*		
Argon	29	2.87	4.70	10.9	50–323	[26]
	12	4.10	3.51	14.2	90–326	[56]
Average	*41**	*3.23*	*4.35*	*11.9*		
Krypton	29	4.00	13.92	9.70	50–3273	[26]
	18	6.83	16.93	3.69	199–6000	[56]
	1	4.90	8.94	3.30	293	[123]
Average	*48**	*5.08*	*14.95*	*7.31*		
Xenon	29	6.27	13.9	16.2	50–3273	[26]
	27	6.02	16.9	13.1	194–15000	[56]
	1	1.06	7.40	17.2	293	[123]
Average	*57**	*6.06*	*15.2*	*14.8*		
Overall	237**	5.03	8.73	10.7		

*Number of data points for each noble gas. **Total number of data points for all noble gases.

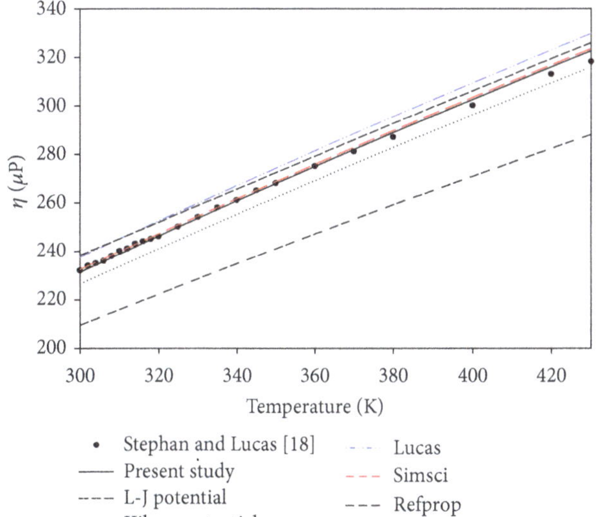

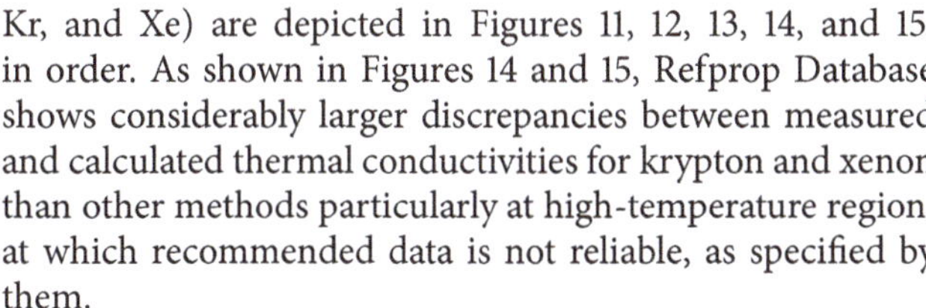

FIGURE 10: Comparison of measured and calculated viscosities for Xe.

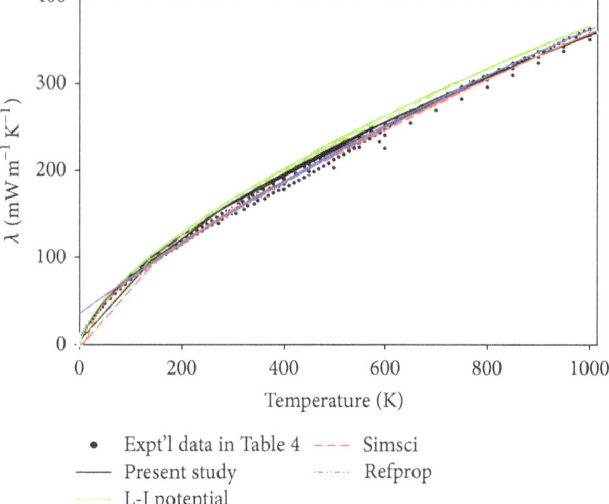

FIGURE 11: Comparison of measured and calculated thermal conductivities for He.

Kr, and Xe) are depicted in Figures 11, 12, 13, 14, and 15, in order. As shown in Figures 14 and 15, Refprop Database shows considerably larger discrepancies between measured and calculated thermal conductivities for krypton and xenon than other methods particularly at high-temperature region, at which recommended data is not reliable, as specified by them.

A total of 237 self-diffusion coefficient data were next tested. As shown in Table 5, the overall average RMSD$_r$ value of 5.03% from all noble gases was obtained by this work, in which the result is in better agreement with experimental data than those of the original Lennard-Jones (12-6) potential 8.73% and of the Fuller method [25] 10.7%. It is indicated that for the helium and neon gas, the proposed method is less

TABLE 6: Deviations between experimental and predicted second cross-virial coefficients of noble gas mixtures.

Mixtures	Number of points	Average RMSD in B_{12} (cm^3 mol^{-1}) Present study	L-J	Dymond et al.	Tsonopoulos	ΔT (K)	Reference
Mixtures of noble gases							
Helium-neon	18	2.05	5.82	0.94	9.40	15–323	[2]
	23	0.66	0.54	1.72	10.6	273–3273	[26]
Helium-argon	21	2.03	0.92	0.44	6.85	90–773	[2]
	23	4.90	5.71	6.37	14.8	273–3273	[26]
Helium-krypton	15	5.66	3.49	0.99	1.84	90–323	[2]
	23	8.54	12.0	12.0	15.9	273–3273	[26]
Helium-xenon	9	10.9	6.07	0.71	7.14	120–323	[2]
	23	15.8	19.7	23.3	21.5	273–3273	[26]
Average	*155**	*6.15*	*7.12*	*6.75*	*11.9*		
Neon-argon	30	1.68	34.9	1.00	7.72	84–475	[2]
	23	3.15	1.15	6.69	4.71	273–3273	[26]
Neon-krypton	48	5.15	48.6	2.76	14.3	100–475	[2]
	23	3.86	3.29	8.07	6.69	273–3273	[26]
Neon-xenon	28	16.5	39.4	2.61	17.6	162–475	[2]
	23	10.8	7.36	18.9	12.5	273–3273	[26]
Average	*175**	*6.68*	*27.2*	*5.77*	*11.2*		
Argon-krypton	50	3.26	150.3	1.82	2.51	108–695	[2]
	23	1.58	16.7	2.93	4.53	273–3273	[26]
Argon-xenon	5	12.3	125.5	0.29	12.1	173–323	[2]
	23	4.60	23.9	30.0	5.76	273–3273	[26]
Average	*101**	*3.63*	*89.9*	*8.43*	*4.18*		
Krypton-xenon	35	7.34	173.7	1.97	5.76	160–700	[2]
	23	2.34	42.6	3.40	4.44	273–3273	[26]
Average	*58**	*5.36*	*121.7*	*2.54*	*5.24*		
Overall	489**	5.73	45.0	6.25	9.28		

*Number of data points for each gas mixture. **Total number of data points for all gas mixtures.

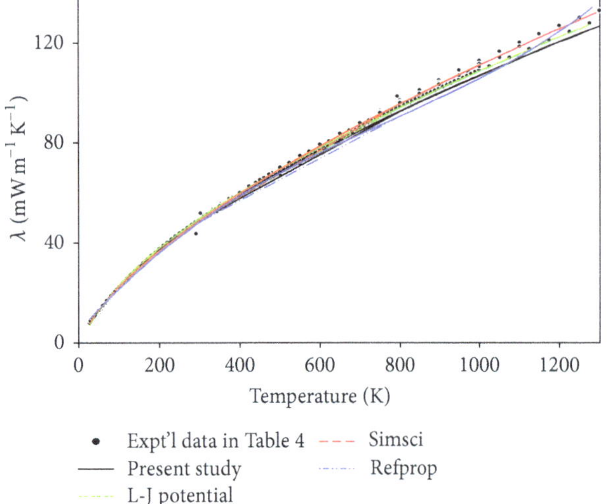

FIGURE 12: Comparison of measured and calculated thermal conductivities for Ne.

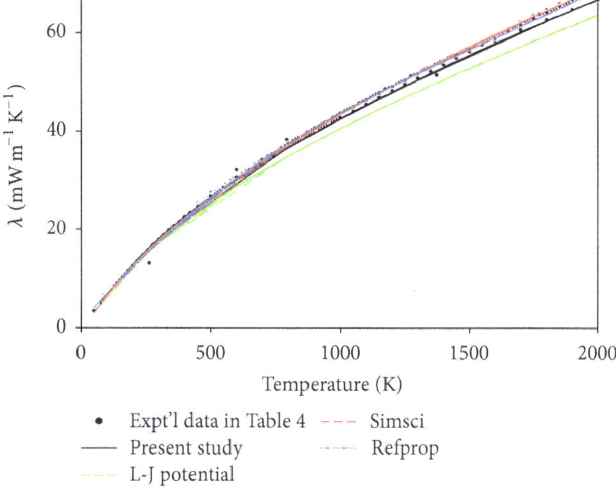

FIGURE 13: Comparison of measured and calculated thermal conductivities for Ar.

accurate than the original Lennard-Jones (12-6) potential. Comparisons of the measured and calculated self-diffusion coefficients for noble gases (He, Ne, Ar, Kr, and Xe) are depicted in Figures 16, 17, 18, 19, and 20, in order.

TABLE 7: Deviations between experimental and predicted mixture viscosities of noble gas mixtures.

Mixtures	Number of points	Average RMSD$_r$ in η_{mix} (%)			ΔT (K)	Reference
		Present study	L-J	Lucas method		
Helium-neon	87	6.24	13.6	8.85	50–3273	[26]
	7	2.40	4.87	3.56	291	[47]
	10	1.93	4.93	4.39	293	[53]
	32	1.75	4.77	8.18	298–973	[36]
	3	2.31	4.88	3.97	293	[40–42]
	10	2.58	5.15	3.80	293–523	[124]
	11	3.36	4.82	6.98	293–523	[55]
	31	2.22	4.96	3.67	20–523	[56]
	162	2.10	8.40	7.49	100–950	[57]
Helium-argon	87	4.49	7.31	4.95	50–3273	[26]
	12	3.38	4.25	3.51	293–303	[125]
	11	3.75	3.66	4.78	291	[47]
	58	3.84	4.07	4.76	72–192	[50]
	40	2.98	6.06	3.63	298–993	[31]
	3	3.52	0.92	6.46	293	[78]
	3	4.09	3.01	5.19	293	[40–42]
	6	3.69	3.12	5.25	293–523	[55]
	16	6.72	7.04	5.47	293–523	[56]
	123	1.29	5.18	3.20	100–1500	[57]
	28	3.14	3.23	4.45	298	[126]
	12	1.26	2.44	4.57	300–1100	[127]
	6	3.62	0.51	7.15	298	[128]
Helium-krypton	87	4.60	9.41	13.4	50–3273	[26]
	9	4.44	7.72	9.63	291	[45, 46]
	16	4.98	5.28	10.5	303	[54]
	40	7.72	8.91	5.78	298–993	[31]
	135	4.07	11.8	4.94	300–1000	[57]
Helium-xenon	87	3.85	9.63	13.9	50–3273	[26]
	10	3.41	6.11	19.2	291	[45, 46]
	18	2.50	8.85	13.6	298–778	[38]
	90	2.34	9.46	12.7	300–750	[57]
Average	1250*	3.52	8.34	8.33		
Neon-argon	87	1.91	3.95	4.23	50–3273	[26]
	6	1.24	2.80	1.60	293	[53]
	36	0.66	2.15	2.85	298–973	[64]
	26	3.87	4.34	4.24	72–291	[77]
	3	1.68	1.48	0.70	293	[78]
	3	1.61	0.95	0.66	293	[40–42]
	9	1.43	1.85	0.70	291	[47]
	12	3.22	3.95	3.56	293–523	[55]
	16	2.56	1.59	1.14	293–523	[56]
	162	0.95	4.28	2.72	100–950	[57]
Neon-krypton	87	2.63	10.3	3.25	50–3273	[26]
	9	1.28	8.04	2.44	291	[45, 46]
	40	0.85	9.63	1.60	298–973	[65]
	126	0.78	10.5	1.55	300–950	[57]

Table 7: Continued.

Mixtures	Number of points	Average RMSD$_r$ in η_{mix} (%) Present study	L-J	Lucas method	ΔT (K)	Reference
Neon-xenon	87	4.70	11.1	3.94	50–3273	[26]
	8	0.85	7.30	6.42	291	[45, 46]
	12	2.46	8.69	3.74	298–773	[38]
	90	4.70	10.2	3.11	300–750	[57]
Average	819*	2.46	8.11	2.95		
Argon-krypton	87	2.69	10.7	3.13	50–3273	[26]
	36	0.71	10.5	1.81	298–973	[64]
	8	2.20	8.40	1.76	291	[45, 46]
	189	0.97	10.8	2.30	300–1300	[57]
Argon-xenon	87	4.07	11.3	2.84	50–3273	[26]
	9	0.64	8.04	4.01	291	[45, 46]
	12	1.87	9.15	1.61	298–773	[38]
	34	2.46	10.5	2.62	173–299	[80]
	189	2.71	13.2	1.42	173–1597	[57]
Average	651*	2.20	11.4	2.23		
Krypton-xenon	87	3.97	14.5	2.71	50–3273	[26]
	9	1.22	11.2	2.19	291	[45, 46]
	12	2.18	14.1	1.18	298–773	[38]
	90	3.52	14.7	2.98	300–750	[57]
Average	198*	3.53	14.4	2.72		
Overall	2918**	2.93	10.6	4.05		

*Number of data points for each gas mixture. **Total number of data points for all gas mixtures.

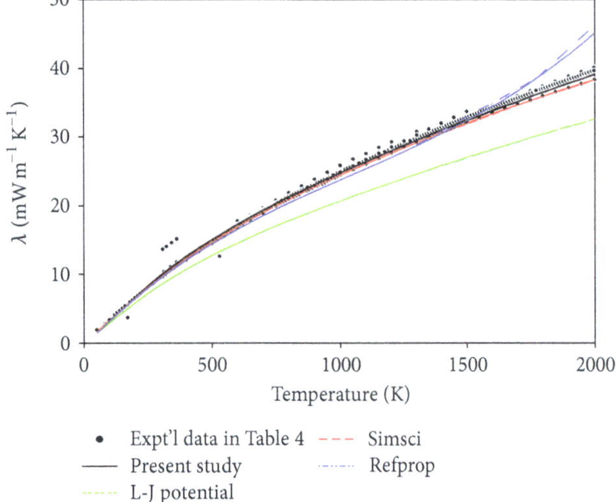

Figure 14: Comparison of measured and calculated thermal conductivities for Kr.

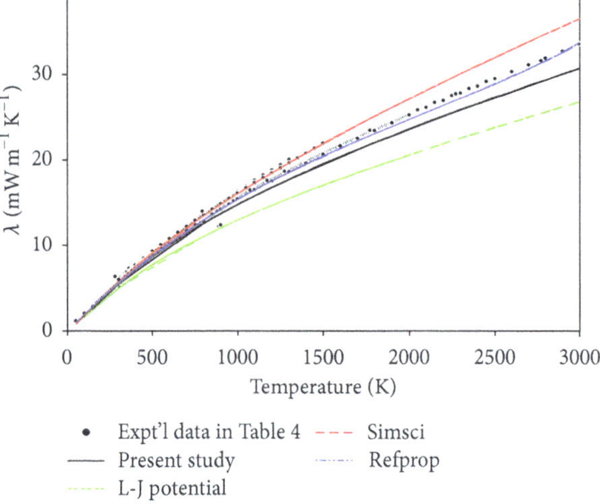

Figure 15: Comparison of measured and calculated thermal conductivities for Xe.

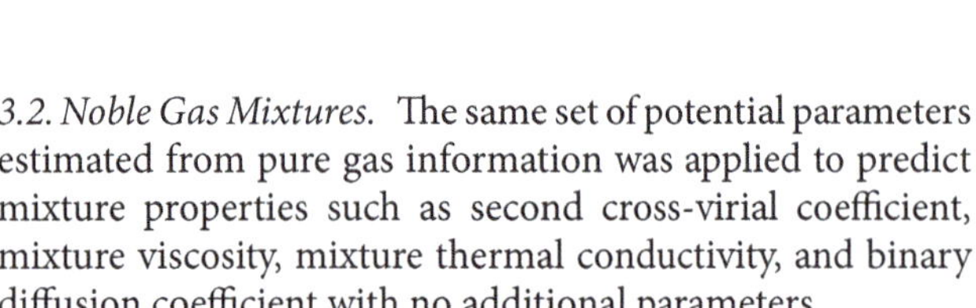

3.2. Noble Gas Mixtures. The same set of potential parameters estimated from pure gas information was applied to predict mixture properties such as second cross-virial coefficient, mixture viscosity, mixture thermal conductivity, and binary diffusion coefficient with no additional parameters.

Table 6 shows that for the second cross-virial coefficient calculations, a total of 489 data of noble gas mixtures taken from the critical compilation of Dymond et al. [17] and from the Kestin et al. [26] were fitted to be overall average RMSD value of 5.73 cm^3 mol^{-1}, while the 6.25 cm^3 mol^{-1} value was obtained by the smoothing functions of Dymond et al. in the same fashion as used in the pure gas calculations. And the 45.0 and 9.28 cm^3 mol^{-1} RMSD values were found by the original Lennard-Jones (12-6) potential and the

TABLE 8: Deviations between experimental and predicted thermal conductivities of noble gas mixtures.

Mixtures	Number of points	Average RMSD$_r$ in λ_{mix} (%)			ΔT (K)	Reference
		Present study	L-J	Wassiljewa[a]		
Helium-neon	32	9.69	12.2	5.98	298–973	[36]
	7	2.30	4.68	1.59	291	[47]
	138	5.62	8.08	3.14	275–1175	[57]
	12	11.2	13.9	7.17	303–363	[100]
	15	8.59	11.3	3.73	350–1050	[104]
	66	7.65	10.1	4.28	50–973	[26]
	7	12.6	12.2	6.08	298	[107]
	8	7.14	7.62	6.27	302–793	[110]
Helium-argon	32	15.2	14.4	10.8	298–993	[31]
	20	8.61	9.59	12.1	291	[47]
	4	12.9	12.2	8.31	273	[56]
	138	9.00	7.94	5.22	275–1025	[57]
	15	10.8	9.46	5.98	350–1050	[104]
	66	8.16	7.22	3.91	50–973	[26]
	9	9.20	8.23	4.99	298	[107]
	12	12.3	11.8	7.54	308–363	[109]
	8	10.6	10.5	11.6	302–793	[110]
	5	10.3	9.99	5.79	311	[111]
	4	9.21	8.76	6.80	273	[129]
Helium-krypton	6	8.47	6.24	3.27	302	[112]
	32	9.82	6.11	4.16	298–993	[31]
	9	5.00	3.12	1.88	291	[57]
	15	9.92	5.27	4.18	350–1050	[104]
	66	8.76	5.81	3.68	50–973	[26]
	9	5.39	3.09	2.29	291	[45, 46]
	16	10.1	9.50	4.52	308–363	[109]
	12	6.23	5.49	4.16	302–793	[110]
Helium-xenon	4	3.95	8.95	1.97	302	[112]
	16	4.57	7.00	3.52	303–363	[100]
	15	1.54	4.05	2.62	350–1050	[104]
	66	7.19	8.93	4.22	50–973	[26]
	10	6.19	10.02	3.61	291	[45, 46]
	8	4.17	8.2	3.43	302–793	[110]
	6	4.35	9.3	2.24	311	[111]
Average	888*	8.11	8.42	4.73		
Neon-argon	32	3.82	4.91	4.91	298–973	[64]
	9	3.59	5.91	5.08	291	[47]
	12	2.87	5.15	4.36	313–363	[113]
	24	3.51	5.70	6.54	350–1500	[104]
	87	2.65	4.62	5.34	50–3273	[26]
	11	0.84	2.23	1.17	311	[115]
	8	0.77	2.64	2.12	298	[107]
	8	5.22	7.33	7.34	302–793	[110]
Neon-krypton	32	1.37	4.94	2.10	298–973	[65]
	198	25.7	32.2	28.1	275–1275	[57]
	12	0.97	8.51	2.26	313–363	[113]
	24	15.8	23.6	19.2	350–1500	[104]
	87	3.35	7.55	2.58	50–3273	[26]
	9	4.04	5.58	1.54	311	[115]
	9	3.85	5.28	2.34	291	[45, 46]
	6	6.10	10.4	6.27	302–793	[110]

TABLE 8: Continued.

Mixtures	Number of points	Average $RMSD_r$ in λ_{mix} (%) Present study	L-J	Wassiljewa[a]	ΔT (K)	Reference
Neon-xenon	9	2.11	4.50	2.18	291	[57]
	12	5.24	8.08	4.49	303–363	[100]
	24	4.52	7.59	5.03	350–1500	[104]
	87	4.41	5.24	2.43	50–3273	[26]
	9	1.90	4.42	2.04	291	[45, 46]
	6	6.00	8.72	6.06	302–793	[110]
Average	715*	9.94	13.7	11.0		
Argon-krypton	32	1.14	10.3	2.42	298–973	[64]
	9	5.65	5.02	3.31	291	[57]
	24	2.11	9.69	2.51	350–1500	[104]
	87	2.73	10.1	3.32	50–3273	[26]
	6	4.70	5.58	2.46	311	[115]
	8	5.77	5.28	3.58	291	[45, 46]
	12	3.52	7.30	1.50	308–363	[109]
	6	3.32	13.2	5.55	302–793	[110]
Argon-xenon	9	2.63	5.59	2.94	291	[57]
	8	10.5	17.1	10.0	311–366	[113]
	24	1.14	8.46	1.38	350–1500	[104]
	87	3.10	8.11	2.34	50–3273	[26]
	9	2.87	5.43	3.19	291	[45, 46]
	6	3.43	10.8	3.14	302–793	[110]
	6	4.05	4.03	4.94	311	[111]
Average	333*	2.97	8.90	2.92		
Krypton-xenon	5	4.42	14.9	3.91	302	[112]
	9	1.87	12.5	1.03	291	[57]
	12	7.14	18.3	7.24	313–363	[113]
	24	2.35	13.7	2.37	350–1500	[104]
	87	5.13	16.0	3.97	50–3273	[26]
	9	1.76	12.4	0.92	291	[45, 46]
	10	6.06	17.5	5.90	302–793	[110]
Average	156*	4.51	15.5	3.75		
Overall	2092**	7.65	10.8	6.50		

*Number of data points for each mixture.
**Total number of data points for all gas mixture mixtures.
[a]Wassiljewa equation [15] with the combinational factor of Mason and Saxena [16].

corresponding states method of Tsonopoulos, respectively. A comparison of measured and calculated second cross-virial coefficients of Ar + Kr mixture is shown in Figure 21.

Prediction results of noble gas mixture viscosities are presented in Table 7. A total of 2918 viscosity data points for all noble gas mixtures were calculated in this paper, indicating that the present study is in better agreement between experimental and calculated data than other methods on a $\%RMSD_r$ criterion: 2.93% by the present model, 10.6% by the original Lennard-Jones (12-6) potential, and 4.05% by the Lucas method. However, it is noted that for krypton and xenon mixture, this work is less reliable to the Lucas method. Figure 22 shows the comparison of measured and calculated viscosities of He + Ne mixture.

Given in Table 8 are the resulting $\%RMSD_r$ values between a total of 2092 measured and predicted mixture thermal conductivity data. Based on the overall average $\%RMSD_r$ value of all noble gas mixtures, 7.65% of the present study is in slightly worse agreement between measured and predicted mixture thermal conductivities than 6.50% of the Wassiljewa equation [15] with the combinational factor of Mason and Saxena [16] and is more feasible to 10.8% of the original Lennard-Jones (12-6) potential. A comparison of measured and calculated thermal conductivities of Ne + Kr mixture is shown in Figure 23.

Included in Table 9 are the resulting $\%RMSD_r$ values between a total of 1240 measured and predicted binary diffusion coefficient data for noble gas mixtures, in which 4.98 $\%RMSD_r$ of the present study is in quite better agreement between measured and predicted binary diffusion coefficients than 7.95 $\%RMSD_r$ of the original Lennard-Jones (12-6) potential and 6.24 $\%RMSD_r$ of the Fuller method.

TABLE 9: Deviations between experimental and predicted binary diffusion coefficients of noble gas mixtures.

Compound	Number of points	Average RMSD$_r$ in D (%)			ΔT (K)	Reference
		Present study	L-J	Fuller		
Helium-neon	26	2.45	3.77	10.6	50–1773	[26]
	7	1.04	3.09	8.31	200–673	[29]
	1	5.20	1.32	12.7	1000	[130]
	5	2.01	6.19	3.52	65–295	[131]
	6	3.52	1.03	9.42	298–973	[36]
	26	9.89	7.35	5.09	200–1250	[132]
	20	5.27	4.93	12.7	297–1268	[121]
	1	14.0	10.6	8.99	293	[133]
	14	2.81	5.29	7.79	273–683	[56]
Helium-argon	29	2.35	2.72	7.04	50–3273	[26]
	7	0.56	0.12	4.91	200–673	[29]
	1	8.91	8.18	0.77	1000	[130]
	3	3.15	4.11	2.93	276–346	[134]
	8	2.78	3.57	4.06	298–498	[135]
	1	1.66	2.67	0.67	298	[136]
	1	0.91	0.08	1.87	298	[137]
	3	3.84	3.71	2.87	287–418	[138]
	6	8.30	7.80	1.68	298–1100	[139]
	1	1.21	2.22	0.17	296	[140]
	1	0.58	1.63	1.11	273	[12]
	1	1.69	2.75	0.01	273	[55]
	2	5.42	4.92	4.23	276–418	[18]
	4	1.37	2.84	2.76	90–400	[131]
	14	2.63	2.50	2.65	251–418	[141]
	17	5.01	5.08	3.39	276–1100	[142]
	8	5.69	5.05	0.65	298–993	[31]
	26	6.02	5.59	2.13	200–1250	[132]
	21	9.66	8.79	6.62	298–1272	[121]
	84	8.65	8.68	5.25	118–4500	[56]
Helium-krypton	29	2.60	7.00	5.51	50–3273	[26]
	7	2.44	5.18	6.12	200–673	[29]
	1	0.88	8.40	0.34	1000	[130]
	4	14.2	17.0	18.6	111–400	[131]
	8	2.04	7.40	4.22	298–993	[31]
	26	5.12	2.86	3.41	200–940	[132]
	21	5.32	12.4	8.11	307–1274	[121]
	4	3.48	4.02	6.69	273–318	[56]
Helium-xenon	29	7.21	13.1	5.35	50–3273	[26]
	7	5.13	11.8	4.80	200–673	[29]
	1	11.0	17.5	2.85	1000	[130]
	4	6.21	12.6	6.93	169–400	[131]
	6	1.69	8.48	3.59	298–778	[38]
	26	9.4	15.6	4.93	200–940	[132]
	21	9.96	16.3	6.65	297–1270	[121]
	8	7.6	13.5	8.01	273–394	[56]
	10	2.2	8.97	2.82	220–400	[143]
	2	2.3	9.1	2.97	273–315	[144]
Average	557*	5.80	7.61	5.72		

TABLE 9: Continued.

Compound	Number of points	Average RMSD$_r$ in D (%)			ΔT (K)	Reference
		Present study	L-J	Fuller		
Neon-argon	29	2.03	3.08	5.38	50–3273	[26]
	7	1.64	2.83	2.25	200–673	[29]
	1	3.38	6.84	1.17	1000	[130]
	4	2.60	3.14	3.36	90–400	[131]
	8	0.76	3.61	2.50	298–973	[64]
	5	5.42	1.66	4.03	90–473	[145]
	4	2.77	1.72	2.10	273–318	[146]
	26	1.86	3.76	1.51	200–940	[132]
	20	5.89	7.29	4.91	297–1274	[121]
	1	3.21	7.39	7.77	293	[147, 148]
	19	3.15	3.25	3.52	194–680	[56]
	5	5.47	0.79	1.06	277–365	[149]
Neon-krypton	29	2.06	10.4	6.99	50–3273	[26]
	7	2.31	10.4	8.25	200–673	[29]
	1	0.58	12.3	4.19	1000	[130]
	4	4.37	9.32	6.83	111–400	[131]
	4	6.11	7.25	6.32	273–318	[146]
	8	2.75	9.80	5.78	298–973	[65]
	26	3.73	13.0	8.44	200–940	[132]
	20	5.72	9.45	5.31	269–1270	[121]
	4	6.11	7.25	6.32	273–318	[56]
Neon-xenon	29	5.49	11.8	6.50	50–3273	[26]
	7	4.82	11.9	7.49	200–673	[29]
	1	6.22	13.3	3.79	1000	[130]
	4	4.09	11.1	6.82	169–400	[131]
	6	2.03	9.38	4.24	298–773	[38]
	26	6.97	7.81	7.58	200–940	[132]
	21	8.96	15.5	9.15	297–1270	[121]
	4	1.14	8.37	4.82	273–318	[56]
	5	5.59	12.6	9.16	275–362	[149]
Average	335*	4.15	8.18	5.73		
Argon-krypton	29	3.49	7.91	10.0	50–3273	[26]
	7	2.37	6.02	4.28	200–673	[29]
	1	0.08	8.05	0.37	1000	[130]
	1	0.33	7.03	0.26	273	[18]
	4	2.26	4.42	5.99	169–400	[131]
	8	8.60	5.25	19.2	77–600	[150]
	9	4.01	7.89	3.39	199–473	[142]
	5	5.85	8.91	5.25	273–473	[145]
	4	3.85	3.69	2.98	273–318	[146]
	8	1.87	8.82	3.42	298–973	[64]
	11	2.23	4.79	2.89	200–400	[151]
	26	9.30	15.3	10.6	200–940	[132]
	21	2.88	9.59	4.20	300–1274	[121]
	13	3.51	7.89	3.37	199–407	[56]
Argon-xenon	29	5.48	8.62	11.5	50–3273	[26]
	7	2.46	6.39	6.68	200–673	[29]
	1	7.99	13.0	3.23	1000	[130]

Table 9: Continued.

Compound	Number of points	Average $RMSD_r$ in D (%) Present study	L-J	Fuller	ΔT (K)	Reference
	2	5.72	6.25	6.34	195–378	[18]
	4	2.43	5.43	10.6	169–400	[131]
	4	3.16	6.25	6.88	194–378	[152]
	6	1.86	6.41	3.16	298–773	[38]
	26	5.57	9.95	4.59	200–940	[132]
	21	4.18	7.51	5.35	298–1272	[121]
	19	3.79	7.11	9.07	173–1597	[153]
	13	3.93	6.96	5.83	194–394	[56]
Average	279*	4.48	8.42	7.16		
Krypton-xenon	29	4.46	7.05	12.1	50–3273	[26]
	7	1.38	4.42	7.54	200–673	[29]
	1	4.70	10.5	4.63	1000	[130]
	4	1.38	3.27	12.2	169–400	[131]
	6	1.10	4.72	2.88	298–773	[38]
	22	7.00	11.1	7.52	297–1270	[121]
Average	69*	4.49	7.70	9.27		
Overall	1240**	4.98	7.95	6.24		

*Number of data points for each mixture. **Total number of data points for all gas mixtures.

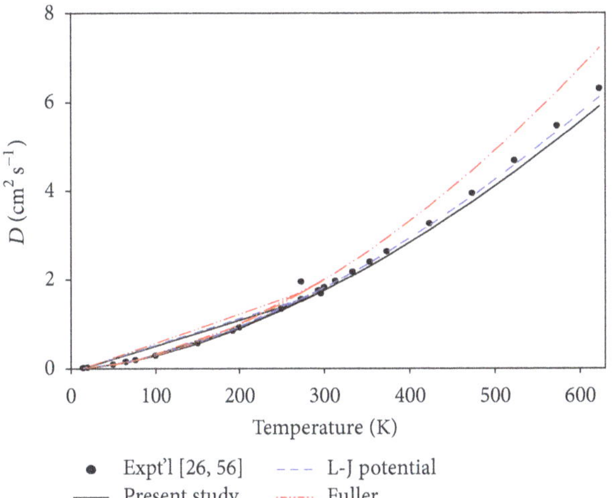

Figure 16: Comparison of measured and calculated self-diffusion coefficients for He.

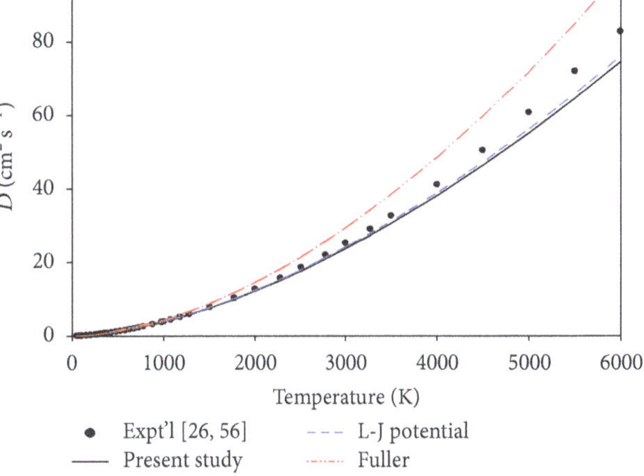

Figure 17: Comparison of measured and calculated self-diffusion coefficients for Ne.

A comparison of measured and calculated binary diffusion coefficients for He + Ne mixture is shown in Figure 24.

4. Conclusions

The three-parameter Lennard-Jones (12-6) potential function has been empirically modified by introducing a temperature-correction parameter to the reduced temperature T^* for the calculation of the thermodynamic property (second virial coefficient) and dilute transport properties (viscosity, thermal conductivity, and diffusion coefficient) of noble gases (He, Ne, Ar, Kr, and Xe) and their binary mixtures. Separately for each species, a single set of three potential parameters (σ, ε, and τ) is estimated when the second virial coefficient and viscosity data are regressed together within the experimental errors. Obtained potential parameters are used to reproduce second virial coefficient and viscosity data and in all following predictions of other properties like thermal conductivity and diffusion coefficient. Noble gas mixture properties are calculated with the same set of parameters as well.

For the second virial coefficient calculations of pure noble gases, the three-parameter Lennard-Jones (12-6) potential proposed in this paper is quite comparable to Dymond's correlations and produces more accurate results than the original two-parameter Lennard-Jones (12-6) potential, the

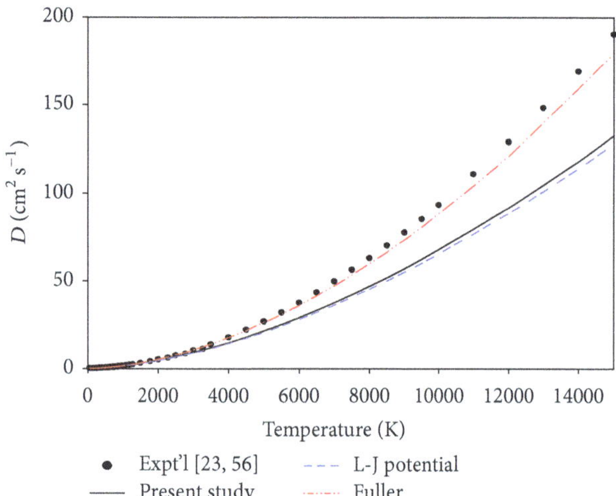

FIGURE 18: Comparison of measured and calculated self-diffusion coefficients for Ar.

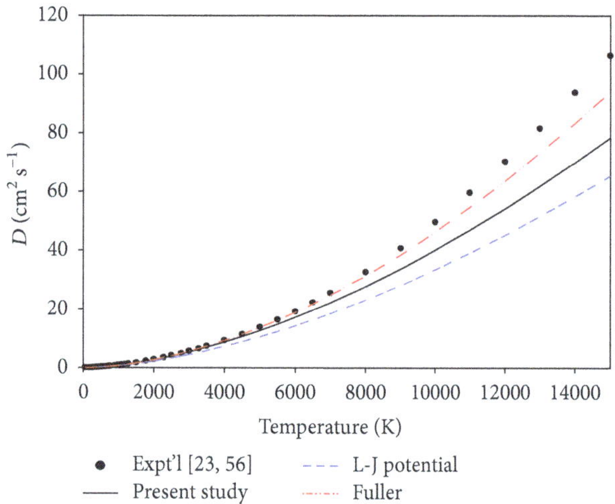

FIGURE 19: Comparison of measured and calculated self-diffusion coefficients for Kr.

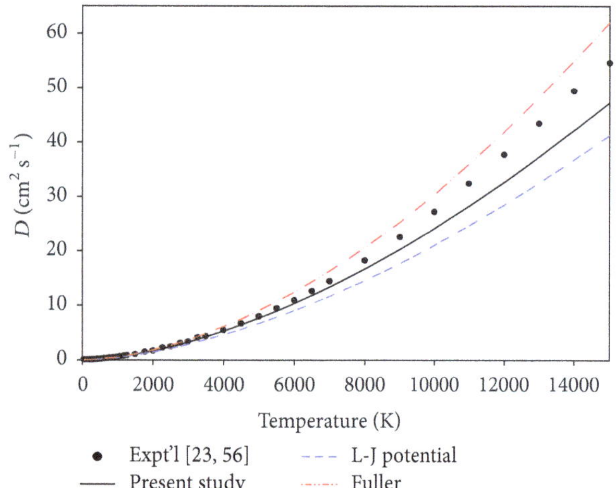

FIGURE 20: Comparison of measured and calculated self-diffusion coefficients for Xe.

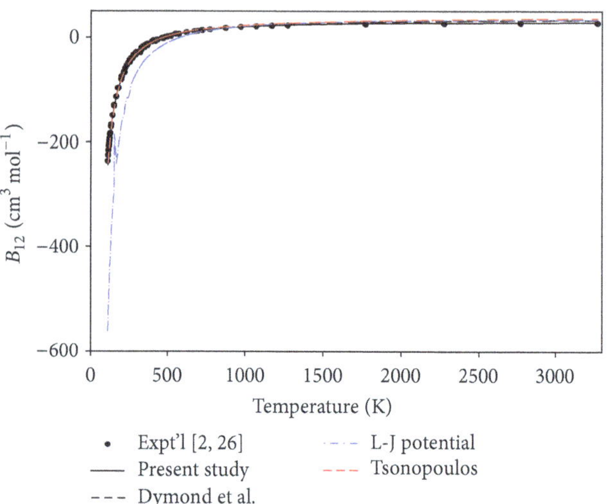

FIGURE 21: Comparison of measured and calculated second cross-virial coefficients for Ar + Kr mixture.

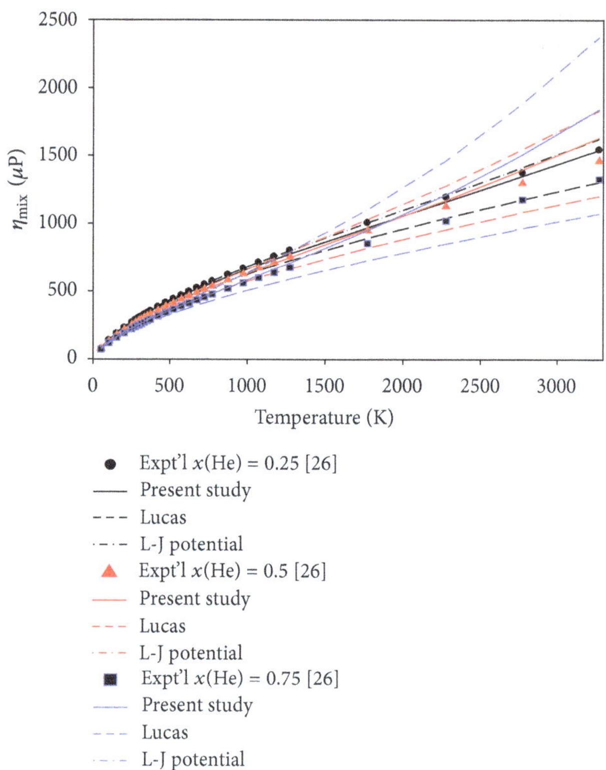

FIGURE 22: Comparison of measured and calculated viscosities for He + Ne mixture.

Kihara potential with group contribution concept, and the Tsonopoulos correlations. For the viscosity calculations, the proposed model agrees better with the observed and calculated data than the original Lennard-Jones (12-6) potential, the Kihara potential with group contribution method, the Lucas method, the Simsci Database, and the Refprop Database. Agreement between experimental and calculated thermal conductivities obtained by the proposed model

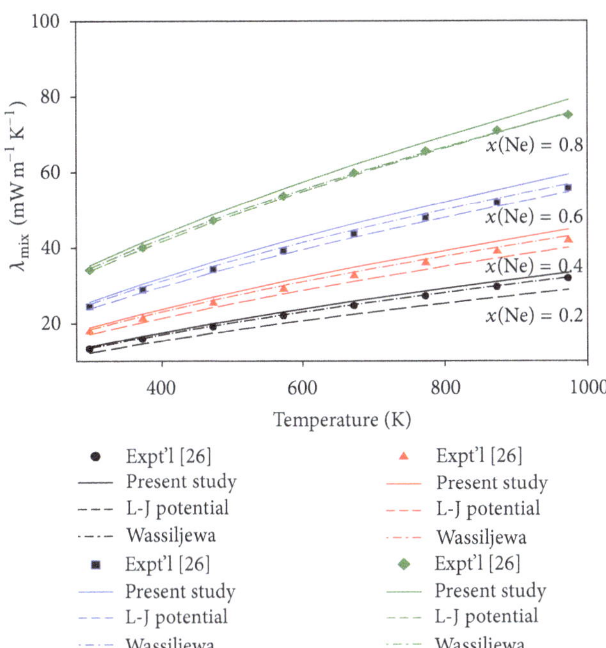

Figure 23: Comparison of measured and calculated thermal conductivities for Ne + Kr mixture.

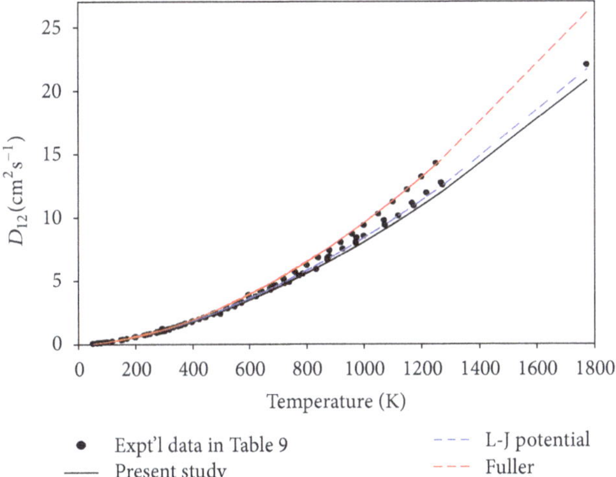

Figure 24: Comparison of measured and calculated binary diffusion coefficients for He + Ne mixture.

is somewhat less accurate than the Simsci Database, but compares very well with the original Lennard-Jones (12-6) potential and with the Refprop Database. Calculation of self-diffusion coefficients shows that this work is in better agreement with experimental data than those of the original Lennard-Jones (12-6) potential and of the Fuller method, and that for helium gas and for neon gas results of the proposed method is less accurate than the original Lennard-Jones (12-6) potential.

For mixture property predictions, the same set of potential parameters is applied with no additional parameters. Second cross-virial coefficient data calculated by the present study is less feasible to those of Dymond's correlations and is in better agreement with the observed data than the original Lennard-Jones (12-6) potential and the corresponding states method of Tsonopoulos. The present study is in better agreement between experimental mixture viscosity data than the original Lennard-Jones (12-6) potential and the Lucas method, except for the mixture of kr + Xe. The present study is in somewhat worse agreement between measured and calculated mixture thermal conductivities than the Wassiljewa equation with the combinational factor of Mason and Saxena and is more accurate than the original Lennard-Jones (12-6) potential. However, thermal conductivity of the He + Ne and Kr + Xe mixtures was reproduced better with the Wassiljewa equation than with the proposed method. The present study is in appreciably better agreement between the observed and calculated binary diffusion coefficients of noble gases mixtures than the original Lennard-Jones (12-6) potential and the Fuller method.

In this work, the empirical approach of adding a temperature-correction parameter to the reduced temperature in the Lennard-Jones (12-6) potential function has been tested with good success for the calculations of thermodynamic and transport property of noble gases and their binary mixtures in dilute gas region. Application of this approach to other substances such as polyatomic, polar gases will be tested in the near future.

Nomenclature

A:	Combinational factor in (18)
A^*:	Defined by (13), $\Omega^{(2,2)*}(T^*)/\Omega^{(1,1)*}(T^*)$
B:	Second virial coefficient (cm^3 mol^{-1})
D:	Diffusion coefficient (cm^2 sec^{-1})
k_B:	Boltzmann constant, $1.3806488{\cdot}10^{-23}$(JK^{-1})
M:	Molecular weight (gram mol^{-1})
N_A:	Avogadro constant, $6.022{\cdot}10^{23}$(mol^{-1})
P:	Pressure (bar)
R:	Distance between molecular centers of molecules 1 and 2 (Å)
RMSD$_r$:	Root-mean-square deviation (cm^3 mol^{-1})
%RMSD$_r$:	Percent relative root-mean-square deviation, relative (%)
T:	Absolute temperature (K)
T^*:	Reduced temperature, $k_B(T-\tau)/\varepsilon$
U:	Intermolecular potential function
y:	Mole fraction in the gas phase
X_η:	Defined by (10)
Y_η:	Defined by (11)
Z_η:	Defined by (12).

Greek Letters

ε: Depth of the potential well [J]
η: Viscosity (μP)
λ: Thermal conductivity (mW m^{-1} K^{-1})
σ: Collision diameter (Å)
τ: Reduced temperature-correction parameter

$\Omega^{(1,1)*}$: Collision integral for diffusion coefficient
$\Omega^{(2,2)*}$: Collision integral for viscosity.

Acknowledgment

This paper was supported by the Konyang University Research Fund, in 2012-2013.

References

[1] T. H. Chung, M. Ajlan, L. L. Lee, and K. E. Starling, "Generalized multiparameter correlation for nonpolar and polar fluid transport properties," *Industrial & Engineering Chemistry Research*, vol. 27, no. 4, pp. 671–679, 1988.

[2] J. P. O'Connell and J. M. Prausnitz, "Advances in thermophysical properties at extreme temperatures and pressures," in *Proceedings of the 3rd Symposium of Thermophysical Properties*, ASME, New York, NY, USA, 1965.

[3] S. Chapman and T. G. Cowling, *Th Mathematical Theo y of Non-Uniform Gases*, Cambridge University Press, New York, NY, USA, 3rd edition, 1970.

[4] J. O. Hirschfelder, C. F. Curtiss, and R. B. Bird, *Molecular Theory of Gases and Liquids*, Wiley, New York, NY, USA, 1954.

[5] J. H. Dymond and B. J. Alder, "Pair potential for Argon," *Th Journal of Chemical Physics*, vol. 51, no. 50, pp. 309–320, 1969.

[6] L. S. Tee, S. Gotoh, and W. E. Stewart, "Molecular parameters for normal fluids. Lennard-Jones 12-6 Potential," *Industrial and Engineering Chemistry Fundamentals*, vol. 5, no. 3, pp. 356–363, 1966.

[7] L. S. Tee, S. Gotoh, and W. E. Stewart, "molecular parameters for normal fluids. kihara potential with spherical core," *Industrial and Engineering Chemistry Fundamentals*, vol. 5, no. 3, pp. 363–367, 1966.

[8] J. E. Lennard-Jones, "On the determination of molecular fields. II. from the equation of state of a gas," *Proceedings of the Royal Society A*, vol. 106, no. 738, pp. 463–477, 1924.

[9] T. Kojima, *A Study on measurement for speed of sound in R125, R143a, and the mixture and precise determination of thermodynamic properties base on intermolecular potential model [M.S. thesis]*, Keio University, Yokohama, Japan, 2001.

[10] K. Ichikura, Y. Kano, and H. Sato, "Importance of third virial coefficients for representing the gaseous phase based on measuring PVT-properties of 1,1,1-Trifluoroethane (R143a)," *International Journal of Thermophysics*, vol. 27, pp. 23–38, 2006.

[11] B. E. Poling, J. M. Prausnitz, and J. P. O'Connell, *Th Properties of Gases and Liquids*, McGraw Hill, New York, NY, USA, 5th edition, 2004.

[12] P. D. Neufeld, A. R. Janzen, and R. A. Aziz, "Empirical equations to calculate 16 of the transport collision integrals $\Omega^{(l,s)*}$ for the lennard-jones (12-6) potential," *Th Journal of Chemical Physics*, vol. 57, no. 3, pp. 1100–1102, 1972.

[13] R. Brokaw, "Predicting transport properties of dilute gases," *Industrial & Engineering Chemistry Process Design and Development*, vol. 8, no. 2, pp. 240–253, 1969.

[14] H. J. M. Hanley and M. Klein, "Application of the m-6-8 potential to simple gases," *The Journal of Physical Chemistry*, vol. 76, no. 12, pp. 1743–1751, 1972.

[15] A. Wassiljewa, "Wärmeleitung in Gasgemischen," *Physikalische Zeitschrift*, vol. 5, p. 737, 1904.

[16] E. A. Mason and S. C. Saxena, "Approximate formula for the thermal conductivity of gas mixtures," *Physics of Fluids*, vol. 1, no. 5, pp. 361–369, 1958.

[17] J. H. Dymond, K. N. Marsh, R. C. Wilhoit, and K. C. Wong, *Virial Coeffici ts of Pure Gases and Mixtures*, vol. 4 of *Landolt-Börnstein: Numerical Data and Functional Relationships in Science and Technology*, part 21A&B, Springer, Hidelberg, Germany, 2002.

[18] K. Stephan and K. Lucas, *Viscosity of Dense Fluids*, Plenum Press, New York, NY, USA, 1979.

[19] IMSL, *IMSL STAT/LIBRARY: Regression*, IMSL, Houston, Tex, USA, 1994.

[20] S.-K. Oh, "An extension of the group contribution method for estimating thermodynamic and transport properties. Part III. Noble gases," *Korean Journal of Chemical Engineering*, vol. 22, no. 6, pp. 949–959, 2005.

[21] C. Tsonopoulos, "Empirical correlation of second virial coefficients," *AIChE Journal*, vol. 20, no. 2, pp. 263–272, 1974.

[22] E. W. Lemmon, M. L. Huber, and M. O. McLinden, "NIST standard reference database 23," Reference Fluid Thermodynamic and Transport Properties (REFPROP), version 9.0, National Institute of Standards and Technology, 2010.

[23] K. Lucas, *Phase Equilibria and Fluid Properties in the Chemical Industry*, Dechema, Frankfurt, Germany, 1980.

[24] Simsci Database, Pro/II with Provision Software, http://iom invensys.com/EN/Pages/SimSci-Esscor_ProcessEngSuite_PROII .aspx.

[25] E. N. Fuller, P. D. Schettler, and J. C. Giddings, "new method for prediction of binary gas-phase diffusion coefficients," *Industrial & Engineering Chemistry Research*, vol. 58, no. 5, pp. 18–27, 1966.

[26] J. Kestin, K. Knierim, E. A. Mason, B. Najafi, S. T. Ro, and M. Waldman, "equilibrium and transport properties of the noble gases and their mixtures at low density," *Journal of Physical and Chemical Reference Data*, vol. 13, no. 1, 75 pages, 1984.

[27] A. Arteconi, G. Di Nicola, G. Santori, and R. Stryjek, "Second virial coefficients for dimethyl ether," *Journal of Chemical & Engineering Data*, vol. 54, no. 6, pp. 1840–1843, 2009.

[28] A. A. Richcardo and M. N. da Ponte, "Second virial coefficients of mixtures of Xenon and lower hydrocarbons. 1. Experimental apparatus and results for Xe + C_2H_6," *Th Journal of Physical Chemistry*, vol. 100, no. 48, pp. 18839–18843, 1996.

[29] D. R. Lide, *CRC Handbook of Chemistry and Physics*, CRC Press, New York, NY, USA, 90th edition, 2010.

[30] J. Kestin and J. Yata, "Viscosity and diffusion coefficient of six binary mixtures," *Th Journal of Chemical Physics*, vol. 49, no. 11, pp. 4780–4791, 1968.

[31] A. S. Kalelkar and J. Kestin, "Viscosity of He-Ar and He-Kr binary gaseous mixtures in the temperature range 25–720°C," *Th Journal of Chemical Physics*, vol. 52, no. 8, pp. 4248–4261, 1970.

[32] J. Kestin, E. Paykoç, and J. V. Sengers, "On the density expansion for viscosity in gases," *Physica*, vol. 54, no. 1, pp. 1–19, 1971.

[33] J. Kestin, S. T. Ro, and Ber. Bunsenges, "The viscosity of nine binary and two ternary mixtures of gases at low density," *Berichte der Bunsengesellschaft für Physikalische Chemie*, vol. 20, no. 1, pp. 20–24, 1974.

[34] J. Kestin, S. T. Ro, and W. A. Wakeham, "Viscosity of the binary gaseous mixture Helium-nitrogen," *Th Journal of Chemical Physics*, vol. 56, no. 8, pp. 4032–4036, 1972.

[35] J. Kestin, S. T. Ro, and W. A. Wakeham, "Viscosity of the noble gases in the temperature range 25–700°C," *Journal of Chemical Physics*, vol. 56, no. 8, 6 pages, 1972.

[36] J. Kestin, S. T. Ro, and W. A. Wakeham, "Viscosity of the binary gaseous mixtures He–Ne and Ne–N_2 in the Temperature Range 25–700°C," *Journal of Chemical Physics*, vol. 56, no. 12, 6 pages, 1972.

[37] J. Kestin, H. E. Kalifa, S. T. Ro, and W. A. Wakeham, "The viscosity and diffusion coefficients of eighteen binary gaseous systems," *Physica A*, vol. 88, no. 2, pp. 242–260, 1977.

[38] J. Kestin, H. E. Khalifa, and W. A. Wakeham, "The viscosity and diffusion coefficients of the binary mixtures of Xenon with the other noble gases," *Physica A*, vol. 90, no. 2, pp. 215–228, 1978.

[39] J. Kestin and W. A. Wakeham, "The viscosity and diffusion coefficient of binary mixtures of nitrous oxide with He, Ne and CO," *Berichte der Bunsengesellschaft für Physikalische Chemie*, vol. 87, no. 4, pp. 309–311, 1983.

[40] M. Trauntz and K. F. Kipphan, "Die Reibung, Wärmeleitung und diffusion in gasmischungen. IV. Die reibung binärer und ternärer edelgasgemische," *Annalen der Physik*, vol. 394, no. 6, p. 743, 1929.

[41] M. Trauntz and R. Zink, "Die Reibung, Wärmeleitung und diffusion in gasmischungen XII. Gasreibung bei höheren temperaturen," *Annalen der Physik*, vol. 399, no. 4, p. 427, 1930.

[42] M. Trauntz and H. Zimmermann, "Die Reibung, Wärmeleitung und diffusion von gasmischungen XXX. Die innere reibung bei tiefen temperaturen von wasserstoff, helium und neon und binären gemischen davon bis 90,0° abs. herab," *Annalen der Physik*, vol. 414, no. 2, p. 189, 1935.

[43] J. Kestin and W. Leidenfrost, "The viscosity of Helium," *Physica*, vol. 25, no. 1–6, pp. 537–555, 1959.

[44] K. Kestin and W. Leidenfrost, *Th rmal and Transport Properties of Gases and Liquids*, McGraw-Hill, New York, NY, USA, 1959.

[45] E. Thornton, "Viscosity and thermal conductivity of binary gas mixtures: Xenon-Krypton, Xenon-Argon, Xenon-Neon and Xenon-Helium," *Proceedings of the Physical Society*, vol. 76, no. 1, p. 104, 1960.

[46] E. Thornton, "Viscosity and thermal conductivity of binary gas mixtures: Krypton-Argon, Krypton-Neon, and Krypton-Helium," *Proceedings of the Physical Society*, vol. 77, no. 6, p. 1166, 1961.

[47] E. Thornton and W. A. D. Baker, "Viscosity and thermal conductivity of binary gas mixtures: Argon-Neon, Argon-Helium, and Neon-Helium," *Proceedings of the Physical Society*, vol. 80, no. 5, p. 1171, 1962.

[48] H. Iwasaki and J. Kestin, "The viscosity of Argon-Helium mixtures," *Physica*, vol. 29, no. 12, pp. 1345–1372, 1963.

[49] A. B. Rakshit and C. S. Roy, "Viscosity and polar-nonpolar interactions in mixtures of inert gases with ammonia," *Physica*, vol. 78, no. 1, pp. 153–164, 1974.

[50] A. O. Rietveld, A. Van Itterbeek, and G. J. Van Den Berg, "Measurements on the viscosity of mixtures of Helium and Argon," *Physica*, vol. 19, no. 1-12, pp. 517–524, 1953.

[51] A. O. Rietveld, A. Van Itterbeek, and C. A. Velds, "Viscosity of binary mixtures of hydrogen isotopes and mixtures of He and Ne," *Physica*, vol. 25, no. 1–6, pp. 205–216, 1959.

[52] A. E. Schuil, "A note on the viscosity of gases and molecular mean free path," *Philosophical Magazine Series 7*, vol. 28, no. 191, p. 679, 1939.

[53] J. Kestin and A. Nagashima, "Viscosity of Neon-Helium and Neon-Argon mixtures at 20° and 30°C," *Th Journal of Chemical Physics*, vol. 40, no. 12, pp. 3648–3654, 1964.

[54] J. Kestin, Y. Kobayashi, and R. T. Wood, "The viscosity of four binary, gaseous mixtures at 20° and 30°C," *Physica*, vol. 32, no. 6, pp. 1065–1089, 1966.

[55] H. Landolt and R. Börnstein, *Landolt-Börnstein Physikalisch-chemische Tabellen*, Springer, Berlin, Germany, 1923.

[56] N. B. Vargaftik, *Tables on the Thermophysical Properties of Liquids and Gases*, Hemisphere, Washington, DC, USA, 2nd edition, 1975.

[57] K. Stephan and T. Heckenberger, *Th rmal Conductivity and Viscosity Data of Fluid Mixtures*, vol. 10 of *Chemistry Data Series*, Dechema, Frankfurt, Germany, 1988.

[58] E. Bich, J. Millat, and E. Vogel, "The viscosity and thermal conductivity of pure monatomic gases from their normal boiling point up to 5000 K in the limit of zero density and at 0.101325 MPa," *Journal of Physical and Chemical Reference Data*, vol. 19, no. 6, article 1289, 17 pages, 1990.

[59] W. A. Wakeham, A. Nagashima, and J. V. Sengers, *Measurement of the Transport Properties of Fluids*, IUPAC Chemical Data Series no. 37, Blackwell Scientific, London, UK, 1991.

[60] A. Bejan and A. D. Kraus, *Heat Transfer Handbook*, chapter 2, John Wiley & Sons, New York, NY, USA, 2003.

[61] G. C. Maitland and E. B. Smith, "Critical reassessment of viscosities of 11 common gases," *Journal of Chemical & Engineering Data*, vol. 17, no. 2, pp. 150–156, 1972.

[62] G. F. C. Rogers and Y. R. Mayhew, *Thermodynamic and Transport Properties of Fluids*, Basil Blackwell, Oxford, UK, 3rd edition, 1981.

[63] IAEA, *The mophysical Properties of Materials for Nuuclear Engineering: A Tutorial and Collection of Data*, IAEA, Vienna, 2008.

[64] J. Kestin, W. Wakeham, and K. Watanabe, "Viscosity, thermal conductivity, and diffusion coefficient of Ar-Ne and Ar-Kr gaseous mixtures in the temperature range 25–700°C," *Th Journal of Chemical Physics*, vol. 53, no. 10, pp. 3773–3780, 1970.

[65] J. Kestin, S. T. Ro, and W. A. Wakeham, "Viscosity of the binary gaseous mixture Neon–Krypton," *Journal of Chemical Physics*, vol. 56, no. 8, article 4086, 6 pages, 1972.

[66] A. O. Rankine, "On the variation with temperature of the viscosities of the gases of the Argon group," *Zeitschrift für Physik*, vol. 11, p. 497608, pp. 745–762, 1910.

[67] A. G. Clarke and E. B. Smith, "Low-temperature viscosities of argon, krypton, and xenon," *The Journal of Chemical Physics*, vol. 48, no. 9, pp. 3988–3991, 1968.

[68] R. A. Dawe and E. B. Smith, "Viscosities of the inert gases at high temperatures," *Th Journal of Chemical Physics*, vol. 52, no. 2, pp. 693–703, 1970.

[69] R. S. Edwards, "The effect of temperature on the viscosity of Neon," *Proceedings of the Royal Society A*, vol. 119, no. 783, pp. 578–590, 1928.

[70] A. van Itterbeek and O. van Paemel, "Viscosity of liquid deuterium," *Physica*, vol. 7, no. 3, p. 208, 1940.

[71] R. Wobser and F. Müller, "Die innere reibung von gasen und dämpfen und ihre messung im Höppler-viskosimeter," *Kolloid-Beihefte*, vol. 52, no. 6-7, pp. 165–276, 1941.

[72] H. Johnston and E. R. Grilly, "Viscosities of carbon monoxide, Helium, Neon, and Argon between 80° and 300°K. coefficients of viscosity," *Th Journal of Physical Chemistry*, vol. 46, no. 8, pp. 948–963, 1942.

[73] J. Kestin and W. Leidenfrost, "An absolute determination of the viscosity of eleven gases over a range of pressures," *Physica*, vol. 25, no. 7-12, pp. 1033–1062, 1959.

[74] G. P. Flynn, R. V. Hanks, N. A. Lemaire, and J. F. Ross, "Viscosity of Nitrogen, Helium, Neon, and Argon from −78.5° to 100°C

below 200 Atmospheres," *Journal of Chemical Physics*, vol. 38, no. 1, article 154, 9 pages, 1963.

[75] J. Kestin and J. H. Whitelaw, "A relative determination of the viscosity of several gases by the oscillating disk method," *Physica*, vol. 29, no. 4, pp. 335–356, 1963.

[76] N. J. Trappeniers, A. Botzen, H. R. Van Den Berg, and J. Van Oosten, "The viscosity of Neon between 25°C and 75°C at pressures up to 1800 atmospheres. Corresponding states for the viscosity of the noble gases up to high densities," *Physica*, vol. 30, no. 5, pp. 985–996, 1964.

[77] A. O. Rietveld and A. van Itterbeek, "Measurements on the viscosity of Ne-A mixtures between 300 and 70°K," *Physica*, vol. 22, no. 6-12, pp. 785–790, 1956.

[78] M. Trautz and H. E. Binkele, "Die Reibung, Wärmeleitung und diffusion in gasmischungen. VIII. die reibung des H_2, He, Ne, Ar und ihrer binären gemische," *Annalen der Physik*, vol. 397, no. 5, pp. 561–580, 1930.

[79] B. A. Younglove and H. J. M. Hanley, "The viscosity and thermal conductivity coefficients of gaseous and liquid Argon," *Journal of Physical and Chemical Reference Data*, vol. 5, no. 4, article 1323, 15 pages, 1986.

[80] I. A. Barr, G. P. Matthews, E. B. Smith, and A. R. Tindell, "Intermolecular forces and the gaseous viscosities of Argon-Xenon mixtures," *Journal of Physical Chemistry*, vol. 85, no. 22, pp. 3342–3347, 1981.

[81] J. M. Hellemans, J. Kestin, and S. T. Ro, "Viscosity of the binary gaseous mixtures of nitrogen with Argon and Krypton," *Th Journal of Chemical Physics*, vol. 57, no. 9, pp. 4038–4042, 1972.

[82] J. Kestin and S. T. Ro, "The viscosity and diffusion coefficients of binary mixtures of nitrous Oxide with Ar, N_2, and CO_2," *Berichte der Bunsengesellschaft für Physikalische Chemie*, vol. 86, no. 10, pp. 948–950, 1982.

[83] J. Kestin and W. A. Wakaham, "The viscosity of three polar gases," *Berichte der Bunsengesellschaft für Physikalische Chemie*, vol. 83, no. 6, pp. 573–576, 1979.

[84] A. van Itterbeek and O. van Paemel, "Measurements on the viscosity of Argon gas at room temperature and between 9° and 55°K," *Physica*, vol. 5, no. 10, pp. 1009–1012, 1938.

[85] P. Gray and A. O. S. Maczek, "The thermal conductivities, viscositites, and diffusion coefficients of mixtures, containing two polar gases," in *Proceedings of the 4th Symposium on Th rmophysical Properties*, pp. 380–391, ASME, New York, NY, USA, 1968.

[86] C. F. Bonilla, S. J. Wang, and H. Weiner, "The viscosity of steam, heavy-water vapor, and Argon at atmospheric pressure up to high temperatures," *Transactions of the American Society of Mechanical Engineers*, vol. 78, pp. 1285–1289.

[87] H. Iwasaki, J. Kestin, and A. Nagashima, "Viscosity of Argon-ammonia mixtures," *Th Journal of Chemical Physics*, vol. 40, no. 10, pp. 2988–2995, 1964.

[88] J. Hilsenrath, *Tables of Th rmal Properties of Gases*, vol. 564 of *National Bureau of Standards Circular*, US Government Printing Office, Washington, DC, USA, 1955.

[89] J. M. Hellemans, J. Kestin, and S. T. Ro, "The viscosity of oxygen and of some of its mixtures with other gases," *Physica*, vol. 65, no. 2, pp. 362–375, 1973.

[90] E. F. May, R. F. Berg, and M. R. Moldover, "Reference viscosities of H_2, CH_4, Ar, and Xe at low densities," *International Journal of Th rmophysics*, vol. 28, no. 4, pp. 1085–1110, 2007.

[91] V. Vasilesco, "Recherches expérimentales sur la viscosité des gaz aux températures élevées," *Annales de Physique*, vol. 20, pp. 292–334, 1945.

[92] H. J. M. Hanley, "the viscosity and thermal conductivity coefficients of dilute Argon, Krypton, and Xenon," *Journal of Physical and Chemical Reference Data*, vol. 3, no. 3, article 619, 24 pages, 1973.

[93] H. J. M. Hanley, R. D. McCarty, and W. M. Haynes J, "The viscosity and thermal conductivity coefficients for dense gaseous and liquid Argon, Krypton, Xenon, Nitrogen, and Oxygen," *Journal of Physical and Chemical Reference Data*, vol. 3, no. 4, article 979, 39 pages, 1974.

[94] X. Wang, J. Wu, and Z. Liu, "Viscosity of gaseous HFC245fa," *Journal of Chemical & Engineering Data*, vol. 55, no. 1, pp. 496–499, 2010.

[95] R. Kiyama and T. Makita, "The viscosity of carbon dioxide, ammonia, acetylene, Argon and oxygen under high pressures," *Review of Physical Chemistry of Japan*, vol. 22, pp. 49–58, 1952.

[96] T. Makita, "The viscosity at pressures to 80o kg/cm^2 up of Argon, nitrogen and air," *Review of Physical Chemistry of Japan*, vol. 27, no. 1, pp. 16–21, 1957.

[97] M. Hongo, "Viscosity of Argon and of Argon-ammonia mixtures under pressures," *Th Review of Physical Chemistry of Japan*, vol. 48, no. 2, pp. 63–71, 1979.

[98] J. Kestin, H. E. Khalifa, and W. A. Wakeham, "The viscosity of gaseous mixtures containing Krypton," *Th Journal of Chemical Physics*, vol. 67, no. 9, pp. 4254–4259, 1977.

[99] M. Trautz, "Die reibung, Wärmeleitung und diffusion in Gasmischungen XXI. Absoluter η-Wirkungsquerschnitt, molekulartheoretische bedeutung der kritischen temperatur und berechung kritischer drucke aus η," *Annalen Der Physik*, vol. 407, no. 2, p. 198, 1932.

[100] J. M. Gandhi and S. C. Saxena, "Thermal conductivity of binary and ternary mixtures of Helium, Neon and Xenon," *Molecular Physics*, vol. 12, no. 1, p. 57, 1967.

[101] S. C. Saxena, M. P. Saksena, R. S. Gambhir, and J. M. Gandhi, "The thermal conductivity of nonpolar polyatomic gas mixtures," *Physica*, vol. 31, no. 3, pp. 333–341, 1965.

[102] H. Ziebland, "commission on physicochemical measurements and standards," *Pure & Applied Chemistry*, vol. 53, no. 10, pp. 1863–1877, 1981.

[103] R. W. Powell, C. Y. Ho, and P. E. Liley, "Thermal conductivity of seleted materials," National Standard Reference Data Series NBS 8, US Department of Commerce, 1966.

[104] S. C. Saxena, "Transport properties of gases and gaseous mixtures at high temperatures," *High Temperature Science*, vol. 3, pp. 168–188, 1971.

[105] S. C. Saxena, S. Mathur, and G. P. Gupta, "The thermal conductivity data of some binary gas mixtures involving nonpolar polyatomic gases," *Defence Science Journal*, vol. 16, supplement, pp. 99–112, 1966.

[106] N. B. Vargaftik, L. P. Filippon, A. A. Tarzimanov, and E. E. Totskii, *Handbook of Th rmal Conductivity of Liquids and Gaseous*, CRC Press, Boca Raton, Fla, USA, 1994.

[107] W. Van Dael and H. Cauwenbergh, "Measurements of the thermal conductivity of gases. II. Data for binary mixtures of He, Ne and Ar," *Physica*, vol. 40, no. 2, pp. 173–181, 1968.

[108] W. G. Kannuluik and E. H. Carman, "The thermal conductivity of rare gases," *Proceedings of the Physical Society B*, vol. 65, no. 9, p. 701, 1952.

[109] R. S. Gambhir and S. C. Saxena, "Thermal conductivity of binary and ternary mixtures of Krypton, Argon and Helium," *Molecular Physics*, vol. 11, no. 3, pp. 233–241, 1966.

[110] E. A. Mason and H. Von Ubisch, "Thermal conductivities of rare gas mixtures," *Physics of Fluids*, vol. 3, no. 3, pp. 355–361, 1960.

[111] S. C. Saxena, "Thermal conductivity of binary and ternary mixtures of helium, argon and xenon," *Indian Journal of Physics*, vol. 31, pp. 597–606, 1957.

[112] H. von Ubisch, "The thermal conductivities of mixtures of rare gases at 29° and 520°," *Arkiv för Fysik*, vol. 16, pp. 93–100, 1959.

[113] S. Mathur, P. K. Tondon, and S. C. Saxena, "Thermal conductivity of binary, ternary and quaternary mixtures of rare gases," *Molecular Physics*, vol. 12, no. 6, pp. 569–579, 1967.

[114] C. Y. Ho, R. W. Powell, and P. E. Liley J, "Thermal conductivity of the elements," *Journal of Physical and Chemical Reference Data*, vol. 1, no. 2, article 279, 143 pages, 1972.

[115] B. N. Srivastava and S. C. Saxena, "Thermal conductivity of binary and ternary rare gas mixtures," *Proceedings of the Physical Society B*, vol. 70, no. 4, p. 369, 1957.

[116] K. Schäfer, "Transport Phenomena in the temperature range up to 1100 degrees," *Dechema Monograph*, vol. 32, pp. 61–73, 1959.

[117] K. C. Hansen, L. H. Tsao, T. M. Aminabhavi, and C. L. Yaws, "Gaseous thermal conductivity of hydrogen chloride, hydrogen bromide, boron trichloride, and boron trifluoride in the temperature range from 55 to 380°C," *Journal of Chemical and Engineering Data*, vol. 40, no. 1, pp. 18–20, 1995.

[118] R. S. Gambhir and S. C. Saxena, "Thermal conductivity of the gas mixtures: Ar-D_2, Kr-D_2 and Ar-Kr-D_2," *Physica*, vol. 32, no. 11-12, pp. 2037–2043, 1966.

[119] L. Sun and J. E. S. Venart, "Thermal conductivity, thermal diffusivity, and heat capacity of gaseous Argon and nitrogen," *International Journal of Thermophysics*, vol. 26, no. 2, pp. 325–372, 2005.

[120] L. Sun, J. E. S. Venart, and R. C. Prasad, "The thermal conductivity, thermal diffusivity, and heat capacity of gaseous Argon," *International Journal of Thermophysics*, vol. 23, pp. 357–389, 2002.

[121] C. J. Zwakhals and K. W. Reus, "Corrections to the Smoluchowski equation in the presence of hydrodynamic interactions," *Physica C*, vol. 100, no. 2, pp. 251–265, 1980.

[122] W. Groth and E. Sußner, "Selbstdiffusionsmessungen III. Der selbstdiffusionskoeffizient des neon (self-diffusion measurments III. The coefficient of self diffusion of neon)," *Zeitschrift für Physikalische Chemie*, vol. 193, pp. 296–300, 1944.

[123] W. Groth and P. Harteck, "Die selbstdiffusion des Xenons und des Kryptons," *Zeitschrift für Elektrochemie*, vol. 47, no. 2, pp. 167–172, 1941.

[124] M. P. Saksena and S. C. Saxena, "Viscosity of multicomponent gas mixtures," *Physica A*, vol. 31, p. 18, 1965.

[125] H. Iwasaki and J. Kestin, "The viscosity of Argon-Helium mixtures," *Physica*, vol. 29, no. 12, pp. 1345–1372, 1963.

[126] D. J. Richardson, G. Mason, B. A. Buffham, K. Hellgardt, I. W. Cumming, and P. A. Russell, "Viscosity of binary mixtures of carbon monoxide and Helium," *Journal of Chemical and Engineering Data*, vol. 53, no. 1, pp. 303–306, 2008.

[127] S. C. Saxena and T. K. S. Narayanan, "Multicomponent viscosities of gaseous mixtures at high temperatures," *Industrial & Engineering Chemistry Fundamentals*, vol. 1, no. 3, pp. 191–195, 1962.

[128] J. W. Buddenberg and C. R. Wilke, "Calculation of gas mixture viscosities," *Industrial & Engineering Chemistry Research*, vol. 41, no. 7, pp. 1345–1347, 1949.

[129] J. Wachsmuth, "Über die wärmeleitung von gemischen zwischen argon und helium," *Physikalische Zeitschrift*, vol. 7, p. 235, 1908.

[130] E. A. Mason and T. R. Marrero, "Gaseous diffusion coefficients," *Journal of Physical and Chemical Reference Data*, vol. 1, no. 1, 116 pages, 1972.

[131] R. J. J. Van Heijningen, J. P. Harpe, and J. J. M. Beenakker, "Determination of the diffusion coefficients of binary mixtures of the noble gases as a function of temperature and concentration," *Physica*, vol. 38, no. 1, pp. 1–34, 1968.

[132] P. S. Arora, H. L. Robjohns, and P. J. Dunlop, "Use of accurate diffusion and second virial coefficients to determine (m, 6, 8) potential parameters for nine binary noble gas systems," *Physica A*, vol. 95, no. 3, pp. 561–571, 1979.

[133] M. Trautz and W. Muller, "Die reibung, wärmeleitung und diffusion in gasmischungen III. Die korrektion der bisher mit der verdampfungsmethode gemessenen diffusionskonstanten," *Annalen der Physik*, vol. 414, no. 4, pp. 333–352, 1935.

[134] J. N. Holsen and M. R. Strunk, "Binary diffusion coefficients in nonpolar gases," *Industrial and Engineering Chemistry Fundamentals*, vol. 3, no. 2, pp. 143–146, 1964.

[135] S. L. Seager, L. R. Geertson, and J. C. Giddings, "Temperature dependence of gas and vapor diffusion coefficients," *Journal of Chemical & Engineering Data*, vol. 8, no. 2, pp. 168–169, 1963.

[136] A. T. Hu and R. Kobayashi, "Measurements of gaseous diffusion coefficients for dilute and moderately dense gases by perturbation chromatography," *Journal of Chemical & Engineering Data*, vol. 15, no. 2, pp. 326–335, 1970.

[137] A. P. Malinauskas, "Gaseous diffusion. The systems He–Ar, Ar–Xe, and He–Xe," *Journal of Chemical Physics*, vol. 42, no. 1, article 157, 4 pages, 1965.

[138] R. A. Strehlow, "The temperature dependence of the mutual diffusion coefficient for four gaseous systems," *Journal of Chemical & Engineering Data*, vol. 21, no. 12, article 2101, 6 pages, 1953.

[139] R. E. Walker and A. A. Westenberg, "Molecular diffusion studies in gases at high temperature. III. results and interpretation of the He—a system," *Journal of Chemical Physics*, vol. 31, no. 519, 4 pages, 1959.

[140] J. C. Giddings and S. L. Seager, "Method for the rapid determination of diffusion coefficients. theory and application," *Industrial and Engineering Chemistry Fundamentals*, vol. 1, no. 4, pp. 277–283, 1962.

[141] S. C. Sexena and E. A. Mason, "Thermal diffusion and the approach to the steady state in gases: II," *Molecular Physics*, vol. 2, no. 4, p. 379, 1959.

[142] E. N. Fuller, P. D. Schettler, and J. C. Giddings, "New method for prediction of binary gas-phase diffusion coefficients," *Industrial & Engineering Chemistry*, vol. 58, no. 5, pp. 18–27, 1966.

[143] M. Keil, L. Danielson, and P. J. Dunlop, "On obtaining interatomic potentials from multiproperty fits to experimental data ," *Journal of Chemical Physics*, vol. 84, no. 14, article 296, 14 pages, 1991.

[144] A. P. Malinauskas, "Gaseous Diffusion. The Systems He–Ar, Ar–Xe, and He–Xe," *Journal of Chemical Physics*, vol. 42, no. 1, article 156, 4 pages, 1964.

[145] K. Schafer and K. Schuhman Z, "Zwischenmolekulare kräfte und temperaturabhängigkeit von diffusion und selbstdiffusion in Edelgasen," *Zeitschrift für Elektrochemie*, vol. 61, no. 2, pp. 246–252, 1957.

[146] B. N. Srivastava and K. P. Srivastava, "Mutual diffusion of pairs of rare gases at different temperatures," *Journal of Chemical Physics*, vol. 30, no. 984, article 984, 7 pages, 1959.

[147] W. Sutherland, "The viscosity of gases and molecular force," *Philosophical Magazine*, vol. 36, pp. 507–631, 1893.

[148] W. Sutherland, "The attraction of unlike molecules.-I. The diffusion of gases," *Philosophical Magazine*, vol. 38, no. 230, pp. 1–19, 1894.

[149] V. P. S. Main and S. C. Saxena, "measurement of the concentration diffusion coefficient for Ne-Ar, Ne-Xe, Ne-H_2, Xe-H_2, H_2-N_2 and H_2-O_2 gas systems," *Applied Scientifi Research*, vol. 23, pp. 121–133, 1971.

[150] A. E. Humphreys and E. A. Mason, "Intermolecular forces: thermal diffusion and diffusion in Ar–Kr ," *Physics of Fluids*, vol. 13, no. 1, article 65, 6 pages, 1970.

[151] P. J. Dunlop and C. M. Bignell, "Diffusion and thermal diffusion in binary mixtures of methane with noble gases and of Argon with Krypton," *Physica A*, vol. 145, no. 3, pp. 584–596, 1987.

[152] L. Andrussow and T. F. Schatzki, "Diffusion coefficients of the systems Xe–Xe and A–Xe," *Journal of Chemical Physics*, vol. 27, no. 5, article 1049, 6 pages, 1957.

[153] E. A. Mason and T. R. Marrero, "The diffusion of atoms and molecules," *Advances in Atomic and Molecular Physics*, vol. 6, pp. 155–232, 1970.

Permissions

The contributors of this book come from diverse backgrounds, making this book a truly international effort. This book will bring forth new frontiers with its revolutionizing research information and detailed analysis of the nascent developments around the world.

We would like to thank all the contributing authors for lending their expertise to make the book truly unique. They have played a crucial role in the development of this book. Without their invaluable contributions this book wouldn't have been possible. They have made vital efforts to compile up to date information on the varied aspects of this subject to make this book a valuable addition to the collection of many professionals and students.

This book was conceptualized with the vision of imparting up-to-date information and advanced data in this field. To ensure the same, a matchless editorial board was set up. Every individual on the board went through rigorous rounds of assessment to prove their worth. After which they invested a large part of their time researching and compiling the most relevant data for our readers. Conferences and sessions were held from time to time between the editorial board and the contributing authors to present the data in the most comprehensible form. The editorial team has worked tirelessly to provide valuable and valid information to help people across the globe.

Every chapter published in this book has been scrutinized by our experts. Their significance has been extensively debated. The topics covered herein carry significant findings which will fuel the growth of the discipline. They may even be implemented as practical applications or may be referred to as a beginning point for another development. Chapters in this book were first published by Hindawi Publishing Corporation; hereby published with permission under the Creative Commons Attribution License or equivalent.

The editorial board has been involved in producing this book since its inception. They have spent rigorous hours researching and exploring the diverse topics which have resulted in the successful publishing of this book. They have passed on their knowledge of decades through this book. To expedite this challenging task, the publisher supported the team at every step. A small team of assistant editors was also appointed to further simplify the editing procedure and attain best results for the readers.

Our editorial team has been hand-picked from every corner of the world. Their multi-ethnicity adds dynamic inputs to the discussions which result in innovative outcomes. These outcomes are then further discussed with the researchers and contributors who give their valuable feedback and opinion regarding the same. The feedback is then collaborated with the researches and they are edited in a comprehensive manner to aid the understanding of the subject.

Apart from the editorial board, the designing team has also invested a significant amount of their time in understanding the subject and creating the most relevant covers. They scrutinized every image to scout for the most suitable representation of the subject and create an appropriate cover for the book.

The publishing team has been involved in this book since its early stages. They were actively engaged in every process, be it collecting the data, connecting with the contributors or procuring relevant information. The team has been an ardent support to the editorial, designing and production team. Their endless efforts to recruit the best for this project, has resulted in the accomplishment of this book. They are a veteran in the field of academics and their pool of knowledge is as vast as their experience in printing. Their expertise and guidance has proved useful at every step. Their uncompromising quality standards have made this book an exceptional effort. Their encouragement from time to time has been an inspiration for everyone.

The publisher and the editorial board hope that this book will prove to be a valuable piece of knowledge for researchers, students, practitioners and scholars across the globe.

List of Contributors

Mehdi Taghdiri, Mahmood Payehghadr, Reza Behjatmanesh-Ardakani and Homa Gha'ari
Department of Chemistry, Payame Noor University, P.O. Box 19395-3697, Tehran, Iran

Duminda A. Gunawardena and Sandun D. Fernando
Biological and Agricultural Engineering Department, Texas A&M University, College Station, TX 77843, USA

Abhishek Khanna
Department of Physical Sciences, Indian Institute of Science Education and Research Mohali, Sector 81, Manauli, Mohali, Punjab 140306, India
Elite Course Theoretical and Mathematical Physics, Ludwig Maximillian University, D-80333 Munich, Germany

Ramandeep S. Johal
Department of Physical Sciences, Indian Institute of Science Education and Research Mohali, Sector 81, Manauli, Mohali, Punjab 140306, India

B. Nagarjun
Department of Physics, G.V.P. College of Engineering (A), Visakhapatnam 530048, Andhra Pradesh, India

A. V. Sarma
Department of Physics, Andhra University, Visakhapatnam 530003, Andhra Pradesh, India

G. V. Rama Rao
Department of Physics, DAR College, Nuzvid 521201,Andhra Pradesh, India

C. Rambabu
Department of Chemistry, Acharya Nagarjuna University, Guntur 522510, Andhra Pradesh, India

V. V. Ryazanov
Institute for Nuclear Research, Prospect Nauki 47, Kiev 252028, Ukraine

Hiroshi Abe, Tomohiro Mori and Yusuke Imai
Department of Materials Science and Engineering, National Defense Academy, Yokosuka 239-8686, Japan

Yukihiro Yoshimura
Department of Applied Chemistry, National Defense Academy, Yokosuka 239-8686, Japan

Maninder Kumar and V. K. Rattan
University Institute of Chemical Engineering and Technology, Panjab University, Chandigarh 160014, India

M. V. Rathnam and Devappa R. Ambavadekar
Physical Chemistry Research Laboratory, B.N. Bandodkar College of Science, Th ne 400601, India

M. Nandini
Department of Chemistry, Dr. P.R.Ghogrey Science College, Deopur, Dhule 424005, India

Enrique Hern´andez-Lemus
Computational Genomics Department, National Institute of Genomic Medicine, Perif´erico Sur 4809, Col. Arenal Tepepan, Delegaci´on Tlalpan, 14610 Mexico City, DF, Mexico

Arvind R. Mahajan and Sunil R. Mirgane
P. G. Department of Chemistry, Jalna Education Society's R. G. Bagdia Arts, R. Benzonji Science College, S. B. Lakhotia Commerce, Jalna, Maharashtra 431203, India

Ouldouz Nourani Zonouz and Mehdi Salmanpour
Department of Mechanical Engineering, Marvdasht Branch, Islamic Azad University, Marvdasht 73711-13119, Iran

Yukihiro Yoshimura and Naohiro Hatano
Department of Applied Chemistry, National Defense Academy, Yokosuka, Kanagawa 239-8686, Japan

Yusuke Imai and Hiroshi Abe
Department of Materials Science and Engineering, National Defense Academy, Yokosuka, Kanagawa 239-8686, Japan

Osamu Shimada and Tomonori Hanasaki
Department of Applied Chemistry, Ritsumeikan University, Kusatsu, Shiga 525-8577, Japan

R. J. Yadav
Department of Mechanical Engineering, MIT College of Engineering, Pune-411028, India

A. S. Padalkar
Department of Mechanical Engineering, Flora Institute of Technology, Pune-412205, India

Daming Gao, Hui Zhang, Hong Sun, Dechun Zhu and Jianjun Shi
Department of Chemistry and Materials Engineering, Hefei University, Anhui, Hefei 230022, China
Sino-German Research Center for Process Engineering and Energy Technology, Anhui, Hefei 230022, China

Peter Lücking
Sino-German Research Center for Process Engineering and Energy Technology, Anhui, Hefei 230022, China
Department of Engineering, Jade University of Applied Science, 26389 Wilhelmshaven, Germany

Hong Chen and Jingyu Si
Department of Chemistry and Materials Engineering, Hefei University, Anhui, Hefei 230022, China

Kh. Abdul Maleque
Department of Mathematics, American International University-Bangladesh, House-23, 17, Kamal Ataturk Avenue, Banani, Dhaka 1213, Bangladesh

Tadeusz Michałowski
Faculty of Engineering and Chemical Technology, Technical University of Cracow, 31-155 Krak´ow, Poland

Agustin G. Asuero
Department of Analytical Chemistry, The University of Seville, 41012 Seville, Spain

B. Vasu and V. R. Prasad
Department of Mathematics, Madanapalle Institute of Technology and Science, Madanapalle 517325, India

O. Anwar Bég
Department of Engineering and Mathematics, Sheffield Hallam University, Room 4112, Sheaf Building, Sheffield S1 1WB, UK

A. Malvandi
Department of Mechanical Engineering, Amirkabir University of Technology (Tehran Polytechnic), 424 Hafez Avenue, P.O. Box 15875-4413, Tehran, Iran

F. Hedayati
Department of Mechanical Engineering, Islamic Azad University, Sari Branch, Sari, Iran

G. Domairry
Department of Mechanical Engineering, Babol Noshirvani University of Technology, Babol 47148-71167, Iran

J. A. de Freitas Pacheco
Laboratoire Cassiopée-UMR 6202, Observatoire de la C^ote d'Azur, University of Nice-Sophia Antipolis, BP 4222, 06304 Nice, Cedex 4, France

M.Weclas and J. Cypris
Institute of Vehicle Technology (IFZN), Faculty of Mechanical Engineering, Georg Simon Ohm University of Applied Sciences, Nuremberg, Kesslerplatz 12, 90489 Nuremberg, Germany

T. M. A. Maksoud
Faculty of Advanced Technology, University of Glamorgan, Pontypridd CF37 1DL, UK

Seung-Kyo Oh
Pharmaceutics and Biotechnology Department, Konyang University, 121University Road, Nonsan, Chungnam 320-711,Republic of Korea

www.ingramcontent.com/pod-product-compliance
Lightning Source LLC
Chambersburg PA
CBHW080625200326
41458CB00013B/4515

* 9 7 8 1 6 3 2 4 0 5 0 7 4 *